MANUFACTURING TECHNOLOGY RESEARCH

LEAN MANUFACTURING

IMPLEMENTATION, OPPORTUNITIES AND CHALLENGES

Manufacturing Technology Research

Additional books and e-books in this series can be found
on Nova's website under the Series tab.

MANUFACTURING TECHNOLOGY RESEARCH

LEAN MANUFACTURING

IMPLEMENTATION, OPPORTUNITIES AND CHALLENGES

FRANCISCO J. G. SILVA
AND
LUÍS CARLOS PINTO FERREIRA
EDITORS

Copyright © 2019 by Nova Science Publishers, Inc.

All rights reserved. No part of this book may be reproduced, stored in a retrieval system or transmitted in any form or by any means: electronic, electrostatic, magnetic, tape, mechanical photocopying, recording or otherwise without the written permission of the Publisher.

We have partnered with Copyright Clearance Center to make it easy for you to obtain permissions to reuse content from this publication. Simply navigate to this publication's page on Nova's website and locate the "Get Permission" button below the title description. This button is linked directly to the title's permission page on copyright.com. Alternatively, you can visit copyright.com and search by title, ISBN, or ISSN.

For further questions about using the service on copyright.com, please contact:
Copyright Clearance Center
Phone: +1-(978) 750-8400   Fax: +1-(978) 750-4470   E-mail: info@copyright.com.

NOTICE TO THE READER

The Publisher has taken reasonable care in the preparation of this book, but makes no expressed or implied warranty of any kind and assumes no responsibility for any errors or omissions. No liability is assumed for incidental or consequential damages in connection with or arising out of information contained in this book. The Publisher shall not be liable for any special, consequential, or exemplary damages resulting, in whole or in part, from the readers' use of, or reliance upon, this material. Any parts of this book based on government reports are so indicated and copyright is claimed for those parts to the extent applicable to compilations of such works.

Independent verification should be sought for any data, advice or recommendations contained in this book. In addition, no responsibility is assumed by the Publisher for any injury and/or damage to persons or property arising from any methods, products, instructions, ideas or otherwise contained in this publication.

This publication is designed to provide accurate and authoritative information with regard to the subject matter covered herein. It is sold with the clear understanding that the Publisher is not engaged in rendering legal or any other professional services. If legal or any other expert assistance is required, the services of a competent person should be sought. FROM A DECLARATION OF PARTICIPANTS JOINTLY ADOPTED BY A COMMITTEE OF THE AMERICAN BAR ASSOCIATION AND A COMMITTEE OF PUBLISHERS.

Additional color graphics may be available in the e-book version of this book.

Library of Congress Cataloging-in-Publication Data

ISBN: 978-1-53615-725-3
Library of Congress Control Number:2019943457

Published by Nova Science Publishers, Inc. † New York

CONTENTS

Foreword		vii
	Maria João Viamonte	
Acknowledgments		ix
Chapter 1	Lean Thinking across the Company: Successful Cases in the Manufacturing Industry *Rui Borges Lopes, Leonor Teixeira and Carlos Ferreira*	1
Chapter 2	From the Factory Floor to New Product Development: Development and Implementation of a Lean Assessment Tool *T. Welo*	31
Chapter 3	Benefits and Challenges of Lean Manufacturing in Make-to-Order Systems *Oladipupo Olaitan, Anna Rotondo, John Geraghty and Paul Young*	57
Chapter 4	Sustaining Lean in Organizations through the Management of Tensions and Paradoxes *Malek Maalouf, Peter Hasle, Jan Vang and Imranul Hoque*	79
Chapter 5	The Impact of 5S + 1S Methodology on Occupational Health and Safety *Joana P. R. Fernandes, Radu Godina, Carina M. O. Pimentel and João C. O. Matias*	101
Chapter 6	Value of Real-Time Data for Cycle Time Optimisation at Wet Tools *A. Rotondo, J. Geraghty and P. S. Young*	123
Chapter 7	The Eighth Waste: Non-Utilized Talent *M. Brito, A. L. Ramos, P. Carneiro and M. A. Gonçalves*	151
Chapter 8	Quality and Safety Continuous Improvement through Lean Tools *Gilberto Santos, J. C. Sá, J. Oliveira, Delfina G. Ramos and C. Ferreira*	165

Chapter 9	Lean Manufacturing Applied to the Production and Assembly Lines of Complex Automotive Parts *Conceição Rosa, Francisco J. G. Silva, Luís Pinto Ferreira and J. C. Sá*	189
Chapter 10	SMED Applied to Composed Cork Stoppers *Eduardo Sousa, F. J. G. Silva, Carina M. O. Pimentel and Luís Pinto Ferreira*	225
Chapter 11	Lean Production in the Portuguese Textile and Clothing Industry: The Extent of Its Implementation and Role *Laura Costa Maia, Anabela Carvalho Alves and Celina Pinto Leão*	255
Chapter 12	Lean Manufacturing Applied to a Complex Electronic Assembly Line *F. J. G. Silva, Andresa Baptista, Gustavo Pinto and Damásio Correia*	285
Chapter 13	Karakuri: The Application of Lean Thinking in Low-Cost Automation *Stephanie D. Nascimento, Milena B. Alves, Julia O. Morais, Laryssa C. Carvalho, Robson F. Lima, Ricardo R. Alves and Robisom D. Calado*	335
Chapter 14	Lean and Ergonomics: How to Increase the Productivity Improving the Wellbeing of the Workers – A Case Study *J. Santos, F. J. G. Silva, G. Pinto and A. Baptista*	355
Chapter 15	Measurement of the Level of Implementation of Sociotechnical and Ergonomic Practices and Lean Production Practices: Considerations from a Systematic Review Process *E. P. Ferreira, J. Schmitt, L. G. L. Vergara, D. F. de Andrade and G. L. Tortorella*	385
Chapter 16	Lean Manufacturing and Industry 4.0: Facing New Challenges from a Shop-Floor Perspective *Antonio Sartal and Helena Navas*	407
Chapter 17	Lean Healthcare in a Cancer Chemotherapy Unit: Implementation and Results *T. M. Bertani, A. F. Rentes, M. Godinho Filho and R. Mardegan*	425
About the Editors		459
Index		461
Related Nova Publications		467

Foreword

The Lean Manufacturing subject has become deeply rooted in companies around the world because of the gradual way it can be introduced in companies and the way it can optimize their operation, generating value-added for the company and, simultaneously, leading to better working conditions for the employees. In order to make this a reality, special attention to detail is needed, eliminating all sources of waste that exists throughout the entire production or services chain, from conception to delivery to the customer. Although very widespread, this subject continues deserving further studies and publications, as it helps greatly to increase the flexibility of the companies' productive systems or services' quality. Indeed, the challenge for rapid response to market needs and the reduction of inventories leads Lean tools to be a real asset in pursuing these goals.

This book aims to contribute to the extension of the implementation of these tools, through the presentation of several case studies explained in detail, as well as to promote a discussion about future developments of these tools. Thus, it is intended to make it easier to apply several of the main Lean tools used in the industry, as well as extending this implementation to the services sector. Based on the experience of these scientists that contributed to the accomplishment of this book, the sequence of the chapters is intended to stimulate new practitioners to promote new applications, describing several cases where the tools have been properly used, contributing to the improvement of the companies' performance and benefiting the workers in questions related to health at work.

The Editors hope that this book can definitely contribute to the broadening of the application of these tools in industry and services sectors, making products more competitive and improving the working conditions of employees.

Maria João Viamonte, PhD
President, ISEP

ACKNOWLEDGMENTS

First of all, the Editors need to thank the Nova Science Publishers for giving us this opportunity to conduct the edition of this book, giving us also the opportunity to contact authors from different countries, and getting a much wider scope on how the subject of this book is treated, bringing together diverse approaches and experiences.

The Editors also wish to thank the Authors for their extremely valuable contributions through the different chapters of this book, because this experience has been translated into an extension of the contacts with researchers who have shown to be a real asset for this work. The Editors also emphasize the commitment of the Authors in meeting the stipulated dates and contribute to the quality of the present work. It was really amazing to note that Authors were always available to hear and help us during this process.

The Editors also want to give a special thanks to the help provided by our colleague Gustavo Pinto, who played a crucial role in the formatting and final revision of each chapter, revealing his usual high professionalism. Here we can also express our thanks to our colleague Andresa Baptista for her support in completing one of the chapters.

The Editors also want to thank Ronny Gouveia for the most accurate review of some chapters, with the usual care.

The Editors also thank the Department of Mechanical Engineering of ISEP, where they are working, due to the available time given to manage this work.

Last, but not least, the unconditional support of the President of ISEP, who gives us the necessary conditions to carry out works like this.

In: Lean Manufacturing
Editors: F. J. G. Silva and L. C. Pinto Ferreira
ISBN: 978-1-53615-725-3
© 2019 Nova Science Publishers, Inc.

Chapter 1

LEAN THINKING ACROSS THE COMPANY: SUCCESSFUL CASES IN THE MANUFACTURING INDUSTRY

*Rui Borges Lopes[1], Leonor Teixeira[2] and Carlos Ferreira[2],**

[1]Department of Economics, Management,
Industrial Engineering and Tourism (DEGEIT),
Center for Research and Development in Mathematics and Applications (CIDMA),
University of Aveiro, Aveiro, Portugal
[2]Department of Economics, Management, Industrial Engineering and Tourism (DEGEIT), Institute of Electronics and Informatics Engineering of Aveiro (IEETA),
University of Aveiro, Aveiro, Portugal

ABSTRACT

Lean thinking is a well-known and well-established management philosophy in the production environment. Having had significant success in reducing wastes, its principles and tools can now be found applied in several industrial sectors with good results. Outside the production environment, however, literature is scarce concerning its applicability and outcome. This work presents several case studies, of companies of different sectors, where both inside and outside production areas wastes were identified and acted upon, namely, in Logistics and Production, Information Flow, and Ergonomics. Results suggest that Lean can be equally applicable and beneficial in these areas, potentially making it a viable philosophy across the company.

Keywords: lean, production, logistics, information flow, ergonomics

---

* Corresponding Author's E-mail: carlosf@ua.pt.

1. INTRODUCTION

Technological advances in the several areas of organizations, coupled with the best business practices they can access, have provided organizations with a set of tools which enables them to improve production processes and increase throughput while at the same time reducing costs. Despite this, the pressure over today's manufacturing industry is always increasing, highly influenced by external factors. One of the major challenges industries face is how they can adjust to this ever-moving environment while still being able to produce in the most efficient way. In this context it is important to adopt principles and practices which allow them to react to market dynamics.

Continuous improvement (named *Kaizen*) is currently widely accepted as one of these practices which can help to identify problems and find solutions to tackle them. Lean Thinking, Lean Manufacturing, or just Lean, aims to continuously improve production systems through the elimination of the unnecessary, i.e., what does not add value to the product (called Wastes or *Muda*), regardless of the nature of the processes involved.

Having emerged in Japan, after the second World War, it comprises a set of principles and techniques, which now are familiar names in most production environments (*Kanban*, *Heijunka*, *Jidoka*, etc.), greatly inspired by the Toyota Production System (TPS) (Toyota Motor Corporation 1998). As a consequence of successful applications in different industrial sectors, today it represents the backbone of important management tools, to eliminate *Muda* as overproduction, waiting, transporting, over processing, inventory, motion and defects, proposing several Lean principles to achieve a Lean organization, i.e., inducing the Lean Thinking.

Although Lean has originated in the production departments of automotive industries, more specifically in the shop floor, currently its underlying management principles and tools are successfully used in other areas/processes within organizations. Organizations are typically structured around a set of operations, characterized by the workstations to process the product in the shop floor, and a set of services, which include support processes such as management, logistics, accounting, human resources, research and development, as well as information and corresponding technology. Although the set of operations establishes the core business of an organization, the associated processes and support services represent important pillars to achieve the organization's goals, and thus are suitable to similar procedures of waste elimination.

In this work, several successful cases are reported detailing how Lean was adapted and used in these other departments/services in companies of such diverse sectors as consumer goods packaging, footwear, cork, automotive, security and safety systems, visual communication solutions, ceramics and assembly systems. After more than a decade of achieving relevant improvements in the processes of several companies implementing Lean practices (methods and tools), we argue that Lean is equally applicable and beneficial in almost all departments, potentially making it a viable management philosophy across the company and inducing a Lean Thinking daily attitude.

The remainder of this work is organized in the following way. Section 2 provides a theoretical background for the main Lean methods and tools and their application in different industrial areas. Afterwards, in Section 3, the case studies are detailed: firstly, the characteristics of the companies where the case studies took place are presented; then the methodology and tools used are shown, followed by the main results and findings. The last section summarises the main conclusions and puts forward some future research avenues.

2. THEORETICAL BACKGROUND

Nowadays, in order to increase the efficiency of the processes that are the basis of increasing productivity, increasing quality and reducing organizational costs, one of the most widely used management practices in the world is Lean. Although the concept proliferated in the 1980s when Toyota began to raise the attention of the global community with the efficiency and consistency of its production processes, the book "The Machine that Changed the World" published in 1990 by Womack, Jones and Roos strongly contributed to consolidate this philosophy in organizations, since it describes a simple and easy-to-apply set of tools that, when implemented, can contribute to helping organizations to increase their results (Womack, Jones, and Roos 1990). Another related concept is the TPS, based on value creation and at the same time reducing waste (Ohno 1988). One of the basic principles of Lean Thinking is precisely the creation of value for the client, which can be effectively the end customer, or simply the element of the organization that is downstream of the process in question. In fact, the concepts of value and waste can be found at any level of the organization, from the most elementary process to the organization as a whole. Hence, the application of the Lean practices can cover not only the processes, but also the industrial sectors, the chain that integrates the different industrial sector on the organization or even the organization as a whole (Figure 1).

Figure 1. Different organizational levels where Lean practices can be applied (adapted from (Keyte and Locher 2016; Jones and Womack 2011; Rother and Shook 2003; Rother and Harris 2001).

2.1. Value-Added Activities and Wastes

Value-added (VA) activities in a company or in a supply chain, according to Hines and Taylo (2000), contribute to keeping consumers satisfied or correspond to the activities for which customers are willing to pay. Waste, or *Muda* in Japanese, means any activity that consumes resources and time and does not add any kind of value, usually resulting in increased costs associated with the product and/or service (Jasti and Kodali 2015). However, although waste should be reduced or eliminated, some authors argue for two types of waste: (i) pure waste, i.e., completely expendable activities, such as stoppages, breakdowns, trips or unnecessary meetings; (ii) necessary waste, i.e., necessary activities such as quality inspections and setups, which, while not adding value, are necessary and cannot be completely eliminated (Womack and Jones 2003).

According to Liker (2004), pure wastes which are to be targeted for reduction or elimination, can form the acronym DOWNTIME:

- **D**efects: Efforts caused by rework or scrap, or even incorrect information;
- **O**verproduction: Production that is more than is required downstream or by clients or produced before it is needed;
- **W**aiting: Wasted time waiting for the next step in a process;
- **N**on-utilized talent: Underutilizing people's talents, creativity, know-how, experience and skills;
- **T**ransportation: Unnecessary movements of products and materials;
- **I**nventory: Excess products and materials not being processed;
- **M**otion: Unnecessary action and movements by people;
- **E**xtra work/**E**xcess Processing: More work or higher quality than is required by the customer.

In order to identify the aforementioned wastes, problems and associated root-causes, as well as propose solutions in a sustainable way supported by continuous improvement, a set of methods and tools can be considered. These methods are addressed in the next subsection and are often used in the context of Lean.

2.2. Some Lean Methods and Tools

This section presents some tools categorized according to the most essential methods of the Lean approach, following the classification by Belekoukias, Garza-Reyes, and Kumar (2014): Just-In-Time (JIT), Autonomation, Total Productive Maintenance (TPM), Value Stream Mapping (VSM) and *Kaizen*. JIT represents a method which states that an organisation should produce the right item at the right time and aims to reduce inventories, space utilisation and possible wastes (Womack and Jones 2003). TPM is considered a method that allows the optimisation of the activities related with corrective, preventive and predictive maintenance (Brah and Chong 2004). Autonomation, also known as *Jidoka*, is a method that aims to reduce quality defects (Baudin 2007). VSM is a method that visually identifies and measures waste resulting from inefficiencies at various resources as information, time, money, space, people, machines and material, (Rother and Shook 2003). Finally, *Kaizen* represents an approach for focusing on the elimination of wastes through continuous and incremental improvement of processes (Imai 2012). Table 1 summarizes the most well-known Lean tools, categorized by five methods of Lean manufacturing addressed by (Belekoukias, Garza-Reyes, and Kumar 2014).

Table 1. Some of the Lean manufacturing most used methods and tools

JIT	TPM	Autonomation	VSM	Kaizen
Kanban Visual control Visual management Levelled production/Heijunka Supermarket	5S Single minute exchange of die (SMED)	Visual control System/Andon Poka-yoke - Jidoka	Spaghetti diagram	Spaghetti diagram Kanban 5 Whys Ishikawa diagram 5S VSM Standardization

adapted from Belekoukias, Garza-Reyes, and Kumar (2014)

These tools are presented as follows:

- *Kanban* is a tool to control the flow of materials, people and information in the *Gemba* often used in pull system operations. The *Kanban* system is based on the principle that the consumption of components triggers a request for replenishment, focusing on the production of small pieces (Hammarberg and Sunden 2014).
- *Visual Control/Visual Management* represents an information display tool that aims to support operators with visual signs (Shimbun 1995). Often a set of standards, graphs, colour systems, space delimitations are used in order to facilitate interpretation by the operator about possible deviations. An important benefit of this tool is that it enables the management and control of processes, taking advantage of the graphical representation (Bateman, Philp, and Warrender 2016).
- *Levelled production/Heijunka* which means levelling production, aims to reduce the fluctuations in production, namely situations of underused capacity or overburden, to have the best use of the available capacity (Hüttmeir et al. 2009). It consists in establishing a continuous flow of production, reducing stock levels and having increased stability of the processes.
- *Supermarket* or inventory buffer is an area where inventory is organized according to some conditions, namely: the split of products by type of components; first-in-first-out principle; and a pull system (using for example *Kanban* cards). Supermarkets are typically located among processes and allow a better visual control for the logistics operator as well as helping avoid overproduction (Kerber and Dreckshage 2011; Battini, Boysen, and Emde 2013).
- *5S* represents a tool that allows reducing or even eliminating wastes and activities without added value (Jiménez et al. 2015; Omogbai and Salonitis 2017). This tool is often used to improve and maintain the organization of the job and the five 'S' helping to remove items that are no longer needed on a workplace (Sort - *Seiri*), to organize the items to optimize efficiency and flow (Straighten - *Seiton*), to clean the area in order to more easily identify problems (Shine - *Seiso*), to implement colour coding and labels to stay consistent with other areas (Standardize - *Seiketsu*) and to develop behaviours that keep the workplace organized over the long term (Sustain - *Shitsuke*) (Jiménez et al. 2015).
- *Single Minute Exchange of Die (SMED)* allows reducing machine setup time ensuring fast tool change. It is defined as the minimum amount of time required to

change from one type of activity to another (setup time), considering the last piece in conformity, manufactured in the previous lot, to the first conforming piece of the next batch (Shingo 1985).
- *Error Proofing: Poka-Yoke* aims to have an error proof process, ensuring that proper conditions exist prior to initiating any process. It performs a detection function, eliminating defects in the process as soon as possible. According to (Ghinato 1998), this tool is fundamental in zero defect quality control, which aims the total elimination of defects through the identification and elimination of their root causes.
- *Jidoka* also known as Autonomation, means automation with intelligence or with a human touch. It consists of providing man and machine with the autonomy to interrupt production whenever an abnormal situation has been found or when the scheduled production has been reached (Ghinato 1998).
- *Spaghetti* diagram is a visual representation of the physical flow of materials and people throughout the processes, detailing the transportation distance and tracing (walking patterns of people, shuttling back and forth of materials), and waiting times (Wilson 2010).
- *5Whys* is a tool to identify the root cause of a problem and act upon it. Correctly performed has a lot of depth and breadth and can be viewed as both corrective as well as preventive action (Ohno 1988).
- *Ishikawa Diagram* also known as Cause-Effect diagram is a tool where causes are identified to get to the root of a specific problem, by analysing all the factors that contributed to the source of the problem. To identify the root-causes, Gwiazda (2006) proposes the 6M method, where the causes can be derived from Manpower, Machinery, Materials, Method, Mother-nature (Environment) and Measurement.
- *Value Stream Map* is considered an effective method/tool to identify waste in the value chain and aims to analyse and design flows at the system level across multiple processes. It represents a technique used to document, analyse and improve the flow of information or materials required to produce a product or service for a customer (Tyagi, Choudhary, et al. 2015; Forno et al. 2014). The VSM-based value chain mapping is divided into three levels: at the top of the map, the information flow is present; in the centre of the map, the flow of materials; and at the bottom of the map, the distances travelled taking into account a timeline.
- *Standard Work (Standardization)* is a Lean tool to ensure that improvements implemented are to be sustained, setting how activities are to be performed (Womack and Jones 2003).

Although these concepts and tools have emerged on the shop-floor associated with production processes, the concept of waste can exist in any human or non-human activity, so the application of Lean to other areas can also provide strong benefits. The following subsections address some applications of Lean, namely in logistics and production (traditional areas), information flow and ergonomics (non-traditional areas).

2.3. Lean Applied across the Company

2.3.1. About Lean in Logistics

Logistics plays a critical role in most companies as it is often what enables them to stay competitive or gain competitive advantage (Rahman 2006). Improvements in logistics lead to reduced costs, improved resource utilization and more efficient systems (Beamon and Ware 1998). Companies must therefore focus on continuously monitoring and improving their logistic systems, with its key components being (Rushton, Croucher, and Baker 2017) transport, information and control; inventory; packaging and unitization; and storage, warehousing and material handling.

To that end, Lean provides a fitting set of principles and tools. As with production processes, it is important to first identify the wastes in the context of logistics. These are in some ways equivalent to the ones found in production (Sutherland and Bennett 2007). Table 2 summarizes the eight Lean wastes that are possible to find on Logistics processes.

Having started in production, the most natural entry point of Lean concepts into logistics is internal logistics, which concerns managing logistics activities within the boundaries of the company (i.e., among the different production areas).

Some of the activities in internal logistics are often regarded as non-value added (NVA) – mostly the ones concerning movement of components or products. Following the Lean philosophy, this makes them target to elimination or to be minimized and done in the most efficient way possible. Some of the tools that can be used to help in this regard are supermarkets, visual management, *Heijunka* and *Kanbans* (Coimbra 2013).

Table 2. The eight Lean wastes (DOWNTIME) on Logistics

Lean waste	Characterization
Defects	Activities causing reworks, unnecessary adjustments or returns (e.g., billing errors, inventory discrepancies, and damaged/defective/wrong/mislabeled products)
Overproduction	Delivering products before needed or demand information overproduction (caused by requesting larger quantities than needed or before needed)
Waiting	Delays between logistics activities (e.g., time between truck arrival and its loading/unloading, delay between receiving customer's order information and beginning fulfilling that order)
Non-utilized talent	Concerning, for example, the know-how and experience of logistics operators, warehouse responsible, or salesperson
Transportation	Unnecessary movements of products and materials
Inventory	Logistics activities resulting in added inventory in the wrong locations or above requirements (e.g., early deliveries, inventory in the wrong distribution centre)
Motion	Unnecessary movement and actions by people (e.g., extra walking, reaching, or stretching to reach products in warehouses due to poor arrangement or ergonomic design)
Extra work/Excess Processing	Extra space or suboptimal use of space (e.g., less than full truckloads, boxes not used to full capacity or inefficiently filled, inefficient use of warehouse space)

2.3.2. About Lean in Information Flows

Information is one of the most valuable organizational resources, which requires an adequate process of management. It has become a valuable asset, compared to production, material and financial resources; therefore, in a globalized society information it is a precious resource. However, to be useful, it should be relevant, complete, accurate and up-to-date, and at same time, obtained economically. Thus, having good quality information, reliable, in the right quantity and at the right time is a differentiating factor and an added value, while the lack of information originates loss of competitive opportunities. According to (Hicks 2007) the way information is organized, in a format that is easy to visualize and use, can add value to flow and to the end-user (decision-maker). This author also sustains that it is difficult to align complex systems with the organization, which can origin a significant detrimental effect on the organization and its performance. Therefore, the capacity of an Information System to store and manage data to produce useful information is the key to gain competitive advantage.

An information process begins with a set of data (raw material), then these data are stored, processed and organized (transformation) and the results are delivered to customers (information for decision-makers). Thus, also in the information processes it is possible to identify waste (Hicks 2007; Ibbitson and Smith 2011; Blijleven, Koelemeijer, and Jaspers 2017). Table 3 summarizes the eight Lean wastes that are possible to find on the information processes.

Table 3. The eight Lean wastes (DOWNTIME) on information flow

Lean waste	Characterization
Defects	Inappropriate actions based on inaccurate/lack of information which can lead to poor decision-making
Overproduction	Generation and maintenance of excess non-value-added information
Waiting	Time and resources used to identify information that is not easily available, which usually results in people waiting for the correct information
Non-utilized talent	Human-resources needed to work in non-optimized information processes due to the existence of other wastes
Transportation	Movement of information which does not add value to the user who receives it
Inventory	Legacy databases and file archives that provide more information than those that consumer needs at a given moment in time
Motion	Unnecessary movement to seek information due to improper organization and representation of information
Extra work/Excess Processing	Activities undertaken to overcome a lack of information that may include creating or searching for additional information

Although it is a less common approach, there are already some studies that show the benefits of application of Lean thinking in the information processes (Pereira and Teixeira 2017; Soares and Teixeira 2014; Hicks 2007; Teixeira, Ferreira, and Sousa-Santos 2019; Ré and Teixeira 2018). Goodman (2012) also applied the Lean and Six Sigma methodology in a pharmaceutical industry to access, analyse and present information publically available; Khodambashi (2014) evaluated an IS in clinical process applying the VSM and the A3 method; Soares and Teixeira (2014) conducted a study based on SIPOC (Suppliers, Inputs, Process, Outputs and Customer) and PDCA (Plan, Do, Check and Act) methodologies to improve a set of indicators in an industrial company. Chookittikul, Busarathit, and Chookittikul (2008) used DMAIC (Define, Measure, Analyse, Improve, Control) and SIPOC to improve and control the quality of graduates in a university.

2.3.3. About Lean in Ergonomics

In order to have an efficient and effective manufacturing environment, it is important to guarantee good working conditions for all employees in terms of workplace health and safety. According to Hendrick (2003), ergonomics is a discipline that can increase savings and productivity in organizations, reduce worker injuries and absenteeism, and increase morale. This discipline considers different ergonomic risk factors that can affect workers while performing their tasks at work, classifying those in physical/mechanical factors, and in psychosocial factors (Koukoulaki 2014). Thus, ergonomic principles represent valuable practices to prevent worker tiredness and strains, which can lead to work-related musculoskeletal disorders (WMSDs) and/or work-related neurovascular disorders (WNVDs). It should be noted that psychosocial exposure, usually associated with stress and mental disorders, can also lead to physical/mechanical disorders. Repetitive motion injuries, cumulative trauma disorders, carpal tunnel syndrome, tension neck syndrome, low back pain, and soft tissue disorders represent some of the most common injuries related with WMSDs (Costa and Vieira 2010). Often these injuries are associated with workloads, repeated movements, handling of large loads, lack of ergonomics in the workstations, bad postures, and obstructed passages, among other causes that can increase ergonomic risks and cause accidents.

Concerning the ergonomics and safety risks, Aqlan et al. (2014) distinguish those two terms and state that the former are visible hazards whereas the latter can be invisible ones. Consequently, the authors highlight the importance of integrating Lean with ergonomics issues for effective assessment and elimination of lean wastes and ergonomic risks which normally have impact on safety risks. In fact, high ergonomic risk can reflect the existence of one or more lean wastes. For example, according to Aqlan et al. (2014) the poor design of tools and awkward postures can lead to wasted motions that increase the time to perform tasks and reduce the quality of work. Another example is related with the defects that can create fatigue to operators and, consequently, increase the overtime.

Table 4 summarizes the eight Lean wastes (DOWTIME) that are possible to find on ergonomics area.

Table 4. The eight Lean wastes (DOWNTIME) on the Ergonomics

Lean waste	Characterization
Defects	Activities that cause wrong and extra movements
Overproduction	Additional work that can lead to operator overload and consequently to fatigue
Waiting	Injuries that can lead to the stops working
Non-utilized talent	Human-resources needed to work in some operations due to absence of an injured worker
Transportation	Poor design of workstation that can increase certain movements
Inventory	--
Motion	Unnecessary movement that can cause tiredness and awkward posture
Extra work/Excess Processing	Unnecessary movements with excessive weight that can cause injury

Some studies indicate that the integration of Lean manufacturing and ergonomics issues could improve work environments by minimizing both lean wastes and consequently minimizing the ergonomic risks that can cause WMSDs (Aqlan et al. 2014; Aqlan et al. 2013; Santos, Vieira, and Balbinotti 2015; Saurin and Ferreira 2009). However, related with WNVDs, Koukoulaki (2014) argues that Lean production can generate some conditions associated with time pressure that affect mental workload and thus can contribute to the increase of ergonomic risks.

3. CASE STUDIES SHOWING LEAN APPLIED ACROSS THE COMPANY

3.1. Introduction and Companies' Contextualization

The cases mentioned in this work concern projects with the duration of 6 to 9 months using an action research approach, where one of the members of the research team was within the company taking part of data collection, analysis, development of solutions, implementation and follow-up. All the companies in these case studies are located in Portugal, although some of them belong to major international groups.

The main characteristics of these companies are detailed as follows and summarized in Table 5.

Table 5. Characteristics of the companies where the case studies took place

Company	Location	Size	Sector	Source
A	North region	Small	Assembly systems	(Oliveira 2017)
B	North region	Large	Consumer goods packaging	(Ferreira 2017)
C	Centre region	Large	Ceramics	(Millheiro 2011)
D	Centre region	Large	Visual communication solutions	(Ferreira 2014)
E	Centre region	Large	Cork (floor and wall coverings)	(Lopes 2013)
F	Centre region	Large	Cork (capsules)	(Pereira 2015)
G	Centre region	Large	Footwear	(Ferreira 2015)
H: H.1 & H.2	Centre region	Large	Automotive	(Rolo 2018) (Teixeira 2018)
I	Centre region	Large	Security and safety systems	(Gomes 2018)

Company A started its activity 10 years ago, in the north of Portugal, with the goal of filling a void in the Portuguese market concerning Lean solutions for workstations, logistics trains, supermarkets, warehouses and logistics. Over the years the company found the same opportunities in other countries and started exporting its products. It currently has around 50 employees, and can be considered a small enterprise.

Company B is part of a major group in the consumer goods packaging and contract manufacturing industry. Located in the north of Portugal, it was founded in the 1960s with the intent of producing cans for cookies. Over the years the company grew and diversified its business, starting to produce plastic containers in the 1980s. The major group has a sales volume of around 475 million euro employing over 2800 employees worldwide. Currently the company has its production divided into packaging (metallic or plastic) and filling (mostly sprays).

Company C was created nearly 200 years ago to produce glass and crystal, specialized in porcelain, and is today the largest Iberian group (and sixth worldwide) in tableware and giftware. With more than 1700 employees manufacturing porcelain, earthenware, glass and hand-made glass, it has a sales volume of around 85 million euro, where the external market represents 64% of the turnover.

Company D, founded in 1979, produces and markets visual communication products, exporting 99% of its production to around 80 countries. It occupies an area of 25,000 m2, has more than 500 employees and a turnover of more than 50 million euro.

Companies E and *F* are part of a major group that initiated the activity in the cork area around 1870. Nowadays Company E is the world leader in the production and distribution of cork flooring coatings and cork with wood, with a market share of 65%. It has a mix of about 2000 products, an annual productive capacity of 10 million m2, occupies an area of 120,000 m2, has 650 employees, and a turnover that reached 100 million euro. Company F is focused in the production of cork capsulated stoppers, with a workforce of around 85 employees, producing more than 800 thousand cork capsules per day.

Company G is part of a footwear manufacturing multinational based in southern Denmark, which was founded in 1963. The company started production in Portugal in 1984, is in the centre of Portugal, and currently has more than 1200 employees. This was the first factory of the group outside Denmark, and the choice was due to Portugal being a major player in the footwear industry.

Company H is one of the several factories of a major automotive group. This factory is in the centre region of Portugal and has assets valued in around 400 million euro. The company has around 1300 workers, occupies an area of 300,000 m2 and produces two main types of components for its group: gearboxes and engine components. Gearboxes represent the largest business volume of the company and the company exports all its production to the other factories in the group.

The group *Company I* is in appeared in the year 1886 in Germany. The group is currently present in 4 continents and more than 50 countries, employing approximately 282,000 people worldwide. It is divided into three main business areas: (i) automotive technology; (ii) industrial technology; and, (iii) consumer goods and industrial technology. Company I is an industrial unit that belongs to the business area of consumer goods and industrial technology, and is located in the centre of Portugal.

In the following sections, the case studies are presented according to the area of the company where they took place, namely, Logistics and Production in Section 3.2, Information Flow in Section 3.3, and Ergonomics in Section 3.4. For each case, problem contextualization and main objectives are provided, followed by the methodology, main results and conclusions.

3.2. Cases of Lean Applied to Logistics and Production

3.2.1. The Case of Company A

Problem contextualization and objectives - the company needed to control the stock of the produced components as several problems were identified, namely, stock-out of some items and excess stock in items with low turnover. The stock-out resulted in an increase in work in progress (WIP) and often compromised the lead time of the products. The main objective was therefore to reduce the number of times the items were out-of-stock in order to reach a goal set by the top management: less than one stock-out per month. With this objective in mind a *Kanban* system was to be implemented in a Supermarket of components.

Methodology, results and conclusions - due to its area of activity the company is quite aware of Lean philosophies and tools; nevertheless its rapid growth did not allow fully implementing them indoors. The company follows a methodology composed of four steps when addressing Lean projects. Firstly, the project is defined in order to understand its scope, the business case indicating the main goal, risk and stakeholder analysis; the output is a project charter. Then, an analysis phase is performed, where data is collected, solutions are devised, and a timeframe is defined, resulting in a project board made visible to all. Then follows an implementation stage where pilot testing occurs and, if the solution is found suitable, it is employed. Finally, there is a control and maintenance phase, where standards, training and auditing ensure that the solution is maintained.

In the analysis phase a Value Stream Design (VSD) was made to better understand how materials and information were flowing. Through the VSD it was possible to understand the complexity of the exchange of information to ascertain if an item was already in stock or needed to be produced. Additionally, based on the monthly sales information, a criterion was defined to classify components into make-to-stock (MTS) or make-to-order (MTO).

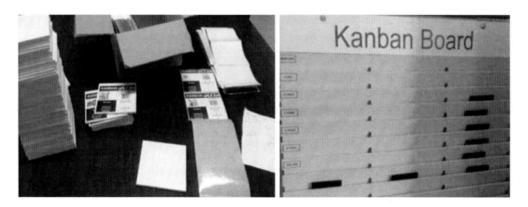

Figure 2. Creating Kanban cards (left) and Kanban board (right).

In the implementation stage the number of *Kanbans* for each component to be produced was determined (Figure 2). Moreover, the layout of the supermarket was drawn, locating the components with higher turnover in the most ergonomic positions and the remaining in higher or lower positions allowing for a more effective picking process. Finally, the information flow and the procedures concerning supply to the production lines were redefined. In the control and maintenance phase it was possible to measure the gains from this project and ensure the newly defined standards were being followed. The number of stock outs was reduced to meet the goal of the administration (on average less than one per month) and there was a reduction of stock around 7% for MTO components and around 50% for MTS components.

3.2.2. The Case of Company B

Problem contextualization and objectives - the company has an extensive experience in Lean, having its own methodology, adjusted to the organizational culture. This methodology is composed of four pillars with a strong focus in continuous improvement: Leaders, Support,

Project and Daily. Leaders concerns the top management, which defines the company's strategies and goals, provides support and leadership and ensures all are involved. Support is composed of the continuous improvement teams and aims to support and train the teams in the pillars Project and Daily. Project represents the disruptive improvements in processes or equipment, encompassing diversified teams. Finally, Daily concerns the daily activities in continuous improvement and how to properly maintain them. This last pillar is further divided into 4 levels, denoting the maturity level of each sector in the company (1 is lower, 4 is higher).

The main goal was to improve the performance of the logistics teams, making them reach level 4 in the Daily pillar of their Lean methodology. This would allow teams to be more autonomous in identifying problems and finding proper solutions, thus contributing to improving the efficiency of their work: supply the production lines.

Methodology, results and conclusions - the internal logistics sector is composed of three teams responsible for supplying the production lines and moving the finished products to the warehouse. The teams started out in level 1 of the Daily pillar, as teams were generally unorganized and uncoordinated, the daily boards used to gather information relevant for their work were incomplete and outdated, leading to being disregarded. Moreover, daily meetings rarely occurred and were generally considered unproductive. Therefore, despite the company experiencing a relatively high Lean maturity level these teams did not follow the rest of the company in this regard. Several of the Lean wastes could be found in this sector, such as, lack of organization in the workspaces, improper location of components and finished products, and no synergy between the teams (leading to unnecessary movements).

To tackle these issues, firstly, two key performance indicators (KPI) were defined to allow measuring the efficiency of the supply to the production lines: line stoppages time and number of supplies per hour.

Based on the data of the previous two years it could be concluded that there was a considerable variation in the time of stoppages. The main factors were: high load level of the production lines; flaws in the definition of the production schedule; experience of the workers supplying the lines; and time lost searching for misplaced tools. Moreover, the total time for line stoppages was averaging 140 minutes per week, which was considerably high when compared with the defined target of 60 minutes per week.

Concerning the number of supplies per hour, the movements of workers were analysed during the course of several weeks. It comprised mostly movements to bring the raw materials from the warehouse to production lines. The time required to reach the different lines was similar, taking on average 2 minutes. A target of 8 of these trips per hour was defined for this indicator; the currently only 6 were being done on average. If the target is reached the company can reduce the number of workers while improving the performance of the teams. To better understand the required movements for the trips a Spaghetti diagram was drawn. This diagram allowed to identify the larger flows and showed unnecessary movements.

Following the company's methodology, concerning the Daily pillar, initially, the teams were organized, and a team leader was elected. The election of leaders with a good knowledge of the process and with leadership skills was encouraged. Then all the teams had a training period to better understand the methodology and the boards were reorganized to meet the standards. The level 2 of the Daily pillar was achieved after several hands-on workshops allowed implementing 5S in their workplaces, warehouses and vehicles. Level 3 was reached after good practices were standardized and embedded in the daily activities of workers. For the main tasks performed, the workers jointly reached several ways to perform them, from which the best was chosen using an impact/effort matrix; the best way became the standard. Level 4 was not reached during the duration of the project and the necessary steps to reach it were postponed to a later date.

Regarding this Lean implementation, it was observed that the teams became more autonomous and efficient, having improved the performance of their daily activities. Concerning the two KPIs defined for this project, total stoppages time was reduced for two of the three main types of products in 83% for one and 17% for the other, having reached the target of under 60 minutes per week (average of 57 minutes in the last six weeks analysed in the case study). The number of supplies per hour did not reach the defined target (staying on average in 7 per week); however, it could be observed that the workers were still highly motivated and committed to reach this goal.

3.2.3. The Case of Company C

Problem contextualization and objectives - this project was carried out in a work centre of isostatic pressing, working 24 hours a day and 7 days a week and allowed to reduce costs inherent to activities that do not add value. This was achieved through the implementation of a KPI of overall efficiency, namely the Overall Equipment Effectiveness (OEE), which monitors the process in terms of availability, performance and quality. This information, made available through visual management to the employees, has also potentiated a proposal for intervention in the unavailability of the equipment, through TPM. The methodology involved the elaboration of a VSM, which allowed to understand the interactions between the different stages of the manufacturing process and contributed to create an equipment efficiency indicator, outlining the improvement plan leading to the goals and established targets of the centre.

Methodology, results and conclusions - a data collection of the typologies of equipment stoppage events in the *Gemba* through the SAP as well as a Pareto analysis allowed to focus the intervention in the classes cleaning of the work station, change of forms and failure with maintenance intervention. This pointed out which Lean tools to use: 5S, SMED and preventive maintenance. Regarding the monitoring and impact of the OEE, Visual Management was used through a table that includes information on 5 relevant aspects: (i) daily logging and weekly OEE evolution and monitoring of production; (ii) daily monitoring of production with indication of the reasons and times of stops; (iii) daily and weekly monitoring of the quality of the products conformation; (iv) showcase of faulty parts, with reference to the responsible team, indicating the percentage of refused items in the final choice, and (v) area of suggestions.

The 3 steps of SMED (separation of internal operations from external ones, conversion of internal operations to external ones, and reduction of internal operations) were applied to 2 tools (forms) that initially had a mean change time of 1.5 hours. As a final result, there was a reduction of about 70% of that time. For a consistent evaluation of the improvements introduced by SMED (reduction of set-up time) the strategy of every part every interval (EPEI) was used to quantify the reduction of the production batches size. This allows to produce the product mix of each series which satisfies the demand of the customer without creating waste with the excess stock (scenario 1), the possibility of increasing the quantity produced (scenario 2) or an intermediate solution. Scenario 1 led to a reduction of the average lot from 6000 to 2000 units, rotating 150 references 18 times a year, instead of the previous 6 times. Scenario 2, was an unwanted alternative for the company since it was producing very close to takt-time, was evaluated by simulation and revealed an increase of OEE by 5%.

In the stop due to maintenance intervention, 5S enabled to shorten the periods of intervention for maintenance. In the screening phase (*Seiri*) the red label technique was used; in the storage and cleaning (*Seiton* and *Seiso*) cars of components and boxes properly labelled were used; in the standardization and discipline (*Seiketsu* and *Shitsuke*) for the tools an adequate cabinet was acquired, and a standardization dossier was elaborated. The evaluation was performed based on the comparison of the intervention time (before and after) in the most frequent failure (oil leakage), with a median time decrease of 2 hours to 1 hour. Also, the intervention time of 50% of the faults (that varied between 1 hour and 4 hours) was reduced to the interval of 0.5 hour to 1 hour.

Regarding the stop to clean the work station the new defined sequence of tasks led to a reduction of this period of 8 hours/week, allowing the stopping of 1 equipment during a full shift/week to carry out the important preventive maintenance. This way the productivity was not harmed. Moreover, the OEE showed an overall increasing trend, together with a decrease in its weekly variability.

3.2.4. The Case of Company D

Problem contextualization and objectives - the objective of the project was to analyse and improve the processes and methods of supplying plans (primary components of cork, felt, ceramic, magnetic, etc.) and to increase the productivity of a pallet packaging machine. Previously the supply was done with a stacker, with individual routes to each workstation of the section, in a total of 16 tours/day and 370 meters/tour. The supply request was made orally by the employee, resulting in: (i) a large and unnecessary stock of plans along the line and/or (ii) lack of plans at the workstations. Concerning the pallet packaging machine, a detailed analysis revealed several unproductive times, a non-balanced line and a disorganized warehouse. Thus, in a continuous improvement environment (*Kaizen*), a JIT system was implemented adopting the Pull paradigm (with *Kanbans*) and visual communication frameworks (*Heijunka box*) to coordinate and support the workflow in using a logistic train (*Mizusumashi*). The parameters of the pallet packaging machine were tuned according to the balanced line.

Methodology, results and conclusions - the data collection to the supply needs of each workstation, allowed sizing the *Mizusumashi*, defining routes and with the elaboration of sequencers and the respective *Heijunka box* and *Kanban* creating a Pull system. There was a 50% improvement in the distance travelled by the *Mizusumashi* (Figure 3) and a saving of 57% per shift in the number of tours. Also, the number of pallets transported per tour decreased by 36%. The accumulated stock in some lines decreased by 83%, and the stock in warehouse was also reduced. Concerning the pallet packaging machine, an Ishikawa diagram revealed the causes of non-productive time. An analysis of these times showed that the bottleneck was the delay of removal of the pallets. The process was re-designed; the machine parameters tuned and considering the obtained takt-time production goals of each shift were established. It was possible to save 1 of the 3 operators and machine productivity increased by 23%.

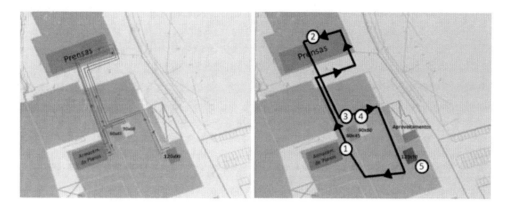

Figure 3. Supplying with the stacker (left) and with the Mizusumashi (right).

3.3. Cases of Lean Applied to Information Flow

3.3.1. The Case of Company E

Problem contextualization and objectives - the project aimed at analysing and optimizing the information flow across the organization through the application of multiple Lean tools (*Kaizen*, 5Whys, VSM, Standard Work) to eliminate waste and create value, with emphasis on the departments of BackOffice, Planning and Expedition. The proposed solutions include the improvement of the Enterprise Resource Planning (ERP) sessions, the definition of responsibilities, processes and the creation of new work methods, which allowed an alignment of the organization with its strategy, leading to an increase in its efficacy and efficiency.

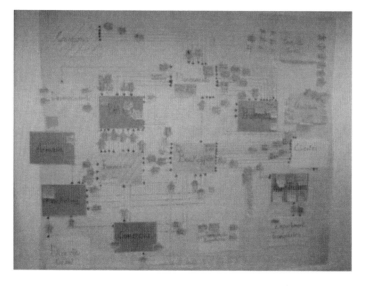

Figure 4. Map of inter-departmental relations.

The approach involved four phases:(i) mapping the flow information across the organization, Figure 4, identifying the areas associated with the main sources of waste; (ii) detailed mapping of the current situation of the areas identified in the previous phase; (iii) analysis of the resulting map and development of proposals for improvement based on Lean principles; (iv) development of an action plan for its implementation.

Methodology, results and conclusions - the elaboration of a VSM allowed obtain an overview of the information flow of the company to identify and document the main errors and wastes and the areas where they occurred. This survey led to the focus on the areas of BackOffice, Planning and Expedition, mapping the current situation, and recording time and frequency of occurrence of actions/tasks. These quantitative indicators, collected from the ERP and by direct observation in the departments, were used to evaluate the impact of the improvement proposals; their analysis supported the transition step between the mapping of the current situation and the mapping of the future situation.

The data analysis involved: (i) a Pareto analysis, which detected the most frequent and highest impact errors; (ii) an adaptation of Breakdown Analysis (graphs and levelling tables) and the *Yamazumi* graph to assess the impact on the current workflow and detect *Mura, Muri* and *Muda*.

The development and implementation of the proposals, validated by top-level management, included the development of: (i) a simplified input interface for production logging and typing in the ERP (which significantly decreased process times and errors); (ii) standardized work that defines responsibilities, ensures a better flow of information, improved process quality and productivity and a convergence of working methods between BackOffice and Planning. As a result, BackOffice achieved reductions of 20% to 50% in the volume of tasks, errors and work on hold, and 60% in the total management time spent. The Planning area reduced 30% to 70% in the time associated with production registration and line sequencing, issuing *Kanbans* and component planning. The Expedition area reduced the amount of errors allowing the saving 1 hour/day.

Finally, it should be noted that, following an initial resistance to change by the employees, their involvement, a detailed explanation of the objectives, and the gradual verified improvement resulting from the implemented changes, increased the necessary trust to establish the continuous improvement environment (*Kaizen*) of the information flow in the organization, paving the way to the reduction of the resource utilization rate, the decrease of lead times and the increase of service level.

3.3.2. The Case of Company F

Problem contextualization and objectives - the process of this unit includes the reception of the cork stopper and consequent transformation, as well as the reception and/or production of capsules, which originate the capped stopper, when integrated in the stopper. TSU currently has more than 300 customers in over 50 countries, with a strong growth in the market for spirits. This implies a great variability in production and projects due to the specificities of the orders, often with great variety and in small quantities (e.g., the capsules of the most expensive whisky in the world in a limited edition of 3 luxury bottles). This product diversity, which in turn implies variability in production, contributes strongly to the existence of wastes in the processes, making them less efficient.

To overcome some of the causes of these wastes, it was decided to implement an Information System (IS) to manage all the data related to the products, their specificities and respective production parameterizations, while supporting the decision support in the factory floor, with a rationale based on experience. The major challenge in this project was the engineering process of requirements on the factory floor, together with a profile of users who operated daily in the process and without knowledge in tools and techniques of software engineering. Moreover, and faced with a set of requirements only present in each employee's tacit knowledge and experience, the requirements elicitation process has thus become a rather difficult task. Considering that the employees were involved in a *Kaizen* process, the analysis and development of the IS was conducted using a set of Lean tools/methodologies.

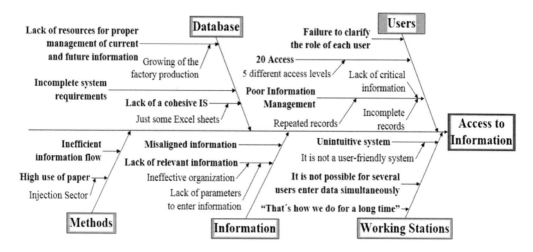

Figure 5. Ishikawa diagram to identify the causes of problems related to access to information (adapted from (Pereira and Teixeira 2017)).

Methodology, results and conclusions - the development of the IS was based on the DMAIC methodology of Six Sigma (Kwak and Anbari 2006) integrated in the systems development life cycle (SDLC) (Whitten and Bentley 2016). A rationale of problem identification and solution based on the PDCA cycle underpinned this approach. Thus, at an early stage, data were collected based on interviews and observation methods. To analyse the current state and identify the main points to act, a VSM was made to better understand how information related to the process under study was flowing inside the company. With this method it was possible to know and map the entire process and, consequently, identifying activities that did not add value (contributing to waste). One of the main identified problems was the access to information, which was analysed in detail based on an Ishikawa diagram, Figure 5.

Considering the process and the result reported in the Ishikawa diagram, the seven wastes in terms of information were identified:

- Defects: there is no clarification of the role of each user; unintuitive system.
- Overproduction: information that is not used as some repeated fields; multiple Excel files with the same information.
- Waiting: the lack of information slows down the production; the need to use multiple files to cross information.
- Transportation: lack of structured databases resulting in movements of unnecessary information.
- Inventory: there are a lot of files that can be unified into one.

- Motion: it is not possible to have multiple users entering data simultaneously; users prefer to ask information to others than using the actual system.
- Extra work: users need to copy the information to other files.

It should be noted that the waste non-utilized talent is naturally present and associated with all the other 7, since the need to reduce them inevitably implies the use of human resources, which could be used in activities that add value. To summarize this phase, the 5W2H tool for planning work and demonstrating the existing problem was applied. This analysis revealed the inefficiency of the information flow and the misalignment between the available information and the required information, as the main causes of the problem. It contributed, also, to a better understanding of the problem and the requirements gathering based on the tacit knowledge and experiences of the collaborators. In order to validate the requirements, they were converted into scenarios and the 5Whys technique was used along with informal interviews.

Considering the data collected on the previous phase, the solution was conceptualized using the Unified Modeling Language (UML) notation (Whitten and Bentley 2016). In this phase, and in order to identify the relevant information, the 5S tool was applied to the model, as follows: (i) Sort, separate the information, maintaining the necessary (with value) and discarding the unnecessary (waste); (ii) Straighten, structuring information with value and establish the different links (achieved with a UML class diagram); (iii) Shine, create mechanisms to ensure that there is no waste; (iv) Standardize, develop standards to maintain and control the first 3 Ss using a proper user manual; and (v) Sustain, bet on training to discipline users to follow the implemented standards.

After this, in the transition phase the abstract model represented with UML notation was converted into a model closer to reality. Therefore, the IS design was done using a mock-up prototyping tool (ForeUI©). The operational model developed includes a set of interfaces that will become part of the system. The interfaces were done in order to get closer to reality and had different iterative cycles (PDCA) to incrementally incorporate suggestions from users to develop a user-friendly system. Finally, the technological solution was developed using the Microsoft Access tool and the Visual Basic language.

In this study some Lean techniques were applied in the development process of a IS, and results confirmed the huge potential of Lean thinking, increasing activities with added value and reducing waste in the process. With this experience, the authors also conclude the importance of the involvement of key-users in processes like this, and the importance of finding requirements gathering and validation tools aligned with the users' knowledge. First it is necessary to understand their needs and then integrate them in the development of the IS to get positive results. Finally, it is essential to internalize that being Lean is always a work in progress, a continuous improvement, requiring the involvement and commitment of all stakeholders.

3.3.3. The Case of Company G

Problem contextualization and objectives - this case occurred in the Global Shoe Costing (GSC) department responsible for the products costing of the whole group. This department was set up in 2015; until then the company had used a decentralized costing system in which each production unit agreed with the parent company (based in Denmark) a production price for its various items.

From 2015, with the creation of the GSC, the costing system became centralized, and it is the responsibility of this department to set the prices of the products for the remaining units. For this reason, since its origin, this department has been strategic within the group. Given the potential expansion of this department and its importance, there was a need to standardize the processes and procedures associated with the work of its human resources. It should be noted that since it is a cost department, the main tasks carried out are, by their nature, management activities supported by a set of technological tools. In this context, the main objective of this project was to improve the costing process of GSC, using a set of Lean tools.

The following specific objectives were defined: (i) identification of wastes associated with the activities and processes of the GSC, also exploring their causes; (ii) identification of VA activities and NVA activities within GSC processes; (iii) identification of opportunities for improvement in information management processes; and finally, (iv) the presentation of concrete proposals for tools and methodologies aimed at reducing the waste associated with the processes.

Methodology, results and conclusions - to reach the goal a set of Lean tools was adopted, emphasizing at an early stage of the project the use of the 5W2H method. This method highlighted relevant issues of the process and made it possible to gain a deeper understanding of the What, the Why, the Who, and contributed to the following approach to the problem resolution (How). Regarding the main objective and identified problem (i.e., lack of efficiency of the costing process), and in order to know its main root causes an Ishikawa diagram was done.

In this diagram the root causes were grouped according to the model 6M, explained by Gwiazda (2006), and no causes were attributed to the management. These causes were identified through an iterative approach of questions based on the technique of the 5Whys, according to Tyagi, Cai, et al. (2015), aiming to help establish the cause-effect relationship, allowing more easily understand what contributes to the occurrence of problems. With this analysis, it was concluded that the main root cause of the problem was that the department was recent and, as such, had not yet consolidated and standardized processes. To know the activities of the process, the actors that worked upstream and downstream of each activity, as well as the flows in terms of inputs and outputs of each activity, a SIPOC analysis was performed. This analysis, besides contributing to the mapping of the processes, allowed knowing the main actors and the type of tasks that each executed, and allowed to identify activities of NVA, arriving to a list of 10 wastes:

- Time spent in organizing cost requests (extra work).
- Preparation of unnecessary costing (overproduction).
- Bill of Material sent with errors or lack of information (extra work/defects).
- High number of manual processes (defects).
- Repetition of operations in all costing processes (defects).
- Lack of historical records (defects / waiting).
- Excessive time in the elaboration of costs for ongoing articles (defects).
- High number of exchange of emails or phone calls (transportation).
- Existence of cost errors (defects).

- Inefficiency associated with clutter in computers (inventory).

Although the 8th waste – non-utilized talent – is not identified, it exists and is associated with all the other 7 types of waste, as the need to reduce them necessarily involves the use of human resources that could be used in activities that add value.

Knowing the list of waste and focusing on improving the efficiency of the GSC process, improvement actions were proposed, which in turn were implemented according to a list of priorities that was based on a cost/benefit analysis of each proposal. This analysis had the participation of the direction of the department. Among the several implemented proposals, it is worth mentioning the creation of some simple computer based solutions that integrated the process and allowed to automate certain tasks until then manual, as well as the application of the 5S in the own computers and in the shared areas of the computer system, to make intuitive use of documents, in the vast majority of documents collaborative work.

From the obtained results, the following facts stand out: (i) the department can now act on a pull perspective, something that did not happen prior to this project; (ii) the time of some processes and consequently the number of processes awaiting resolution have been reduced substantially; (iii) a large part of the waste associated with the rework was eliminated; (iv) takt-time guidelines are being followed, aligning production to demand; (v) tools have been created capable of providing objective information, minimizing decision-making based on unsupported information; (vi) it contributed to the monitoring of employee performance and to the gradual decrease in the number of human errors; (vii) *Poka-yoke* mechanisms, capable of alerting the user to a possible existence of errors in the process, were implemented and (viii) some phases of the process were automated, allowing the channelling of resources for VA activities.

As a theoretical contribution, and based on the present experience, it can be concluded that Lean, although originating in traditional areas such as production, can contribute very positively to less tangible areas such as services and particularly information management.

3.4. Cases of Lean Applied to Ergonomics

3.4.1. The Case of Company H.1

Problem contextualization and objectives - being part of a major automotive group, Company H is highly experienced in Lean practices, also having its own methodology with a lot of similarities with the Toyota Production System. Considering the company's motto (safety is a priority) and given the existence of some ergonomic problems in some jobs, it was decided to modify the supply process of the lines. The project delt with the improvement of the product flow in the assembly of gearboxes, from their physical storage to the supply in the assembly lines, minimizing the logistic imperfections and the issues of low ergonomic index in the workstations; due to this last issue, the goal was to achieve the values of the ergonomic indices stipulated by the factory (maximum of 20% NOK).

Methodology, results and conclusions - an ABC analysis identified gearboxes as group A (16% production and 82% revenue). It was decided, concerning ergonomics, to improve the supply of the production lines based on moving parts on rolling bases towed by a forklift, a charlotte or an AGV, acting as a *Mizusumashi*. These tours, with a frequency of 2, 8 and 24 hours, are called *tournées*. In this project, 6 of the 11 *tournées* of the internal logistics department (DIL) were analysed and improved. The study of the supply to the production lines involved a data analysis concerning ergonomics at the points of loading and unloading and sampling of times of provisioning of the packages in the rolling base and corresponding *tournée*. Results revealed that only 20-30% of the supply was OK with respect to ergonomics and the total tour time could be reduced, with emphasis on the preparation and loading time.

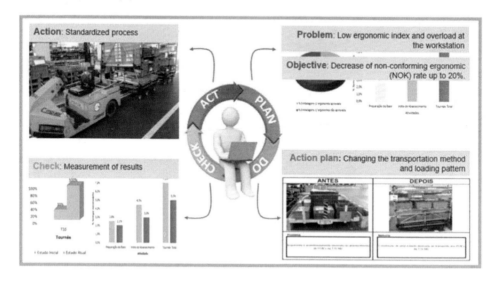

Figure 6. PDCA cycle for a *tournée*.

The use of a PDCA cycle, Figure 6, allowed to significantly improve those problems. Ergonomics rose to the range of 96% to 100% OK and the preparation and loading times of the bases reduced 30% - 50%. There was an aggregation of two tours, resulting in a yearly decrease of 792 turns to the assembly lines (reduction of movement), increasing the safety and freeing space to park the rolling bases. A standardized supply plan has also been drawn up for all tours including: components to be loaded, routes to be carried out, location of the Supermarket in the warehouse, maximum capacity and autonomy and quantity of packaging to be supplied at the place of supply. The Picking/Kitting diagram, the elaboration of a Waste Walk and recording of times revealed the non-use of 2 to 4 hours per day of 4 rolling bases. Solutions pointed to the changing of the layout and the use of AGV. The estimated saving was 30,000 Euro per year.

3.4.2. The case of Company H.2

Problem contextualization and objectives - one of the main goals of the project was to improve the ergonomics of the tasks required for supplying the production lines. The KPIs used to reach this goal were: total distance covered by the operators and an ergonomic indicator which measured the number of boxes being moved above a maximum weight (this weight would vary according to the height where the boxes were moved from/to).

Methodology, results and conclusions - firstly, the supply to the production lines was studied and all the processes mapped, with detailed information concerning unitary operations done by logistics personnel. The flow of materials from the warehouses to the production lines was analysed, allowing having an accurate idea of the amount of materials moved to the lines, the intervals between supplies, and the routes taken. This was done with the help of Spaghetti diagrams, interviews with the operators, and analysing historical data.

The first step was to reduce the distance covered by operators. To this end, all the NVA activities during the transportation were identified and targeted for elimination (e.g., checking the production schedule several times, making dedicated trips to bring empty boxes to the warehouses, etc.). Moreover, it was uncovered that the most efficient routes were not being taken, also leading to unnecessary movements. New routes were therefore drawn and made standard, which allowed reducing the number of different routes from 11 to 9 and reduce the overall distance travelled in 19%, which amounts to less 3,500 meters per day. Visual management was also employed to reorganize movable racks, assigning dedicated colours to them, with dedicated tracks and easily interpretable labels; considered to have a good impact in the well-being of the operators (avoiding accidents and making the routes less stressful).

Other ergonomic aspects were also studied, namely, the weight being transported by the operators (as this was considered the most important ergonomic factor). Moving boxes with less than 15 kg was considered acceptable, between 15 kg and 25 kg was considered a significant effort, and over 25 kg was considered a huge effort. Moreover, the frequency and height at which they were moved also influenced the ergonomics of the task: a significant effort was considered acceptable if the posture was considered correct (height of the movements between 700 mm and 1100 mm) and with low frequency (performed less than 10 times per hour). Overall, 80% of the boxes being moved were within the acceptable range, 13% needed a significant effort, and 7% required a huge effort.

To meet these goals, the weight of the boxes was changed, by introducing in the Supermarket a new Work standard that ensured the reduction of the number of components per box to meet the ergonomic limits. This allowed to increase the percentage of acceptable movements to 96%, reduce movements requiring significant effort to 4%, and eliminate movements requiring a huge effort.

Concluding, this Lean project allowed reducing the distance covered and the weight being carried by logistics operators, thus improving the ergonomics of the tasks in internal logistics.

3.4.3. The Case of Company I

Problem contextualization and objectives - the increasing market competitiveness and volatility have become a challenge for organizations, causing high levels of stress for workers, which in most cases end in Work-related Musculoskeletal Disorders (WMSD). To tackle this problem, which also has a strong implication on productive performance, companies increasingly tend to look for mechanisms to increase the safety of their workers, while increasing satisfaction and promoting higher productivity.

This case aimed to conduct a study in occupational ergonomics, based on the application of a set of tools, some of them from the Lean philosophy in a warehouse, more specifically in the area of repacking, trying to eliminate or minimize the risk factors that cause WMSD at those workstations.

Five specific objectives, which guided the methodology used, were defined: (i) ergonomic evaluation of repacking tasks; (ii) identification of tasks that foster the development of WMSD; (iii) identification of wastes in the process; (iv) implementation of ergonomic improvements; and finally, (v) implementation of improvements in the process under study.

Methodology, results and conclusions - to reach the defined objectives, the following steps were followed: (i) in a first phase, ergonomic evaluation methods to identify the risk factors and potential wastes associated with the movements of material manipulation by the operators; (ii) in a second phase, and in order to eliminate these wastes and consequently implement mechanisms to improve the associated processes, while creating the desired working conditions, a set of Lean tools.

To study the jobs in the repacking area, as well as the tasks that are performed by the employees operating in those posts, direct observation was carried out, together with informal interviews. This allowed understand the problems and consequently to select the most appropriate methods to obtain a holistic and correct evaluation of the tasks. Three different methods were applied: the NIOSH method (Waters et al. 1993) to evaluate the acyclic tasks; the Bosch method to evaluate the cyclical tasks; and finally, to complete the evaluation of the previous methods, the REBA method (Hignett and McAtamney 2000) was used to evaluate the body postures of the collaborators and in what ways they contribute to the development of WMSD.

At a later stage of the project, the Nordic Musculoskeletal Questionnaire (NMQ) (Nkuorinka et al. 1987) was used to study the symptomatology of warehouse employees.

The obtained results allowed identifying a set of risk factors to which also a set of wastes was associated. Moreover, and to assess the severity of these identified risk factors, they were validated by the employees, using the 5Whys technique. The set included: (i) unnecessary displacement to reach the materials; and (ii) inadequate geometry of the workstations to the employees' statures, which contributed to more unnecessary movements. These factors led to the employees having more material near the workstation than they needed, contributing to the disorganization of the work space while increasing the risk factors of WMSD.

Thus, a set of improvement proposals was defined, some of which were implemented during the project, namely the optimization of the repacking area, the Standardization of certain procedures, and the organization of workspaces based on 5S and Visual Management.

The post-implementation results seemed to be encouraging, with the improvement of workspaces, especially in what can be seen from visual management. Based on a survey made to the employees of those workstations, a reduction of wasted time required for the execution of the tasks was also confirmed, due to the reduction of the operations carried out, as well as an increase in user satisfaction and a decrease in the number of users' risk factors of WMSD.

CONCLUSION AND FUTURE WORK

Over the years Lean has played a significant role in finding and eliminating inefficiencies or wastes in the shop floor. The methods and tools used in the context of Lean have enabled many companies to improve production processes and increase throughput, while reducing costs. Moreover, they have often instilled a continuous improvement culture in those companies.

Having originated in the shop floor, Lean has found its way into other areas/processes within organizations. Efficiency in these other areas, such as management, logistics, accounting or human resources, is equally important. Therefore, waste elimination can also be targeted in these areas using Lean's methods and tools, once the necessary adjustments are made.

This work presents several successful cases where Lean was adapted and used across some areas of companies of such diverse sectors as consumer goods packaging, footwear, cork and automotive, to name a few.

Table 6. Lean methods and tools used and main results obtained on the companies

Company	Main methods and tools used	Waste reduced or eliminated
A	Kanban; Supermarket; VSD	D-O-N-I-M-E (Logistics)
B	Spaghetti diagram; 5S; Standardization	D-W-N-T-I-M-E (Logistics)
C	Pareto analysis; 5S; SMED; OEE	D-O-W-N-M (Production)
D	Kaizen; Kanban; Heijunka; Mizusumashi; Ishikawa diagram	O-W-T-I-M-E (Logistics)
E	Kaizen; 5Whys; Standardization; VSM; Pareto analysis; Breakdown analysis; Yamazumi; graph; Kanban	D-O-W-N-T-I-M-E (Inform. flow)
F	DMAIC; VSM; Ishikawa diagram; 5W2H; 5Whys; 5S; PDCA	D-O-W-N-T-I-M-E (Inform. flow)
G	5W2H; Ishikawa diagram; SIPOC; 5S; Poke-yoke	D-O-W-N-T-I-M-E (Inform. flow)
H.1	ABC analysis; Mizusumashi; PDCA; Standardization; Supermarket	D-O-W-T-M-E (Ergonomics)
H.2	Spaghetti diagram; Visual management; Supermarket; Standardization	D-O-W-N-T-M-E (Ergonomics)
I	5Whys; Standardization; Visual management; 5S	D-O-N-T-M-E (Ergonomics)

Results show that significant gains could be achieved in the various case studies. In the cases from Logistics and Production the following could be observed: overall less stock and stock outs, reduced total stoppages time, lower changeover times, OEE increased, and the distance travelled by logistics operator and number of tours was significantly reduced. In the cases focusing in information flow, among the several gains, we highlight the following: reduction of the volume of tasks, less time spent on management activities, less area needed for the different support activities, less waste, and gradual decrease in the number of human errors. Finally, in the cases where Lean was applied to ergonomics: less movement, less weight carried by operators, activities overall causing less burden to workers, risk factors identified, and a general rise of the ergonomic indicators and standards of each company.

Table 6 summarizes the main lean methods and tools used in the ten cases studies and the pure waste that was reduced or eliminated.

These case studies lead us to conclude that Lean is equally applicable and beneficial in almost all departments and activities, potentially making it a viable management philosophy across the company.

For future work it is suggested to extend the application of Lean to other less traditional areas and following up on the case studies reported in this work to know if the gains obtained are maintained or even improved.

ACKNOWLEDGMENTS

This work was supported in part by the Portuguese Foundation for Science and Technology (FCT - Fundação para a Ciência e a Tecnologia), through CIDMA - Center for Research and Development in Mathematics and Applications, within project UID/MAT/04106/2013 and through IEETA - Electronics and Informatics Engineering of Aveiro Research Unit - in the context of the project PEst-OE/EEI/UI0127/2014.

REFERENCES

Aqlan, F., S. Lam, S. Ramakrishnan, and W. Boldrin. 2014. "Integrating Lean and Ergonomics to Improve Internal Transportation in a Manufacturing Environment." *IIE Annual Conference and Expo 2014*, 3096–3101.

Aqlan, F., S. Lam, M. Testani, and S. Ramakrishnan. 2013. "Ergonomic Risk Reduction to Enhance Lean Transformation." *IIE Annual Conference and Expo 2013*, 989–97.

Bateman, N., L. Philp, and H. Warrender. 2016. "Visual Management and Shop Floor Teams – Development, Implementation and Use." *International Journal of Production Research* 54 (24): 7345–58. doi:10.1080/00207543.2016.1184349.

Battini, D., N. Boysen, and S. Emde. 2013. "Just-in-Time Supermarkets for Part Supply in the Automobile Industry." *Journal of Management Control* 24 (2): 209–17. doi:10.1007/s00187-012-0154-y.

Baudin, M. 2007. *Working with Machines: The Nuts and Bolts of Lean Operations with Jidoka*. Cambridge: Productivity Press.

Beamon, B., and T. Ware. 1998. "A Process Quality Model for the Analysis, Improvement and Control of Supply Chain Systems." *Logistics Information Management* 11 (2): 105–13. doi:10.1108/09576059810209991.

Belekoukias, I., J. Garza-Reyes, and V. Kumar. 2014. "The Impact of Lean Methods and Tools on the Operational Performance of Manufacturing Organisations." *International Journal of Production Research* 52 (18): 5346–66. doi:10.1080/00207543.2014.903348.

Blijleven, V., K. Koelemeijer, and M. Jaspers. 2017. "Identifying and Eliminating Inefficiencies in Information System Usage: A Lean Perspective." *International Journal of Medical Informatics* 107: 40–47. doi:https://doi.org/10.1016/j.ijmedinf.2017.08.005.

Brah, S., and W. Chong. 2004. "Relationship between Total Productive Maintenance and Performance." *International Journal of Production Research* 42 (12): 2383–2401. doi:10.1080/00207540410001661418.

Chookittikul, J., S. Busarathit, and W. Chookittikul. 2008. "A Six Sigma Support Information System: Process Improvement at a Thai University." In *Fifth International Conference on Information Technology: New Generations (Itng 2008)*, 518–23. doi:10.1109/ITNG.2008.212.

Coimbra, E. 2013. *Kaizen in Logistics and Supply Chains*. USA: McGraw-Hill Education.
Costa, B., and E. Vieira. 2010. "Risk Factors for Work-Related Musculoskeletal Disorders: A Systematic Review of Recent Longitudinal Studies." *American Journal of Industrial Medicine* 53: 285–323. doi:10.1002/ajim.20750.
Ferreira, J. 2017. "Implementação de Um Processo de Melhoria Contínua Na Área Da Logística Interna Na Colep." [Implementation of a continuous improvement process in the internal logistics of Colep] University of Aveiro, Portugal (MSc Thesis [in Portuguese]).
Ferreira, R. 2015. "Princípios Lean Aplicados À Gestão de Informação: Um Caso Prático No Departamento de Custeio Do ECCO." [Lean principles applied to information management: A pratical case in the costs department of ECCO] University of Aveiro, Portugal (MSc Thesis [in Portuguese]).
Ferreira, T. 2014. "Metodologias Lean Na Logística Interna Da Bi-Silque." [Lean methodologies in the internal logistics of Bi-Silque] University of Aveiro, Portugal (MSc Thesis [in Portuguese]).
Forno, A., F. Pereira, F. Forcellini, and L. Kipper. 2014. "Value Stream Mapping: A Study about the Problems and Challenges Found in the Literature from the Past 15 Years about Application of Lean Tools." *The International Journal of Advanced Manufacturing Technology* 72 (5): 779–90. doi:10.1007/s00170-014-5712-z.
Ghinato, P. 1998. "Quality Control Methods: Towards Modern Approaches through Well Established Principles." *Total Quality Management* 9 (6): 463–77. doi:10.1080/09544129883398.
Gomes, J. 2018. "Avaliação E Melhoria Das Condições Ergonómicas de Trabalho Num Armazém – Um Caso Prático Na Bosch Security Systems." [Evaluation and improvement of the ergonomic working conditions in a warehouse – A pratical case in Bosch Security Systems] University of Aveiro, Portugal (MSc Thesis [in Portuguese]).
Goodman, E. 2012. "Information Analysis: A Lean and Six Sigma Case Study." *Business Information Review* 29 (2): 105–10. doi:10.1177/0266382112450767.
Gwiazda, A. 2006. "Quality Tools in a Process of Technikal Project Management." *Journal of Achievements in Materials and Manufacturing Engineering* 18 (1–2): 439–42.
Hammarberg, M., and J. Sunden. 2014. *Kanban in Action*. 1st ed. Greenwich, CT, USA: Manning Publications Co.
Hendrick, H. 2003. "Determining the Cost - Benefits of Ergonomics Projects and Factors That Lead to Their Success." *Applied Ergonomics* 34 (5): 419–27. doi:10.1016/S0003-6870(03)00062-0.
Hicks, B. J. 2007. "Lean Information Management: Understanding and Eliminating Waste." *International Journal of Information Management* 27 (4): 233–49. doi:10.1016/j.ijinfomgt.2006.12.001.
Hignett, S., and L. McAtamney. 2000. "Rapid Entire Body Assessment (REBA)." *Applied Ergonomics* 31 (2): 201–205. doi:10.1016/S0003-6870(99)00039-3.
Hines, P., and D. Taylo. 2000. *Going Lean*. Cardiff: Lean Enterprise Research Centre - Cardiff Business School.
Hüttmeir, A., S. de Treville, A. van Ackere, L. Monnier, and J. Prenninger. 2009. "Trading off between Heijunka and Just-in-Sequence." *International Journal of Production Economics* 118 (2): 501–7. doi:10.1016/j.ijpe.2008.12.014.
Ibbitson, A., and R Smith. 2011. *The Lean Information Management*. Toolkit. Ark Group.

Imai, M. 2012. *Gemba Kaizen: A Common Sense Approach to Continuous Improvement Strategy*. 2nd ed. New York: McGrawHill Professional.

Jasti, N., and R. Kodali. 2015. "Lean Production: Literature Review and Trends." *International Journal of Production Research* 53 (3): 867–85. doi:10.1080/00207543.2014.937508.

Jiménez, M., L. Romero, M. Domínguez, and M. Espinosa. 2015. "5S Methodology Implementation in the Laboratories of an Industrial Engineering University School." *Safety Science* 78: 163–72. doi:https://doi.org/10.1016/j.ssci.2015.04.022.

Jones, D., and J. Womack. 2011. *Seeing the Whole Value Stream*. 2nd ed. Cambridge: Lean Enterprise Institute.

Kerber, B., and B. Dreckshage. 2011. *Lean Supply Chain Management Essentials : A Framework for Materials Managers*. New York: CRC Press.

Keyte, B., and D. Locher. 2016. *The Complete Lean Enterprise: Value Stream Mapping for Administrative and and Office Processes*. Boca Raton, FL: CRC Press.

Khodambashi, S. 2014. "Lean Analysis of an Intra-Operating Management Process - Identifying Opportunities for Improvement in Health Information Systems." *Procedia Computer Science* 37 (1877): 309–16. doi:10.1016/j.procs.2014.08.046.

Koukoulaki, T. 2014. "The Impact of Lean Production on Musculoskeletal and Psychosocial Risks: An Examination of Sociotechnical Trends over 20 Years." *Applied Ergonomics* 45 (2 Part A): 198–212. doi:10.1016/j.apergo.2013.07.018.

Kwak, Y., and F Anbari. 2006. "Benefits, Obstacles, and Future of Six Sigma Approach." *Technovation* 26 (5–6): 708–15. doi:10.1016/j.technovation.2004.10.003.

Liker, J. 2004. *The Toyota Way: 14 Management Principles from the World's Greatest Manufacturer*. New York: McGraw-Hill.

Lopes, R. 2013. *Ferramentas Lean E Gestão Da Informação Na Indústria Corticeira*. [*Lean tools and information management in the cork industry*] University of Aveiro, Portugal (MSc Thesis).

Millheiro, P. 2011. *Aplicação de Metodologias Lean Nas Prensas Isostáticas Da Vista Alegre*. [*Application of lean methodologies in the isostatic presses of Vista Alegre*] University of Aveiro, Portugal (MSc Thesis).

Nkuorinka, I., B. Jonsson, A. Kilbom, H. Vinterberg, F. Biering-Sørensen, G. Andersson, and K. Jørgensen. 1987. "Standardised Nordic Questionnaires for the Analysis of Musculoskeletal Symptoms." *Applied Ergonomics* 18 (3): 233–237. doi:10.1016/0003-6870(87)90010-X.

Ohno, T. 1988. *Toyota Production System: Beyond Large-Scale Production*. Portland: Productivity Press.

Oliveira, C. 2017. *Implementação de Um Sistema Kanban Para a Logística Interna Na 4Lean*. [*Implementation of a kanban system for the internal logistics of 4Lean*] University of Aveiro, Portugal (MSc Thesis).

Omogbai, O., and K. Salonitis. 2017. "The Implementation of 5S Lean Tool Using System Dynamics Approach." *Procedia CIRP* 60: 380–85. doi:https://doi.org/10.1016/j.procir.2017.01.057.

Pereira, J. 2015. *Aplicação Dos Princípios Lean No Desenvolvimento de Um Sistema de Informação: Um Caso Prático Na Industria Corticeira*. [*Application of lean principles in the development of an information system: A practical case in the cork industry*] University of Aveiro, Portugal (MSc Thesis).

Pereira, J., and L. Teixeira. 2017. "Contribution of Lean Principles in the Information Systems Development: An Experience Based on a Practical Case." In *Engineering Systems and Networks: The Way Ahead for Industrial Engineering and Operations Management (Lecture Notes in Management and Industrial Engineering)*, edited by M. Amorim, C. Ferreira, M. Junior, and C. Prado, 117–25. Springer International Publishing. doi:10.1007/978-3-319-45748-2_13.

Rahman, S. 2006. "Quality Management in Logistics: An Examination of Industry Practices." *International Journal of Supply Chain Management* 11 (3): 233–40. doi:10.1108/135985406610662130.

Ré, M., and L. Teixeira. 2018. "Information Systems in the Context of Industry 4.0: A Lean Approach to Information Flows for the Calculation of Indicators." In *Proceedings of the Portuguese Association for Information Systems Conference*, 19 pgs (in press). APSI/PTAIS.

Rolo, F. 2018. *Melhoria de Fluxo de Abastecimento Das Linhas de Montagem de Caixas de Velocidades Na Indústria Automóvel*. [*Improvement of the supply flow of gearboxes assembly lines of the automotive industry*] University of Aveiro, Portugal (MSc Thesis).

Rother, M., and R. Harris. 2001. *Creating Continuous Flow*. Cambridge: Lean Enterprise Institute, Inc.

Rother, M., and J. Shook. 2003. *Learning to See: Value Stream Mapping to Add Value and Eliminate MUDA*. 1st ed. Cambridge: Lean Enterprise Institute.

Rushton, A., P Croucher, and P. Baker. 2017. *The Handbook of Logistics and Distribution Management - Understand the Supply Chain*. 6th ed. New York: Kogan Page.

Santos, Z., L. Vieira, and G. Balbinotti. 2015. "Lean Manufacturing and Ergonomic Working Conditions in the Automotive Industry." *Procedia Manufacturing* 3: 5947–54. doi:10.1016/j.promfg.2015.07.687.

Saurin, T., and C. Ferreira. 2009. "The Impacts of Lean Production on Working Conditions: A Case Study of a Harvester Assembly Line in Brazil." *International Journal of Industrial Ergonomics* 39 (2): 403–12. doi:10.1016/j.ergon.2008.08.003.

Shimbun, N. 1995. *Visual Control Systems*. Productivity Press.

Shingo, S. 1985. *A Revolution in Manufacturing: The SMED System*. Cambridge: Produtivity Press.

Soares, S., and L. Teixeira. 2014. "Lean Information Management in Industrial Context: An Experience Based on a Practical Case." *International Journal of Industrial Engineering and Management* 5 (2): 107–114.

Sutherland, J., and B. Bennett. 2007. "The Seven Deadly Wastes of Logistics: Applying Toyota Production System Principles to Create Logistics Value." *CVCR White Paper #0701*. www.lehigh.edu/~inchain .

Teixeira, L., C. Ferreira, and B. Sousa-Santos. 2019. "An Information Management Frmework to Industry 4.0: A Lean Thinking Approach." In *Human Systems Engineering and Design*, edited by T. Ahram, W. Karwowski, and R. Taiar, Vol. 876, 1063-1069. Switzerland: Springer Nature. doi:10.1007/978-3-030-02053-8_162.

Teixeira, M. 2018. *Melhoria Do Processo de Abastecimento Das Linhas de Produção de Componentes Mecânicos Na Renault Cacia*. [*Improvement of the supply process of mechanical components assembly lines in Renault Cacia*] University of Aveiro, Portugal (MSc Thesis [in Portuguese]).

Toyota Motor Corporation. 1998. *The Toyota Production System: Leaner Manufacturing for a Greener Planet*. Tokyo: TMC Public Affairs Division.

Tyagi, S., X. Cai, K. Yang, and T. Chambers. 2015. "Lean Tools and Methods to Support Efficient Knowledge Creation." *International Journal of Information Management* 35 (2): 204–14. doi:10.1016/j.ijinfomgt.2014.12.007.

Tyagi, S., A. Choudhary, X. Cai, and K. Yang. 2015. "Value Stream Mapping to Reduce the Lead-Time of a Product Development Process." *International Journal of Production Economics* 160: 202–12. doi:https://doi.org/10.1016/j.ijpe.2014.11.002.

Waters, T., V. Putz-Anderson, A. Garg, and L. J. Fine. 1993. "Revised NIOSH Equation for the Design and Evaluation of Manual Lifting Tasks." *Applied Ergonomics* 36 (7): 749–776. doi:10.1080/00140139308967940.

Whitten, J. L., and L. Bentley. 2016. *Systems Analysis and Design Methods*. 7th ed. Irwin: McGraw-Hill.

Wilson, L. 2010. *How to Implemente Lean Manufacturing*. New York: McGraw-Hill.

Womack, J., and D. Jones. 2003. *Lean Thinking: Banish Waste and Create Wealth in Your Corporation*. 2nd ed. New York: Simon & Schuster, Inc.

Womack, J., D. Jones, and D. Roos. 1990. *The Machine That Changed the World*. New York: Free Press.

In: Lean Manufacturing
Editors: F. J. G. Silva and L. C. Pinto Ferreira
ISBN: 978-1-53615-725-3
© 2019 Nova Science Publishers, Inc.

Chapter 2

FROM THE FACTORY FLOOR TO NEW PRODUCT DEVELOPMENT: DEVELOPMENT AND IMPLEMENTATION OF A LEAN ASSESSMENT TOOL

T. Welo[*]
NTNU – Department of Mechanical and Industrial Engineering, Norway

ABSTRACT

In today's competitive business climate, many firms have attempted to expand the Lean concept to new Product Development (PD). Yet there exists a number of definitions of Lean when this concept is applied to PD, where the artefact is information and knowledge rather than a physical object. Its practical applicability remains unknown as there exist only a few dependable case studies of Lean being applied in PD. We hypothesize that the potential benefits of Lean PD can only be realized once the concept is contextualized and made scalable to the actual business environment. In this chapter, we develop a Lean PD maturity framework for identifying gaps between current capabilities and those deemed necessary in the future. The overall research objective is to identify differences between companies as to how they assess Lean PD capability gaps within their operational contexts. The framework consists of a three-level hierarchical model with 6 key components, 22 characteristics and 66 capabilities or practices. The framework has been tested in four global knowledge-intensive product manufacturing companies having R&D operations in Norway, using a continuous descriptive five-level maturity grid method. The results show that to which degree PD practices are 'project-driven' or 'process-driven' largely determine which components (or capabilities) where the companies identify having the main improvement potential. It is concluded that the framework has proven its capability as a practical assessment tool, which can be used in other business contexts to identify maturity gaps as a starting point for Lean PD transformation activities.

Keywords: new product development, lean concept, implementation, framework, capabilities, practices, contextual assessment, comparative case study

---

[*] Corresponding Author's E-mail: torgeir.welo@ntnu.no.

1. INTRODUCTION

Nowadays manufacturers, customers and consumers operate globally in highly competitive marketplaces. This trend is expected to continue at increased strength in the future. Business counterstrategies such as low-cost-country sourcing may provide immediate benefits in terms of cost reductions and expanded market presence, but does not guarantee sustaining competitiveness (Hines,1998). For example, many automakers have been struggling financially for quite some time, despite their early outsourcing efforts. In a broader perspective, there are many examples showing that cost reductions for instance by developing global vehicle architectures, cutting labor and development costs, increasing outsourcing and subcontracting, leveraging stronger supplier competition, establishing new purchasing methods, etc. do not guarantee long-term profits, or even the future existence of a company, according to Kahn et al. (Kahn, 2017).

One key question is then how a product manufacturing firm can become a future winner in the market place, particularly if it is primarily based in a high-cost country? The only meaningful answer to this question is to improve its capability in inventing, developing and producing new products that offer increased new value to customers. Hence, companies must strengthen their innovation capabilities such that attractive products that are more attractive reach the marketplace earlier than products offered from competitors—before improved technology is available and before the market changes. Innovation represents in general terms a match between an unmet need or problem (customer want), its solution (technology and thing), the human knowledge needed, and the commercially successful use (of the solution) in the marketplace. Huthwaite (Huthwaite, 2007) defines innovation as "the process of creating new (customer) value with a minimum waste". In other words, the ability of bringing ideas into viable concepts along with the further process of developing and launching new products define a company's innovation capability and its long-term viability in the marketplace.

The application of concurrent engineering practices (Prasad, 1996) over the past two decades has played a major role in drifting new Product Development (PD) from being an intangible, almost mystical activity belonging to the product engineering group, to becoming a concern that involves the entire enterprise. There is currently no longer a clear distinction between an innovation process and a PD process, other than the latter usually exclude high-risk efforts such as basic research, see Marxt and Hacklin (Marxt, 2005). Therefore, bridging the gap between the research phase and the development phase, sometimes referred as the 'valley of death' (Markham, 2010), is critical to the outcome of any innovation effort. For example, Toyota's ability to manage risk by effectively integrating the research/concept phase (Kentou) and the conventional development phase is possibly one of the most important factors to their history of success (Welo, 2011). Another success factor—nowadays when all significant players have introduced Lean in their manufacturing system, with different level of inspiration from the Toyota's Production System (TPS)—is Toyota's Product Development System (TPDS), commonly referred to as Lean Product Development (LPD). However, manufacturing and PD are indeed very different matters—so are TPS and TPDS—and it is of great importance to know to what extent, how, and with what potential Lean principles can be applied to help improve PD operations in Western companies, which are fundamentally different from Toyota in terms of operation, strategy and, even more importantly, culture.

In this chapter on Lean Product Development, focus is placed on discussing the fundamentals of PD and innovation as a basis, before aiming to take the lean concept from the factory floor to PD. The objective is to develop and pilot a lean PD capability-maturity assessment tool, taking into account the fundamentals of PD and the essential differences between PD and manufacturing, as well as the context of the company.

The reminder of this chapter is organized as follows: Section 2 discusses the theoretical fundamentals associated with product innovation in preparation of the application of Lean in PD (LPD). It elaborates on the differences between manufacturing and product development, and gives a brief overview of the history of Lean along with a comparison of LPD and some classical PD theories. Section 3 presents a contextual model for LPD. Section 4 discusses an LPD capability-maturity assessment tool developed with basis in our current understanding of LPD. The tool is built as a hierarchical system of components, characteristics and sub-characteristics. Section 5 presents results from the application of the tool in four global companies, which operate in different industries and contexts. Finally, Section 6 gives the conclusion.

2. THEORY: LEAN IN THE PERSPECTIVE OF NEW PRODUCT DEVELOPMENT FUNDAMENTALS

2.1. Some Definitions of Innovation, Product Development and Value

Innovation is according to Carlson and Wilmot (Carlson, 2006) defined by as: "*the successful creation and delivery of a new or improved product or service in the marketplace. (...). Innovation is the process that turns an idea into value for the customer and results in sustainable profit for the enterprise.*" New Product Development (NPD) can be defined as: '*the collective activities, or system, that a company uses to convert its technology and ideas into a stream of products that meet the needs of customers and the strategic goals of the company*" (Kennedy, 2008). The overall goal of a company's product development efforts is thus to maximize the overall productivity from its overall product portfolio, aiming to maximize the financial return from its new product (R&D) investments, see e.g., Cooper and Edgett (Cooper, 2005). This implies maximizing the expected commercial value of every new product in the market place, while minimizing the cost and time invested, within the constraints of resources and investment provided.

Although the definitions of innovation and PD may appear confusingly similar, the former usually involves the entire time line from the very first ideation, invention, commercialization and deployment, including use, service and end-of-life in a cradle-to-cradle perspective, see Figure 1. Following from the definition above, any innovation must result in a successful financial outcome for the owners. On the contrary, new solutions to problems that fail to result in targeted financial returns remain inventions but are strictly not innovations. The PD phase usually covers a shorter time line from program approval through production and product launch, as indicated by the colored area in the figure. PD is sometimes considered the collective company efforts undertaken once there exists a 'customer' for a planned product; i.e., usually after the so-called fuzzy front-end (scoping) and before deployment in the market place. Overall, in more popular terms, it can be stated that PD is the multi-disciplinary problem-solving efforts conducted to bridge the gap between customer needs and a new solution that satisfy those needs.

Creating customer value in today's highly competitive markets has proven to be difficult. As many as 70 to 90% of new-product introductions fail in the marketplace. Many of these failures are caused by lack of understanding of customer and user needs and values (Welo, 2012); (Nakagawa, 2009); (Carlson 2006); (Gordon, 2006); (Sanders, 1992). According to Carlson and Wilmot (Carlson, 2006), the most successful companies are the ones that focus on customers with a shared language and tools for understanding customer value, as well as a systematic process of creating customer value, as illustrated in Figure 2.

Product innovation is the frontier to sustain and improve competitiveness, according to Elverum and Welo (Elverum, 2015). In this regard, understanding customer value is the most important principle; and this is also the first principle in the lean philosophy. Customer value can be defined as "the benefits that a customer explicitly or implicitly ascribes to a product, relative to its price" (Browning, 2003). Historically, early lean PD efforts were usually motivated from the desire to reduce time-to-market and production cost through focus on lead time and production cost, see Figure 2 (right). Lean efforts in this cost-lead time perspective usually focus on eliminating waste rather than creating additional customer benefits. Research by Gautam et al. indicate that this type of Lean (PD) strategy is more successful in companies targeting incremental innovations than in companies targeting more radical innovations (Gautam, 2008).

Competitiveness can only be sustained by extending a company's PD efforts into the domain of differentiation through offering additional customer benefits, as indicated by the third 'innovation' dimension in Figure 2. This dimension can further be decomposed into two sub-dimensions: *product performance* (quantitative characteristics) and *product meanings* (qualitative characteristics) Gudem, 2013), see Figure 3, which is made with inspiration from Verganti (Verganti, 2009).

This model illustrates that a viable product innovation strategy should potentially leverage a customer-value proposition that target both these dimensions. Particularly in the case of consumer products, identifying latent user needs and use these to add new meanings to a product is a frontier for the success. In many new product cases, adding meanings are more important than adding features or improving performance characteristics. This is illustrated by the blue arrows in the figure, showing that a marginally improved technology can either lead to an incremental innovation, or generate a radical innovation if combined with new meanings or experiences. Apple and Harley-Davidson are two company examples that have demonstrated how important the dimension 'meanings' is to their financial success (Oosterwal, 2010).

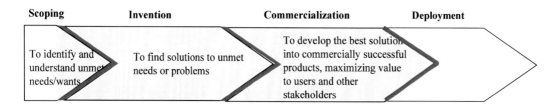

Figure 1. Time line for innovation process with a typical PD process indicated with the shaded area.

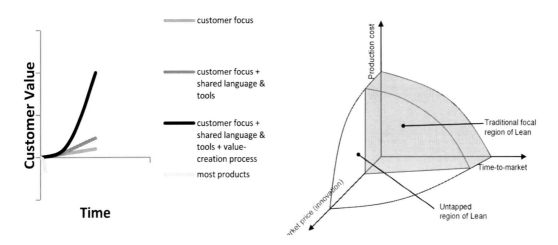

Figure 2. (a) Strategies to create customer value (Carlson and Wilmot, 2006); (b) Three dimensions of new-product development in competitive businesses.

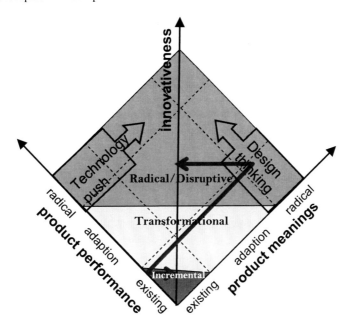

Figure 3. Innovativeness as represented by the two dimensions product performance and product meanings.

Leveraging product innovation could be done by following an outside-in-inside-out strategy, hence combining 'Design thinking' with 'Technology push', see Figure 3. The former is based on the assumption that identifying and transforming needs into ideas and product concepts that ultimately satisfy those needs, in the so-called fuzzy front end of a company's innovation efforts, is associated with great ambiguity. Therefore, the early stage of the innovation process represents the best opportunity for creating viable product concepts before other factors, such as reliability, risk migration, cost, investment levels, etc., restrict the solution space, see e.g., Sobek et al. (Sobek, 1999). Any innovation effort—lean or not lean—must leverage perceived product benefits, including features, functions, performance, attributes, properties, etc. as well as meanings, experiences and emotions associated with a product in everyday life (Verganti, 2009).

In summary, defining customer value essentially means understanding the underlying factors associated with different product benefits, realizing that these are continuously changing as technology, markets and competitive situation change. Moreover, important customer needs are typically implicit or tacit in nature (Olsen, 2011), and are in many cases difficult to capture. Hence, one side of the Lean coin, 'doing the right thing', is closely related to understanding customer value, whereas the other side of the Lean coin, process efficiency, is closely related to 'doing things right' to reduce cost. When taking the Lean concept to PD to gain competitiveness in the market place, the approach have to target both these sides of the Lean coin.

2.2. Some Fundamental Differences between Manufacturing and NPD

Some of the typical key characteristics that separate PD and manufacturing are summarized in Table 1. To the very basic, the overall objective of manufacturing is to use material and by machineries or manually convert it into a physical product that meets financial and strategic goals of the company. In PD, on the other hand, the overall objective is to make a recipe of useful information that reduces the risk of producing a new product to an acceptable level for the company.

Perhaps the most striking difference between manufacturing and PD is the conception of value. Manufacturing deals with physical products, which enable value to be separated from waste by systematically following the product throughout the process route, asking three essential questions according to Fiore (Fiore, 2005):

(1) Does the specific operation physically change the product (in the direction to solve customer needs)?
(2) Is the customer willing to pay for the change—does s/he need the specific feature accommodated in the operation?
(3) Is the operation done correctly first time (without rework)?

Answering 'yes' to all these three questions, the operation is by definition value-added. In PD, however, the work product is essentially information turned into knowledge, which means that value must be assigned to the system of generating data, conversion into information, transformation into (re)usable knowledge that solves the problem in question or helps the company solve future problems (strategic value). Unlike manufacturing, where value can be optimized by eliminating unnecessary activities (waste), value in PD is mainly related to providing the right information when needed as input to activities conducted to transform the design into a product that meet customer needs.

Waste elimination methods and related thinking from manufacturing are generally not very useful in PD. As an example, consider an engineer calculating stiffness of a (partially) constrained aluminum beam that, say, is going into an automotive frame assembly. The design of the cross section of the beam can be broken down into smaller and smaller subsets of activities, all the way down to a detail level where the engineer lifts her hand, grabs a pen, moves her arm, rests her elbow on the desk, grabs paper, starts drawing a concept, checks meeting part con figurations, define model, searches for beam equations, checks load requirement, and so on, until arriving at the final formula, $\delta = \varphi \cdot PL^3/EI$, where δ represent the allowable deflection, EI is the flexural rigidity, L is free length, φ is a constant (depending on constraints of meeting parts) and P is the applied load requirement.

Table 1. Characteristics that separate product development from manufacturing (Welo, 2011)

	Characteristics of Manufacturing	Characteristics of Product Development
Objective	To use material and (by machineries or manually) make them to a physical product that meets financial and strategic goals	To make a recipe of useful information that collectively reduce the risk of producing a new product
Value-added (First Lean Principle)	Clear, simple definitions of waste and value: proven tools available	No precise definitions to separate value from waste: no proven tools
	High processing-to-total time (i.e., 'value added contents')	Low processing-to-total time (i.e., 'value added contents')
	Value notion mainly linked to quality, delivery and costs	Value notion includes multiple user desires, preferences and meanings
	Waste easily visible; may be approached at micro process level	Most wastes not easily visible; have to be approached at system level
	Waste stems mainly from doing unnecessary activities	Waste stems mainly form doing activities with wrong input
Product/ Inventory	Physical object	Information, knowledge
	Visible	Invisible
	One place at a time	Multiple locations at a time
	Product requirements are constraints (or limitations); 'must haves'	Product requirements represent opportunities (value); 'could haves'
Tasks	Outcomes are predictable and fixed	Outcomes are unpredictable, floating
	Less tasks are better: actions that do not add value are to be removed	Less tasks not always better: apparent waste at action level may generate value at system level
	Duration is consistent (within tight limits)	Duration is non-consistent (and unpredictable)
	Done by machines and then people	Done by people and then machines
People	Practitioner and factory floor people with basic education level	Engineers and designers with college or university degree
	Disciplined and used to comply with procedures and methods made by others	Hesitant to unreservedly follow rules and procedures
	Humble belief that reuse (what has worked in the past) saves time and efforts, and reduce risk exposure	Reuse makes work dull and erode 'signature' from new products: i.e., not-invented-here syndrome
	Team culture that promotes the safe and well-established, avoiding risk	Individuals that promote entrepreneurship, creativity and risk taking
Process	Operational process within the stringent definition of a process (set of activities that creates an output from different inputs)	A process in a broader context—and partly a network—with 'floating' input, activities and output
	Simple and linear with a number of repetitive, sequential steps dealing with physical objects	Complex and non-linear, generating, transforming and integrating information
	Mostly independent operations (no or low degree of 'adaptivity')	Interrelated operations (high degree of 'adaptivity')
	Duration of seconds and minutes	Duration of months and years
	Variability prohibited (waste)	Variability a necessary means to add value, especially in front end
	Requirements are fixed throughout all process steps	Some requirements are fixed but many change during the process (sometimes the entire context)

Using a stop watch time study approach, as would be the case in Value Stream Mapping (VSM) of a manufacturing process, to identify when the engineer actually did something valuable (i.e., writes on the sheet of paper = process time) and compare it with the time it took from somebody 'ordered' the task until the dimensions and drawings of the beam were communicated to the internal customer (i.e., cycle time) would reveal a very low productivity; i.e., in lean terms, a process full of waste. Improving each micro-operation by remove 'waste' would obviously increase productivity. However, does it matter much to the overall value creation if the process time was 2 hours or 30 minutes? Most likely the beam calculation was not on the critical path of the project anyway. What would truly generate waste at system (business) level in this example, however, is if the engineer failed to use the right value of φ in the calculation model. Then, the project manager would kick off tooling with the wrong part dimensions, other parts (and tooling) in the assembly would be affected, the prototype assembly would possibly fail the stiffness test, the problem would bounce back to engineering, new validation calculations would be needed, tool modifications, and so on. Hence, the problem was not that the engineer worked 'inefficiently' but rather that she ended up giving wrong information as input to the internal customer (the project manager) using the information in the next stage of the PD process.

Once the correctly designed part goes into the production line, however, lack of value would be identified solely from answering 'no' to one of the three questions (1-3) listed above. For example, the end customer did not seem to really need a datum hole, which was put on the drawing, since he can use one of the bolt holes instead ('customer not willing to pay for hole'). Alternatively, the grinding operation at the ends of the beam turned out to be unnecessary since the beam is only going to be handled by robots in the assembly operation ('does not physically alter the product'). Furthermore, once in production, a pile of beams stacked at the end of the production line with bolt holes missing due to a piercing tool braking constantly, would be the ultimate proof of rework ('not done correctly first time'). Hence, question 3 would be answered by a 'no'. This simple example may illustrates some of the basic differences between the perceptions of value and waste in manufacturing and PD.

Another important factor that separates production from PD is the characteristic of product requirements. In production, requirements are solely constraints that must be met to be able to ship a good part to the customer ('must haves' according to the Kano model, see Figure 4 (Kano, 1984)). For 'must haves' (also denoted 'dissatisfiers'), the most positive outcome that can be achieved is to ship a product with a quality that the customer expects. Even if the product characteristics exceed the minimum requirements, for example, by providing a better dimensional capability than specified on the drawing, the customer would expect production releasing dimensional control in the process and pass on the savings in terms of cost reductions. In the case of PD, product requirements are opportunities, which are denoted 'could haves' in the Kano model. By successfully implementing a 'could have' feature you could have a delighted customer, who did not expect to get a product with this feature or function. The worst outcome you could achieve by not implementing it, would be a customer that did not notice that the feature or function was missing. The 'should haves' are typically features whose degree of implementation satisfies customers to an increasing degree; i.e., the more, the better (more value).

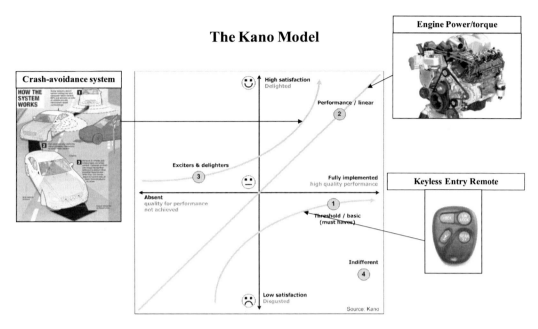

Figure 4. The Kano model showing three product feature examples, including keyless entry remote ('Dissatisfier'); engine power ('Satisfier'); crash avoidance system ('Delighter').

In summary, since the value potential in production is fixed, waste elimination is important. On the other hand, the value potential in product development is nearly unlimited. In PD, therefore, maximizing value is more important than eliminating waste, especially if the latter is attacked at micro operation level.

Comparing the basic tasks typically done in production and PD, the former has fixed outcomes (a product according to quality specification), duration (cycle time) and the less operations you need to make the product, the better. PD is an exploration and learning process, which means that the outcomes and duration are unpredictable from the very beginning, which is exactly the reason why the company invests in it. PD is very much about defining problems and closing gaps by generating information and knowledge, and activities are directed as learning takes place. This is in contrast to the thinking behind a pre-scheduled governance process with fixed tasks for completion, such as Stage Gate (Cooper, 2005).

One other important issue that separates PD from production is that doing less is rarely better in PD. If a PD team for different reasons arrives at a single concept solution early and uses all its efforts to iterate on this concept to product specifications, this is likely to generate a suboptimal outcome for end customers, internal customers and other stakeholders. On the other hand, if the team keeps the solution space open and thoroughly explores the potential of a set of different concepts, this is more likely to result in an optimal solution—although this involve front-loading more resources. The additional efforts needed to explore the design space to find a more optimal solution, however, would result in design iterations and multiple concepts that were not going into production. According to 'Lean production thinking', these iterations are wasteful activities since the value potential is assumed fixed; i.e., any final solution would cost the same, provide the same benefits, sell in the same volumes, capture the same markets, etc. In a Lean PD perspective, however, the investments made upfront in exploring the design and solution spaces would result in a more desirable product, which ultimately could result in better financial performance. Moreover, the additional concepts and iterations not directly used in the new product represent sources for organizational learning and knowledge generation; i.e., providing 'strategic value' for the company.

Improving operations, this being production or PD, involves changing the company culture and with this comes changing the behavior of individuals along with formal procedures. While management is the driver for such changes, the practical deployment has to take place by the employees. In this connection, practitioners and factory floor people are usually disciplined people who are used to follow procedures and methods made by others. Their believes and behaviors are commonly based on experience, and team culture promotes the safe and well-established, avoiding risks by knowing what has worked in the past. Any improvement effort that would save time and reduce risk is greatly appreciated among production workers. These elements, along with the methodological nature of lean production, has made lean implementation in production environments such a big success. On the other hand, Product engineer usually have a college, MSc or PhD degree. They are so-called knowledge workers. Rather than being told by others, they are used to establish methods at their own. They are typically more hesitant to unreservedly follow rules and procedures than production workers. From school they are used to solve problems and create solutions by combining algorithmic and heuristic work, see e.g., Martin (Martin, 2009). Creativity and entrepreneurship—and with this comes risk taking—remain a part of the culture in PD environments. Their job and reward scheme are to create something new; putting their own 'design signature' on new products create pride and self-esteem among design engineers.

PD is a process in a broader context—more a network than a process—with 'floating' input, activities and output. Applying lean manufacturing methods, ones that are developed for a linear, simple and repetitive process with a physical product to which value and waste can be assigned, to a dynamic and complex PD process is not possible without performing significant adjustments, both in terms of methodology and mindset. One of the major differences in terms of process characteristics is production focusing on repeatability and robustness, with variability being the enemy. In PD, however, variability is a necessary means to generate value—without variability (and risks), the potential for creating new value and innovation vanishes.

2.3. Lean from the Manufacturing Floor to the Product Developer's Desk

Although the origin of Lean may be seen as the Japanese response to the crude oil crisis in the 1970s (Schoenberger, 1982), the concept has a significantly longer history through the Toyota Production System (TPS), see Haque and Moore (Haque, 2002). TPS represents a practical demonstrator of lean processes and has become the benchmark for competitive manufacturing throughout the world. Toyota has also been the original inspiration to the applications of a more systematic approach to improve efficiency and effectiveness in product development, also denoted Lean.

Development of innovative products combined with improved production techniques made the Japanese industry develop quickly after the World War II. Japanese companies adapted and optimized at that time already known manufacturing techniques and principles, many of them originated in USA and Europe. They developed the techniques to a new level and put them into operational practice to surpass the US mass production concepts, making

Japan become one of the great economic nations. One of the main features of Japanese production concepts is their focus on developing optimal systems for multi-item, small lot production (flexible automation) after experiencing that only a minor part of the output actually reached the volume output characteristics of mass production. The present goal of a modern manufacturing plant is the unmanned factory, which is more easily achieved in connection with few-item mass production than with multi-item small lot production. A flexible manufacturing system (FMS) refers to an automated production system, which connects machines, presses and/or machining centers by conveyor belts or unmanned material handling systems such as industrial robots. The unmanned flexible production system is possible, except in the preparation of raw materials and shipment of the final product. It permits a 24-hour operation, and can be applied in dirty work environments or where dangerous operations take place. The success of using robotics in connection with flexible production systems in Japan was associated with the early general acceptance of the work unions.

Flexible automation (systems for multi-item, small lot production), Total Quality Control (TQC) and the Kanban system were key success factors. The Kanban system was initiated in Japan and developed further by Toyota. The system is based on two criteria: a) just-in-time and b) self-actuation where the responsible worker stops the line and remove the cause of the problem. In Kanban systems, work-in-progress is minimized by the workers performing the required work on the part provided by the proceeding work station at the necessary time. The main challenges associated with the Kanban system is related to employer-employee cooperation, people rotation, training and adaptability to order-fluctuations.

The Kanban system and other manufacturing techniques were commonly applied for production of high-quality, compact-size products, integrating mechanical technology and electronics (Hitomi,1985). This (production) technology-product-driven transformation along with many other more cultural factors made Japanese companies outperform many of their Western competitors. In Japan, the educational level of factory workers is relatively high and there is relatively little difference in terms of status and wages between workers and managers. Employees and employers also have a mutual desire for the advancement of production and improvement in operations. Moreover, worker's self-management is commonly used and so-called quality control circles are practiced where workers analyze inferior parts from quality and productivity standpoints. The quality circle concept is also referred to as total quality control (TQC).

One other success factor is the close cooperation between design and production engineers in Japanese firms, conducting R&D activities on the products from a system optimization perspective; i.e., performing a successive sequence of R&D, design and manufacturing within a framework of integrated production process. Also, lifetime employment by seniority rule and the Japanese wage system create skilled workers and loyalty within the work group. These factors related to PD, manufacturing technology and people (culture) combined with long term focus have been key for making Japanese companies competitive over the past several decades.

As mentioned above, Lean manufacturing is an operational process management strategy derived mainly from TPS in the 1980s (Womack, 1990); (Karlsson, 1996); (Womack, 1996); (Baines, 2006), focusing on waste reduction in the factory. The term Lean Production was introduced by researchers of MIT in the late 1980s, describing an approach that used less of everything and did it faster and cheaper than traditional production techniques. Womack et al.

(Womack, 1990) published their famous book The Machine that Changed the World, which covers a set of techniques under a common heading used to explain the success of Japanese auto manufacturer in the 1980s. Many of the techniques were already well known at that time but looking at lean as part of an overall production strategy was new-and somewhat confusing since there was no detailed description as to how to apply it in practice.

Lean methods (in manufacturing) are often linked with the Six Sigma methodology (Karlsson, 1996) because of their emphasis on reducing variability. Lean is generally considered as a set of tools that assists in the identification and elimination of waste, the improvement of quality and production time, as well as cost reduction, as discussed above.

The term Lean was first coined by MIT student John Krafcik in his 1988 article 'Thriumph of the Lean Production System' (Krafcik, 1988) based on his thesis. It is noteworthy there is no equivalent or translation of Lean in the Japanese language. In the mid-1990s, the Lean notion was extended to the Lean Automotive Factory and later to the Lean Factory, focusing on costs, quality and delivery, with major contributions from researches at the University of Michigan (Sobek, 1999); (Morgan, 2002); (Liker, 2003); (Liker, 2006); (Ward, 2007). In the late 1990s, the lean concept was extended to the Lean Enterprise, and during the past decade it has been introduced to new areas such as PD, engineering, design, software development (agile) and accounting. Over a 20-years period, the focus of lean had drifted from waste elimination to cost, quality and delivery and then further into customer value. More recently, the Lean notion has also been used within management and leadership, healthcare and hospitals, military and even other public and governmental organizations.

2.4. Comparison of Traditional PD and Lean PD

Traditional PD and Lean PD (LPD) are not directly comparable since their overall goal and scope are somewhat different. The former describes a systematic approach of well-defined steps guiding engineers to create a product that solves a given technical or functional problem. The latter, on the other hand, defines a direction to make engineering processes more effective and efficient, improving company outcomes. LPD describes how people, processes and technology shall operate and be managed to help a company become more productive by 'pulling' customer value up the value chain. Lean is more a philosophy, or mindset, rather than a detailed methodology for solving engineering problems, according to Welo (Welo, 2011). Traditional 'schools of PD thought' explain the steps that have to be conducted and what has to be done during these steps, whereas LPD describes the working philosophy associated with the PD process. However, LPD and traditional PD are not contradictory in any respect, as it can be claimed that lean complements traditional methods by adding managerial factors such as effectiveness and waste reduction (e.g., people, money, rework). Table 2 summarizes some key goal characteristics of both approaches.

Based on the above discussion, one can ask in which way traditional PD approaches are lean? Analyzing altogether six different traditional schools of PD thought and one integrated PD approach—ones that are commonly referred as benchmarks in traditional PD—in the context of lean, the findings are summarized Table 3. Here a set of selected lean components are correlated with traditional PD approaches. When a lean component is indicated with an 'x', it is a part of the traditional PD approach, and vice-versa. The chosen lean components represent a broad selection, which are essentially based on Morgan and Liker (Morgan, 2006).

Table 2. Characteristics of traditional PD and lean PD

Goals of Traditional Product Development	Goals of Lean Product Development
Gives specific 'work instructions' to mainly engineers at detail level	Gives visionary and directional strategies for the entire company at system level with PD being the core component
Methodology that provides engineers with tools for solving a wide range of technical problems, and developing and designing products	A company-wide PD system aimed at maximizing value to the customer or user, within the constraints of value to other stakeholders
Focusing on developing the best technical solution (high quality) with basis in engineering excellence	Focusing on using an effective process to develop an overall optimal (customer) solution from a system perspective, including operational and strategic management
Use of knowledge and ideas to create solutions for technical problems	Effective capturing and reuse of knowledge and ideas for increased learning, and to develop solutions with highest possible value in the eyes of the customer
Can solve unknown problems and improve existing products; i.e., offering methodologies for both	Strong basis in known processes with predictable outcome (continuous improvement), minimizing technical risk within PD, i.e., after program definition
Follows parallel or sequential processes, aiming to solve the task as well as possible	Follows parallel processes, aiming to solve the task fast with effective use of resources

Table 3. Correlation between of traditional product development and components in lean product development

Lean Principle	Roden-acker	Tjalve	Pahl, et al.	Roth	Ehrlen-spiel	Hubka	Hein
Continuous control of requirements	-	x	x	x	(x)	X	x
Front load of the PD process	-	-	x	x	x	-	x
Understanding the customer	-	(x)	x	-	x	x	x
Integrate customer and supplier in complete development	-	-	-	-	-	-	-
Parallel processes	-	x	-	-	(x)	x	x
Increase standardization, reduce variation	-	x	x	x	x	x	(x)
Continuous improvement of product	x	(x)	x	x	x	x	x
Continuous improvement of process	(x)	-	x	-	-	-	x
Capturing and reuse of knowledge and experience	x	(x)	x	x	(x)	(x)	x
Capturing past knowledge in checklists	(x)	-	-	x	x	(x)	-
Short and precise knowledge capture	-	-	-	x	-	-	(x)
Early include all different departments	-	-	(x)	(x)	x	-	x
Learning Cycle	x	x	x	(x)	x	x	x
Set-based concurrent engineering	-	-	-	(x)	-	x	-
Solving the roots of problems	(x)	-	x	x	x	-	x

Rodenacker's approach (Rodenacker, 1970) is one of the earlier ones in systematic engineering design, and is still widely applied today. Rodenacker aims finding solutions for the cause-effect relations by following a stepwise methodology through logical, physical, and structural working principles. He uses a learning cycle similar to PDCA including the steps: information retrieval, information processing, information output, and checking. Capture, reuse and extension of knowledge, which are important for continuous improvement, all are parts of Rodenacker's approach.

Tjalve's main contribution (Tjalve, 1979) to the design methodology is variation of form. Product solutions and alternatives are developed by systematically varying size, number, structure and shape of the design elements. Tjalve uses a learning cycle, called 'product synthesis', similar to the learning cycle in lean. He proposes that the criteria vary from phase to phase and have an increasing number of details, based on information from the former step. This essentially reflects the lean principles, learning and continuous improvement.

Pahl et al. (Pahl et al., 2007) provide a linear, holistic and systematic engineering design process to help design engineers find solutions for products by the use of different tools. They suggest that a PD methodology should save time, reduce work load, speed-up understanding and help maintain active interest. Further, they want the different functions concerned with development of a product to collaborate early, which in accordance with lean principles. Problems should be detected early and clearly defined in the requirement list together with customer needs. Pahl et al. refer to a learning cycle, similar to the LAMDA cycle in Lean. They interpret the design process as a dynamic control process that continues until the information (content) has reached a level for optimum solution. Here, it should be noted that many lean approaches follow a similar strategy.

Roth (Roth, 2000) introduces design catalogues for engineers. 'Effects', 'effect owners', materials, etc. are systematically structured in catalogues, which make knowledge capture and reuse simple, providing the design engineers a set of standard solutions and recommendations. Roth states that it is important to define the correct problem statement early and to attack problems at the root cause. He does not explicitly use expressions such *customer* or *customer value*, which are important drivers in LPD. However, customer (value) may still be considered as part of his approach since customer satisfaction is mandatory for the success of a product. Roth applies engineering catalogs, which are essentially similar to knowledge-briefs within LPD (Kennedy, 2008); (Sobek, 2008). Experiences, standards, and former product solutions can be documented in a visual, engineering-friendly way by both approaches. The catalogues, which give fast and clear overview of alternatives, represent a knowledge-based approach. Catalogues can be adapted to the design process of a certain company, and they can also be extended. An additional core component of lean is standardization and the use of check lists. For instance, standard tables and check lists are used for the gathering requirements, and these can be adjusted and extended to meet new challenges. In LPD, a similar approach is employed by alternative concepts such as *house of quality and quality function deployment* (QFD).

Ehrlenspiel (Ehrlenspiel, 1985) discusses the influence of engineering design on product costs, including life-cycle costs. He proposes a number of opportunities to reduce product cost by correct selection of design features, production methods, materials, and good collaboration between different departments inside a company. Cost reduction opportunities lie in standardization of products, which is part of lean, by for instance using modular product concepts with standard parts or assemblies and customer-specific adaption of parts and assemblies. Ehrlenspiel uses *value analysis* to identify unnecessary costs, aiming to determine which product functions are absolutely necessary to accommodate the task that has to be accommodated to satisfy the customer, which can be associated with reduction of waste, meaning lean design. This methodology is also consistent with *value engineering* (Welo, 2011). Further, Ehrlenspiel encourages close communication between teams and short lines of communication, which supports the pull concept in lean. However, his approach is more focused and is guiding engineers to use cost reduction methods in detail, whereas LPD to a more extent considers problems associated with the overall PD system.

Hubka et al. (Hubka, 1988) introduce a theory for technical systems, including transformations (functions), organs (e.g., functional interfaces) and parts (components), where the organs represent the link between two components or one component and the user. Hubka proposes an approach similar to SBCE in Lean. The evaluation at the end of each phase is based on the status, the experience and learning of previous work, and the customer specifications. This resembles the lean principles of continuous learning, reuse of knowledge, and focus on customer value.

Hein and Andreasen (Hein,1988) introduce an approach that considers PD in a broader perspective, the so-called integrated product development (IPD). This is a more holistic approach, which includes engineering design, production, marketing, and organization. IPD seeks to integrate methodologies used in different departments of a company toward common goals, procedures, and attitudes. The customer is of key importance since s/he ultimately decides whether the product becomes a success or not. Focus is not just on the product itself but the entire execution environment, which is necessary to make the product successful in the market place. Hence, IPD makes a step forward from engineering design methodology towards LPD and product management (PM).

3. A Lean Product Development Model

One of the main challenges to move lean practices to other functional areas, such as PD, is the differences between the nature of a production process and that of a much more dynamic PD process. Anyone who is looking into the great body of knowledge that has been created to establish a common strategy for LPD would be surprised by the great variety of interpretations of the concept, implementation strategies as well as practical methods and tools. Although the different LPD schools of thought cover a large landscape, there is at least one common characteristic: i.e., very few well-documented success stories from outside Toyota.

Therefore, there is a need to develop a LPD framework with a broader applicability. Existing models in the literature have a tendency to focus on just a few single components. However, as an example, introducing 'set-based concurrent engineering' and a 'chief engineer' will not automatically improve PD performance if, say, resources are spent on project(s) with the lower potential. In addition, some models define a large number of interrelated (sub)components, sometimes denoted 'principles', without being mutually exclusive. For example, 'simultaneous engineering' and 'work leveling', in which improving simultaneous engineering capabilities will generally lead to better predictability of resource needs within different functions—and vice-versa. Since a general principle has "to stand the test of time and application", its practical usability is limited unless it is filled with more detailed content. The consensus among researchers is that lean, particularly when being applied to PD, is more of a mindset than a toolkit; see e.g., Womack and Jones (1996).

In the following, a brief description of the different components of a proposed LPD framework will be described. The developed model consists of six core components with different (sub-) characteristics. The present model is believed to represent a suitable compromise between being generic and specific at the same time, aiming to keep an overall system perspective with minimum interdependencies see e.g., Hoppman et al. (Hoppman, 2011) between the different components. A car wheel may be used as a metaphor to illustrate the model, see Figure 5. The wheel consists of two core components which have multiple design constraints and limited flexibility, including the hub (understanding customer value) and the rim/tire (knowledge and learning capabilities). By contrast, the design freedom for the spoke arrangement (the four other core components) to fulfill the functional lay-out of a wheel is much more flexible. This means that different numbers, patterns and spoke shapes may serve the same purpose of providing sufficient structural rigidity and integrity for transferring loads, while keeping the rim round and in the right position relative to the hub. Thus, different vehicles ('businesses') used on different roads ('markets') would require different wheel integrity. However, every wheel would need a round rim/tire and a central hub to provide its main functions. In the (figurative) context of LPD, therefore, the focus on customer value and knowledge management is key and indisputable, as has been discussed above, whereas inclusion of the other four core components is subjected to a great amount of flexibility and customization, depending on context and the needs of each individual company. A detailed interpretation of the six core components and the underlying characteristics associated with this model are given elsewhere.

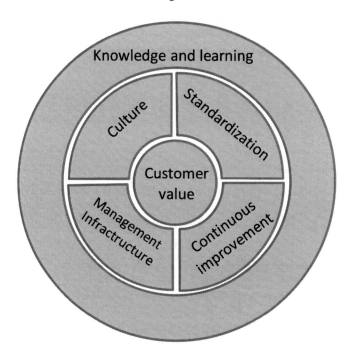

Figure 5. A knowledge-based model for the application of the lean concept in PD (Welo, 2011).

4. A LEAN MATURITY ASSESSMENT TOOL: PROCESS AND STRUCTURE

A questionnaire for assessment of lean practices in PD, with basis in the model presented above, has been developed. The main structure of the tool includes the characteristics associated with each LPD dimension and underlying characteristics. Each individual component of the LPD model is divided into from two to five underlying characteristics, all together 22 (Table 4). Each underlying characteristic (heading) is divided further into three sub-characteristics assigned with situational descriptions on a scale from 1 to 5, to which the assessment of each one is done. An example of the structure of one assessment sheet is shown in Figure 6. The rating includes both an estimate of the current state (C) state and the desired state (D) based on the business conditions and context of each company. The situational description serves as a guide for making an objective choice of the current and desired states based on three practices (sub-characteristics). It is worth noting that the main purpose is to identify gaps, $G = D - C$, rather than the 'score' (C). Based on the assessment, an overall gap (G) for each of the sub-characteristics can be estimated. The assessment tool was developed from literature studies, discussions with industrial companies as well as own hypothetical reasoning, experience and thoughts, following the layout of Nightingale et al. (Nightingale, 2002).

CUSTOMER FOCUS: ROLES AND VALUES		Underlying question: To what extent does the company work with the customer to understand current and future customer needs?			
Component		Description of the situation			Assessment rating
	Low (1)	Intermediate (3)	High (5)	Current	Future
Perceived role of customer:	Company internal preferences and priorities have usually greater impact on design choices and product specifications than customer needs.	Customer requirements, needs and preferences are generally prioritised but management and design teams sometimes miss important aspects due to insufficient field research in early phases.	'Customer first' is a core value in the company. Every employee views customer satisfaction as the primary driver for future company success, and works continuously to improve customer value.	c_1	f_1
Collaboration with customer in NPD:	Customers/end-users does not have an integrated role in the NPD process. The company has a short-term relationship with individual customers. The company only aims to barely meet predetermined specs and no efforts are made to establish directional product targets together with the customers.	A few customers are viewed as long-term partners and these are to some extent integrated in the NPD process. The bulk part of customers (groups) does not participate actively in the PD process, though.	The customer has a central, integrated role in the company's NPD process to create insight in true customer needs and wants. The company has a very good understanding of their products *from a user perspective*, and uses various methods to continuously extend user knowledge.	c_2	f_2
Integration of customer in continuous improvement of product quality:	Complaints or warranty claims are usually solved at contractual or legal level. These are usually considered as individual quality issues, more than opportunities to better understand *why* the customer is dissatisfied.	The most frequent and serious complaints and warranty claims are assigned to task groups, appointed at management level, resulting in a suboptimal learning process in the organization.	Complaints and warranty claims are used systematically as a source of knowledge for developing better and more desirable new products, as well as improving existing ones.	c_3	f_3
		Average score, C and F:		$\frac{1}{3}\Sigma c_i$	$\frac{1}{3}\Sigma f_i$
		Average gap, (G):		\multicolumn{2}{c}{$G = F - C$}

Figure 6. An exhibit of an assessment sheet for underlying characteristic, value and role of customer in LPD, associated with the core component 'customer focus'. Cross-referencing underlying characteristic #1 in Table 4.

Table 4. Summary of questionnaire used to identify areas for introducing lean in PD

Component	Characteristic	Main question to be answered (sub char.)	Rating
Customer Value	1. Role and Values	What role has the customer in the company's strategy and practices?	1 2 3 4 5 C: ○ ○ ● ○ ○ D: ○ ○ ○ ○ ●
	2. Interface between customer and E&D	How do customer desires and requirements reach design engineers?	...
Knowledge	3. Knowledge Value Stream	Rate role of knowledge in terms of capturing new markets and growing the business.	...
	4. Knowledge ownership and management	Is knowledge ownership defined, and is capturing process systematic managed?	...
	5. Cross-functional knowledge Flow	Assess practices for transferring knowledge between functional departments	...
	6. Set-Based Concurrent Engn.	To what extent is front loading and SBCE used in design and knowledge generation?	...
Management Infrastruct	7. Resource planning and management	Do functional departments and projects get the resources they need when needed?	...
	8. Product & portfolio management	Is there a systematic approach to prioritize projects with resource allocation?	...
Organization (Stabilization)	9. Communication between org. levels	Rate the communication practice and info. flow between organizational levels?	...
	10. Manufacturing's role in PD	What role (authority and responsibility) does manufacturing take in projects?	...
Management	11. Supplier's role in PD/company	What is the role of key suppliers and how are they utilized in the organization?	...
Standardization	12. Standardization of the PD process	Assess the product development process from its focus on quality of deliverables?	...
	13. Standardization for flexibility	Does your company standardize skill sets for flexibility in resource management/staffing?	...
	14. Design strategy	Is there a design strategy (reuse), and is it integrated as a part of the design practice?	...
	15 Standardization of problem Solving	Assess company's process for solving problem at the root cause and org. learning.	...
Continuous Improvement	16. Continuous Improvem. in PD	Is CI and systematic waste elimination in PD deeply rooted in company philosophy	...
	17. PD Productivity Measurements	Asses the way company actively uses metrics and productivity measures in PD	...
Culture	18. Trust, respect, and responsibility	To what extent are trust, respect, and responsibility core values in the organization?	...
	19. Fact-based decision making	Rate culture to make fact-based decisions in the organization at all levels?	...
	20. Creativity and entrepreneurship	Is creativity encouraged, valued and part of in product and technology strategy?	...
	21. Digital Tools in product D&E	Assess the role tools in achieving business and PD improvement goals?	...
	22. Simple and visual communication	To what extent is use of visual communication anchored in the culture?	...

The main objective of using the assessment in a case study is twofold: (a) to identify and select pilot areas for implementation of lean PD practices in companies; (b) to generate knowledge as a basis for academic learning and further studies into details associated with the different prioritizations. The assessment is conduced through a full-day workshop and included people from key functional areas in PD, typically 6-10 people. After introducing the LPD model and its associated practices, as well as discussing the characteristics, the questionnaire is completed on an individual basis. Then, the results are collected and processed into a format suitable for further discussion. In case there are significant variations between the individual assessment ratings, these form the basis for a discussion within the assessment team, aiming to arrive at a common assessment score. Prioritization of improvement areas is done collectively based on (i) evaluation of gaps, (ii) importance to business performance and (iii) efforts required to close those gaps.

5. CASE STUDY: THE IMPACT OF BUSINESS CONTEXT ON LPD GAP IDENTIFICATIONS

5.1. Contextual PD Dimensions

The business setting, or context, in which a company operates may be classified through four main dimensions, including strategy, financing, marketing and operations. Each of these dimensions can be further divided into a myriad of subcategories. Seen from the perspective of the actual process for developing products, however, it will be just as common to consider two other context-specific dimensions; namely, the degree to which PD activities are 'process-driven' or 'project-driven'. The former (dimension 'process-driven) is characterized through repetitive steps, stability, robustness, linearity and predictability, where most of the value is embedded in achieving the same repetitive outcome without variation every time. The extreme of this dimension (i.e., 100% process-driven) is represented by a linear production process. The dimension 'project-driven', on the other hand, is characterized through variability, uniqueness, newness and non-repetitiveness, where most of the value is embedded in differentiation aspects of the outcome from the PD activities. To the very extreme of this scale, (i.e., 100% 'project-driven'), one find PD activities in a typical entrepreneurial start-up. Such a firm may have only one project, little or no technology basis to which knowledge or infrastructure is tied up, relying much on its (outside) network to solve the problems necessary to create a viable product for the market place. It may be assumed that the characteristics of the PD process in any firm will reflect a combination of these two (almost) diametrically opposed dimensions, as illustrated in Figure 7.

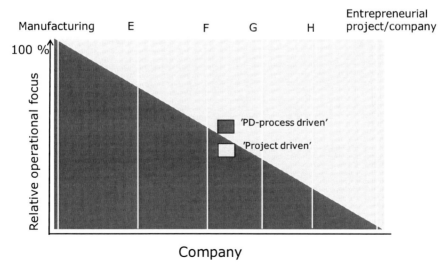

Figure 7. Model showing two diametrically opposed dimensions, reflecting to which relative degree the NPD operation is process or project driven.

50   T. Welo

An LPD assessment, as described above, has been conducted with four global companies (E, F, G and H) with R&D operations in Norway, whose characteristics represent a major part of the scale made up by the combination of the two suggested dimensions.

5.2. Results from Assessment

The assessment showed that Company E—an automotive supplier—whose NPD environment is mostly process-driven, has the higher capability gaps (G) within the assessment characteristics 'Cross-Functional Knowledge Flow', 'Resource Planning and Management', 'Manufacturing Integration' and 'Trust, Respect and Responsibility' all showings a gap of 3.0 between current (C) and desired state (D). The company demonstrated high potential in strengthening focus on knowledge generation and reuse, with multi-disciplinary approach being the major challenges ('Cross functional knowledge flow'). Company E also scores low on 'Resource Planning and Management', where the main challenges were 'Frontloading' and 'Integrating manufacturing early in the PD phases', since its manufacturing units typically do not prioritize assigning manufacturing people to participate in PD activities before the project gets closer to launch. The assessment team claimed that the lack of involvement typically results in iterations/loops, fire fighting and high launch costs, which could exceed the cost of adding more resources early (front-loading). The assessment also revealed that company E is a top-controlled organization, whose reaction to 'problems' is typically seen in terms of more controls and more frequent meetings with focus on check points.

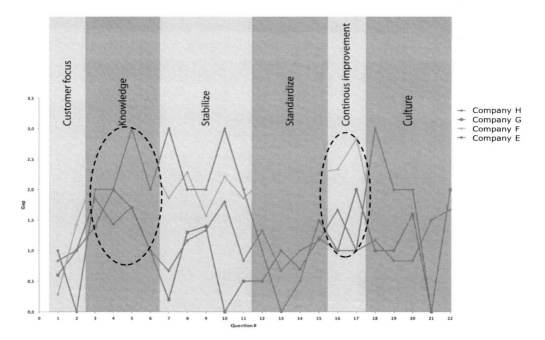

Figure 8. Result summary from assessment of four companies with different business environments, showing assessed gaps associated with characteristics of practice within each of the six LPD components (nos. are cross-referencing Table 4 above).

Figure 8 shows the overall findings from assessments made with the four companies. Here the vertical (y) axis depicts the gaps identified for different capability characteristics, which were evaluated by the multi-functional group in each company. The results show significant differences in terms of how each company evaluates where they have the larger mismatch (or gap) between current and desired future capabilities. For example, companies G and E assess their gap within component Stabilization (Management, infrastructure and organization) very differently. The same comments apply to companies E and F in theway they assess component Standardization.

A common characteristic, associated with this group of companies is that they seem to have the overall larger Lean PD capability-maturity gaps within the components 'Knowledge' and 'Continuous Improvement', as indicated by the broken-line ovals in the figure. This may be explained by the fact that very few companies have made sufficient commitments and related practices in becoming a learning organization, in which organizational knowledge is considered being a value stream in its own. Moreover, continuous improvement efforts are dependent on having a stable new PD environment in place, in combination with systematic efforts to standardize routine operations and processes. It is interesting to note that those companies that score high in current capabilities (low gap) in 'Stabilization' and 'Standardization' also score high (low gap) in 'Continuous Improvement'—and vice-versa—although only partially reflected in the results of company E.

Company F, which is in the defense and aerospace industry, is somewhat more project-driven than company E. Company F shows the higher gaps within components 'Continuous improvements in PD' with underlying characteristic 'Productivity Measurements in Product Development' showing a gap as high as 2.8 between desired state (D) and current state (C). Hence, this company has a high potential in implementing methods for measuring the productivity in new PD and thereby track progress toward defined company goals. The characteristic 'Cross-Functional Knowledge Flow' was another capability with high capability gap ($G = 2.6$). This result may be related to the company experiencing significant business growth prior to the assessment, causing information to be more scattered, hard to find and difficult knowing who knows what.

Several systems such as *Engineering Wikipedia* and *Sharepoint* have been implemented. However, different branches use different systems, and there is a lack of standards on how to capture knowledge and quality assure information. Another challenge is that strategic (long term) research is being conducted in the technology centers, while (short term) research necessary to complete each project is being conducted within the project itself. The result is that insufficient research is being done, and the knowledge is poorly communicated throughout the organization. All capability gaps within the categories 'Standardization' and 'Knowledge' are larger than 2.0. 'Customer focus' and 'Culture' are the two categories showing the lower gap for company F.

Company G—a supplier in the subsea, offshore and oil industry—is even more project-oriented in its PD operation than company F. The assessment shows that the higher gaps exist within core component 'Knowledge creation and reuse' with characteristic 'Knowledge ownership and management' shows a gap (G) as high as 2.0 between the desired state (D) and current states (C). This result may be traced back to the fact that the company is very much project-driven, and the functional areas have limited impact since there is less focus on strategic organizational knowledge development through long-term research. Therefore, this company demonstrates a large improvement potential in strengthening its knowledge transformation and organizational learning capabilities, by improving knowledge flow between projects as well as defining strategic internal projects to position the company long term. Company G should also evaluate if functional mangers should have a stronger ownership to in-house knowledge standards within their area of responsibility. 'Cross-functional knowledge flow' is another area where the practices in this company show a large gap ($G = 1.7$). Also, all the gaps associated with component 'Knowledge' were 1.0 or higher. The company demonstrates high improvement potential within characteristic 'Defining productivity measures in PD', which showed a maturity gap of 2.0 for component 'Continuous Improvement'. The third obvious improvement area is 'Culture' as represented by characteristic 'Simple and visual communication', showing a gap of 2.0. The company rates itself as a quite customer-oriented organization with gaps less than 1.0. This result may be related to the fact that the company operates in a Engineer-to-Order/B2B context, and is thus driven by strong customers and product specifications.

Company H—a total supplier within engineering and product innovation—is located far to the right in Figure 7. This indicates that most of the value from its new PD processes is embedded in its ability to create variable outcomes in terms of newness and uniqueness from project to project. This means that standardization has to take place at higher levels; i.e., process, working modes and quality system, rather than at product, component and solution level. The higher gaps were identified within component 'Knowledge', where the characteristic 'Learning and knowledge value stream' was assessed to have a gap of 2.0.

The company (H) seems to have a large improvement potential in strengthening the knowledge value stream by improving flow between projects and defining more strategic projects, in addition to customer financed PD projects. The assessment also indicated that this company should open the 'bulkheads' between different departments as well as defining clearer ownership to knowledge. The latter is closely linked to leveraging knowledge standards and organizational learning. The gaps associated with all four characteristics (3-6) within component Knowledge were 1.2 or higher. The assessment ranked also 'Integrating manufacturing in product development' with high gap. This may be seen in relation to the company's business model of offering PD services without owning the product, and their role in customer projects. However, to build long term relationship with customers the company enable customer satisfaction by integrating manufacturing concerns more into the PD process. Like the other companies in this case study, company H demonstrated a relatively weak basis for systematically improve its PD operations. This may also be related to the fact that the company to a large degree is a project-driven organization. In addition, since the business model is built on developing advanced new product solutions for its customers, being mostly paid by development contracts or by the hour, the incentives for continuous improvement are less obvious than in more process-driven organizations where profits comes from product sales. However, seeing this together with the need for initiating own strategic projects to strengthen the knowledge fundament, leveraging continuous improvements in project execution, could be a means to fee up time for more value-adding activities within company H.

CONCLUSION

This chapter discusses the fundamentals associated with the application of the Lean concept in product development. It has demonstrated that the original concept for manufacturing needs significant modifications before it can be applied in PD. The main reasons for this has been discussed in broad terms in this chapter.

A model for Lean PD has been proposed, including the following components (dimensions): customer value; standardization; stabilization, continuous improvement; knowledge; and culture. The model has been used as basis for developing a practical tool for assessing Lean capability maturity in PD. The assessment scheme divides the six components into 22 characteristics and 66 sub-characteristics. The tool is contextual in the sense that it allows assessing different business environments at an individual basis, using a relative scoring scale for current and future Lean PD capabilities.

The practical use of the assessment tool has been demonstrated by surveying PD capabilities in four different companies that operates in different industries (automotive, defense and aerospace, subsea and oil, and product engineering and innovation). The results show that there are significant differences between the companies as to how they rank their Lean PD capability maturity. In this connection, two important contextual dimensions were identified—process-driven and project-driven. These can be used to explain many of the differences observed between the companies, motivating the need for contextual Lean PD implementation strategies. More specifically, the results show that the main differences between the four companies are within the components 'Knowledge (and learning)' and 'Continuous improvement'. Based on the results obtained and discussed above, it is concluded that the Lean concept is only applicable in PD once it is modified to represent the fundamentals associated with product innovation and the company context. The developed assessment tool has through testing in four companies proven to be useful in identifying capability maturity gaps as a basis for a transition into leaner PD practices.

REFERENCES

Baines, T., H. Lightfoot, G. M. Williams, and R. Greeough. 2006. "State-of-the-Art in Lean Design Engineering: A Literature Review on White Collar Lean." *Journal of Engineering Manufacture* 220 (9), 1539-1547. doi: 10.1243/09544054JEM613.

Browning, T. R. 2003. "On customer value and improvements in product development processes." *Systems Engineering*, 6 (1), 49–61. doi: 10.1002/sys.10034.

Carlson, C.R., and W. W. Wilmot. 2006. *Innovation: The Five Disciplines for Creating What Customers Want*. Crown Business, New York. ISBN-13: 978-0307336699.

Cooper, R.G, and S. J. Edgett. 2005. *Lean, Rapid and Profitable New Product Development*. Product Development Institute. BookSurge. ISBN-13: 978-1439224601.

Ehrlenspiel, K. 1985. *Kostengünstig Konstruieren: Kostenwissen, Kosteneinflüsse, Kostensenkung. Konstruktionsbücher* [Cost-effective design: cost knowledge, cost effects, cost reduction. construction Books]. Band 35. Springer, Heidelberg. ISBN-13: 978-3540139980.

Elverum, C.W., and T. Welo. 2015. "On the use of directional and incremental prototyping in the development of high novelty products: Two case studies in the automotive industry." *J. of Engineering and Technology Management* 38, 71-88. doi: 10.1016/j.jengtecman.2015.09.003.

Fiore, C. 2005. *Accelerated Product Development – Combining Lean and Six Sigma for Peak Performance*. The Productivity Press, NY, USA. ISBN-13: 978-1563273100.

Gautam, N. and N. Singh. 2008. "Lean product development: maximizing the customer perceived value through design change (redesign)" *Int. J. Production Economics* 114, 313–332. doi: 10.1016/j.ijpe.2006.12.070.

Gordon, J. 2006. "Returning Insight To The Consumer." *Stagnito's New Products Magazine*. December 2006.

Gudem, M., T. Welo, L. Leifer, and M. Steinert. 2013. "Redefining customer value in lean product development design projects." *J. of Engineering, Design and Technology* 11(1), 71-89. doi: 10.1108/17260531311309143.

Haque, B., and M. J. Moore. 2002. "Characteristics of lean product introduction." *Int. J. Automotive Technology and Management* 2 (3–4), 378–401. doi: 10.1504/IJATM.2002.002096.

Hein, L., and M. M. Andreasen. 1985. *Integreret produktudvikling. [Integrated product development]*. Jernets Arbejdsgiver-forening, Copenhagen.

Hines, P., N. Rich. 1998. "Outsourcing competitive advantage: the use of supplier associations." *International Journal of Physical Distribution & Logistics Management* 28 (7), 524-546. doi: 10.1108/09600039810247489.

Hitomi, K. 1985. "The Japanese way of manufacturing and production management" *Technovation* 3, 49–55. doi: 10.1016/0166-4972(85)90036-7.

Hoppman, J., E. Rebentisch, U. Dombrowski, and T. Zahn. 2011. "A Framework for Organizing Lean Product Development." *European Engn. Management Journal* 23 (1), 3-15. doi: 10.1080/10429247.2011.11431883.

Hubka, V., M. Andreasen, W. E. Eder. 1988. *Practical Studies in Systematic Design*. Butterworth & Co., London. ISBN-13: 978-0408014205.

Huthwaite, B. 2007. *The Rules of Innovation, Institute for Lean Innovation*. Michigan. ISBN-13: 978-0971221048.

Kano, N., S. Nobuhiku, T. Fumio, and T. Shinichi. 1984. "Attractive quality and must-be quality." *Journal of the Japanese Society for Quality Control* (in Japanese) 14 (2), 39–48., doi: 10.1007/978-3-7908-2380-6_20.

Karlsson, C. and P. Åhlström. 1996. "The difficult path to lean product development." *J. Prod. Innov. Management* 13, 283–294. doi: 10.1016/S0737-6782(96)00033-1.

Khan, S. T., S. S. Raza, and S. George. 2017. "Resistance to Change in Organizations: A Case of General Motors and Nokia." *International Journal of Research in Management, Economics and Commerce*, 7 (1), 16-25. ISSN 2250-057X.

Kennedy, M. N., K. Harmon, and E. Minnock. 2008a. *Ready, Set, Dominate: Implement Toyota's Set Based Learning for Product Development*. Oaklea Press, Richmond, VA. ISBN-13: 978-1511659659.

Krafcik, J. F. 1988. "Triumph of the lean production system." *Sloan Management Review* 30 (1), 41–52. doi: 10.4236/ajibm.2012.22004 24,406.

Liker, J. K. 2003. *The Toyota Way – 14 Management Principles form the World's Greatest Manufacturer*. McGraw-Hill, NY, USA. ISBN-13: 978-0071392310.

Liker, J. K. and J. M. Morgan. 2006. "The Toyota way in services: the case of lean product development." *Academy of Management Perspectives*, 20 (2), 5–20. doi: 10.5465/amp.2006.20591002.

Markham, S. K., S. J. Ward, and L. Aiman-Smith. 2010. "The Valley of Death as Context for Role Theory in Product Innovation." *The Journal of Product Innovation Management*, 27 (3), 402-417. doi: 10.1111/j.1540-5885.2010.00724.x.

Martin, R., *The Opposable Mind*. Harvard Business School Publishing, MA, ISBN-13: 978-1-4221-3977-6.

Marxt, C, and F. Hacklin. 2005. "Design, product development, innovation: all the same in the end? A short discussion on terminology." *Journal of Engineering Design*, 16 (4), 413-421. doi: 10.1080/09544820500131169.

Morgan, J.M. 2002. *High Performance Product Development: A Systems Approach to a Lean Product Development Process*. PhD Thesis, University of Michigan, Ann Arbor, MI.

Nakagawa, T. 2009. "*Running the numbers on innovation success and failure,*" Blog – Tech, [online]. Available at: <http://www.straight.com/article-242874/running-numbers-innovation-success-and-failure> [Accessed 07 June 2010].

Nightingale, D. J., and J. H. Mize. 2002. "Development of a lean enterprise transformation maturity model." *Information Knowledge Systems Management* 3 (1), 15-30.

Olsen, T. O., and T. Welo. 2011. "Maximizing product innovation through adaptive application of user-centered methods for defining customer value." *J. of Technology Management and Innovation*, Vol. 6, Issue 4, pp 172-191. doi: 10.4067/S0718-27242011000400013.

Oosterwal, D. P. 2010. *The Lean Machine*. AMACOM, American Management Association, N. Y. ISBN-13: 978-0814413784.

Pahl, G., W. Beitz, J. Feldhusen, and K. H. Grote. 2007. *Engineering Design. A systematic Approach*, 3rd Edn. Springer, London. ISBN-13: 978-1-4471-6025-0.

Pink, D. 2009. *Drive: The Surprising Truth about What Motivates Us*. Riverhead Books. ISBN-13: 978-1594484803.

Prasad, B. 1996. *Concurrent Engineering Fundamentals*. Prentice-Hall, Upper Saddle River, NJ, USA. ISBN-13: 978-0133969467.

Rodenacker, W. G. 1970. *Methodisches Konstruieren. Konstruktionsbücher. [Methodical Construction. Construction Books]* Band 27. Springer, Berlin. ISBN-13: 978-3-662-08721-3.

Roth, K. 2000. *Konstruieren mit Konstruktionskatalogen. [Construct with Design Catalogs]* Band 1-2. Springer. ISBN-13: 978-3-540-67142-8.

Sanders, E. B.-N. 1992. "Converging Perspectives: Product Development Research for the 1990s." *Design Management Journal*, 3(4), 49-54.

Schoenberger, R. J. 1982. *Japanese Manufacturing Techniques: Nine Hidden Lessons in Simplicity*. Collier Macmillan, NY, USA. ISBN-13: 978-0029291009.

Sobek, D. K., A. C. Ward, and J. K. Liker. 1999. "Toyota's principles of set-based concurrent engineering." *Sloan Management Review*, 40 (2), 67–82.

Sobek, D. K., A. Smalley. 2008. *Understanding A3 Thinking*. Productivity Press, Boca Raton. ISBN-13: 978-1563273605.

Tjalve, E. 1979. *Systematic Design of Industrial Products. Institute for Product Development.* Technical University of Denmark, Newnes-Butterworths, London. ISBN-13: 978-8798136019.

Verganti, R. 2009. *Design-Driven Innovation: Changing the Rules of Competition by Radically Innovating What Things Mean*. Harvard Business Press, Boston. ISBN-13: 978-1422124826.

Ward, A. 2007. *Lean Product and Process Development*. The Lean Enterprise Institute, Cambridge, MA. ISBN-13: 978-1934109137.

Welo, T. 2011. "On the application of lean principles in Product Development: A commentary on models and practices." *International Journal of Product Development* 13(4):316 – 343. doi: 10.1504/IJPD.2011.042027.

Welo, T., T. O. Olsen, M. Gudem. 2012. "Enhancing Product Innovation through a Customer-centered Lean Framework." *International Journal of Innovation and Technology Management*, 9 (6), 250041 (28 pages). doi: 10.1142/S0219877012500411.

Womack, J. P., D. T. Jones, and D. Roos. 1990. *The Machine that Changed the World: The Story of Lean Production*, Harper Perennial, NY, USA. ASIN: B007ZHTZ8Y.

Womack, J. P. and D. T. Jones. 1996. *Lean Thinking: Banish Waste and Create Wealth in Your Corporation*. Free Press, NY, USA. ISBN-13: 978-0743249270.

In: Lean Manufacturing
Editors: F. J. G. Silva and L. C. Pinto Ferreira

ISBN: 978-1-53615-725-3
© 2019 Nova Science Publishers, Inc.

Chapter 3

BENEFITS AND CHALLENGES OF LEAN MANUFACTURING IN MAKE-TO-ORDER SYSTEMS

Oladipupo Olaitan[*], Anna Rotondo, John Geraghty and Paul Young

Enterprise Process Research Centre,
School of Mechanical and Manufacturing Engineering, Dublin City University,
Dublin, Ireland

ABSTRACT

In make-to-order systems, a key expectation from the implementation of lean manufacturing principles is to cut down on waste, improve flow and, therefore, the responsiveness of the system to fulfilling customer orders. Smooth flow is best achieved in such systems through flexibility of the system's resources, in the form of multi-skilled workers and adaptable machines that can pro-cess any product type and minimise setup times. Flexibility of the system layout is also needed to support variable product routings. These capabilities are achieved through a combination of cross-trained workforce, reconfigurable machines and cellular system layout etc.

However, research works have shown that increased flexibility also increases the complexity of the system's coordination. Therefore, it is important that trade-offs are made between the level of flexibility designed into a system and the coordination effort required for its effective operation. In this chapter, the benefits and challenges of flexibility are discussed, with focus on two case study companies. Issues such as workers' forgetting of skills and the extra sequencing and sched-uling of resources that might result from labour flexibility are discussed. Finally, the potentials of the evolving concepts of Industry 4.0, such as Smart Operators, Data Analytics and Machine Learning, in overcoming some of these issues are outlined.

Keywords: lean manufacturing, make-to-order, resource flexibility, smart operators, Industry 4.0.

---

[*] Corresponding Author's E-mail: oladipupo.olaitan2@mail.dcu.ie.

1. INTRODUCTION

In implementing lean manufacturing, a key expectation is to cut down on waste, improve flow and, therefore, the responsiveness of the system to fulfilling customer orders. However, this is not often easily achievable in make-to-order (MTO) systems because of variation in customer orders, whose variable processing and routing requirements place uneven demands on the system resources. Such system must rely on the flexibility of the system's resources, in the form of multi-skilled workers and adaptable machines to process any product type and minimise setup times. Flexibility of the system layout is also needed to facilitate the manufacture of products that have variable processing and routing requirements. These capabilities are achieved through a combination of cross-trained workforce, flexible and/or reconfigurable machines, flexible system layout, etc.

However, research works have shown that increased flexibility also increases the complexity of a system's coordination. Therefore, it is important that trade-offs are made between the level of flexibility designed into a system and the coordination effort required for its effective operation. In this chapter, the benefits and challenges encountered for the different levels of flexibility in a system's labour, manufacturing system and layout are discussed, with focus on two case companies. These will highlight issues commonly identified with flexibility, such as forgetting of skills and quality issues, and the need for extra coordination effort. These issues are going to be explored from the perspective of the two case companies, in combination with insights from literature. Both companies are MTO but with different types of manufacturing layout and levels of flexibility designed into their systems.

Finally, the potential impacts of recent advances in the Industry 4.0 paradigm in providing answers to some of the issues raised in the chapter are discussed. The possibility to achieve full labour flexibility without the attendant issues through the smart operators of Industry 4.0 is previewed. Specifically, it will be discussed how the use of augmented and virtual reality to aid worker tasks can eradicate the need for hard-learning of skills. It will be shown that there are significant opportunities that could facilitate the implementation of lean manufacturing in MTO systems, which are generally considered difficult candidates for its implementation.

1.1. Background: Lean Manufacturing in MTO Systems

The setup of a manufacturing system for lean manufacturing aims to ensure that as minimal resources as possible can be used in the production of variety of products, mainly through a U-Shaped layout of the machines. To facilitate the switch between product types, the workers are cross-trained to process different product types on different machines. While this is the ideal situation for achieving smooth flow of products through the system, it necessitates a significant amount of decisions on how it should be set up, followed by ongoing, day-to-day coordination decisions to ensure the process runs smoothly. And, this is evident in the volume of research topics that have touched on how to best take these decisions. The initial set up decisions have been covered in research topics such as:

1. System design and layout, which involve the design of the machines and the decision on how to set up the machines (Koren and Shpitalni 2010, 130-41; Correia et al. 2018, 667-8),
2. The cross-training configuration, which determines the number of skills to train each worker on, and the number of workers to trained per skill (Brusco and Johns 1998, 499-515; Bobrowski and Park 1993, 257-68),
3. The shift configuration, which involves determining the number of shifts to have, the shifts' durations and the number of workers to have per shift (Campbell and Diaby 2002, 9-20; Campbell 2011, 1038-47),
4. On the other hand, the coordination decisions have been covered in research topics that involve:
5. Planning the assignment of workers to machines (Liu, Wang, and Leung 2016, 162-79; Sayın and Karabatı 2007, 1643-58),
6. Taking decisions on the duration of time for which each worker should work on a machine before switching to another machine, and to which machine the worker should be switched (Berman, Larson, and Pinker 1997, 158-72),

Naturally, the coordination decisions are dependent on how the system is set up initially.

2. LABOUR FLEXIBILITY: BENEFITS AND CHALLENGES

Labour flexibility is increasingly adopted as an operational strategy to generate flexible capacity within a company and secure its competitiveness and survival in extremely dynamic manufacturing environments (Abrams and Berge 2010, 522). Labour flexibility can refer to a flexible adaption of labour capacity (i.e., number of temporary workers, total working hours, etc.) to fluctuating demand (Askar et al. 2007, 250). More frequently, it is intended as workforce versatility, which is the ability of workers to perform different tasks. It induces a dynamic view of workers skills (Attia, Duquenne, and Le-Lann 2014, 4548) and can be achieved through cross-training, or multiskilling (Abrams and Berge 2010, 522). A flexible workforce fosters the realization of more robust (Abrams and Berge 2010, 528; Sennott, Van Oyen, and Iravani 2006, 542) and efficient (Chen and Askin 2006, 89) production systems. From a wider perspective, the benefits of resource flexibility in terms of productivity have been investigated in the context of resource-constrained systems, including in acyclic networks (Vairaktarakis 2003, 726), cellular manufacturing systems (Cesani and Steudel 2005, 571-91), dual-resource constrained systems (Xu, Xu, and Xie 2011, 309-18), and in understaffed assembly lines (Downey and Leonard 1992, 469-83). The general conclusion of such studies is that a flexible workforce can generate substantial productivity improvements, especially when the complexity of systems and/or tasks increases (Gomar, Haas, and Morton 2002, 108; Mc creery and Krajewski 1999, 2056), and that the most benefits are realised by means of incremental flexibility. It however points out that over-skilling can have neglectable impact on productivity, i.e., having workers learn more than two skills (Gomar, Haas, and Morton 2002, 107). Nevertheless, workers' flexibility provides many other benefits not necessarily related to production efficiency. Flexibility can increase workers' motivation and responsibility (Croci, Perona, and Pozzetti 2000, 244), reduce workers' alienation (Downey and Leonard 1992, 469) and improve production quality, especially because it can help avoid repetitive tasks that can cause ergonomic effects can cause fatigue, boredom, or repetitive stress. Additionally, it can facilitate learning (Hopp and Oyen 2004, 922), aid communication and coordination (Buzacott 2004, 216), and improve organizational culture (Herzenberg, Alic, and Wial 2000, 41).

On the other hand, these benefits, and especially the increased productivity gained by reducing idle times due to more flexible assignment options, may conceal potential reductions in workers' efficiency due to deviations from skill specializations (Abrams and Berge 2010, 524; De Bruecker et al. 2015, 9). Frequent worker transfers across tasks can negatively impact productivity through transaction costs (Schultz, McClain, and Thomas 2003, 81-92) and learning curve effects (Gomar, Haas, and Morton 2002, 104). The effects of learning/forgetting behaviours on production system performance are well-documented (Nembhard and Osothsilp 2005, 576-87; Nembhard 2001, 1955-68; Shafer, Nembhard, and Uzumeri 2001, 1639-53; Heimerl and Kolisch 2010, 3759-81). For instance, Shafer, Nembhard, and Uzumeri (2001, 1645) highlight the importance of capturing heterogeneity and worker's specific learning/forgetting rates in a multi-skilled workforce in order to avoid underestimating productivity. The introduction of learning curve models in workers assignment approaches fosters a more efficient production planning (Anzanello and Fogliatto 2011, 573).

Negative psychological implications of cross-training should also be considered as workers may prefer depth of knowledge to breadth of knowledge, and cross-training might be seen as a strategy to depreciate their expertise, create confusion on compensation rights (Hopp and Oyen 2004, 922) and reduce their opportunities for career path development (Abrams and Berge 2010, 525). Schultz, McClain, and Thomas (2003, 90) also highlights that flexibility can affect the motivation to work faster/slower based on performance feedback and direct comparison of productivity among workers that share a task.

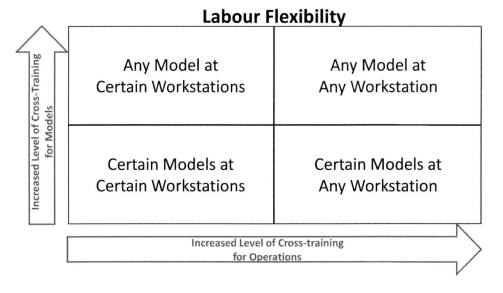

Figure 1. Dimensions and levels of workforce flexibility.

Labour flexibility can be achieved in the two dimensions shown in Figure 1, which are in terms of the ability of workers to operate multiple machines and/or process different product models. The four possible extreme levels of the dimensions of flexibility of workers will be discussed in the following sub-sections. The discussions will be based on a combination of the insights from existing research and observations from two case companies. These two companies have been selected because they provide examples of a job shop and an assembly line in which the flexibility of labour is required to manufacture different product models. To facilitate the understanding of the two case companies when they are referenced in subsequent sections, their manufacturing systems are first described.

2.1. Case System 1: Assembly Line

The first case is a medical device fabrication company which manufactures include cardiac rhythm management (CRM) devices, such as pacemakers and defibrillators, and cardiovascular stents. Some of the lines are fully manned, having a worker assigned to each workstation. For some other lines, productivity is constrained by the availability of both workstations (i.e., some workstations are shut down during specific shifts) and workers (Jie Xu et al. 2015, 309). In these lines, the number of workers allocated is generally less than the number of machines operating; therefore, there is a reliance on the ability to dynamically switch the workers, who are cross-trained, between the machines to achieve production targets. In terms of the overall system's routing flexibility, some lines are fully dedicated to specific product models, while other lines or, in some instances, sections of lines are shared between different products, as shown in Figure 2.

From researching this system, challenges have been observed for the different aspects and levels of flexibility characterising these assembly lines. These observations will be discussed in later parts of this section, with reference to the relevant literature available.

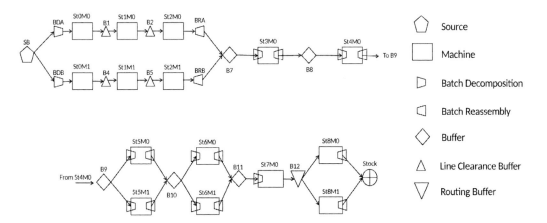

Figure 2. Assembly line layout.

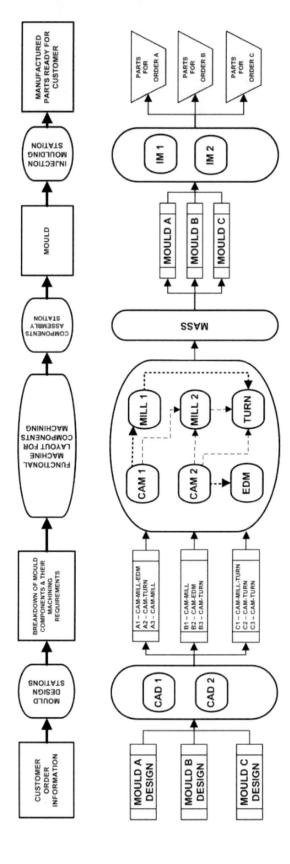

Figure 3. Job shop layout.

2.2. Case System 2: A Job Shop

The second case company is a rapid tooling, rapid prototyping company, which also does small series production. The manufacturing system can be classified as a shop because successive products have different machining and routing requirements. The company's activities involve the design and creation of moulds and subsequent prototype and small series part production. As illustrated in Figure 3, the processing of each customer order starts with a Computer Aided Design (CAD) of its mould, followed by CNC programming (CAM), Milling (MILL), Turning (TURN) and Electric Discharge Machining (EDM) of the components constituting the mould design, before the components are assembled at an Assembly station (MASS) into the mould which is then used for series part production at the Injection Moulding machine (IM).

The machines that are used to carry out these operations are setup in a functional layout. But the company relies on the flexibility of its workers to bridge the flow of parts across the different functional departments. Although this brings desired benefits, its attendant challenges are highlighted in the following subsections.

For completeness, each one of the four extreme levels of labour flexibility shown in Figure 1 is going to be discussed, including those that do not directly apply to the two case companies.

2.3. Certain Models at Certain Workstations

This is the case in which lines/workstations are dedicated to specific product models and no transfer of workers is allowed across workstations. It consists of a configuration characterised by no flexibility, with respect to versatility. Interestingly, this case has not been observed in the assembly lines or in the job shop considered for analysis. This situation would be typical of Taylor-like organisations where workers have to perform a task of a highly repetitive nature (Zoethout, Jager, and Molleman 2006, 344). The frequent repetition of task cycles is likely to generate highly specialised workers, and the time required to complete a task will tend to decrease as the worker gains expertise and identifies the most efficient way to conduct the task. Productivity inefficiencies will tend to emerge in the line as workers that do not keep up with the production pace will be easily identified through the build-up of work in process (WIP).

Training programmes can be immediately defined, as skill gaps will be readily identifiable. Training efficiency will also be high in these systems, as trainings will be conducted by highly specialised workers and will focus on specific skills. This type of assembly lines also can benefit from low organisational challenges as there is no need to coordinate workers and decide on effective task rotation or assignment strategies, which means that the case coordination costs incurred from the time/effort required to allocate a task can be considered null (Zoethout, Jager, and Molleman 2006, 346).

However, advances in expertise enabled through task specialisation may be counter-acted upon by reductions in motivation. Factors such as boredom and alienation may play a fundamental role on productivity performance which can quickly deteriorate if no recovery action (i.e., through task rotation) is taken (Zoethout, Jager, and Molleman 2006, 357). The highly repetitive nature of tasks can also impact the quality of the work performed as attention levels may drop (Costa, Silva, and Campilho 2017, 4044). This is probably the main reason why this type of system configuration is not adopted in the company considered, especially in the assembly system in which production quality is of paramount importance in the medical device industry.

Another limitation of this type of systems is that they require considerable design efforts, as the line must be perfectly balanced or else, the lack of flexibility will lead to WIP buffering at potential bottlenecks with obvious consequences on workers' performance (i.e., idle times caused by blockages at workstations upstream bottlenecks will be regularly observed). It is difficult to ensure a balanced line because of the variability which results from the stochastic nature of manual operations, the presence of heterogeneous workers or from variable takt times that are due to fluctuating demands. Capacity buffering is generally needed to absorb the variability of task/workstation processing times (Sennott, Van Oyen, and Iravani 2006, 542). As a result, it often happens that the available production capacity remains largely underutilised than in more flexible flowlines (Koren et al. 1999, 528). Some level of flexibility is then required to better exploit the productivity potential, and also avoid that absenteeism or resignation of specialised workers would have dramatic effects on line productivity. Sennott, Van Oyen, and Iravani (2006, 542) observe that a modest amount of flexibility can be sufficient to reap significant benefits and suggest the use of a floating worker to address productivity issues in assembly lines usually operated by specialists. As observed in the two case companies, the floating role is usually covered by experienced line supervisors and is used to troubleshoot workstations where issues are experienced (i.e., replace an absent worker, or a worker during break time, quality issues, etc.).

2.4. Any Model at Certain Workstations

In this case, worker flexibility refers to the ability of workers to process different product models at the same workstation. In the assembly lines considered, this situation is rarely observed and would represent an exception of the fully flexible case (see Section 0). Indeed, it can happen that, due to compelling production targets to be met, sections of lines that are generally shared between different product models and usually managed using self-organising approaches by a limited number of workers, are assigned additional workers so that each single workstation is operated. This configuration would not be sustainable in the manufacturing environment considered as rotation policies (see Section 0) would still trigger workers' movements to different workstations for safety and quality reasons.

However, at workstations where this applies, the workers need to take decisions on when to switch between the processing of the different product models. For such workstations, more detailed production targets, which are disaggregated into product models and time slots, are generally provided by the planning department to assist workers in product selection decisions. However, it was observed in the real lines that the suggested schedule is usually overlooked, and more complex dynamics based on informal communication between workers around WIP requirements are used to drive product selection decisions. Experience also plays a fundamental role in this process as workers will tend to anticipate potential production issues and assign priorities to product models based on historical evidence. The product selection decisions are also necessarily influenced by the type of production. For instance, CRM devices are generally produced by single units and more flexibility on the switching frequency is allowed. Conversely, for cardiovascular stents, batch production is adopted, and a line clearance policy applies. This policy, which is commonly observed in the medical device industry, consists of ensuring a physical segmentation of the line with respect to production batches in order to prevent the accidental mixing of components and production units belonging to different batches. In this case, the switching frequency is indirectly constrained by the batch size. Moreover, changeover operations become lengthier and more critical.

The impact that product variety has on the operations and performance of a manufacturing system cannot be neglected (Nazarian, Ko, and Wang 2010, 36), and an explicit consideration of the inter-task times, necessary for setup changes, tools changes or part repositioning, is critical to support more accurate manufacturing lines' design. Usually, standard buffer times are used to capture sequence-dependent inter-task times in production line design models (Becker and Scholl 2006, 706–9). As a consequence, line re-balancing must be regularly performed (Nazarian, Ko, and Wang 2010, 44). Design issues in these flexible lines also involve decisions on the opportune level of machine/workstation flexibility that is required to accomplish the desired product variety; highly flexible machines are generally more capital intensive and less reliable than less flexible machines (Koren et al. 1999, 527). The adoption of flexible machines in production lines should be confined to workstations where product variety impacts the tasks' repeatability or recurrence (Nazarian, Ko, and Wang 2010, 36). In this way, capital investments can be minimised, and the challenges related to flexible machine operations are avoided. This practice is followed in the real-system observed, where only sections of manufacturing lines are equipped with flexible machines which are shared between product models. Product variety within a model can be more easily managed within less flexible lines/workstations as it mainly affects the dimensional features of the units, which are within millimetric precision; processing steps and equipment required to process these product variants do not change.

The possibility of switching from a model to another introduces task variety in this type of assembly lines and positively impacts workers' performance. Although product variety may lead to reduced levels of specialisation with respect to the line configuration with no flexibility, motivation will always be higher, and the effects of boredom or lack of focus may be reduced in this context. This is achieved without the need for workers to be re-assigned to different workstations, which ensures that the inefficiencies caused by travel times and hand-over operations are nullified and coordination costs are minimal.

In this type of systems, training is also quite efficient as similarities between the skills required to perform the same process steps on different product models will be exploited.

2.5. Certain Models at Any Workstation

This is the case of lines dedicated to specific product models and workers that are trained on different tasks and can be reassigned to different workstations. This is a typical system configuration in mass-production assembly lines and certainly a configuration commonly observed in the two companies referenced in this chapter. For a job shop, this can happen in cases where specific workers are assigned to oversee the processing of each customer order from start to finish. And, this could be the case because the workers have previous experience working with a customer, which would facilitate their communication with the customer during the processing of their order.

For the assembly line, due to technological issues and strict production flow regulations, the lines are not perfectly balanced. Also due to the natural stochasticity of manual processing, the processing times required to complete tasks at different workstations are not necessarily equal (i.e., asynchronous flowline). However, in order to minimise the balance delay and also to reduce manufacturing costs, workers are partially cross-trained and the number of workers assigned to the assembly lines is less than the number of machines operating in them (Downey and Leonard 1992, 469). Cross-training programs are developed at this company based on a "3 tasks to 3 workers" concept, whereby the aim is to train workers on up to 3 tasks and have at having at least 3 workers able to work on any given workstation. Training targets for each worker are established by the training department that develops a versatility matrix for the entire workforce based on long-term strategic capacity plans. The aim of this staffing policy is to ensure that the available workforce is sufficiently flexible to respond to worker shortage issues caused by absenteeism/resignation and/or strategic management decisions on staffing economies. Workers may also be occasionally transferred to other production lines based on both staffing requirements and production priorities, which can further limit the daily availability of workers on specific assembly lines. The main challenge in this type of system is to identify worker assignments to workstations that would ensure the balancing of the line (i.e., avoid blockage or starvation at the various workstations) and ensure the fulfilment of productivity targets in a production environment that constantly changes (i.e., the system configuration changes every time workers are reassigned).

Rotation rules are in place in the observed lines to ensure the implementation of good manufacturing practice in the medical device industry. Three types of rules are considered: flexing, safety and quality rules. Flexing rules suggest the tasks on which a worker can freely rotate, and this involves grouping tasks which are similar or logically connected so that a worker assigned to a specific task can autonomously decide when to start working on another task belonging to the same group based on WIP considerations and/or personal motivation. Safety rules are implemented to avoid that the effects of tiredness and lack of focus caused by the highly repetitive nature of a task would impact production yield. These rules mainly apply to critical tasks and prescribe the maximum amount of time for which a worker can perform that task.

An extension of the safety rules also implies that after operating on a critical task, the worker cannot be transferred to "high-risk" tasks, where "high-risk" refers to both the task complexity and the impact that the operation performed can have on the quality of the units produced. Quality rules apply to inspection tasks and prescribe that workers that have been involved in critical production tasks cannot be transferred to inspection workstations. Basically, the aim of these rules is to avoid that a worker may inspect units that he/she has produced. Beside these rules, no formal coordination policy is followed at this company and the line supervisor is made responsible for critical assignment decisions.

A worker coordination policy defines the logic used to assign workers to task over time (Hopp and Oyen 2004, 924). For instance, at the beginning of each shift, the production supervisor assigns workers to workstations based on the number of workers and machines available. Subsequently, considerations on factors such as the WIP distribution in the line, production targets, and workers' certification maintenance are made by the supervisor to decide on potential workers' re-assignment during a shift. A rule of thumb generally adopted at the company is that in case of substantial (and unusual) worker shortage, the supervisor assigns workers to the first workstations in the line so that WIP is built at the intermediate workstations. Towards the end of the shift or during the following shift, workers are then moved to the last workstations to re-balance the line's WIP. It was observed that the implementation of this rule and its effect on production are dependent on the supervisors' policy for taking the decision, experience and his/her promptness in making the assignment decisions based on the line status. As a result, it is clear that the success of running a system with this level of labour flexibility depends on its worker coordination policy. There is significant volume of literature on worker coordination policies and cross-training skill patterns, which are both key elements of a framework developed by Hopp and Oyen (2004, 927) for classifying workforce flexibility and evaluating its advantages from both a strategic and tactical perspective. Various worker coordination policies are suggested in the literature. For instance, Moving Worker Modules (MWM) applications are concerned with systems, especially assembly lines, characterized by more machines than workers. Typical implementations of MWM systems are Bucket Brigade systems, where a worker processes sequential tasks of a job until he/she is blocked by a downstream worker or the downstream worker takes the job from him/her, and the Toyota sewn-product management system, where each worker is restricted to a zone of machines and decides whether to perform tasks at an overlapping workstation based on buffer level status (Sennott, Van Oyen, and Iravani 2006, 544). Other coordination policies include those applicable to systems with floating workers where a small number of generalists is used to perform the most urgent tasks based on the system requirements (Sennott, Van Oyen, and Iravani 2006, 542). Zoned worksharing where workers are organised in groups which are assigned to "zones" or tasks that can overlap (such as the D-skill chaining, the Toyota sewn-product management system, or the Bubcket Brigade system) or are disjoint. Additionally, there are pick and run policies, which are similar to the practice in the job shop system, where each single worker is fully cross-trained and follows a product from start to finish or its team variant, where team dynamics are exploited to create motivation and generate production efficiencies (Hopp and Oyen 2004, 932). Hopp and Oyen (2004, 935) compare various worksharing policies under different dimensions, including training, pre-emption, multi-tasking, collaboration, transition, etc.

Downey and Leonard (1992, 469) discuss the assignment of fully cross-trained workers in an assembly line with fewer workers than stations. Idle workers working at starved or blocked workstations are allowed to move to unattended stations selected based on an index of potential production. Using simulation to develop regression models for cost structures, they investigate the impact of the number of workers assigned to a line and the size of intermediate buffers on average inventory, workers' movements and total direct labour time per production unit. The assignment approach used by Downey and Leonard (1992, 469-83) is similar to the assignment practice observed in the real assembly lines analysed here; this approach has also many commonalities with the assignment strategies used in Dual Resource Constrained (DRC) systems, which are systems where productivity is constrained by the availability of both workstations and workers. While DRC systems concurrently address workers' assignment and job dispatching, dynamic worker assignments uniquely characterize these systems since the number of available workers is always less than the number of machines operating in the system (Cesanì and Steudel 2005, 572). The commonalities of the assignment strategies suggested in the DRC literature and the assignment practice used by the supervisors managing the assembly line analysed here consist of the use of "when" and "where" rules. These rules are followed to ensure efficient and effective workforce re-assignment in real-time and maximize manufacturing productivity (J Xu, Xu, and Xie 2011, 309). "Where" rules concern the selection of machines to assign to different workers whereas "when" rules define the time when re-assignment is allowed (Ammar, Pierreval, and Elkosentini 2013, 639). "Where" rules can consist of static rules (i.e., they do not use information on the system status); for example, queue-length-based rules (i.e., rules that use real-time information on the WIP levels) and workload-based rules (i.e., rules that use information on the cumulative processing times for the jobs in a queue) (Hopp, Tekin, and Van Oyen 2004, 89). "When" rules are generally categorized into centralized and decentralized rules. Centralized rules allow a worker to be reassigned to another machine as soon as the current job is completed whereas decentralized rules allow workers to transfer when the queue at the machine currently assigned is empty (Ammar, Pierreval, and Elkosentini 2013, 640). J Xu, Xu, and Xie (2011, 313) argue that "when" rules are more important and have more effect on the overall system performance than the "where" rules.

The very dynamic nature of systems characterised by workers' availability constraints makes the definition of effective coordination policies a challenging task. For instance, even the decision on how frequently the workers should be re-assigned (i.e., event-based decision or regular time intervals) is not trivial and may have considerable effects on productivity. Dynamic decisions on re-assignment points should be preferred and guided by considerations on typical processing times, line configuration, WIP distribution, motivational and behavioural effects. In the assembly lines observed, excluding the workstations where self-organisation is allowed, it appears that workers' transfers tend to be minimised and coincide with production breaks (i.e., breakfast, lunch, etc.), which helps to avoid the loss of productivity resulting from task/workstation switching, which include travel times from a workstation to another, breaking of working rhythm, hand-over operations (i.e., informational transition), set-up operations (i.e., physical transition, such as wearing protective gear, system log-in, etc.), forgetting effects. However, it is interesting that when these effects are neglected, more frequent re-assignments generally determine productivity increases as WIP imbalance issues are avoided.

Skill retention also becomes a complex task to manage when high levels of cross-training are implemented. For instance, in the assembly lines observed, the lines supervisor is faced with the challenging task of ensuring the workers maintain their proficiency level on the tasks on which they are trained, in terms of both production efficiency and quality yield, by means of effective task rotation. This would entail keeping track of the amount of time spent on different tasks for each worker; although a manufacturing execution system captures detailed production data, there is currently no visualisation means to make information on the workers' assignment available to the supervisor. Hence, workers' skill retention is very much dependent on the supervisor ability to consider this aspect while making assignment decisions.

2.6. Any Model at Any Workstation

Full flexibility can be reached in the assembly lines in terms of the possibility of processing different models at a workstation but also the ability to reassign workers to different workstations (where different models are produced). This case is typically observed in sections of lines where sub-assemblies that are common to different product models are processed. Generally, the difference between these sub-assemblies is minor and concern aesthetic or dimensional aspects of the assembled components. For instance, batteries for pacemakers and defibrillators of different sizes are assembled and tested in a multi-product line segment; likewise, laser marking operations on pacemakers' aluminium casings are performed in a sub-assembly line section shared between different pacemaker models. The high flexibility level characterising these sub-lines is also enabled by the availability of flexible machines at various process steps that can be easily reconfigured to process different product models. For instance, different programmes are available on the laser marking machine and can be selected by a worker based on the units to be processed.

In this system configuration, benefits and challenges deriving from the two dimensions of worker flexibility considered here (i.e., product models and operations) are multiplied with respect to the one-dimensional flexibility counterpart (see previous sub-sections). The main advantage in this configuration lies in the possibility of availing of dynamic capacity at each process step as workers can be assigned to critical workstations whenever required and flexibly work on different products based on production priorities and WIP requirements. This generally ensures that production targets are met, or at least critical production priorities are satisfied. Workers' idle time is also minimised, and the line capacity better utilised.

On the other hand, supervisors are faced with considerable organisational challenges. At the companies analysed, the lack of well-defined worker coordination policies generates a tendency to adopt self-organising approaches within a production team (i.e., set of workers assigned to a line or line segment) that are difficult to control and can conceal production inefficiencies. For instance, in one of the assembly lines considered for analysis, 4 workers are assigned to 8 workstations and different variants of a product model are processed. The production flow in this line is dictated by an automated workstation (i.e., injection moulding of pacemakers' caps) that is characterised by considerably longer processing times than at any other workstation in the line. The line is operated so as to ensure sufficient WIP levels at the automated step in the early hours of a production shift. In the second half of a shift, workers are typically assigned to downstream workstations so that the remaining processing steps can be completed on the production units and production targets can be reached. Although there are task rotation rules provided by industrial engineers, these rules are loosely followed (i.e., except quality and safety rules which are always obeyed) and, in practice, the team of workers assigned to this line tends to adopt a self-organising approach. Based on interviews conducted among the workers operating this line, it clearly emerged that task rotation and workers' assignment to workstations, while taking WIP considerations into account, is mainly based on more personnel related factors such as expertise and motivation. Motivation mainly refers to the interest expressed by a worker in performing a task and also to the likelihood for a worker to become bored of performing the same task if task rotation is not applied with an adequate frequency. The impact of these factors (i.e., motivation and expertise) along with task complexity and variety on the performance of teams consisting of specialists and generalists is investigated by Zoethout, Jager, and Molleman (2008, 84-5). Both groups studied (i.e., generalists and specialists) are fully cross-trained with heterogenous skill levels; however, generalists follow pre-defined coordination policies whereby they are forced to regularly practice their skills whereas specialists are given the freedom to self-organise. The assignment dynamics of self-organising processes are captured in an agent-based simulation model where motivation and expertise are modelled as factors impacting both autonomous assignment decisions and task performance (i.e., the time required to complete a task). It is found that, on one hand, the self-organising dynamics (i.e., for the specialists) enable the team to reshape itself according to the demands of the system; on the other hand, the group of generalists demonstrate that a well-ordered assignment structure can generate good task performance with low motivation issues. Specialists tend to perform best when there is no task variety as they reach higher level of expertise in a limited amount of skills and stabilise their motivation by regular task rotation whereas generalists outperform specialists for low or moderate task variety. For high levels of task variety the two groups perform similarly. Coordination costs and more complex team structures (i.e., number of workers and heterogeneity between workers) are however not considered in the study. The switching frequency between tasks also tends to follow motivational reasons; whereas, switching between products is more subject to WIP considerations.

Training and coordination costs substantially increase in this type of configuration, and inefficiencies resulting from a lack of control on coordination policies and transitional costs can also be observed. For this reason, the implementation of this type of configuration is quite limited in case company and mainly driven by technological reasons (i.e., asynchronous flow caused by the presence of a long process step).

In summary, different levels of labour flexibility provide benefits and challenges that cannot be neglected while analysing training programmes and worker coordination policies. Trade-offs should be considered to balance negative and positive effects. It is also important to consider that cross-training policies should be adapted to the particular system under investigation, without neglecting behavioural traits of the workforce employed (Abrams and Berge 2010, 527). In later sections, the impact that the evolving technologies of Industry 4.0 will have on this perception are going to be discussed. But first flexibility from the manufacturing system design and layout perspectives are going to be discussed in the next section (Section 0).

3. MANUFACTURING SYSTEM FLEXIBILITY: BENEFITS AND CHALLENGES

Although the chapter focusses mainly on manufacturing systems that rely on labour flexibility for the overall system flexibility, it is still worth examining how manufacturing systems with automated operations achieve system flexibility. As observed for labour, a similar notion of dimension and different levels of flexibility apply to such automated systems, and these are mainly achieved through the design of the machines and the system layout. For machine design, three levels of flexibility can be identified in dedicated manufacturing systems (DMS), flexible manufacturing systems (FMS) and reconfigurable manufacturing systems (RMS), which have different levels of flexibility to process different product types. Each one of the manufacturing systems offers different levels of flexibility that are similar to that offered by different levels of cross-training of labour for different product models. As shown in Figure 4, the DMS process only fixed model types but at high production volumes, the FMS are capable of processing different product models at lower production volumes than the DMS (Mehrabi, Ulsoy, and Koren 2000, 406), while the RMS offers flexibility in both product variety and volume (Koren et al. 1999, 528), which is why it appears across the four extremes. Each one of them is going to be briefly discussed in the following subsections.

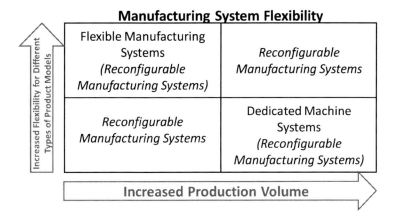

Figure 4. Levels of flexibility of manufacturing systems.

3.1. Dedicated Machine Systems

The DMS represents the oldest form of manufacturing equipment design, which is characterised by high production volume but low flexibility in the variety of products that they can produce. This was so because machines then were fully mechanised, which made them adequate to support the mass production of low variety of products.

However, with these systems, new product types could only be introduced into the system by either purchasing entirely new machines or making costly, and sometimes, irreversible changes to existing ones. This made companies to limit the range of their concurrent product offerings. New products were introduced mainly as replacements to existing ones, and products that have high variation in their demand volume were phased out to avoid having to keep redundant capacity for their production. However, increase in consumer demand for customised products made it inevitable that a quick and low-cost approach be found for adapting systems to simultaneously produce different product types.

3.2. Flexible Manufacturing Systems

The emphasis of the FMS is on the ability to process different product types through the same line. It usually consists of expensive, general-purpose computer numerically controlled (CNC) machines that are supported by other forms of automation to achieve a system that is versatile to processing range of products with different processing and routing requirements (Koren et al. 1999, 527). The CNC machines are controlled by computer programmes and equipped with variety of tools that can be altered or interchanged for different types of machining operations. In addition, the overall system achieves its flexibility through an automated material handling system that can dynamically connect parts to the machines that will satisfy their processing requirements satisfied.

Therefore, while FMS relies on the flexibility of machines to process different types of products, it relies also on the overall system ability to provide flexible routing of different part types to machines in the system according to their respective requirements. As such, even with limited machine-level flexibility, a FMS is set up such that there are machines to satisfy all processing requirements, and an intelligent conveyor system can connect parts to the machines that can satisfy their processing requirements.

Even though the FMS provide the required flexibility to produce different product types, the CNC machines used in them are costlier than the machines used in DMS because the CNC machines are usually overdesigned to have all the functionality that might be needed (Mehrabi, Ulsoy, and Koren 2000, 403; Koren et al. 1999, 528), which makes the FMS uneconomical in some cases. The costliness of the FMS is also worsened by the fact that they are unable to reach the same level of production volume that the dedicated manufacturing systems are capable of. Therefore, it is difficult for them to produce sufficient volume of products to justify their extra cost. However, the FMS's cost of keeping the same set of capacity always might be justifiable in high mix manufacturing systems which usually involve the manufacture of the same known set of products that have predictable and consistent demand volumes.

3.3. Reconfigurable Manufacturing Systems

As defined in (Koren et al. 1999, 529), the underlying principle of RMS is to design the system from the outset to enable a quick adjustment to production capacity and functionality within a part family in response to sudden changes in market or regulatory requirements. And, these are achieved through a design for rapid change in the system's structure, hardware and software components. Prior to this, the focus was on the customisation of products to meet evolving customer preference. However, it was soon realised that satisfying new customer preference through the customisation of products also necessitates that the systems needed to manufacture the products must be adaptable to achieve the quick production of the custom products (Koren et al. 1999, 527).

The RMS offers the flexibility needed to customise manufacturing systems in response to the changes in functionality needed to process new product types and to adjust to the variations in the demand volumes of existing product types (Mehrabi, Ulsoy, and Koren 2000, 405-6; Koren et al. 1999, 528), i.e., adaptability and scalability, respectively. These two features are offered in the RMS through its ability to integrate new functions and technologies into existing systems to process new product types or its ability to tweak the existing system to reduce or increase the product volume of existing product types. Although the adaptability was to some extent present in the FMS, the scalability was a breakthrough for the RMS, most especially because the production volume can be used to offset the extra cost of its costlier machines.

3.4. Manufacturing System Layout

A manufacturing system's layout has the most impact on the operational dimension of flexibility, in that a system layout could facilitate or hamper the application of workers' skills at multiple workstations. Most important in the layout is the proximity of machines to one another, which ensures that the cost of movement can be reduced and that the workers can be self-coordinating. For instance, if the costs involved in moving workers between machines were considered, the workstations must be accessible to one another to minimise the cost of moving workers between workstations. These movement costs have been shown to weaken the benefits that cross-trained teams offer (Hottenstein and Bowman 1998, 167-8). Therefore, to derive the benefits of having flexible, cross-trained workers, the layout of the system must facilitate the process of taking the decisions involved in coordinating the workers' movement. Another benefit of workstations being accessible to one another is that it supports the workers to self-coordinate their tasks effectively, which means that they will be able to monitor the current state of the other workstations/departments and quickly identify when it is necessary to move to a new workstation, without the intervention of a supervisor.

Possible layout configurations vary between having all operations completed in one location (i.e., a fixed position layout) and having them spread out between sequential steps (i.e., a product/line layout), and these provide varying levels of support for system flexibility. As with most job shops, the case system that is referenced in this chapter operates a functional layout in which similar machines are grouped into the same departments. But, to reduce the abruptness that such layout would have on the flow of parts between the departments, the company has cross-trained workers who are able to operate machines in any of the departments. This approach is labour efficient, as it ensures that workers can move across departments to operate machines. Similarly, the dedicated lines for the assembly of different product types are arranged in parallel to one another to facilitate workers in recognising the need to switch between workstations when inventory begins to build up at any workstation.

However, this poses the challenges of deciding on where to assign workers and coordinating their transfer between the workstations (in the job shop) and lines (in the assembly line). And, in cases where the workers are not perfectly interchangeable (Bobrowski and Park 1993, 267; Hottenstein and Bowman 1998, 159), the decision is further complicated by the need to determine what worker should work on what machines.

4. CURRENT IMPLICATIONS FOR LEAN MANUFACTURING IN MAKE-TO-ORDER SYSTEMS

Lean manufacturing, especially in MTO systems, relies significantly on the flexibility of labour to ensure a smooth flow of parts through its favoured U-Shaped cellular layout. And, the most desirable level of labour flexibility for it would be to have workers that are fully cross-trained with skills for Any Model at Any Workstation in the cell. However, as discussed in Section 0, this level of flexibility comes with extra coordination challenges. As a result, despite its benefits, companies are cautious of the implementation of this level of labour flexibility. It is often the case that only a small set of workers are cross-trained to have this level of full flexibility, and they are usually selected after years of demonstrating unique ability to retain skills and self-manage their work. These fully cross-trained workers are then available to complement the level of flexibility provided by the partially cross-trained workers. Additionally, as is the case in the two case systems, production supervisors also possess this level of full cross-training because they have had the opportunity to work in different departments on the way to becoming supervisors, and so have acquired the skills to operate many of the machines and process variety of product types. But because they are not regular workers, they cannot be counted on as regular parts of the production labour.

4.1. Outlook: Impact of Industry 4.0 and Other Evolving Technologies

In this section, the potential impact of the Smart Operator concept of Industry 4.0 in supporting the use of fully cross-trained workers in a system without the attendant issues of extra coordination efforts, as described in Section 0, is discussed. Industry 4.0 seeks to take advantage of high speed computing to interconnect manufacturing entities, collect data about their interactions, analyse the data and re-use the knowledge gained from the analysis to influence the future interactions of the systems entities towards better achieving the system's overall goals and objectives. Because human workers will be among the interconnected system entities, its interaction with the machine entities is elevated through augmented or virtual reality to a layer that corresponds to that of the non-human entities. At this level, the worker becomes "Smarter" and able to receive real time information to support its interaction with the other entities.

The smart operator is a key part of Industry 4.0, and it can play a key role in supporting companies to overcome some of the challenges earlier identified in this chapter. The most important potential for labour flexibility is that the smart operators will not have to hard-learn any skills, which means full cross training for product models and workstations can be achieved without concerns for issues such as forgetting or poor quality. However, it does not directly eradicate the extra coordination that has been shown to result from full cross-training. In fact, it might require that more advanced and dynamic scheduling would need to be done at a detailed level that gives more specifics on the schedule that workers/machines should follow (Longo, Nicoletti, and Padovano 2017, 144-59; Vidoni and Vecchietti 2015, 326-38; Upasani et al. 2017, 1-14; Ivanov et al. 2016, 386-402). Similarly, the scheduling system would need to be dynamic because there will be more information being received in real-time about the system status, to which the schedule can be adjusted in response. However, the data from the interconnectedness of the system entities can be used to facilitate the coordination decisions. For instance, machine learning algorithms can be applied on the data from worker activities and used to support the decisions on the assignment of workers to machines and their rotation to meet the somewhat complex objectives of ensuring that workers make prolonged use of their skills for proficiency, while also applying as many of their skills as possible in order not to forget them. Those tools can provide support for making assignment and rotation decisions that would suit each worker's performance at respective workstations. Worker performance can be monitored and analysed in real time, with the analysis results then used to determine when to move workers and to which workstation, in a way that satisfies immediate production objectives and long-term retention of skill sets. Most importantly, if smart operators will not have to hard-learn skills, then their assignment or rotation decisions can even become less complicated, without the need to worry about them forgetting some of their skill sets.

In addition to the coordination of the shop floor, the improved interconnectivity of the system's entities can be used to improve the workload balancing on the shop floor as shown in (Li et al. 2017, 1855-64). The digitalised technologies can, for example, support the creation of dashboard displays through which workers can monitor the state of all workstation and use this information to facilitate their decision on when to move to a new workstation and the workstation to which to move, i.e., the when and where decisions.

The benefits of smart operators extend beyond labour-reliant systems, such as the two case systems described here, because, even RMS that rely on reconfigurable machines for flexibility would still need human intervention for their reconfiguration. The system reconfiguration involves following a set of instructions to reconfigure the machine for the processing of different types of products. Because the instructions can vary for different product types, they can be as complicated for workers as needing different skills to process different product types. As such, smart operators can significantly increase the speed at which such systems can be reconfigured and extend the number of possible reconfigurations built into them. For instance, the SOPHOS_MS system proposed in (Longo, Nicoletti, and Padovano 2017, 144-59) can interactively support workers with instructions on how to carry out a wide variety of tasks. The implication of this could be that there would be no need to worry about workers forgetting a skill set, which would mean that the sole focus can be on boosting short term productivity when taking decisions on when and where to assign workers, and not on how to ensure they retain all their skill sets. The workers will, however, need to be more flexible and demonstrate adaptive capabilities (Longo, Nicoletti, and Padovano 2017, 144-59).

If the above benefits can be achieved with a Smart Operator, then it should facilitate the implementation of lean manufacturing principles in MTO systems, which are otherwise not very suitable for it. This is because it would mean that lean manufacturing can operate in such MTO systems with its favoured U-Shaped cellular layout and fully-rely on labour flexibility to ensure a smooth flow of parts through the system.

CONCLUSION

In this chapter, the benefits and challenges of implementing lean manufacturing in MTO systems have been discussed, particularly from the perspective of its reliance on labour flexibility for its effectiveness in accomplishing the use of as minimal resources as possible to produce an MTO system's high variety of products. The analysis of two case systems in combination with insights from literature show that while increased flexibility is good for system efficiency, it requires extra coordination efforts to derive the full benefits. Some of the potential complications of increased labour flexibility were highlighted, including a more complex decision-making process, increased likelihood of workers forgetting some of their skills, as well as possible quality and ergonomic issues. However, it is also outlined in this chapter that with advances, such as Smart Operators, Data Analytics and Machine Learning, which have evolved with Industry 4.0, it would be possible to achieve a maximum level of labour flexibility without the attendant problems that were highlighted in Section 0.

Specifically, it is discussed how the use of augmented and virtual reality to aid workers in performing tasks can eradicate the need for hard-learning of skills. It is shown that these would facilitate the implementation of lean manufacturing in MTO systems, which are generally considered difficult candidates for its implementation.

REFERENCES

Abrams, Carmen, and Zane Berge. 2010. "Workforce Cross Training: A Re-Emerging Trend in Tough Times." *Journal of Workplace Learning* 22 (8): 522–29.

Ammar, Achraf, Henri Pierreval, and Sabeur Elkosentini. 2013. "Workers Assignment Problems in Manufacturing Systems: A Literature Analysis." In *Industrial Engineering and Systems Management (IESM), Proceedings of 2013 International Conference On*, 637–43.

Anzanello, Michel J, and Flavio S Fogliatto. 2011. "Learning Curve Models and Applications: Literature Review and Research Directions." *Internation Journal of Industrial Ergonomics* 41: 3–21.

Askar, Gazi, Thomas Sillekens, Leena Suhl, and Decision Support. 2007. "Flexibility Planning in Automotive Plants." *Management Logistischer Netzwerke*, 235–55.

Attia, El Awady, Philippe Duquenne, and Jean Marc Le-Lann. 2014. "Considering Skills Evolutions in Multi-Skilled Workforce Allocation with Flexible Working Hours." *International Journal of Production Research* 52 (15): 4548–73.

Becker, Christian, and Armin Scholl. 2006. "A Survey on Problems and Methods in Generalized Assembly Line Balancing." *European Journal of Operational Research* 168 (3): 694–715.

Berman, O., R. C. Larson, and E. Pinker. 1997. "Scheduling workforce and workflow in a high volume factory." *Management Science* 43 (2): 158-72.

Bobrowski, Paul M, and Paul Sungchil Park. 1993. "An evaluation of labor assignment rules when workers are not perfectly interchangeable." *Journal of Operations Management* 11 (3): 257-68.

Bruecker, Philippe De, Jorne den Bergh, Jeroen Beliën, and Erik Demeulemeester. 2015. "Workforce Planning Incorporating Skills: State of the Art." *European Journal of Operational Research* 243 (1): 1–16.

Brusco, Michael J, and Tony R Johns. 1998. "Staffing a multiskilled workforce with varying levels of productivity: An analysis of cross-training policies." *Decision Sciences* 29 (2): 499-515.

Buzacott, John A. 2004. "Modelling Teams and Workgroups in Manufacturing." *Annals of Operations Research* 126 (1–4): 215–30.

Campbell, G. M. 2011. "A two-stage stochastic program for scheduling and allocating cross-trained workers." *Journal of the Operational Research Society* 62 (6): 1038-47.

Campbell, Gerard M., and Moustapha Diaby. 2002. "Development and evaluation of an assignment heuristic for allocating cross-trained workers." *European Journal of Operational Research* 138 (1): 9-20.

Cesanì, Viviana I, and Harold J Steudel. 2005. "A Study of Labor Assignment Flexibility in Cellular Manufacturing Systems." *Computers & Industrial Engineering* 48 (3): 571–91.

Chen, Jiaqiong, and Ronald G Askin. 2006. "Throughput Maximization in Serial Production Lines with Worksharing." *International Journal of Production Economics* 99 (1–2): 88–101.

Correia, Damásio, FJG Silva, RM Gouveia, Teresa Pereira, and Luís Pinto Ferreira. 2018. "Improving manual assembly lines devoted to complex electronic devices by applying Lean tools." *Procedia Manufacturing* 17: 663-71.

Costa, RJS, FJG Silva, and RDSG Campilho. 2017. "A novel concept of agile assembly machine for sets applied in the automotive industry." *The International Journal of Advanced Manufacturing Technology* 91 (9-12): 4043-54.

Croci, F, M Perona, and A Pozzetti. 2000. "Work-Force Management in Automated Assembly Systems." *International Journal of Production Economics* 64 (1–3): 243–55.

Downey, Barbara Spangler, and M S Leonard. 1992. "Assembly Line with Flexible Work-Force." *The International Journal of Production Research* 30 (3): 469–83.

Gomar, Jorge E., Carl T. Haas, and David P. Morton. 2002. "Assignment and Allocation Optimization of Partially Multiskilled Workforce." *Journal of Construction Engineering and Management* 128 (2): 103–9.

Heimerl, Christian, and Rainer Kolisch. 2010. "Work Assignment to and Qualification of Multi-Skilled Human Resources under Knowledge Depreciation and Company Skill Level Targets." *International Journal of Production Research* 48 (13): 3759–81.

Herzenberg, Stephen A, John A Alic, and Howard Wial. 2000. *New Rules for a New Economy: Employment and Opportunity in Postindustrial America*. Cornell University Press.

Hopp, Wallace J, and Mark P Oyen. 2004. "Agile Workforce Evaluation: A Framework for Cross-Training and Coordination." *IIE Transactions* 36 (10): 919–40.

Hopp, Wallace J, Eylem Tekin, and Mark P Van Oyen. 2004. "Benefits of Skill Chaining in Serial Production Lines with Cross-Trained Workers." *Management Science* 50 (1): 83–98.

Hottenstein, Michael P, and Sherry A Bowman. 1998. "Cross-training and worker flexibility: A review of DRC system research." *The Journal of High Technology Management Research* 9 (2): 157-74.

Ivanov, Dmitry, Alexandre Dolgui, Boris Sokolov, Frank Werner, and Marina Ivanova. 2016. "A dynamic model and an algorithm for short-term supply chain scheduling in the smart factory industry 4.0." *International Journal of Production Research* 54 (2): 386-402.

Koren, Y., U. Heisel, F. Jovane, T. Moriwaki, G. Pritschow, G. Ulsoy, and H. Van Brussel. 1999. "Reconfigurable-Manufacturing-Systems." *Annals of the CIRP* 48 (2): 537–40.

Koren, Yoram, and Moshe Shpitalni. 2010. "Design of reconfigurable manufacturing systems." *Journal of Manufacturing Systems* 29 (4):130-41.

Li, Di, Hao Tang, Shiyong Wang, and Chengliang Liu. 2017. "A big data enabled load-balancing control for smart manufacturing of Industry 4.0." *Cluster Computing* 20 (2): 1855-64.

Liu, Chunfeng, Jufeng Wang, and Joseph Y. T. Leung. 2016. "Worker assignment and production planning with learning and forgetting in manufacturing cells by hybrid bacteria foraging algorithm." *Computers & Industrial Engineering* 96: 162-79.

Longo, Francesco, Letizia Nicoletti, and Antonio Padovano. 2017. "Smart operators in industry 4.0: A human-centered approach to enhance operators' capabilities and competencies within the new smart factory context." *Computers & Industrial Engineering* 113: 144-59.

Mc creery, John K., and Lee J. Krajewski. 1999. "Improving Performance Using Workforce Flexibility in an Assembly Environment with Learning and Forgetting Effects." *International Journal of Production Research* 37 (9): 2031–58.

Mehrabi, Mostafa G., A. Galip Ulsoy, and Yoram Koren. 2000. "Reconfigurable manufacturing systems: key to future manufacturing." *Journal of Intelligent Manufacturing* 11 (4): 403-19.

Nazarian, Ehsan, Jeonghan Ko, and Hui Wang. 2010. "Design of Multi-Product Manufacturing Lines with the Consideration of Product Change Dependent Inter-Task Times, Reduced Changeover and Machine Flexibility." *Journal of Manufacturing Systems* 29 (1): 35–46.

Nembhard, D. A. 2001. "Heuristic Approach for Assigning Workers to Tasks Based on Individual Learning Rates." *International Journal of Production Research* 39 (9): 1955–68.

Nembhard, D. A., and N. Osothsilp. 2005. "Learning and Forgetting-Based Worker Selection for Tasks of Varying Complexity." *Journal of the Operational Research Society* 56 (5): 576–87.

Sayın, Serpil, and Selçuk Karabatı. 2007. "Assigning cross-trained workers to departments: A two-stage optimization model to maximize utility and skill improvement." *European Journal of Operational Research* 176 (3): 1643-58.

Schultz, Kenneth L., John O. McClain, and L. Joseph Thomas. 2003. "Overcoming the Dark Side of Worker Flexibility." *Journal of Operations Management* 21 (1): 81–92.

Sennott, Linn I, Mark P Van Oyen, and Seyed M R Iravani. 2006. "Optimal Dynamic Assignment of a Flexible Worker on an Open Production Line with Specialists." *European Journal of Operational Research* 170 (2): 541–66.

Shafer, Scott M., David A. Nembhard, and Mustafa V. Uzumeri. 2001. "The Effects of Worker Learning, Forgetting, and Heterogeneity on Assembly Line Productivity." *Management Science* 47 (12): 1639–53.

Upasani, Kartikeya, Miroojin Bakshi, Vibhor Pandhare, and Bhupesh Kumar Lad. 2017. "Distributed maintenance planning in manufacturing industries." *Computers & Industrial Engineering* 108: 1-14.

Vairaktarakis, George L. 2003. "The Value of Resource Flexibility in the Resource-Constrained Job Assignment Problem." *Management Science* 49 (6): 718–32.

Vidoni, Melina C., and Aldo R. Vecchietti. 2015. "A systemic approach to define and characterize Advanced Planning Systems (APS)." *Computers & Industrial Engineering* 90: 326-38.

Xu, J, X Xu, and S Q Xie. 2011. "Recent Developments in Dual Resource Constrained ({DRC}) System Research." *European Journal of Operational Research* 215 (2): 309–18.

Xu, Jie, Edward Huang, Chun-Hung Chen, and Loo Hay Lee. 2015. "Simulation Optimization: A Review and Exploration in the New Era of Cloud Computing and Big Data." *Asia-Pacific Journal of Operational Research* 32 (03): 1550019.

In: Lean Manufacturing
Editors: F. J. G. Silva and L. Carlos Pinto Ferreira
ISBN: 978-1-53615-725-3
© 2019 Nova Science Publishers, Inc.

Chapter 4

SUSTAINING LEAN IN ORGANIZATIONS THROUGH THE MANAGEMENT OF TENSIONS AND PARADOXES

Malek Maalouf[1,], Peter Hasle[1], Jan Vang[1] and Imranul Hoque[2]*

[1]Department of Materials and Production, Aalborg University, Aalborg, Denmark
[2]International Business Centre (IBC), Department of Business and Management, Aalborg University, Aalborg, Denmark

ABSTRACT

In this chapter, we investigate the four types of organizational paradoxes of lean and the different types of strategies for dealing with these paradoxes in three organizations (Financial, Healthcare and Garments). The study of various types of paradoxes in lean has enabled companies to better understand the causes of resistance to lean implementation and, in consequence, take effective actions for facilitating lean implementation. For instance, in dealing with the learning paradox, the garment manufacturer in Bangladesh introduced a range of initiatives to foster the skills development of its workforce including machine operators, line supervisors and middle managers. Similarly, the financial company in Denmark attempted to reduce the tensions emerging from the paradox as employees had stronger attachment to their work identity. Dealing with the paradox often required various sessions of coaching, mentoring and group discussions in order to achieve a new level of understanding and the acceptance of the new role among the employees.

Keywords: paradox, lean implementation, organizational change, tensions

1. INTRODUCTION

Although Eisenhardt and Westcott (1988, 170) argued that the inherent organizational paradoxes in lean are a source of energy that facilitates lean transformation, the literature has

---

[*] Corresponding Author's E-mail: mml@business.aau.dk.

only recently (Maalouf and Gammelgaard 2016, 687-709) and (Maalouf, 2013, 1-292) started developing a systematic approach for managing lean paradoxes. An organizational paradox *"denotes contradictory yet interrelated elements"* and involves *"contradictory, mutually exclusive elements that are present and operate equally at the same time"* (Lewis 2000, 760). For instance, lean and just-in-time practices rely on competing processes and designs, such as increasing employee empowerment as well as adopting controls potentially limiting employee autonomy (Eisenhardt and Westcott 1988, 175). These two opposing yet complementary features of a lean system accentuate structural tensions within organizations (Smith and Lewis 2011, 395). Such structural tensions typically result in paradoxical situations, which emerge as organizations implementing lean and creating a competing design to enhance performance.

Managing structural tensions requires new managerial insights, where managers cannot just rely on tools from 'traditional' operations management. In this context, Poole and van de Ven (1989, 563) propose that researching organizational paradoxes provide a promising opportunity to create richer and more complex management research. A focus on organizational paradoxes, *"moves us away from the concept of organizations as static systems coping with problematic environmental fluctuations through deviation counteracting processes to a concept of organizations as continually dynamic systems that carry the seeds of change within themselves."* (Quinn and Cameron 1988, 82)

In the context of lean implementation, the research has identified a range of organizational paradoxes embedded in lean philosophy (Osono, Shimizu and Takeuchi 2008, 387). Yet, with the exception of Maalouf and Gammelgaard (2016, 687), the research has paid very little attention to the management strategies for dealing with the organizational paradoxes in lean. The investigation of the management strategies for dealing with lean paradoxes is crucial for the successful lean transformation as these strategies aim to reduce employees' resistance to lean implementation (Maalouf and Gammelgaard 2016, 703). This chapter builds on the work of Maalouf and Gammelgaard and further investigates the use of a range of management strategies for dealing with lean paradoxes and their outcomes in a group of selected cases. The cases include a financial company (Denmark), a hospital (Denmark) and a large garment manufacturer (Bangladesh).

The remaining part of the chapter is structured in the following way. In the second and third sections, we define and describe the four types of organizational paradoxes, followed by a presentation of the different strategies used for dealing with the organizational paradoxes in lean. Sections 4, 5, and 6 are dedicated to the analysis of the paradoxes and the management strategies adopted in the three cases. Section 7 contains conclusions and recommendations regarding the perspectives in using organizational paradoxes as a framework for facilitating lean implementation in different contexts.

2. TYPES OF ORGANIZATIONAL PARADOXES

The paradox literature identifies four types of organizational paradoxes: paradoxes of organizing, paradoxes of belonging, paradoxes of learning and paradoxes of performing (Smith and Lewis 2011, 383, Lewis 2000, 765). Table 1 presents a description of each paradox and clarifies the inherent tensions present in each situation.

Table 1. The four types of organizational paradoxes

Type of paradox	Description
Learning	The learning paradox is related to the ability of individuals to assimilate new knowledge that is needed to adjust to variations and change. The paradox usually involves struggle between the old and the new knowledge.
Belonging	The paradox of belonging rotates around tensions of identity and interpersonal relationships that arise between the individual and the collective. These paradoxes emerge because actors strive for both preserving their own identities and maintaining a collective affiliation.
Organizing	This paradox emerges as organizations create competing designs and processes in order to enhance performance. Increasing employee empowerment and creativity as well as adopting formal statistical processes and controls is one example of this paradox.
Performing	The performing paradox is initiated by conflicting demands among different stakeholders. Moreover, organizational change tends to exacerbate the tensions of performing by fostering competing measures of managerial success.

The paradox of learning emerges as individuals struggle between the demands of old and new knowledge, both need to perform their activities. The paradox of learning is made salient under lean transformation, which calls for learning a set of skills and applying these in a flow setting rather than achieving higher levels of technical proficiency in narrower areas of specialization (Womack, Jones and Roos 1990, 71-197). The paradox of belonging rotates around tensions associated with time and effort dedicated to personal goals as opposed to time and effort dedicated to team activities. The paradox of belonging accentuates during lean transformation as employees attempt to make sense of systemic contradictions fostering both autonomy and discipline (Lüscher and Lewis 2008, 233).

The paradox of organizing emerges from competing work designs related to discipline and standardization versus autonomy. In fact, lean systems contain features of both mechanistic and motivational designs. While the mechanistic design is grounded on standardization and efficiency, the motivational design is grounded in greater organizational autonomy, job rotation and teamwork (Adler and Borys, 1996, 61, Cullinane, et al. 2012, 41, Cooney 2002, 1130). Finally, the paradox of performing is initiated as individuals tend to accommodate different and even conflicting performance measures. Indeed, lean accentuates the performing paradox as it entails pursuing multiple dimensions of performance, such as lower costs, superior quality and short delivery time (Nawanir, Teong et Othman 2013, 1019, Kosuge, Modig et Åhlström 2010, 1, Adler, Goldoftas and Levine, 1999, 43).

3. THE MANAGEMENT OF ORGANIZATIONAL PARADOXES

Ford and Backoff (1988, 89) define an organizational paradox as "some 'thing' that is constructed by individuals when oppositional tendencies are brought into recognizable proximity through reflection or interaction." According to this view, organizational members confront and construct environments through their mental frames, which are the cognitive mechanisms that form the context, within which, reality construction and the creation of paradoxes occur (Watzlawick, Weakland and Fisch 2012, 92). Thus, dealing with paradoxes must take into account the mental frames of the individuals involved in organizational change, and the success of change through the management of paradoxes must entail some level of reframing or the creation of new mental frames (Smith and Lewis 2011, 389). Furthermore, the reframing process starts with some trigger or event that unfreezes a particular mental frame (way of understanding a situation) and indicates that this understanding might be changed. To be effective, the challenge to the established mental frames has to be strong because once mental frames are developed, they tend to endure (Bartunek 1993, 322). Maalouf and Gammelgaard (2016, 702-703) give examples of reframing during lean implementation such as standardization which can make sense to employees as opposed to the previous mental framework that standardization is equivalent to rigidity.

Moreover, managerial responses can influence reframing and change in two ways. First, a response can trigger an initial stimulus for challenging existing mental frames by making individuals aware of their paradoxical situations and by establishing conditions and setting directions that enable breaking the vicious circle (Eisenman and Rothenberg 1980, 689). Second, a response can motivate individuals to refrain from constraining the outcome of the process (Bartunek 1993, 322). In similar context, Quinn and Cameron (1988, 10) cite that the effective management of organizations require exploring and balancing contradictions and oppositions. They also note that effective organizations "do not pursue a single set of criteria; rather, they pursue competing, or paradoxical, criteria simultaneously," such as standardization versus autonomy, centralization versus decentralization and short- versus long-term focus. Furthermore, Smith and Lewis (2011, 389) argue that organizations are inherently paradoxical and the opposing yet complementary dualities of paradoxes are embedded in the process of organizing. Moreover, the authors argue that the paradoxes remain latent until they are made salient through social interaction, actors' cognition and organizational change. As a consequence, tensions intensify to the point that organizational actors experience and recognize their effect.

Responding to paradoxes, scholars present two generic strategies: acceptance and resolution. Acceptance assumes that tensions and contradictions can coexist and actors can benefit from the increased understanding of the relationship between the two opposites. In acceptance strategies, actors "play through rather than confront tensions, thereby avoiding potentially disastrous conflicts" (Smith and Lewis 2011, 385). Moreover, acceptance entails that actors confront paradoxes and discuss their tensions, which help constructing a more accommodating understanding of the paradoxical phenomenon (Vince and Broussine 1996, 1). In general, acceptance of the presence of contradictions provides a comfort with tensions and a new understanding of the relationship between opposites, thus, enabling actors to use resolution strategies for dealing with paradoxes. Acceptance strategies entail coaching, mentoring, experimentation and intense involvement of employees, who are intrinsically motivated toward the adoption of new mental frame that accommodates lean tensions and paradoxes (Maalouf and Gammelgaard 2016, 703).

Resolution involves responding to paradoxical tensions by separating, physically or temporarily, the tensions between the two poles of paradox. Resolution strategies also entail finding synergies that accommodate the opposing elements of a paradox. The spatial or temporal separation of the two poles of paradox reduce immediate tensions and help actors identify synergies between opposites by making explicit how one pole of the paradox relates to the other (Poole and van de Ven 1989, 566). For instance, under lean implementation, allocating daily activities and problem solving tasks to different groups of employees is an example of spatial separation, while allocation of a part of an employee's working time to problem solving tasks is an example of temporal separation (Maalouf and Gammelgaard 2016, 695). Separation is likely to create focus among employees that reduces the immediate pressure and helps individuals achieve a better understanding of the paradoxical phenomenon and adopt more sustainable management strategies (Poole and van de Ven 1989, 565-566). Resolution strategy also entails synthesis by creating organizational structures and processes that accommodate opposing elements of a paradox simultaneously. Under lean implementation, synthesis entails solutions that balance standardization and autonomy, such as focusing on the standardization of the repetitive parts of a task so that employees have more time to invest in other more creative tasks (Maalouf and Gammelgaard 2016, 698).

Table 2 presents a summary of the management strategies for each type of paradox. These strategies are either acceptance (Employee involvement, Experimentation, Facilitation, Coaching and Mentoring, Class training and On-the-job training) or resolution strategies (Temporal and Spatial separation; Synthesis, and Goal setting).

Table 2. Management strategies for each type of lean paradox

Type of paradox	Management strategies	
	Acceptance	Resolution
Paradox of organizing	Employee involvement (Maalouf and Gammelgaard 2016, 687-709, Glew, et al. 1995, 395-421, Shadur, Kienzle and Rodwell 1999, 479-503) Experimentation, and trial-and-error (Maalouf and Gammelgaard 2016, 687-709, Rerup and Feldman 2011, 577-610)	Synthesis (Maalouf and Gammelgaard 2016, 687-709, Poole and van de Ven 1989, 562-578)
Paradox of performing		Temporal separation (Maalouf and Gammelgaard 2016, 687-709, Poole and van de Ven 1989, 562-578) Spatial separation (Maalouf and Gammelgaard 2016, 687-709, Poole and van de Ven 1989, 562-578) Synthesis (Maalouf and Gammelgaard 2016, 687-709) Goal setting (Maalouf and Gammelgaard 2016, 687-709, Locke and Latham 2013, 3-15)
Paradox of belonging	Facilitation of group discussions (Maalouf and Gammelgaard 2016, 687-709, Ellinger and Bostrom 1999, 752-771) Coaching and mentoring (Ellinger and Bostrom 1999, 752-771)	
Paradox of learning	Class Training (Maalouf and Gammelgaard 2016, 687-709, Ellinger and Bostrom 1999, 752-771) On the job training (Maalouf and Gammelgaard 2016, 687-709, Ellinger and Bostrom 1999, 752-771) Coaching and mentoring (Ellinger and Bostrom 1999, 752-771)	

In the next section, we present the 3 cases that illustrate the identification of the different types of organizational paradoxes in lean (2 cases from Denmark, one case from Bangladesh). Moreover, in each case we present the repertoire of managerial responses to deal with each paradox and discuss the outcomes. The cases are selected from different geographies and sectors in order to capture the effect of varied internal and external contexts on the creation and management of organizational paradoxes in lean.

4. CASE 1. THE FINANCIAL COMPANY (DENMARK)

The company is one of the biggest financial corporations in Denmark. The company offers a typical range of banking products and services for both Danish and international customers. To increase the efficiency of its operations, the Financial Company decided to implement lean in its transaction-processing operations with a focus on the productivity of the case handling process. The company aimed to increase the productivity of the process by 20%. The increase in productivity was set to compensate for the number of employees going into retirement within the next two or three years and to avoid hiring new employees. That is, the company decided to compensate for the natural reduction of employees through retirement by increasing the productivity of the remaining workforce.

Moreover, the company needed to change its staff promotion policy as part of the implementation strategy. Promotion to team leader position was often based on technical skills and knowledge about the claims handling process. Indeed, many of the team leaders were previously senior case handlers and the company promoted these case handlers to team leaders because of their technical skills and experience in claims handling. This promotion policy raised tensions in the Financial Company during lean implementation, mainly because team leaders were required to take on a new role based on lean flow knowledge and workforce management rather than the traditional technical skills in claims handling.

To get acquainted with lean philosophy, team leaders in the company normally go through lean program training in order to learn and apply lean practices and tools. The lean training program includes the use of lean philosophy and practices to execute an improvement project related to the work area of the team leader. The project has two modules of 18 weeks each. The first module is called the analysis and implementation phase, and the second module the follow-up phase. In the first phase, team leaders are trained and supported by an external lean consultant in order to learn to apply lean practices in their respective projects. In the follow-up phase, the team leader is supposed to take ownership of the process, increasing as the support of the lean consultant is gradually reduced. During the project, on-the-job learning is intense as team leaders get acquainted with lean practices and learn how to apply them in real work context. According to one lean consultant involved in the training:

> "Team leaders have to learn lean and operations management techniques where they plan every single day and balance the work load of the employees... they are expected to become not only technical leaders but also process consultants by making improvements and eliminating the root cause of the problems."

During lean implementation at the Financial Company, the organizing and belonging paradoxes accentuated and increased the resistance of team leaders towards the change. The next two sub-sections explain the emergence and accentuation of these two paradoxes and discuss the strategies used for dealing with them and present the outcomes.

4.1. The Organizing Paradox

The organizing paradox emerges at the company as team leaders are required to implement lean flow and follow lean standards instead of their traditional self-developed way of handling cases and claims. Team leaders resist standards because they believe standards limit their autonomy during case handling and reduce their ability to adjust the claims handling process to different types of claims with varying degrees of complexity. Managers have attempted to promote the acceptance of this paradox by communicating to team leaders that standards are not "sacred." On the contrary, the discourse in the company was that standards can be improved by the users as a better standard is identified. According to one senior director:

> "If an employee identifies an opportunity to improve the process, then he or she should submit his idea to the formal suggestion system. By doing this, the idea will be analyzed by the team and discussed with the team leader who submitted it. By doing so, the idea becomes everybody's project."

Moreover, some team leaders resisted standards because of the risk of embarrassment in case their measured performance was assessed below the average performance of their colleagues. Other team leaders argued that standards can limit their autonomy in searching for all potential sources of errors that led to customers' complaints. However, many team leaders noticed and recognized the benefits of following lean standards for their daily production and on workers' motivation. According to one team leader:

> "My claim handling workers feel that it is good to have standards because when they go home, they can say that it was a good day and they achieved their daily goals. By following lean standards and productivity measurements, one can still achieve the daily goal although there is still a bunch of cases waiting in line for the next day. Moreover, I keep telling my workers that the standard will be followed until we decide to change it… when we find better way of doing things, then we improve the standard."

Furthermore, one team leader explained that by following lean standards, employees could use their creativity to find a better way of doing their work and to improve the existing standards, rather than to change operational procedures on an individual basis or find different ways for handling similar cases or claims. However, explaining the benefits of following standards have not been always effective for convincing employees to adhere to standards. In these cases, acceptance strategies based on involvement and experimentation were used intensively and repetitively by the company in order to deal with this paradox and promote the acceptance of the organizing tensions. More specifically, the confrontation of the organizing paradox entailed discussions of tensions in groups, experimentation and trial-and-error learning, and the direct involvement of the most resistant members of the team in the improvement of standards. The experimentation and trial-and error learning were crucial for breaking the vicious circle of resistance as team leaders could put their own ideas into practice, observe the results, and improve. According to one lean consultant:

> "We take the employees that put the most resistance early on the improvement workshop where he or she can have more influence in the output of the process; in the first day of the workshop they might complain; however at a certain point of the workshop they begin to get engaged in the process and contribute to the improvement effort... they have normally a lot of energy... they begin to see the benefits of the process and come up with a lot of good ideas for improvement; they can be considered change agents because other employees usually listen to them."

However, in some cases, the acceptance strategies were not achieving the intended results and some team leaders were still resisting the new work organization based on lean thinking. In this case, the company adopted a top down push in order to make some employees participate in the improvement process. According to one manager:

> "We communicate to the employees that lean has come to the department and will stay; so you have to decide what you want."

4.2. The Belonging Paradox

The belonging paradox has been noticed frequently in the Financial Company during the interviews. The belonging paradox emerges as team leaders are required to take on a new role during lean transformation and abandon the old role based on technical knowledge. According to the new work design, team leaders are expected to act as process and operations managers rather than firefighters or technical experts for case handling. One manager explained the tension associated with the belonging paradoxes among team leaders:

> "People want to hold on to the old role as firefighters because it has been the source of their prestige within the company; it is about letting go of the old role and embracing the new role; sometimes they suddenly embrace the new role and become good leaders... as soon as they reach some level of understanding... so they become the big advocates of the new role... when they see the effect of the new role and of the new tools on their daily work."

To facilitate the acceptance of the paradox and deal with the challenges of the new role, the follow-up phase of the projects was used by team leaders, mentors and lean consultants as a buffer period for reflection where people consolidate the gains achieved during the implementation phase instead of starting new projects. In the reflection phase, the acceptance of lean tensions and paradoxes has increased as team leaders consolidate their knowledge about what has worked and what has not worked during the conversion to the new role. The acceptance of the paradox has enabled team leaders to take on more challenging aspects of the new roles in relation to the dissemination of the lean mind-set and the use of lean tools in their respective areas when the training period is over. However, the belonging paradox has often required various sessions of coaching, mentoring and group discussions in order to achieve the new level of understanding and the acceptance of the new role among the employees. According to one manager:

> "We invest in coaching and mentoring where an external consultant follows and helps the employee; we also use a maturity model where we assess the development of the employees; however, sometimes we can see that even after many attempts, this employee is not the right man for this new role; so we have to find something else for him elsewhere."

Moreover, one director summarized the management view for dealing with the belonging paradox:

> "First of all, we have to be determined that this is something we want to do... and lean should not be seen as a time-bound project... the project is there to facilitate broader change of behavior and attitude... we tell our employee that we want this, so how can we help you to get on?"

Table 3 summarizes the lean paradoxes, the strategies used to deal with them, the factors influencing the management of paradoxes, and the outcomes of change at the Financial Company.

Table 3. Organizational paradoxes of lean, the management strategies and the outcomes in the Financial Company

Description of paradox	Management strategies and outcomes
The organizing paradox: The organizing paradox emerges in the company as team leaders are required to implement lean flow and follow lean standards instead of their own way of handling cases and claims.	Management of the organizing paradox entailed discussions of the tensions in groups, experimentation and trial-and-error learning, and the direct involvement of the most resistant members of the team in the improvement of standards. *Outcomes:* People realized that standards could help them in their daily activities. However, in some instances, top management had to pressure employees to adapt to lean organization or find other positions in the organization.
The belonging paradox: The belonging paradox emerges as team leaders are required to take on a new role during lean transformation and abandon the old role based on technical knowledge.	Team leaders, mentors and lean consultants used the follow up phase of the projects as a buffer period for reflection where people consolidated the gains achieved during the implementation phase instead of starting new projects. In the follow up phase, various sessions of discussions in groups, coaching and mentoring were used to achieve the new level of understanding and the acceptance of the new role among team leaders. *Outcomes:* Most of the team leaders succeeded in the new role. However, the management had to replace some team leaders that failed to succeed or accept the new role.

5. CASE 2. THE CANCER DEPARTMENT AT A UNIVERSITY HOSPITAL

The university hospital has adopted the lean philosophy as a platform for improving efficiency and work conditions. In order to do so, the university hospital integrated lean practices in its global strategy and dedicated considerable resources to lean implementation. An internal lean consultant unit organized lean implementation into waves of five to six departments – each lasting one year. A wave started with an intensive training course for the lean implementation teams. During and after the course, a consultant from the unit was attached to each department for a period of 6-12 months. The cancer department had 350 employees and included wards, out-patient chemotherapy, radiation therapy, a laboratory, and a palliative section. A steering committee was established with the head nurse, the head consulting doctor, the leader of the lean project group, and the consultant from the lean unit. A lean project group was organized with a nurse as project leader, a consulting doctor, two nurses, a secretary, a lab technician, and a radiologist. A relatively large number of activities were initiated concerning activities such as delivery of chemotherapy medicine, collaboration between lab technicians and the chemotherapy outpatient clinic, handling of blood samples, handling of case records, establishment of kaizen boards, and the reorganization of patient booking in the chemotherapy outpatient clinic. It was also decided to reorganize the ward rounds, but that project never got off the ground.

The change strategy was based on the extensive involvement of concerned staff, who were supported by members of the lean project group and, if needed, the lean consultant. Some smaller changes were initiated by several kaizen workshops over one or two half-days, where representatives of the concerned staff analyzed the problems using value stream mapping and came up with solutions. In other cases, working groups were established to analyze a particular problem and come up with a solution. Several changes, which the involved actors described as successful, were implemented, including higher quality medicine delivery, better track of blood samples and case records, the use of kaizen boards with many implemented suggestions, and reorganization of the work of the lab technicians. However, the reorganization of patient booking in the chemotherapy outpatient clinic turned out, in particular, to be very problematic, and the department was still fighting to get the patient booking back on track when the project ended. The outcomes were experienced differently by the lab technicians and the nurses. The lab technicians were quite satisfied with the results whereas the nurses expressed severe dissatisfaction. The question was therefore to understand why the situation differed so dramatically between the two groups. Part of the explanation could be found in the differences in the change process adopted in the two units.

Starting with the laboratory, they initiated two major changes. The first one focused on blood sampling and intubation of intravenous lines. Previously, the lab technicians were called to the outpatient clinic after arrival of the patient. They then had to identify the patient, search for a vacant couch, do the blood sampling and walk back to the lab. This task constituted a large portion of the technicians' work and they spent considerable time walking from one place to another, and finding and sometimes waiting for vacant couches. They organized a kaizen workshop together with nurses from the outpatient clinic and identified possible solutions, which ended in a decision to reorganize that particular task. In the new procedure, the nurses ask the patients to walk to the lab where there will be one or two lab technicians on duty in a room with a couch. The intubation and blood sampling take only a couple of minutes and the patients save time they would previously have just spent waiting in the clinic.

The other major change was the introduction of kaizen meetings. Once a week the lab technicians organize a standing meeting around a kaizen board where they suggest ways of improving everything related to the lab and the technicians' work. By the time the project ended, they had made 84 suggestions, 68 of which had been implemented. Among others, the suggestions have resulted in more space in the quite congested lab, a more secure supply of material, and higher safety in the handling of blood samples. In the interviews, both the head lab technician and the technicians said that they experienced lean as successful and beneficial for the work environment.

The outpatient clinic showed quite a different picture. The change started successfully with the reorganization of the communication lines between the pharmacy and the clinic, which ensured delivery of medicine on time. However, after that effort, the change faced increasing resistance by the nurses as the department management and the lean implementation group decided to introduce a new IT-program. This program aimed to achieve more systematic planning of patients' admission flow, which would reduce waiting time and increase the efficiency of the treatment facilities. Among other things, the new system would reduce the considerable misalignment between the planning of patient booking and the expected duration of the administration of the chemotherapy. It turned out that the IT-implementation was much more complicated than expected, and for quite some time, patients' booking was still fraught with problems. Nevertheless, the new booking system considerably changed the work of the nurses. Previously, nurses used to book their own patients individually, whereas, with the new IT system, bookings are done by a secretary. Nurses explained that the new system created more work, partly because of the persistent booking problems and partly because they had to channel the bookings through a secretary rather than doing the booking themselves. Parallel to the new booking system, attempts were also made to introduce kaizen meetings in the outpatient clinic, but with little success.

Associated with the above considerations of the new IT system, the nurses had quite negative views of lean, and considered that lean had created more work that deteriorated their work environment. A large group of the nurses tended to interpret the IT-program as an attack on their professionalism, and expressed a serious concern about the negative effects of the new system on the quality of care. That is, the nurses believed that the flexibility of the patient professional care was seriously weakened by the new standardized booking system, as nurses were often unable to accommodate the patients' special needs.

5.1. The Organizing Paradox

The organizing paradox rotates on tensions between the autonomy of the workers and the standardization imposed by the new system. The paradoxical lens helps us make sense of this difference in the reactions of the two groups of employees: the lab technicians, on the one hand, and the nurses on the other. The paradoxical tensions had intensified during the organization of the outpatient process as nurses valued autonomy as opposed to increased standardization. Yet, the organizing paradox had a much attenuated effect on the lab technicians. One explanation is that the content of the work of the lab technicians contained more repetitive and measurable tasks than that of the outpatients section. Indeed, the lab tasks consisted of reasonably standardized tasks with a strong emphasis on safety matters, such as avoiding any chance of mixing or delaying blood samples. As a consequence, lab technicians valued the reorganization of their activities as it increased the quality and safety control of their tasks. As for the outpatient clinic, the nurses strongly valued their work autonomy, which increased the resistance to the standardization imposed by the new system. Indeed, autonomy was more relevant for the outpatient section as it contained varied types of tasks requiring flexible decision making associated with local control. Naturally, the nurses opposed the loss of autonomy imposed by the new system and experienced a reduction in their degree of control regarding the quality of the care to the patients.

The results from this study suggest that lean implementation may challenge the traditional understanding of professionalism in hospitals - at least for nurses. Traditionally, treatment and care have had a strong element of trial-and-error. Usually, a certain treatment is adopted and the result is monitored and adjusted according to the patient's behavior and needs. Collective standards obviously seem to conflict with the possibility for the individual nurse to make her own decisions based on her own professional judgement. In the example given here, the idea was to make patient booking more efficient, in adherence to the operational value (McClean, et al. 2008), whereas the nurses were afraid that the standardized booking would have negative consequences for the experiential value as the patients may feel a lack of concern for their personal priorities.

The management strategies for dealing with the organization paradox included acceptance strategies, such as employee involvement, experimentation, trial and error, and resolution strategy, such as synthesis. While the acceptance strategies seemed to deal effectively with the organizing paradox of the lab technicians, it failed to reduce the tensions of the nurses in the outpatient clinic. In the case of the outpatient clinic, the management used a resolution strategy (synthesis), which entails a solution that accommodates standardization and autonomy. Indeed, the management focused on correcting the constraints and problems of the new IT system, which increased the usability of the system by the nurses. At the same time, management attempted to secure commitment to the new system by promoting intensive involvement of nurses in the improvement of the system. However, the resolution strategy reduced the tension temporarily as nurses and doctors were still resisting the standardization of their activities, which caused the roll out of the new system to stop.

5.2. The Performing Paradox

The organizing paradox in the hospital is intimately connected to the performing paradox and rotates on tensions between cost efficiency (treating the maximum number of patients) and individualized care (the current approach adopted by nurses and doctors). Perhaps the increasing demands for treatment and the growing complexity of care activities associated with government pressure to increase efficiency created a need to develop a new balance between the traditional individual professionalism and collective standards. The lean approach entails the standardization of work tasks, which increases efficiency and enables the treatment of a higher number of patients.

Table 4. Organizational paradoxes, management strategies and outcomes at company 2

Description of paradox	Management strategies and Outcomes
The organizing paradox This organizing paradox accentuated between the need for individual care of patients, on the one hand, and the standardization introduced by the new system. The organizing paradox was also present in the control of the booking system: booking controlled by nurses (autonomy) versus booking controlled by the protocol of the system (standardization).	The management used both acceptance and resolution strategies. The acceptance strategies entailed employee involvement and experimentation, and were effective in reducing the organizing tensions of the lab technicians. However, the resolution strategy (synthesis) were needed to deal with the persistent tensions in the outpatients' clinic. The synthesis entailed solutions to a range of functional problems of the system with increased involvement of nurses and doctors. *Outcomes:* The booking system ended up working as intended with a more efficient flow of patients, and the lab experienced several improvements in efficiency of blood sampling and resource utilization. However, the resistance to standardization remained high among nurses and doctors and, therefore, further implementation of the system came to a stop.
The performing paradox The performing paradox rotates around tensions between the number of treated patients and the quality of the care	The synthesis (resolution strategy) entailed that the nurses can make patients bookings directly in the system, which maintained a certain level of individualized care for nurses. *Outcomes:* With the possibility for nurses to book directly in the system, the initial resistance was reduced. However, the basic tensions between efficiency and individualized care persisted among nurses and doctors.

Within the context of this study, standardization seems more suitable to the activities of lab technicians, whereas it challenges the varied nature of tasks of nurses and doctors, which require higher degrees of autonomy for dealing with special cases. This is a real dilemma because standards increase efficiency and allow the professional to focus on the more creative tasks. However, at the same time standards limit the individual assessments, which are a crucial part of nurses' and doctors' professionalism.

In order to deal with the performing paradox, the management adopted a resolution strategy based on synthesis that can achieve a balance between efficiency and individualized care. As for the increase of efficiency, the use of the new system benefits patients by reducing the waiting time for commencement of treatment as well as during the treatment process. However, in order to secure a certain level of autonomy in the treatment of outpatients, the management enabled the nurses to make the bookings of the patients instead of a secretary, which maintained a certain level of nurses' autonomy. By adding the possibility for nurses to make the outpatients booking, management was able to reduce the tensions temporarily. However, the inherent tensions between efficiency and individualized care persisted as nurses and doctors were still valuing the autonomy of patients' treatment.

6. CASE 3. THE GARMENT MANUFACTURER IN BANGLADESH

The company is one of Bangladesh's leading manufacturers of ready-made garments. As the company was regularly producing at the limits of its capacity, it embarked on an expansion and optimization project in 2013 in order to increase its production capacity and competitiveness. The main clients include international retail chains and warehouses. The holding company employs more than 6,300 workers, the large majority of whom (70%) are women. As the group was regularly manufacturing at the limits of its capacity, it has, since 2010, invested in new machinery and in training. Since the 1980s, the RMG industry in Bangladesh has experienced rapid growth. With the Chinese RMG industry in decline, Bangladesh has emerged as a rapidly growing global producer of garments. Owing to low wages and workforce availability, RMG-export levels in Bangladesh have grown steadily with average annual growth rates above 10%, and many leading international retailers from Europe and the US have adopted Bangladesh as a main sourcing country. In 2013-2014, RMG exports amounted to USD 24.5 billion and accounted for 80% of the nation's export earnings. The RMG industry in Bangladesh employs some 4.2 million, out of Bangladesh's total workforce of about 80 million. About 80% of these workers are women.

In order to remain globally competitive, RMG manufacturers in Bangladesh need to invest in the skills development of their workforce in order to improve productivity and increase sales. Since most of the machine operators are poorly educated, manufacturers are increasingly looking for higher-skilled workers and professionals to increase the productivity of their plants. Moreover, there are only a few dedicated training programs for garment workers at higher skill levels. Given the poor public education and limited vocational garment-specific training, it is important to involve the private sector in fostering skills development. The widest skills gap is in production mid-management, i.e., line supervisors and line managers. As there is no dedicated training institute in the country for the relevant technical skills, the line supervisors and line managers often have a weak understanding of industrial engineering and modern production systems. Given the increasingly sophisticated machines that are used for achieving greater automation and higher productivity, supervisors also need a good technical understanding of the different specialized machines, requiring special on-the-job training. In terms of people management skills, the company has also identified some deficits in leadership and communication skills among its production supervisors and managers.

6.1. Learning Paradox

Learning paradoxes surface as companies attempt to change, adjust and innovate, which involve both building upon existing knowledge and resources in order to improve performance (O'Reilly and Tushman 2008). Line supervisors and production managers need a deep understanding of modern production systems, and technical expertise on the different types of machines, as well as leadership and communication skills. At the level of machine operator, the company can easily recruit a sufficiently large number of workers, who must be trained internally to acquire the required manufacturing skills and to comply with health and safety standards. Regarding unskilled machine operators, the company has a sufficient supply of women available from the villages nearby who can be recruited directly at the factory gate. These women need intensive on-the-job training to reach an acceptable level of quality and efficiency. Moreover, these new workers often lack a proper understanding of health and safety issues and most of them lack a basic school education.

To tackle these skills and education gaps, the company has introduced a range of initiatives to foster the skills development of its workforce focusing mainly on machine operators, line supervisors and middle managers. In this context, the company has introduced training courses for productivity as an integral part of a larger initiative of production improvement. Moreover, the company introduced a separate training station for providing practical training and production courses for established staff without hampering the actual production process. One of the initiatives aimed to radically transform the production process with the help of outside consultants, who have experience in modern manufacturing systems such as lean and agile manufacturing. From 2012 to 2014, the consultants radically transformed operations at the company by introducing specialization in production (critical tasks were allocated to specially trained workers), changing the production layout to more lean flow and eliminate wastes, and by introducing new machines to increase automation.

In addition to these changes in operations, the initiative involved extensive training courses covering the entire workforce. With the help of job descriptions detailing the skill requirements for each position, all staff from machine operators to the managing director had their aptitude for their current position assessed. Dedicated training modules were then provided as appropriate, based on an elaborate training manual: the modules involved 4 weeks of theoretical training for lean production tools (all workers) and seven days for production leadership (managers, line chiefs and supervisors). Additionally, all sewing operators and workers were trained on the job at a separate training station. From the ranks of production managers, more than 100 employees received training related to lean manufacturing. Moreover, almost all production workers received basic theoretical and practical training. To sustain the training initiative, one consultant with much experience in the international garments sector was hired as plant manager, who has enabled the company to provide the needed training to employees and to offer refresher training courses internally.

However, the company soon noticed that an increasing number of highly qualified workers started to leave the company as they could easily find jobs with higher salaries at nearby competitors. As a consequence, the learning paradox accentuated in the company as more investment in training and education, the more skilled workers found jobs elsewhere with higher salaries, usually in competitors nearby. Indeed, the retention of workers is already a major issue in the Bangladesh garment industry. Attrition rates of more than 5% per month are common among garment manufacturers in Bangladesh. There are at least two main reasons for the high attrition rate in this industry. First, a high proportion of employees are migrant workers from other parts of the country, who are likely to return to their home region. Second, a high share of the employees are women, who often stop working once they are married and have children. Moreover, the company often loses women workers who return to their home villages and marry. Then, in order to deal with this learning paradox and increase the retention of workers, the company soon realized that increasing workers skills alone was not sufficient. In order to retain workers, the company had to look outside the shop-floor, reaching out to the families of workers and to the community nearby where workers live. For instance, the company provided childcare for the children of employees and free transport to work. Moreover, the company invested in the local community by improving roads and sewage systems.

Moreover, the company was sourcing most of its fabrics from China. However, in an effort to support the economy of the local community as much as possible, the company started to purchase fabrics from local suppliers. Today, these local purchases account for about 15% of the fabrics bought by the company. Overall, about 20% of all supplies are produced in Bangladesh, including packaging materials such as plastic bags and cardboard boxes. Moreover, the company supports the development of its suppliers through skills transfer, where the purchasers invite pre-selected suppliers to the main production site of the company, and explain to them the quality requirements posed by international buyers and the quality-testing procedures. Once a local supplier is engaged, the company sourcing team continuously interacts with this supplier in case of quality issues and supports them in finding solutions to fix the underlying problems. The purchasers also regularly visit the suppliers' production facilities to provide support and prevent potential problems related to quality and delivery time. Apart from knowledge transfer, the company supports its local suppliers financially by paying invoices right away and sometimes even in advance of delivery. In addition, the company conducts regular audits of its suppliers to ensure compliance with safety and health standards, suggesting specific corrective actions and deadlines for implementation, monitoring progress, and providing training sessions on health and safety and environmental concerns.

6.2. Performing Paradox

The performing paradox rotates around competing and even contradictory demands among different stakeholders. Moreover, organizational change tends to exacerbate the tensions of performing by fostering competing measures of performance. In this company, the performing paradox rotates around tensions between the efforts to increase productivity and remain competitive on the one hand, and the efforts to comply with code of conduct and maintain good occupational health and safety (OHS) conditions for workers, on the other hand.

A look at the Bangladeshi context reveals important features of this performing paradox. Indeed, the garment sector in Bangladesh has been hit by several deadly accidents with thousands of fatalities while producing garments for international brands from Europe and USA. Moreover, this sector is known to be infamous for low wages, unsafe working conditions, long overtime hours, poor safety regulations, inadequate reinforcement of laws and inefficient factory inspections. More specifically, international pressure for safer work conditions has intensified following the collapse of the Rana Plaza building in 2013. As a consequence, the ACCORD on Fire and Building Safety in Bangladesh was signed that year by the majority of apparel brands, retailers and importers, Bangladeshi trade unions, and NGOs. The ACCORD aims to improve safety in the garment sector (fire, building and electrical safety) by means of independent factory inspections, with corrective action plans, as well providing training on fire precautions and operational health and safety.

In addition to these substantial changes in the country's context, Bangladeshi RMG manufacturers have seen their margins under pressure because of international buyers' use of their bargaining power to keep prices low, an increase of minimum wages approved by the government, tougher regulation on health and safety, and large investments to be made in further fire and building-safety measures. Consequently, the company management was putting pressure on workers to increase productivity while adhering to safety rules and procedures. However, in some cases, workers were shortcutting safety procedures especially when the delivery deadline was under risk of delay.

The main actions of the company for dealing with the performing paradox were focused on implementing improvements in the production processes, which can increase productivity and improve OHS conditions simultaneously. Indeed, research on work conditions and lean implementation reveal that the same issues - such as improper workplace design, poor human-machine fit and inappropriate incentives - are responsible for both worsening work conditions and reducing worker productivity. As a consequence, it would be difficult to improve productivity and OHS simultaneously without dealing with these joint safety and productivity issues (Shikdar and Sawaqed 2003). Moreover, there is evidence in the literature that some lean tools have a positive effect on work productivity and on workers' health and safety. For instance, Gapp, Fisher and Kobayashi (2008, p. 567) argue that *"a primary objective of practicing 5S is to maximize the level of workplace health and safety in conjunction with increased productivity."*

In this context, workers would benefit from better working conditions through changes in the production layout that improves both the ergonomic conditions of the workers and their productivity. More specifically, higher desks and ergonomic chairs reduce the strain on workers' backs caused by excessive bending and extreme body movements. Another important strategy for dealing with the performing paradox is based on goal setting. Goal setting at this company meant that the increase in workers' salaries were linked to productivity gains through a transparent efficiency bonus, and to occupational health and safety compliance of all workers in the production line. These improvements in the production line and changes in work practices not only benefit workers' health but also increase efficiency.

More importantly, a central component in simultaneously improving productivity and working conditions for workers is the existence of a well-working industrial relations system. Despite recent regulatory changes and the creation of several factory level trade unions and employers' associations, Bangladesh is in the very early stages of the development of its industrial relations system. The trade unions in Bangladesh have limited knowledge of social dialogue, collective bargaining, and productivity-based salary structures. Moreover, the garment employers are frequently hostile to unions and consider them a source of hassles and problems. Therefore, introducing social dialogue in the garment industry in Bangladesh faces significant challenges, but contains relevant opportunities for creating the conditions for synergies between improved work environment, social dialogue and productivity.

By adopting solutions that improve both OHS and productivity, and introduce new terms to resolve the tension, the main strategy of the company for dealing with the performing paradox is a synthesis of the two poles of the paradox. As such, lean implementation reflects this synthesis as it introduces solutions that benefit productivity and OHS conditions simultaneously. Moreover, the introduction of social dialogue between company, unions and workers is likely to facilitate this synthesis. Indeed, the company was still reluctant to engage in social dialogue activities because - it is believed - that this social dialogue could lead to counterproductive behavior among the employees. Indeed, the company fears that the employees might go on strike if social dialogue was more pronounced, which partially reflects the confrontational strategies used by existing trade unions.

At its core, social dialogue is a process involving workers, employers, trade unions and governments. Social dialogue can involve collective bargaining, Workplace cooperation, and tripartite social dialogue. Collective bargaining takes place between workers' organizations and employers or representative of employers' organization. It commonly deals with wages, working time, and terms of employment, as well as ongoing relations between workers and employers. Workplace cooperation is another form of social dialogue that can be used to improve work conditions and organization, introduce new production methods, and secure a safe working environment. Tripartite social dialogue involves the participation of the government and usually deals with policy issues. The company was still in the initial phase and was implementing some features of collective bargaining and workplace cooperation. Despite some initial benefits and open dialogue between workers and employers, the sustainability of social dialogue was still elusive as it depends on a long-term strategy, which involves political and economic stability and strong representation of workers.

Table 5. Organizational paradoxes, management strategies and outcomes

Description of paradox	Management strategies and Outcomes
The learning paradox This paradox accentuates as the company invests in increasing the skills level of its workers. Yet, as workers become more skilled, they are more likely to find jobs elsewhere.	The company provided both technical and behavioral training to all levels of staff and workers. Moreover, the company helped the families of workers and the local community where workers live. For instance, the company provided childcare for the children of employees and transport between work and home. The company also invested in the local community by improving roads and sewage systems, and started to purchase fabrics from local suppliers in an attempt to boost local economy. *Outcomes:* The company showed a turnover rate of 4%, which is - according to the company - lower than the industry average. The company spotted a new trend among workers as they started to move to houses situated closer to the company.
The performing paradox In this company, the performing paradox rotates around tensions between the efforts to increase competitiveness and productivity, on the one hand, and the efforts to ensure compliance and maintain good occupational health and safety (OHS) conditions for workers, on the other hand.	The main strategy for dealing with this paradox focused on implementing improvements in the production processes benefitting productivity and OHS conditions simultaneously (Resolution strategy: Synthesis). The other strategy was goal setting, which entailed that increases in workers' salaries were linked to productivity gains through a transparent efficiency bonus, and to occupational health and safety compliance of all workers in the production line. These improvements in work practices not only benefited workers' health but also increased efficiency. The company started to introduce features of social dialogue, such as collective bargaining and workplace cooperation. Yet, the effect of these features was still not evident. *Outcomes:* The company reported a lower number of accidents and better OHS conditions than the industry average. According to the company, workers are less likely to skip safety procedures because the compliance with safety procedures will not reduce their productivity, on the contrary, it will boost their bonus and salaries.

Moreover, the role of social dialogue could be relevant for dealing with the learning paradox as well. Indeed, increasing workers skills, without collective negotiation that involves work conditions, salaries and career planning, will most likely increase the turnover rate as these workers are easy targets for competitors. Collective bargaining entails sustainable benefits for employers, workers, and society in general. Collective bargaining can increase the trust associated with the employment relationship between workers and companies, greater motivation and retention, and ultimately higher performance. Better collective bargaining between company and workers can help to reduce the impact of shocks and seasonality on unemployment by enabling adjustments in wages and working time, so that layoffs could be reduced or avoided. Collective bargaining makes it easier to engage in temporary wage or working-time concessions. On the one hand, collective bargaining often reduces transaction costs involved in the negotiation of temporary wages and working-time reductions between workers and companies and facilitate their implementation. On the other, because wage and working-time concessions are coordinated between workers and companies, collective bargaining can make these concessions more acceptable to workers.

In summary, collective bargaining represents a powerful resolution strategy to deal with both the learning and performing paradoxes by achieving a synthesis that accommodates the two opposing poles of the paradoxes. Table 5 contains a summary of the learning and performing paradoxes at the company, the management strategies used to deal with these paradoxes and the related outcomes.

CONCLUSION AND RECOMMENDATIONS

In this chapter, we presented four types of organizational paradoxes of lean representing four different motivations for resisting lean change. We also investigated the different types of strategies for dealing with these paradoxes and the outcomes. Through the investigation of the organizational paradoxes in lean, we sought to add clarity to the processes of lean implementation in three case companies. The identification of the various types of paradoxes in lean has enabled companies to better understand the causes of resistance to lean implementation and, as a consequence, take effective actions for facilitating lean implementation. That is, companies would increase the likelihood of successful lean transformation if they focus on the relevant tensions among employees. For instance, in case 3 (the Garment Manufacturer in Bangladesh), the learning paradox had accentuated as skilled workers were increasingly leaving the company, and in case 1 (The Financial Company), the belonging paradox was increasingly noticed as team leaders had a stronger attachment to their work identity. As such, the paradoxical framework constituted an alternative to the top down approach. More specifically, this study recommends that companies should not rush to action before understanding the different sources of resistance to lean implementation among workers. By not rushing to action through a top down approach, companies are more likely to target the real causes of resistance and increase the likelihood of success associated with lean transformation.

Furthermore, this study adds to previous knowledge on organizational paradoxes in lean by investigating paradoxical tensions in a developing and industrializing country (e.g., Bangladesh). While previous studies in developing countries have mainly revealed the development of three paradoxes in lean (organizing, performing, and belonging) (Maalouf and Gammelgaard 2016), the garment manufacturer case revealed the relevance of the learning paradox as the motivation for workers to leave the company had increased and endangered the success of lean transformation (loss of knowledge). Moreover, the garment manufacturer case emphasized the importance of including external institutional factors within the range of actions used for dealing with paradoxes. For instance, the garment manufacturer in Bangladesh had to reach out to the families and local communities where the workers live in order to deal with the learning and performing paradoxes and avoid the loss of skilled workers.

Through the investigation of the organizational paradoxes in lean, we sought to add clarity to the processes of lean transformation in 3 cases from different sectors and geographies (Financial (Denmark), Hospital (Denmark), and Garment manufacturer (Bangladesh)). This study has increased our understanding of the different motivations of workers, supervisors and managers to resist lean transformation. Consequently, to avoid unexpected negative outcomes, companies must understand the nature of lean paradoxes and their impact on individuals within their organizations. More importantly, dealing with organizational paradoxes in lean is a long-term process, which involves learning, experimentation, and trial and error.

REFERENCES

Adler, Paul S., Barbara Goldoftas, and David I. Levine. "Flexibility Versus Efficiency? A Case Study of Model Changeovers in the Toyota Production System." *Organization Science* 10, n°1 (1999): 43-68.

Adler, Paul S., and Bryan Borys. "Two Types of Bureaucracy: Coercive and Enabling." *Administrative Science Quarterly* 41, n°1 (1996): 61–89.

Bartunek, Jean M. "The Multiple Cognitions and Conflicts Associated with Second Order Organizational Change." In *Social Psychology in Organizations: Advances in Theory and Research*, by J. Keith Murnighan, 322-49. Englewood Cliffs, N.J.: Prentice Hall, 1993.

Cao, Rui, Kong Bieng Chuah, Yiu Chung Chao, Kar Fai Kwong, and Mo Yin Law. "The Role of Facilitators in Project Action Learning Implementation." *Learning Organization* 19, n°5 (2012): 414-27.

Cooney, Richard. "Is 'Lean' a Universal Production System?: Batch Production in the Automotive Industry." *International Journal of Operations & Production Management* 22, n°10 (2002): 1130-47.

Cullinane, Sarah-Jane, Janine Bosak, Patrick C. Flood, and Evangelia Demerouti. "Job Design under Lean Manufacturing and its Impact on Employee Outcomes." *Organizational Psychology Review* 3, n°1 (2012): 41-61.

Cyert, Richard M., and James G. March. "Behavioral Theory of the Firm." In *A Behavioral Theory of the Firm*, by Richard M. Cyert, 1-53. Englewood Cliffs, N.J.: Prentice-Hall, 1992.

Eisenhardt, Kathleen M., and Brian J Westcott. "Paradoxical Demands and the Creation of Excellence: The Case of Just-in-Time Manufacturing." In *Paradox and Transformation: Toward a Theory of Change in Organization and Management*, by Robert E. Quinn and Kim S. Cameron, 169-93. Cambridge, MA: Ballinger, 1988.

Eisenman, H. J., and Albert Rothenberg. "The Emerging Goddess: The Creative Process in Art, Science and Other Fields." *Technology and Culture* 21, n°4 (1980): 689-91.

Ellinger, Andrea D., and Robert P. Bostrom. "Managerial Coaching Behaviors in Learning Organizations." *Journal of Management Development* 18, n°9 (1999): 752-71.

Ford, Jeffrey D., and Robert W. Backoff. "Organizational Change In and Out of Dualities and Paradox." In *Paradox and Transformation: Toward a Theory of Change in Organization and Management*, by Robert E. Quinn and Kim S. Cameron, 81-121. Ballinger: Cambridge, MA, 1988.

Gapp, Rod, Ron Fisher, and Kaoru Kobayashi . "Implementing 5S within a Japanese Context: An Integrated Management System." *Management Decision* 46, n°4 (2008): 565-79.

Glew, David J., Anne M. O'Leary-Kelly, Ricky W. Griffin, and David D. Van Fleet. "Participation in Organizations: A Preview of the Issues and Proposed Framework for Future Analysis." *Journal of Management* 21, n°3 (1995): 395-421.

Karlsson, Christer, and Pär Åhlström. "The Difficult Path to Lean Product Development." *Journal of Product Innovation Management* 13, n°4 (1996): 283-95.

Kosuge, Ryusuke, Niklas Modig, and Pär Åhlström. "Standardization in Lean Service: Exploring the Contradiction." *Proceedings of the 17th International Annual EurOMA Conference*. Porto, 2010. 1-10.

Lewis, Marianne W. "Exploring Paradox: Toward a More Comprehensive Guide." *The Academy of Management Review* 25, n°4 (2000): 760-76.

Locke, Edwin A., and Gary P. Latham. "Goal Setting Theory." In *New Developments in Goal Setting and Task Performance*, by Edwin A. Locke and Gary P. Latham, 3-15. New York, NY: Routledge, 2013.

Lüscher, Lotte S., and Marianne W. Lewis. "Organizational Change and Managerial Sensemaking: Working through Paradox." *Academy of Management Journal* 51, n°2 (2008): 221-40.

McClean, Sally, Terry Young, Dave Bustard, Peter Millard, and Maria Barton. "Discovery of Value Streams for Lean Healthcare." *4th International IEEE Conference Intelligent Systems*. 2008. 32-38.

Maalouf, M. 2013. Sustaining Lean: Strategies for Dealing with Organizational Paradoxes. PhD thesis. Copenhagen Business School. Copenhagen (1-292). http://hdl.handle.net/10398/8816.

Maalouf, Malek Miguel, and Britta Gammelgaard. "Managing Paradoxical Tensions during the Implementation of Lean Capabilities for Improvements." *International Journal of Operations and Production Management* 36, n°6 (2016): 687-709.

Nawanir, Gusman, Lim Kong Teong, and Siti Norezam Othman. "Impact of Lean Practices on Operations Performance and Business Performance: Some Evidence from Indonesian Manufacturing Companies." *Journal of Manufacturing Technology Management* 24, n° 7 (2013): 1019–50.

O'Reilly, Charles A., and Michael L. Tushman. "Ambidexterity as a Dynamic Capability: Resolving the Innovator's Dilemma." *Research in Organizational Behavior* 28 (2008): 185-206.

Osono, Emi, Norihiko Shimizu, and Hirotaka Takeuchi. "Extreme Toyota : Radical Contradictions That Drive Success at the World's Best Manufacturer." *Journal of Environmental Management* 80, n°4 (2008): 387-93.

Poole, Marshall Scott, and Andrew H. van de Ven. "Using Paradox to Build Management and Organization Theories." *Academy of Management Review* 14, n°4 (1989): 562-78.

Quinn, Robert E., and Kim S. Cameron. *Paradox and Transformation: Toward a Theory of Change in Organization and Management.* Cambridge, MA: Ballinger, 1988.

Rerup, Claus, and Martha S. Feldman. "Routines as a Source of Change in Organizational Schemata: The Role of Trial-and-Error Learning." *Academy of Management Journal* 54, n°3 (2011): 577–610.

Shadur, Mark A., Rene Kienzle, and John J. Rodwell. "The Relationship between Organizational Climate and Employee Perceptions of Involvement: The Importance of Support." *Group and Organization Management* 24, n°4 (1999): 479–503.

Shah, Rachna, and Peter T. Ward. "Lean Manufacturing: Context, Practice Bundles, and Performance." *Journal of Operations Management* 21, n°2 (2003): 129-49.

Shikdar, Ashraf A., and Naseem M. Sawaqed. "Worker Productivity, and Occupational Health and Safety Issues in Selected Industries." *Computers and Industrial Engineering* 45, n°4 (2003): 563–72.

Smith, Kenwyn K., and David N. Berg. *Paradoxes of Group Life: Understanding Conflict, Paralysis, and Movement in Group Dynamics.* San Francisco, Calif.: Jossey-Bass Inc., 1987.

Smith, Wendy K, and Marianne W. Lewis. "Toward a Theory of Paradox: A Dynamic Equilibrium of Organizing." *Academy of Management Review* 36, n°2 (2011): 381–403.

Vince, Russ, and Michael Broussine. "Paradox, Defense and Attachment: Accessing and Working with Emotions and Relations Underlying Organizational Change." *Organization Studies* 17, n°1 (1996): 1-21.

Watzlawick, Paul, John H Weakland, and Richard Fisch. "Change: Principles of Problem Formulation and Problem Resolution (Rev. Ed.)." *The American Journal of Family Therapy* 40 (2012): 92-95.

Womack, James P., Daniel T. Jones, and Daniel Roos. *The Machine that Changed the World: The Story of Lean Production.* New York, NY: Free Press, 1990.

Chapter 5

THE IMPACT OF 5S + 1S METHODOLOGY ON OCCUPATIONAL HEALTH AND SAFETY

Joana P. R. Fernandes[1], Radu Godina[2], Carina M. O. Pimentel[1,2]* and João C. O. Matias[1]

[1]Department of Economics, Management, Industrial Engineering and Tourism (DEGEIT), University of Aveiro, Aveiro, Portugal
[2]Research and Development Unit in Mechanical and Industrial Engineering (UNIDEMI), Department of Mechanical and Industrial Engineering, Faculty of Science and Technology (FCT), Universidade NOVA de Lisboa, Caparica, Portugal

ABSTRACT

The research community agrees there is an interrelation between lean manufacturing system 5S tool and workplace organization, organizational performance, and waste elimination, among others. However, the relationship between the 5S tool and Safety is scarcely explored in the literature and since evidences from practice exist, it is important to turn it into a pertinent research topic. Therefore, the current chapter explores how the implementation of 5S can contribute to the occupational safety conditions through a case study made at an industrial unit of one of the largest European car producers. The study was focused on the company's sorting process. Through this case study it could be demonstrated that apart from other quite well-known benefits of 5S, it is also essential to ensure the occupational safety. This conclusion is supported by a risk assessment analysis applied right before and right after the 5S implementation in which a decrease of 64% of the risk quantification could be reached.

Keywords: lean manufacturing, 6S, occupational safety, risk assessment, automotive industry, case study, sorting process

---

* Corresponding Author's E-mail: carina.pimentel@ua.pt.

1. INTRODUCTION

In a competitive world, all enterprises are constantly trying to reinvent themselves and adopt good practices that would allow them to overcome various obstacles that they could face along the way. Lean Manufacturing (LM) is one of the most widespread implemented practices adopted by the manufacturing sector (Marodin et al. 2018, in press). Lean is considered by several authors as a technique for reducing waste, but in fact, Lean maximizes product value by diminishing waste (Sundar, Balaji, and Kumar 2014, 1875). The adoption of Lean manufacturing entails the practice of certain tools and methodologies like: Kaizen, Pull system, Visual management, Heijunka, 5S, Six Sigma, among others (Alhuraish, Robledo, and Kobi 2017, 327).

Although the application of Lean principles brings numerous benefits to many organizations, nowadays the optimization of the productive process is not enough for a company to be considered as an example of success. There are other aspects to consider. Ensuring optimal Occupational Health and Safety1 (OHS) conditions (at work) has been one of the top priority issues for many enterprises, thus exceeding "quality" or "performance" as indicators of success (Shea et al. 2016, 293). Occupational safety can be defined as "the set of methods that aim the prevention of work accidents, through the evaluation and control of occupational risks" (Sutton 2017, 381). In the workplace, it is necessary to have full conditions in order for employees to be able to safely perform their duties, not only on safety (professional accidents prevention: occupational safety) but also on hygiene (professional injuries prevention: occupational hygiene hazards) and healthcare (professional healthcare: occupational medicine and nursing) (Coelho and Matias 2005, 414). Thus, there is some evidence showing that Lean manufacturing has a positive impact on decreasing the number of health and safety hazards (Babur, Cevikcan, and Durmusoglu 2016, 89).

In order to check if Lean manufacturing principles, and in particular 5S, have an impact in the occupational safety (in this work only on safety), a case has been studied in an industrial unit of one of the largest automotive industry enterprises operating in Portugal, and specifically in the sorting section of this industrial unit. One of the most famous Lean manufacturing tools is 5S which is a simple 5 step program that leads to an optimized work environment. Thus, the goal of this study is to verify if 5S + 1S has an impact on occupational safety in this particular enterprise.

In an initial phase, relevant topics to this research will be discussed. After the description of the sorting area, a risk assessment is performed in order to identify the existent hazards as well as the present safety level. Then, the 5S methodology is implemented and a new risk assessment is performed in order to assess whether 5S had any impact or not and to what extent this impact was significant to achieve 5S + 1S.

This paper is organized as follows. In section 2 the occupational safety is addressed. Lean manufacturing's main advantages and the lean thinking are addressed in section 3. Section 4 is dedicated to the research methodology. The case study is thoroughly analyzed in section 5. Finally, the conclusions are drawn in section 6.

---

[1] Concerning the acronym, there is no unanimous designation. We use OHS - Occupational Health and Safety which includes Health, Hygiene, and Safety at work.

2. OCCUPATIONAL SAFETY

As stated in section 1, OHS (Occupational Health and Safety) addresses the preservation and protection of human resources in the workplace. It is a discipline that aims to eliminate and reduce the risks that people are exposed while they carry out their professional activities (Amirah et al. 2013, 182).

ACT (Working Conditions Authority) is the Portuguese Entity whose mission is to promote and improve working conditions all over the country. According to this entity, 113 fatal accidents occurred in 2017 (see Figure 1) in Portugal (ACT 2017).

Occupational safety has been increasingly gaining focus from researchers and policymakers. Nowadays, it is possible to find several studies and well-based research on this topic (Clarke 2006, 413) (Geldart et al. 2010, 562) (Hale, Borys, and Adams 2015, 112). In Srinivasan et al. (2016, 364) it is affirmed that two known methods for improving safety exist: reactive/proactive or predictive. Usual safety statistics, such as number of incidents, workplace accidents, and absenteeism due to accidents, are reactive measures of safety since they measure the level of safety of a place after the incidents have occurred. On the contrary, the safety setting of an organization is considered to be a predictive measure.

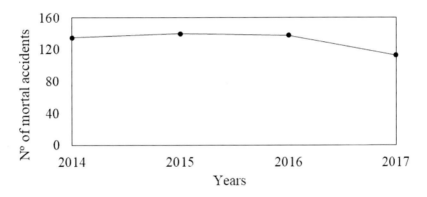

Figure 1. Number of work-related mortal accidents between 2014 and 2017 in Portugal.

2.1. Hazards and Risks

The terms "hazard" and "risk" are often confused and used for the same purpose, yet this practice is incorrect since both concepts have different meanings. The risk is defined as the combination of the consequences of a hazard (accident in safety evaluation) with the probability of occurring (Coelho and Matias 2015, 414) (Marzocchi et al. 2012, 553). Risks result from the interaction between man and environment. The risk exists at all levels of human activity, in private or professional life, being always present (Nicola and McCallister 2006, 180).

In (Kaplan 1997, 408) the question that is asked is "what is risk?" and this question is further divided into 3 main questions: What can go wrong? What is likely to happen? If so, what are the consequences? Hazard is defined as the source, agent or situation whether biological, chemical or physical, which is capable of causing disease or injury if not controlled (Oyarzabal and Rowe 2017, 3). The hazard represents the source while the risk includes the likelihood that this source resulting in actual loss or injury.

In (Yared and Abdulrazak 2018, 555) an example of our daily life that helps us to clarify these concepts is presented. Let us suppose the following situation: somebody finds himself preparing dinner. After putting the pot on the stove, he/she decides to watch television while waiting. However, once distracted by the television, he/she forgets the pot and his kitchen starts to burn. In this situation, we can conclude that the hazard is the pot being on the stove without any vigilance and the risk is the interaction of the probability of the stove without any vigilance and the fire consequences.

2.2. Risk Assessment

A risk assessment is a pro-active method to detect, analyze, and manage every potential risk faced by an enterprise (Lau and Lai 2012, 666). The risk assessment is designed to assess all the undesirable risk scenarios that exist and can lead to real accidents that affect human health as well as the environment around them (Abdo et al. 2018, 175). The risk is the hazard/severity combination associated with it and the likelihood that this hazard will cause injury. The risk assessment is the measurement and quantification of the variables that cause the risk. A risk assessment can be qualitative or quantitative (Nicola and McCallister 2006, 179).

Risk analysis is a process that helps identifying and managing potential problems that could undermine key initiatives or projects. In order to perform a risk analysis, one must first identify the possible threats he/she is facing, and then estimate the likelihood that these threats will materialize (Santos and Santos Nunes 2017, 350). Risk analysis can be performed using qualitative, quantitative, and mixed tools. According to (Pansiri and Jogulu 2011, 689) the mixed tools are the most indicated for studies such as the current research.

Several methods are used for the risk analysis. Some are more detailed while others are more generic and superficial. In Purdy (2010, 884) it is stated that a process of risk assessment and analysis is the result of three steps: the first one being the identification of risk scenarios, the second one being the analysis of the probability of occurrence, and the third one being the effect analysis. Based on such data it can be obtained the level of the risk that is given depending on the scenario in order to see if it is acceptable or not. If it is not acceptable, it is necessary to implement measures to correct such a deviation.

2.3. Industrial Vehicles

The use of motorized industrial vehicles increases the likelihood of a possible accident. Even though, it helps to save time and increases productivity. Balancing safety with productivity is a challenge for many industrial domains. Nonetheless, there are expectations that in the future there will be more reliable and secure industrial equipment (Horberry 2011, 552).

An accident at work may occur due to multiple causes, such as not using appropriate Personal Protective Equipment (PPE), employee distraction, and not following traffic rules, among others. Yet, the main cause is the use of machines and vehicles.

The use of forklifts has increased in many industrial units. Forklifts offer numerous benefits as they help increase productivity, reduce manual handling, and are quite easy to use. However, despite all these positive aspects, forklifts are a potential security threat, especially when used near pedestrians (Horberry et al. 2004, 575).

Horberry et al. (2004, 576) developed a forklift safety demonstration project at two manufacturing sites. These sites were chosen due to a high number of pedestrians, vehicular traffic (especially forklifts) and because the production staff were operating very close to the forklifts. Some principles have been proposed to reduce the risks associated with the use of forklifts in these sites:

1) Eliminating the problem in order to always avoid the contact of the pedestrian with forklifts;
2) Implement barriers (physical or temporal) between forklifts and pedestrians;
3) Improve warnings, visibility and traffic rules where pedestrians and forklifts intersect.

The final area to focus on in order to create a safer environment for forklift operation is the evaluation of the workplace. Only by constantly evaluating the workplace it is possible to determine the objectives and identify the areas of training and further improvements (Milanowicz, Budziszewski, and Kędzior 2018, 99).

3. LEAN THINKING

The "Lean" concept came through the Toyota Production System (TPS) developed by Taiichi Ohno in the mid-1940s. This production system is based on a set of principles such as "cost reduction through waste elimination" (Lander and Liker 2007, 3681) and the recognition that only a small fraction of the overall effort and time of the production means added value to the final product (Melton 2005, 662). TPS was the strategy adopted by Ohno to overcome the difficulties of World War II. Due to the scarcity of resources (material, financial, human), TPS was forced to focus on waste reduction as its main strategy (Behrouzi and Wong 2011, 389).

Lean thinking is a dynamic, knowledge-driven, customer-focused process which allows all the employees to continually eliminate waste and create value (Caldera, Desha, and Dawes 2017, 1547). Thus, lean thinking turns out to be a way of thinking and acting and it involves rethinking the way human resources are led, managed and developed (Pessôa and Trabasso 2017, 45). The aim of lean thinking is to encourage workers to use their full capacity to improve their own work and their involvement in the work of others (Chiarini 2012, 30).

The Lean thinking concept encompasses five basic principles (Randhawa and Ahuja 2017, 336):

- Identify the value: create value for the customer;
- Value chain: identify activities that do not add value and only mean waste;
- Optimize the value chain: create value through a continuous flow without interruptions (waiting times and manufacturing defective products);

- Pull production system: produce only what the customer asks for;
- Perfection: to put into practice a constant continuous improvement of the process by eliminating activities that do not add any type of value and that are wasteful in order to optimize the said process.

The adoption of a Lean philosophy in an organization can be initiated by the introduction of basic Kaizen techniques and tools such as the application of standards, elimination of waste or application of 5S (Ramis-Pujol and Suárez Barraza 2010, 390). In Figure 2 the Toyota Production System basics are illustrated.

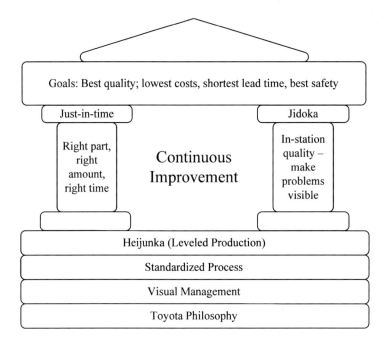

Figure 2. Toyota Production System (adapted from Goubergen and Lambert 2012, 2).

3.1. 5S

5Ss were introduced in Japan at the time TPS was designed to optimize organizational performance. The historical origin of the 5S is essentially based on the principles of Shinto, Buddhism (Self-discipline) and Confucianism (order) related to the Japanese culture (Randhawa and Ahuja 2017, 336).

- Order (Seiri and Seiton) - simplify processes to maximize efficiency and effectiveness by reducing people's workload and human error;
- Cleaning (Seiso and Seiketsu) - maximizing effectiveness by contributing to a healthy life, safety, well-being and enhancing transparency;
- Discipline (Shitsuke) - through education and training in order to improve the quality of life of operators in the workplace and used to rise the standards (Gapp, Fisher, and Kobayashi 2008, 566).

The implementation and practice of 5S are one of the most important steps in being able to achieve continuous improvement. Arunagiri and Gnanavelbabu (2014, 2073) explained that industries face enormous challenges in identifying and adopting the best Lean tools in order to eradicate their problems. In order to quantify and distinguish best practices from the Lean philosophy Arunagiri and Gnanavelbabu (2014, 2073) carried out a study in which 30 different tools were considered, including 5S and popular others, such as Overall Equipment Effectiveness, Value Stream Mapping, 5 Why's, Six Sigma, etc.). Of these 30 tools, its impact, ease of use and flexibility depending on the type of industry and existing production systems were analyzed. From this analysis, 5S ranked first in the ranking thus showing itself as the tool with the highest impact. The 5S methodology is organized in the following steps:

- *Seiri* (Sort) - The first step of 5S refers to keeping only the necessary items in the workplace in appropriate places. The focus is to eliminate everything (items) that is unnecessary from the workplace. These items should be identified with a red label and removed. Items that are used are kept and moved to appropriate places (Jain and Gupta 2015, 74). The organization of the workspace allows the elimination of damaged tools or objects or the removing of the unnecessary items because they are simply no longer needed and can be eliminated. Some advantages of having an organized workspace are stock containment, better use of the workplace, a lower chance of losing working tools (Agrahari, Dangle, and Chandratre 2017, 181).
- *Seiton* (Set in order) - The second step in implementing 5S is to store all the required items in the right place. All employees should be encouraged and motivated to place the working items in the best location and contribute to the visual improvement of the workplace (Agrahari, Dangle, and Chandratre 2017, 181). At this stage, each item can be identified with a label through a color so that it is quick to identify and found. Another storage strategy might be to store similar objects in the same place, attribute names, numbers among other organizing techniques (Jain and Gupta 2015, 75).
- *Seiso* (Clean) - The workplace must be constantly clean and neat. Daily cleaning is necessary for a better workplace (Jain and Gupta 2015, 75) and ends up acting as a motivating factor for the employee since he/she is in a more organized and healthy environment (Gupta and Jain 2014, 25).
- *Seiketsu* (Standardise) – Definition of standards and reference states for the level of cleaning at the workplace (Jiménez et al. 2015, 164). At this stage, importance should be given to the implementation of visual management and rules to ensure the continued practice of the 5S principles. All employees should be involved in the development of standards so to know their responsibilities and duties to perform in their daily routine (Gupta and Jain 2014, 25).
- *Shitsuke* (Sustained discipline) - Discipline means having the ability to do what is expected. However, self-discipline is more than that. While a person being disciplined in a given moment may or may not continue to be, when one is self-disciplined it is a guarantee of the continuity of these actions as a daily routine (Ho and Cicmil 1996, 49).

The emphasis at this stage is the creation of a working environment with good habits. In Liker (2004, 36) it is stated that the fifth S is undoubtedly the most difficult one since in order to maintain discipline practice a combination of several factors is necessary, such as management commitment, adequate training and a culture of continuous improvement from operational level to top management. In most cases, the implementation of 5S is a reasonably simple process. However, it becomes quite difficult to continue with such types of practices for a long period of time (Jain and Gupta 2015, 75). At this stage, the application of standards is essential in order to ensure the maintenance of these good practices. A schematic representation of 5S tools can be found in Figure 3.

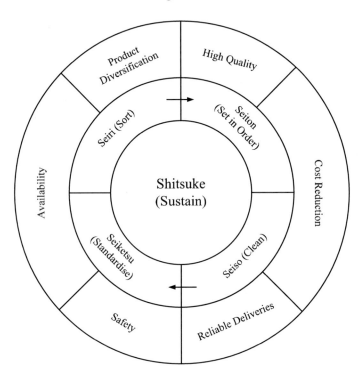

Figure 3. A schematic representation of the 5S tools.

The existing literature regarding 5S prior to this decade speculated a positive impact of 5S on safety. However, such a claim does not seem to be supported by empirical evidence. In Main, Taubitz, and Wood (2008, 38) it is claimed that lean manufacturing and safety are closely related and that any modifications by employing lean tools would have a degree of negative or positive influence on the safety risks. Several studies have addressed the impact that 5S might have on safety. For instance, John E. Becker addressed this issue in (Becker 2001, 31) and stated that 5S can be utilized to make an analysis of the existing procedures and to guarantee that safety is definitely included in the production process. In Shakouri et al. (2016, 2) the authors study the impact of 5S on the safety climate of manufacturing workers operating in the packaging area of a manufacturing industrial unit. In this study, two groups of workers (case and control groups) finalized a safety climate questionnaire preceding the 5S event. The overall safety in the industrial unit considerably improved for the case group. However, the control group remained unaffected during the period of study. The study made in Kiran (2017, 339) confirms that the visual controls of Seiketsu should include work instructions, hazard warnings, cautions, among others and includes an element of safety hazards in the 5S Audit Sheet. A few authors even consider that 5S could be expanded into becoming a crucial part of environmental management systems (Ramis-Pujol and Suárez-Barraza 2012, 80). These authors made a study in which the objective was to research and study the application of 5S in Mexican multinational organizations, with the intention of analyzing and comparing them face to the existing studies (Ramis-Pujol and Suárez-Barraza 2012, 77). In this study, the authors also mention the safety aspect.

3.2. 6S

5S not only simplifies the working environment but also contributes to increased safety (Korkut et al. 2009, 1721). Recently several authors have used the 6S concept, that is, in addition to the well-known 5S terms of this methodology, a new S is added, which is related to safety (Randhawa and Ahuja 2017, 3).

The 6S concept was introduced by Hiroyuki Hirano as a method for reducing waste, improving safety and optimizing productivity (Sari, Suryoputro, and Rahmillah 2017, 1). The implementation of 5S without regarding safety aspects becomes ineffective since safety is or should be the top priority at any working environment. The implementation of safety in 6S has the objective of identifying present hazards/risks, ensuring safe work equipment, no injuries, the existence of fire extinguishers and escape routes. It aims to improve efficiency, quality, and safety and optimize the discipline at work (Sari, Suryoputro, and Rahmillah 2017, 1).

Several authors address the 6S concept in the literature. In Gnoni et al. (2013, 98) the authors draw attention to a 6S Audit that intends to confirm the definitive utilization of the whole Bosch Production System. According to the abovementioned study, the Bosch Production System has successfully improved the traditional Lean Management concepts – resulting from the Toyota Production systems – by making safety management as a central point. The authors Nazarali et al. (2017, 436) proposed an application of a 6S initiative with the objective of improving the workflow for emergency eye examination rooms in which the safety played an important part. A discussion of 6S implementation in a Laboratory of Ergonomy, in the Department of Industrial Engineering of the Islamic University of Indonesia, is addressed in Sari, Suryoputro, and Rahmillah (2017, 1). The authors Vinodh, Arvind, and Somanaathan (2011, 475) even suggest a 7S concept, meaning 5S + 2S with 2S being safety and sustainability. Overall, studies mentioning the 6S concept are quite scarce in the research community.

4. RESEARCH METHODOLOGY

A qualitative methodology, based on a case study strategy, was the selected option to approach the research presented in this chapter. A case study is a history of a past or current phenomenon, drawn from multiple sources of evidence. It can include data from direct observation and systematic interviewing as well as from public and private archives (Karlsson 2016, 167). In terms of scope, according to Yin (2003, 13) a case study is an empirical inquiry that investigates a contemporary phenomenon within its real-life context, especially when the boundaries between phenomenon and context are not clearly evident, allowing an investigation to retain the holistic and meaningful characteristics of real-life events. Benbasat et al. (1987, 370) refer three strengths of using the case study research: 1) the phenomenon can be studied in its natural setting and meaningful, relevant theory generated from the understanding gained through observing actual practice; 2) the case method allows important questions such as why, what and how. These need to be answered with a relatively full understanding of the nature and complexity of the entire phenomenon; 3) the case method lends itself to early, exploratory investigations where the variables are still unknown and the phenomenon not at all understood. On the other side, some of the difficulties of doing case study research are the requirements of direct observation in the actual contemporary situation (cost, time, access hurdles); the need for multiple methods, tools, and entities for triangulation; the lack of controls; and the complications of context and temporal dynamics (Meredith 1998, 444).

5. CASE STUDY

A case-study approach is developed in this section, considering one company from the Portuguese automotive sector. During this case study, the company is not identified due to confidentiality reasons.

5.1. Initial State

5.1.1. Sorting Space Safety

The automotive industry is one of the most demanding quality control industries. All components of a car undergo a rigorous inspection in the sense that only conforming parts are delivered to the customer. Often, it is necessary to make sorting processes in order to make the separation of conforming and non-conforming products (Godina, Matias, and Azevedo 2016, 2).

In the industrial unit of this case study company, the sorting processes are performed by two subcontractors. These subcontractors operate in an area outside the factory where they have to verify and to guarantee the conformity of all the pieces that are expedited to the client.

The sorting itself is the evaluation of parts which implies a high movement of containers. The forklift is the equipment used to make the material handling using stack containers to optimize space. Although the forklift is very effective and easy to use, it also represents a hazard and it can seriously injure an operator or collide with a stack of containers. Another important aspect to consider is that the space itself is quite small for the daily activity that exists, which sometimes leads to the containers having to be stored temporarily outside.

This step began by creating an inventory of every object that was in the area and by assessing what was their necessity. Since the existing space was not enough for an ideal operating site, every unnecessary object removed would increase the useful space. The goal is to delimit a space for every object and to define designated zones with the purpose of maximizing the existing space. More than a few copies of the plan of the existing space were printed, with containers designed to scale, with the purpose of simulating an improved layout and to outline specific zones.

The implementation of the 5S came from the need to maximize the existing space, improve working conditions and, above all, to assess the extent to which the application of Lean tools influences or not the occupational safety. A risk assessment as observed in Sommestad et al. (2016, 199) was performed in order to identify and quantify existing risks. The risk quantification was calculated from the product between the following two variables:

$$risk = probability \times severity = P \times S \qquad (1)$$

In order to quantify the variables from equation (1), qualitative data was converted into a quantitative scale ranging from 1 to 5 as can be seen in Table 1.

The following risks were assessed, as observed in Sommestad et al. (2016, 199):

- Possible fall - Slippery floor with water or dirt (dust). It is likely that an operator might slip and the severity may be slight to moderate;
- Fall of objects – The fall of stored material (an extreme case will be the fall of a container with parts). The likelihood of something like this happening is possible given a large number of containers those are stored and stacked in that space daily and the severity is extreme since it can lead to death.
- Storage, stacking - In the sorting area there is temporary storage of containers and since the space is considerably reduced these are usually stacked. An accident is likely to occur in this regard as there is a serious lack of organization and gravity will be high as an accident of this can cause serious injury.
- Vehicle loading and unloading, vehicle movement, and vehicle circulating in reverse - The vehicles used in the sorting zone are the forklifts responsible for handling containers. Often these enter the space while the operators are concentrated in the evaluation and verification of parts, so the probability of an accident happening is quite possible and the gravity is high.

Table 1. Quantification scale of the of risk

Scale	1	2	3	4	5
Probability (P)	Impossible or remote under normal conditions	Unlikely under normal conditions	50% chance	>50% chance	Very likely
Severity (S)	None or slight	Minimal	Significant	Major	Catastrophic

Table 2. Quantification of the potential hazards

Potential hazards	P	S	Total
Dangerous movement, fall, slippery	3	3	9
Fall of objects	2	5	10
Storage, stacking	3	5	15
Loading and unloading of vehicles	3	4	12
Vehicle moving or circulation	3	4	12
Vehicles circulating in reverse	3	4	12
Total			70

In Table 2 the probability and severity assessment is presented as well as the potential hazards that present the highest scores.

After the risk assessment, it was possible to identify the risks in the sorting zone as well as those that represented a greater danger. Instead of taking actions, for example considering the hazards that had a higher score, an alternative strategy was chosen that consisted in implementing a 5S action in an independent way without taking into account the results found with the risk assessment. Only then is it possible to verify whether a 5S action will actually have an indirect impact on safety and to what extent this influence is significant or not.

5.1.2. 5S Implementation

The first step of 5S implementation was taken through an interview of the employees with the purpose of recording the main difficulties felt by them and also to collect their opinion regarding possible changes that could be implemented. The employees converged on two opinions, the concern for safety due to the presence of the forklift and the location not being ideal since the available space was considered to be insufficient.

During the entire research process, the research team assigned an active role in sorting outsourcing enterprises. It is necessary to recognize that it is not always easy to change and, above all, to maintain the changes over a long period of time until they become a habit again. 5S, in addition to being a tool of simple understanding and execution, often fails for this very reason, people tend to go back to old habits (Keyser, Sawhney, and Marella 2016, 32). Thus, in this study, the involvement of the sorting section employees was considered very important and useful.

Figure 4. Containers stored in the designated sorting space.

The implementation of 5S took about 25 days to complete. In an initial phase, data were collected, in addition to the risk assessment presented previously. Afterwards, a pre-analysis form (as can be seen in Table 3) was prepared for the first three S. The initial state of the sorting area was also recorded through photographs which allow making a visual comparison of the state before and after the implementation of the 5S. In Table 4 it is possible to analyze some of the actions implemented in each step and their impact on the safety level.

Since the existing space was inadequate for every operating action, anything removed would expand the useful space and would allow to rearrange the remaining objects in a way to maximize the existing space. Along with the safety risks related to the abovementioned practice, and since the dimensions of the existing space are quite reduced, it became unmanageable to accumulate and store large containers. All piece containers that were already sorted were also removed from the site. In some areas of the site, several containers were waiting to be sorted or awaiting transportation, either to the warehouse of the factory where the stock is kept or to the return zone to be returned to the supplier. In Figure 4 it is possible to observe how containers were stored in the designated sorting space.

Table 3. A pre-analysis for the first three S

	Criteria under analysis	State	Observations
Seiri (Sort)	Are there unnecessary objects?	Yes	Crane lift
	Several workbenches (one company only has two and the other ten)	No	Workbenches are in poor condition, but cannot be considered damaged.
	Multiple parts out of the flow	No	The sorting ranges that are required are displayed, however, there is scattered information.
	Containers of already sorted parts that are stored on site	No	One of the procedures performed daily by the sorting companies is the accounting for conforming and non-conforming parts, which is a necessary task.
Seiton (Set in Order)	Are there specific locations/spaces for objects?	No	Open sorting place with no defined areas/zones Lack of visual management
	Is the space divided with tape/marks on the floor?	No	Open space without defined zones
	Are there any signs?	No	Lack of signs on traffic rules
Seiso (Clean)	Is the place clean?	No	Difficult to maintain the space clean: constant entry/exit of forklifts, dust release in some jobs
	Is the cleaning material available?	Yes	
	Is there a regular cleaning plan?	No	Somctimes the space is cleaned by a subcontracted company, sometimes the employees clean themselves

Through the analysis of the Table 4, it is possible to conclude that the application of 2S (Set in Order) was the one that brought the most significant impact in terms of security. The delimitation of the areas of the sorting room through the installation of safety barriers has created an area where it became improbable to have any contact between man and machine. In general, it can be assessed that practically all actions had a significant impact on improving the overall safety of the industrial unit. However, the cleaning action of the site did not have a significant impact on safety, which confirms similar results from previous studies, such as in Srinivasan et al. (2016, 375).

Table 4. The 5S and its impact on safety

Step	Actions	Safety Impact
Seiri (Sort)	Elimination of the lifting crane; Elimination of 5 workbenches; Elimination of objects that were on the tables; Disposal of containers that were stored at the sorting site; Identification and disposal of non-flow parts according to type (conforming or non-conforming).	More empty and empty space, making potential dangers transparent and easy to detect Disposal of stored containers has reduced the likelihood of a dropping container; Empty tables have reduced the likelihood of heavy objects falling on someone else;
Seiton (Set in Order)	Delimitation of the various areas of the sorting site; Rearrangement of the office's layout; Implementation of barriers with two meters of height; Application of adhesive tape to the floor in order to define areas; Closing of the sorting zone through a padlock; Identification of accesses through the use of signalling;	Creation of a safety zone where the forklift is prevented from entering; Barriers protect operators from a potential drop of a container; Since the sorting space is sealed with a chain, it prevents an immediate and surprise entrance of a forklift; Warning signs for forklift trucks; Pedestrian crossing information as well as which Personal protective equipment (PPE) to use.
Seiso (Clean)	Space cleaning Replacement of workbenches	There was no significant effect on safety.
Seiketsu (Standardize)	Elaboration of an animation panel and dashboard with the definition of the reference state to be maintained.	Definition of the traffic rules associated with the new disposal site - contact between man and forklift changed from excessive to almost zero.
Shitsuke (Sustained discipline)	Adaptation of the observation of work posts (OWPs) to the sorting area to verify compliance with the 5S principles	Compliance with safety rules

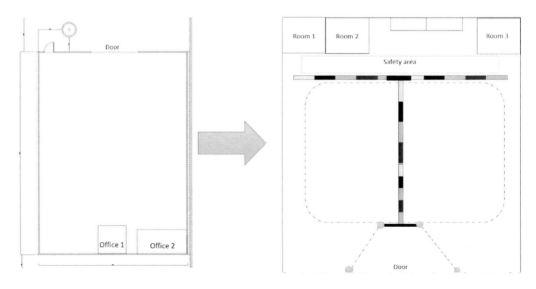

Figure 5. Sorting area - before and after the implementation of 5S.

It is possible to verify if the actions carried out in this study had any impact on safety. These actions, with the aim of demonstrating the safety of the forklift, were applied in compliance with the principles defined by Horberry et al. (2004, 577), such as: the implementation of barriers between forklifts and pedestrians, improvement of warnings and implementation of traffic rules, where pedestrians and forklifts cross. Similar to the results presented by Horberry et al. (2004, 580), this work also managed to drastically reduce the contact between the forklift and the operator, allowing to evolve from a scenario in which this contact was excessive to a scenario of a practically non-existent contact. The reference zone observed in Figure 5 shows how the sorting area should be constantly kept. It is the standard state of this sorting room and it should be respected by all employees all the time. These rules have been added in the information panel so that they would be constantly remembered.

5.2. The Impact of 5S on Safety

In the initial phase of this case study, a risk assessment was performed using the matrix defined by the industrial unit to quantify and obtain a quote for each risk. With this risk assessment, it is possible to make a comparison, quantitatively, of the state before and after implementation of the 5S. Thus, after the 5S action, a new risk assessment was carried out in order to understand whether these actions had or not any impact on the safety of the sorting area.

Table 5. Results of the potential hazards evaluation before and after the implementation of 5S

	Before the Action			After the Action		
Potential hazards	P	G	Total	P	G	Total
Dangerous displacement, fall, slippery	3	3	9	1	3	3
Fall of Objects	2	5	10	1	5	5
Storage, stacking	3	5	15	1	5	5
Loading and unloading of vehicles	3	4	12	1	4	4
Ccirculation of vehicles	3	4	12	1	4	4
Vehicles in reverse	3	4	12	1	4	4
Total			70			25

Right after the 5S action terminated a new risk assessment was executed. This was done exactly like the previous one, through direct observation on the ground. In Table 5 it is possible to observe the quantification of the potential hazards attributed after the 5S action.

The likelihood of risks associated with falls decreased after the 5S action and the space became cleaner and more organized. All hazards associated with the presence of the forklift such as its reverse movement or loading and unloading went from a probability of "probable" to "unlikely" once the physical barriers were created between man and machine so that there would be no contact between them. After the quantification of all hazards a score of 25 points was obtained, a value considerably lower than the starting score, previously to the 5S implementation. In addition, there was also an ergonomic improvement, since the containers to be sorted began to be placed on shelves and installed at an ergonomic level.

Figure 6. Evolution of the screening area before and after the implementation of 5S.

Figure 7. Evolution of the area before and after the implementation of 5S - Security barriers.

Figure 8. An example of an identified and delineated loading zone of the Forklift.

Through all these actions, in an initial phase of implementation of the 5S, it was possible to free some space in the sorting area, thus, making all the potential hazards more visible and detectable. As it can be seen in Table 5, the results indicated that through the application of the 5S tools on the demarcated location it was possible to decrease the total risk by 64% by employing risk assessment. The results also indicate that the use of one of the 5S, namely *Seiton* or Set in Order, was the S that had the most substantial and tangible effect regarding safety. After the successful implementation of 5S+1S, several zones of the sorting site have been bordered by the implementation of safety barriers, in turn resulting in a space in which the probability of having any type of contact between man and machine was heavily decreased. An example of the sorting area, before and after, can be seen in Figures 6 and 7. Finally, in Figure 8 an example of an identified and delineated loading zone of the Forklift can be observed, after the successful implementation of the 5S.

CONCLUSION

Ensuring the safety of all employees has been a major concern shared by many enterprises. Although it is impossible to guarantee zero accidents and there is always a probability that they could occur, conditions can always be created so that the risk is reduced as low as possible. It is also important to stress that security is part of ourselves and our ability to be aware in our daily lives and that our own actions can put us at risk or the people around us.

With this work, it was tried to identify the role of 5S in the safety and to what extent this tool is useful to assure it. 5S is one of the most popular tools of Lean philosophy, being constantly referred to as a tool to create and maintain an organized, clean and effective working environment. Nonetheless, several authors advocate a broader definition in the sense of adding an S – safety in this tool.

The present study aimed to prove and confirm the veracity and feasibility of security integration, demonstrating that the 5S plays a fundamental role and that it is an effective tool for the consolidation of the security as a 5S + 1S = 6S. Process control, setting standards, objects' organization, elimination of obsolete items, and separation of spaces were all actions that led to a reduction in the probability of occurring accidents and injuries. When the quantification of all the risk of potential hazards was performed, a score of 25 points was obtained, a value that is considerably lower than the value before the implementation of 5S (70). The results indicated that through the application of 5S tools on the demarcated location it was possible to decrease the total risk by 64% by employing risk assessment. This allowed creating a safer work environment at the enterprise, which is just as important as increasing productivity, quality, or income. Once an employee feels safe then he/she will contribute positively to the productivity, efficiency, and quality of his/her work.

REFERENCES

Abdo, H., M. Kaouk, J. M. Flaus, and F. Masse. 2018. "A Safety/Security Risk Analysis Approach of Industrial Control Systems: A Cyber Bowtie – Combining New Version of

Attack Tree with Bowtie Analysis." *Computers and Security* 72 (January): 175–95. https://doi.org/10.1016/j.cose.2017.09.004.

ACT. 2017. *Mortal Work Accidents*. 2017. http://www.act.gov.pt/(pt-PT)/CentroInformacao/Estatistica/Paginas/AcidentesdeTrabalhoMortais.aspx.

Agrahari, R S, P A Dangle, and K V Chandratre. 2017. "Implementation of 5S Methodology in the Small Scale Industry: A Case Study." *International Research Journal of Engineering and Technology(IRJET)* 4 (3): 130–37.

Alhuraish, Ibrahim, Christian Robledo, and Abdessamad Kobi. 2017. "A Comparative Exploration of Lean Manufacturing and Six Sigma in Terms of Their Critical Success Factors." *Journal of Cleaner Production* 164 (October): 325–37. https://doi.org/10.1016/j.jclepro.2017.06.146.

Amirah, Noor Aina, Wan Izatul Asma, Mohd Shaladdin Muda, and Wan Abd Aziz Wan Mohd Amin. 2013. "Safety Culture in Combating Occupational Safety and Health Problems in the Malaysian Manufacturing Sectors." *Asian Social Science* 9 (3): 182–91. https://doi.org/10.5539/ass.v9n3p182.

Arunagiri, P., and A. Gnanavelbabu. 2014. "Identification of High Impact Lean Production Tools in Automobile Industries Using Weighted Average Method." In *Procedia Engineering*, 97:2072–80. Elsevier. https://doi.org/10.1016/j.proeng.2014.12.450.

Babur, Ferhat, Emre Cevikcan, and M. Bulent Durmusoglu. 2016. "Axiomatic Design for Lean-Oriented Occupational Health and Safety Systems: An Application in Shipbuilding Industry." *Computers & Industrial Engineering* 100 (October): 88–109. https://doi.org/10.1016/j.cie.2016.08.007.

Becker, J. E. 2001. "Implementing 5s to Promote Safety & Housekeeping." *Professional Safety* 46: 29–31.

Behrouzi, Farzad, and Kuan Yew Wong. 2011. "Lean Performance Evaluation of Manufacturing Systems: A Dynamic and Innovative Approach." In *Procedia Computer Science*, 3:388–95. https://doi.org/10.1016/j.procs.2010.12.065.

Benbasat, Izak, David Goldstein and Melissa Mead. 1987."The case research strategy in studies of information systems." *MIS Quarterly*, 11 (3): 369-386.

Caldera, H. T. S., C. Desha, and L. Dawes. "Exploring the Role of Lean Thinking in Sustainable Business Practice: A Systematic Literature Review." *Journal of Cleaner Production*, 167: 1546-1565. https://doi.org/10.1016/j.jclepro.2017.05.126.

Chiarini, Andrea. 2012. "Lean Thinking." In *From Total Quality Control to Lean Six Sigma: Evolution of the Most Important Management Systems for the Excellence*, edited by Andrea Chiarini, 29–36. SpringerBriefs in Business. Milano: Springer Milan. https://doi.org/10.1007/978-88-470-2658-2_9.

Clarke, Sharon. 2006. "Safety Climate in an Automobile Manufacturing Plant: The Effects of Work Environment, Job Communication and Safety Attitudes on Accidents and Unsafe Behaviour." *Personnel Review* 35 (4): 413–30. https:// doi. org/ 10. 1108/ 00483480610670580.

Coelho, Denis A. and J. C. O. Matias. 2015. *The Benefits of Occupational Health and Safety Standards – Part VI: Management of Occupational Safety and Health of the Handbook of Standards and Guidelines in Ergonomics and Human Factors*, Waldemar Karwowski (Editor), Lawrence Erlbaum Associates, ISBN: 0-8058-4129-6, 413-440, December 2005.

Gapp, Rod, Ron Fisher, and Kaoru Kobayashi. 2008. "Implementing 5S within a Japanese Context: An Integrated Management System." *Management Decision* 46 (4): 565–79. https://doi.org/10.1108/00251740810865067.

Geldart, Sybil, Christopher A. Smith, Harry S. Shannon, and Lynne Lohfeld. 2010. "Organizational Practices and Workplace Health and Safety: A Cross-Sectional Study in Manufacturing Companies." *Safety Science* 48 (5): 562–69. https://doi.org/10.1016/j.ssci.2010.01.004.

Gnoni, M. G., S. Andriulo, G. Maggio, and P. Nardone. 2013. "'Lean Occupational' Safety: An Application for a Near-Miss Management System Design." *Safety Science* 53 (March): 96–104. https://doi.org/10.1016/j.ssci.2012.09.012.

Godina, Radu, João C.O. Matias, and Susana G. Azevedo. 2016. "Quality Improvement with Statistical Process Control in the Automotive Industry." *International Journal of Industrial Engineering and Management* 7 (1): 1–8.

Goubergen, D Van, and Jo Lambert. 2012. "Applying Toyota Production System Principles and Tools at the Ghent University Hospital." *62nd IIE Annual Conference and Expo 2012*, no. Figure 1: 2536–42.

Gupta, Shaman, and Sanjiv Kumar Jain. 2014. "The 5S and Kaizen Concept for Overall Improvement of the Organisation: A Case Study." *International Journal of Lean Enterprise Research* 1 (1): 22. https://doi.org/10.1504/IJLER.2014.062280.

Hale, Andrew, David Borys, and Mark Adams. 2015. "Safety Regulation: The Lessons of Workplace Safety Rule Management for Managing the Regulatory Burden." *Safety Science* 71 (PB): 112–22. https://doi.org/10.1016/j.ssci.2013.11.012.

Horberry, Tim. 2011. "Safe Design of Mobile Equipment Traffic Management Systems." *International Journal of Industrial Ergonomics* 41 (5): 551–60. https://doi.org/10.1016/j.ergon.2011.04.003.

Horberry, Tim, Tore J. Larsson, Ian Johnston, and John Lambert. 2004. "Forklift Safety, Traffic Engineering and Intelligent Transport Systems: A Case Study." *Applied Ergonomics* 35 (6): 575–81. https://doi.org/10.1016/j.apergo.2004.05.004.

Jain, Sanjiv Kumar, and Shaman Gupta. 2015. "An Application of 5S Concept to Organize the Workplace at a Scientific Instruments Manufacturing Company." *International Journal of Lean Six Sigma* 6 (1): 73–88. https://doi.org/10.1108/IJLSS-08-2013-0047.

Jiménez, Mariano, Luis Romero, Manuel Domínguez, and María del Mar Espinosa. 2015. "5S Methodology Implementation in the Laboratories of an Industrial Engineering University School." *Safety Science* 78 (October): 163–72. https://doi.org/10.1016/j.ssci.2015.04.022.

Kaplan, Stan. 1997. "The Words of Risk Analysis." *Risk Analysis*. Wiley/Blackwell (10.1111). https://doi.org/10.1111/j.1539-6924.1997.tb00881.x.

Karlsson, Christer (ed.). 2016. *Research Methods for Operations Management*. 2nd. ed. UK: Routledge.

Keyser, R. S., R. S. Sawhney, and L. Marella. 2016. "A Management Framework for Understanding Change in a Lean Environment." *Tékhne* 14 (1): 31–44. https://doi.org/10.1016/j.tekhne.2016.06.004.

Kiran, D. R. 2017. "Chapter 23 - 5S." In *Total Quality Management*, edited by D. R. Kiran, 333–46. Butterworth-Heinemann. http://www.sciencedirect.com/science/article/pii/B9780128110355000234.

Korkut, Derya Sevim, Nevzat Cakıcıer, E Seda Erdinler, Göksel Ulay, and Ahmet Muhlis. 2009. "5S Activities and Its Application at a Sample Company." *Journal of Biotechnology* 8 (8): 1720–28. https://doi.org/http://dx.doi.org/10.5897/AJB09.145.

Lander, E., and J. K. Liker. 2007. "The Toyota Production System and Art: Making Highly Customized and Creative Products the Toyota Way." *International Journal of Production Research* 45 (16): 3681–98. https://doi.org/10.1080/00207540701223519.

Lau, Henry C.W., and Ivan K.W. Lai. 2012. "A Hybrid Risk Management Model: A Case Study of the Textile Industry." *Journal of Manufacturing Technology Management* 23 (5): 665–80. https://doi.org/10.1108/17410381211234453.

Liker, Jeffrey K. 2004. *The Toyota Way: 14 Management Principles from the World's Greatest Manufacturer.* New York, New York, USA.

Main, Bruce, Michael Taubitz, and Willard Wood. 2008. "You Cannot Get Lean Without Safety Understanding the Common Goals." *Professional Safety* 53 (1). https://www.onepetro.org/journal-paper/ASSE-08-01-38.

Marodin, Giuliano, Alejandro Germán Frank, Guilherme Luz Tortorella, and Torbjørn Netland. 2018. "Lean Product Development and Lean Manufacturing: Testing Moderation Effects." *International Journal of Production Economics*, July. https://doi.org/10.1016/j.ijpe.2018.07.009.

Marzocchi, Warner, Alexander Garcia-Aristizabal, Paolo Gasparini, Maria Laura Mastellone, and Angela Di Ruocco. 2012. "Basic Principles of Multi-Risk Assessment: A Case Study in Italy." *Natural Hazards* 62 (2): 551–73. https://doi.org/10.1007/s11069-012-0092-x.

Melton, Trish. 2005. "The Benefits of Lean Manufacturing: What Lean Thinking Has to Offer the Process Industries." *Chemical Engineering Research and Design* 83 (6 A): 662–73. https://doi.org/10.1205/cherd.04351.

Meredith, Jack. 1998. "Building operations management theory through case and field research." *Journal of Operations Management* 16 (4): 441-454.

Milanowicz, Marcin, Paweł Budziszewski, and Krzysztof Kędzior. 2018. "Numerical Analysis of Passive Safety Systems in Forklift Trucks." *Safety Science* 101 (January): 98–107. https://doi.org/10.1016/j.ssci.2017.07.006.

Nazarali, Samir, Jaspreet Rayat, Hilary Salmonson, Theodora Moss, Pamela Mathura, and Karim F. Damji. 2017. "The Application of a '6S Lean' Initiative to Improve Workflow for Emergency Eye Examination Rooms." *Canadian Journal of Ophthalmology / Journal Canadien d'Ophtalmologie* 52 (5): 435–40.

Nicola, Andrea Di, and Andrew McCallister. 2006. "Existing Experiences of Risk Assessment." *European Journal on Criminal Policy and Research.* Kluwer Academic Publishers. https://doi.org/10.1007/s10610-007-9034-7.

Oyarzabal, Omar A., and Ellen Rowe. 2017. "Evaluation of an Active Learning Module to Teach Hazard and Risk in Hazard Analysis and Critical Control Points (HACCP) Classes." *Heliyon* 3 (4). https://doi.org/10.1016/j.heliyon.2017.e00297.

Pansiri, Jaloni, and Uma D. Jogulu. 2011. "Mixed Methods: A Research Design for Management Doctoral Dissertations." *Management Research Review* 34 (6): 687–701. https://doi.org/10.1108/01409171111136211.

Pessôa, Marcus Vinicius Pereira, and Luís Gonzaga Trabasso. 2017. "Lean Thinking." In *The Lean Product Design and Development Journey: A Practical View*, edited by Marcus Vinicius Pereira Pessôa and Luis Gonzaga Trabasso, 43–53. Cham: Springer International Publishing. https://doi.org/10.1007/978-3-319-46792-4_3.

Purdy, Grant. 2010. "ISO 31000:2009 - Setting a New Standard for Risk Management: Perspective." *Risk Analysis*. Wiley/Blackwell (10.1111). https://doi.org/10.1111/j.1539-6924.2010.01442.x.

Ramis-Pujol, Juan, and Manuel F. Suárez-Barraza. 2010. "Implementation of Lean-Kaizen in the Human Resource Service Process: A Case Study in a Mexican Public Service Organisation." *Journal of Manufacturing Technology Management* 21 (3): 388–410. https://doi.org/10.1108/17410381011024359.

Randhawa, Jugraj Singh, and Inderpreet Singh Ahuja. 2017. "5S – a Quality Improvement Tool for Sustainable Performance: Literature Review and Directions." *International Journal of Quality & Reliability Management* 34 (3): 334–61. https://doi.org/10.1108/IJQRM-03-2015-0045.

Samuel K. Ho Svetlana Cicmil. 1996. "Japanese 5-S Practice." *The TQM Magazine* 8 (1): 44–53. https://doi.org/10.1108/09544789610107261.

Santos, Eduardo Ferro dos, and Letícia dos Santos Nunes. 2017. "Methodology of Risk Analysis to Health and Occupational Safety Integrated for the Principles of Lean Manufacturing." In *Advances in Social & Occupational Ergonomics*, edited by Richard H.M. Goossens, 349–53. Advances in Intelligent Systems and Computing. Springer International Publishing.

Sari, Ad, Mr Suryoputro, and Fi Rahmillah. 2017. "A Study of 6S Workplace Improvement in Ergonomic Laboratory." In *IOP Conference Series: Materials Science and Engineering*. Vol. 277. https://doi.org/10.1088/1757-899X/277/1/012016.

Shakouri, Mahmoud, Isabelina Nahmens, Craig Harvey, Laura Hughes Ikuma, and Siddarth Srinivasan. 2016. "5S Impact on Safety Climate of Manufacturing Workers." *Journal of Manufacturing Technology Management* 27 (3): 364–78. https://doi.org/10.1108/JMTM-07-2015-0053.

Shea, Tracey, Helen De Cieri, Ross Donohue, Brian Cooper, and Cathy Sheehan. 2016. "Leading Indicators of Occupational Health and Safety: An Employee and Workplace Level Validation Study." *Safety Science* 85 (June): 293–304. https://doi.org/10.1016/j.ssci.2016.01.015.

Sommestad, Teodor, Henrik Karlzén, Jonas Hallberg, and Peter Nilsson. 2016. "An Empirical Test of the Perceived Relationship between Risk and the Constituents Severity and Probability." *Information and Computer Security* 24 (2): 194–204. https://doi.org/10.1108/ICS-01-2016-0004.

Srinivasan, Siddarth, Laura Hughes Ikuma, Mahmoud Shakouri, Isabelina Nahmens, and Craig Harvey. 2016. "5S Impact on Safety Climate of Manufacturing Workers." *Journal of Manufacturing Technology Management* 27 (3): 364–78. https://doi.org/10.1108/JMTM-07-2015-0053.

Sundar, R., A. N. Balaji, and R. M. Satheesh Kumar. 2014. "A Review on Lean Manufacturing Implementation Techniques." In *Procedia Engineering*, 97:1875–85. Elsevier. https://doi.org/10.1016/j.proeng.2014.12.341.

Sutton, Ian. 2017. "Chapter 13 - Occupational Safety." In *Plant Design and Operations (Second Edition)*, edited by Ian Sutton, 381–400. Gulf Professional Publishing. http://www.sciencedirect.com/science/article/pii/B9780128128831000139.

Vinodh, S., K. R. Arvind, and M. Somanaathan. 2011. "Tools and Techniques for Enabling Sustainability through Lean Initiatives." *Clean Technologies and Environmental Policy* 13 (3): 469–79. https://doi.org/10.1007/s10098-010-0329-x.

Yared, Rami, and Bessam Abdulrazak. 2018. "Risk Analysis and Assessment to Enhance Safety in a Smart Kitchen." *Fire Technology*, 2018. https://doi.org/10.1007/s10694-017-0696-5.

Yin, Robert. 2003. *Case Study Research: Design and Methods*. 2nd ed. SAGE Publications.

In: Lean Manufacturing
Editors: F. J. G. Silva and L. Carlos Pinto Ferreira
ISBN: 978-1-53615-725-3
© 2019 Nova Science Publishers, Inc.

Chapter 6

VALUE OF REAL-TIME DATA FOR CYCLE TIME OPTIMISATION AT WET TOOLS

A. Rotondo[1,], J. Geraghty[2] and P. S. Young[2]*

[1]Irish Manufacturing Research, Dublin, Ireland
[2]Dublin City University, Dublin, Ireland

ABSTRACT

When solution approaches are developed to facilitate decision processes in real manufacturing systems, fundamental information constraints may apply. The data required by theoretically effective and efficient decision support tools may not be available or accessible, especially for real-time applications; therefore, the possibility of implementing the solutions developed is compromised. In this chapter, the problem of optimising assignment strategies subjected to information constraints is analysed in the context of the semiconductor wet etch process. More specifically, the problem of scheduling a wet station including multiple wet tools, each consisting of an automated handling system, multiple tanks containing etch chemicals or deionized water, and an internal buffer is addressed. Assignment strategies previously developed for a particular wet station are generalised to wet stations characterised by different tool configurations and process recipes. These strategies have been designed to be integrated with data management systems and make use of real time data to minimise the stations' average cycle time. The original strategy developed is based on the concept of assigning a batch to the tool that will complete it the soonest; however, due to data constraints, implementation costs for this strategy would be very high. The alternative strategies suggested progressively enlarge the information domain on which the assignment decisions are based and consider more sophisticated look-ahead logics: one that examines the status of all chemical stations, one that considers the availability of the dryer station, and one that considers the dryer and repour times. The generalisation of these strategies to various wet stations' configurations aims at providing general validity to the concept that the incorporation of more detailed information on the system status in decision support or prescriptive tools enhances the strategies' performance, generates productivity increases, reduces waiting times, and supports improvements to operational policies; in

---
[*] Corresponding Author's E-mail: anna.rotondo@imr.ie.

summary, using data to govern a production system makes the system leaner. The experimental results obtained support this statement as more information intensive assignment strategies outperform data-poor strategies, regardless of the tools' configuration and the recipes processed. The relevance of data-driven simulation as a decision support tool is also demonstrated with reference to wet stations operating in a real semiconductor fabrication facility.

Keywords: wet etch station, scheduling optimization, data-based assignment heuristics, earliest finish

1. INTRODUCTION

The importance of incorporating problem-specific knowledge into the behaviour of any optimisation algorithm is well explicated by the No Free Lunch (NFL) theorems developed by Wolpert and McReady (Wolpert and Macready 1997). These theorems establish that for any optimisation or search algorithm (Wolpert and Macready 1995) any superior performance over one class of problems is offset by performance over another class. In other words, the pre-eminence of a particular algorithm to solve a specific optimisation problem is not extendable to all optimisation problems. Hence, in order to be effective in generating optimal solutions, an algorithm should be tailored to the specific problem analysed and incorporate information on the contingent scenario observed when the solution is generated. This concept can be certainly extended to decision support tools that are designed to be adopted in a specific manufacturing environment to facilitate real-time decisions related to production issues.

A detailed knowledge of the system where the issue is experienced is then fundamental to provide effective decision support. Knowledge is enhanced by information and information "can only be obtained by providing attributes or relevance and purpose to data" (Emblemsvg 2005, 48). The concept of transforming data into knowledge is typical of Data Analytics (DA) (Harding et al. 2006, 969). DA consists of a broad range and combination of analytical techniques and methodologies applied to raw data to support decision making (Chae et al. 2014, 119); data management approaches are used to extract, transform and load (ETL) relevant data that are consequently analysed using data mining and prescriptive approaches. The results are then contextualised in relevant models of the knowledge domain under investigation. The value of DA is widely recognised by the industrial and academic community as real and significant. Big Data is revolutionizing all aspects of our lives ranging from enterprises to consumers, from science to government (Jagadish et al. 2014, 86) . The application of DA and its inherent methodologies covers a broad area of industries and business functions and ranges from business process management (zur Muehlen and Shapiro 2010) to finance (Zhang and Zhou 2004), customer relationship management (Sisselman and Whitt 2007), supply chain management (Trkman et al. 2010; Oliveira, McCormack, and Trkman 2012; Chae and Olson 2013) and manufacturing (Harding et al. 2006; Kusiak 2006; Choudhary, Harding, and Tiwari 2009), among others. As regards manufacturing, a significant growth of DA applications, especially in terms of data mining, has been observed in the automotive and semiconductor industry (Harding et al. 2006). As regards the semiconductor industry, the high automation levels that characterise its production process suggest that big data is available in this sector. Here, big data is being leveraged to adopt a more predictive or even proactive approach to factory control through advanced process control (APC) solutions (Moyne, Samantaray, and Armacost 2016): failure detection is being augmented with failure prediction (Lim, Kim, and Kim 2017); virtual metrology is used to reduce cycle time, avoid metrology delay and improve process capability (Khan, Moyne, and Tilbury 2007; Kang et al. 2011; Cheng et al. 2015); predictive scheduling can lessen capacity constraints; yield prediction with feedback into process control, production scheduling, and maintenance scheduling is being investigated so that fab-wide control can be optimized to yield-throughput objectives (Monzon and Gray 2018; Moyne and Schulze 2010). The complex dynamics of the semiconductor manufacturing process are also likely to generate the so called butterfly effect for which small improvements in one area can have a significant impact on the system's overall performance (Haghighirad, Makui, and Ashtiani 2008, 739); this encourages the implementation of analytics even in small area of the process.

It is evident that in order to benefit from the application of DA techniques, relevant data should be available and accessible. Data lies at the hearth of the fourth industrial revolution as end-to-end digitisation of all physical assets within a value chain and their integration into a digital ecosystem is the focus of this fourth wave of technological advancement (PwC 2016). Sensorisation and Internet of Thing (IoT) communication technologies can be used to generate data from different sources and connect them to decision support tools so that feasible and effective solutions can be developed and easily implemented in the system where decision support is required. However, information constraints may exist and introduce relevant challenges for the development of successful solution approaches since when a limited amount of data is available on a system's current status, the decision will be based on partial information and this could affect the quality of the solutions obtained (Kusiak 2006, 4176).

In this study, the impact of information quality and quantity on the performances of different assignment strategies applied to a wet station operating in a real semiconductor manufacturing facility is investigated. The objective of this study is to show that manufacturing data available in information systems can effectively support optimal assignment decisions in real-time and that the quality of the assignment decision improves with the quality of the data considered during the decision process. In this context, data quality refers to the accuracy of the data available and the amount and type of information that can be extracted from them. Assignment strategies developed to minimise the average cycle time for a specific wet tool's configuration (Rotondo, Young, and Geraghty 2014) have been generalised to more complex production scenarios, from both a tool configuration and operations perspective. These strategies are inspired by an efficient assignment concept based on the shortest completion time (Earliest Finish concept). The original concept would require detailed information on the system's current status; however, due to limited access to the tools' internal database, fundamental information constraints apply. Taking these constraints into account, the strategies developed progressively enlarge the information domain on which the assignment decisions are based and provide results that could be used by production management to identify trade-off's between information costs and cycle time performances.

1.1. Organisation of the Chapter

This chapter continues with a review of the relevant literature on wet etch station scheduling; Section 0 describes the configuration of the wet tool considered in this study and the governing scheduling logic. This logic is captured in the scheduling algorithms described in Section 0 where details on the validation approach are also given. The assignment strategies originally developed are presented in Section 0 and their application to the wet station originally considered is presented in Section 0. The assignment strategies are generalized to more complex tools and recipe configurations in Section 0 and experimental evidence of the robustness of the strategies effectiveness is presented in Section 0. Considerations on the results obtained and conclusions are drawn in Section 0 and Section 0, respectively.

2. Literature Review

Wet stations represent critical production steps for the semiconductor wafer manufacturing process (Aydt et al. 2008, 26). A wet station usually consists of several identical tools that operate in parallel. Batches of wafers made of one or two lots are processed at wet tools as they are transferred and immerged into the various chemical and water tanks available within the tools by an automated material handling system. Being batch chamber tools (Lee 2008, 2128) subjected to peculiar scheduling constraints, wet tools are generally difficult to model (Govind and Fronckowiak 2003, 1401). Simulation approaches have been developed to perform what-if analyses and evaluate the effect of modifications of operational settings on the tools performance (Noack et al. 2008). Simulation has also been coupled with optimisation approaches to identify efficient strategies for operational planning and control of wet-etch tools; maximum waiting time for batching operations (Govind and Fronckowiak 2003, 1403), virtual queue capacity (Govind and Fronckowiak 2003, 1403; Noack et al. 2008, 2196), recipe dedication (Aydt et al. 2008, 32; Te Quek et al. 2007, 2) have been subject of simulation-based optimisation studies on wet tools.

A fundamental feature of wet tools and, more generally, integrated or cluster tools (Niedermayer and Rose 2003, 349) consists of parallel processing; this transforms the inversely proportional relationship between cycle time and throughput (Mauer and Schelasin 1993, 814) so that higher CTs can also generate increases in throughput and reductions of overall run time. The presence of parallel processing and the variety of operations performed at a wet tool enables the improvement of the tools' performance by means of efficient assignment strategies and sequencing optimisation. Mathematical programming approaches (Bhushan and Karimi 2003; Karimi, Tan, and Bhushan 2003; Aguirre, Méndez, and Castro 2011; Zeballos, Castro, and Méndez 2010; Novas and Henning 2012; Castro, Zeballos, and Méndez 2012) and heuristics (Bhushan and Karimi 2004; Geiger, Kempf, and Uzsoy 1997) have been extensively used for solving sequencing and scheduling optimization problems at wet stations. For these problems, as a result of management's suggestions, makespan minimization represents the most common objective function (Zeballos, Castro, and Méndez 2010, 2). This is because makespan reductions lead to throughput increases and reduce the likelihood that wet tools could constitute a constraint to the factory output (Geiger, Kempf, and Uzsoy 1997, 104). Reducing the makespan also supports an increase in tool capacity and, hence, minimizes the number of tools needed, with obvious advantages in terms of occupied clean room floor space (Geiger, Kempf, and Uzsoy 1997, 104). Moreover, decreasing the makespan leads to a lower inventory and contamination and results in greater profits (Bhushan and Karimi 2004, 363). The observed research trend on wet stations sequencing and scheduling optimization is towards the development of approaches able to deliver nearly optimal solutions in a reasonable time for increasingly larger sized problems (Karimi, Tan, and Bhushan 2003, 222; Castro et al. 2011; Castro, Zeballos, and Méndez 2012, 6). More recent studies also focus on the introduction of modelling details, such as those regarding the material handling system, so that more realistic assumptions are considered in the solutions developed (Aguirre, Cafaro, and Méndez 2011, 1822; Lee, Lee, and Lee 2007, 490).

Scheduling at wet tools is also analysed as an integrated problem with scheduling at furnaces. This is due to the fact that furnace operations for diffusion processes usually follow wet tools operations, such as wafer cleaning and etching (Gan et al. 2006, 1822). The most challenging aspects of the integrated scheduling problem consist of the different maximum batch sizes at the two batch tools and the possible presence of wait time constraints between the two processes (Scholl and Domaschke 2000, 274). Efficient schedules for the wet etch operations are fundamental to ensure good performances for the furnace operations (Ham and Fowler 2008, 1016). Optimal dispatching rules (Scholl and Domaschke 2000, 275; Ham and Fowler 2008, 1014) and the impact of furnace upgrades on the upstream flow (Gan et al. 2006, 1822; Yugma et al. 2012, 2121) have been analysed by means of simulation and heuristic optimisation.

As a further effect of parallel processing and recipes' variety, dispatching and assignment strategies also impact cycle time at wet tools (Govind and Fronckowiak 2003, 1403; Te Quek et al. 2007, 2; Noack et al. 2008, 2196). The assignment concept (i.e., Earliest Finish, EF) that has inspired the definition of the assignment strategies illustrated in this paper presents similarities with the concept developed by Hsieh et al. (2002); the fundamental differences between the two assignment approaches concern the underlying equations that support the choice of the optimal tool in the station. Hsieh et al. (2002) adopt an approximated procedure for calculating the processing completion time of a batch at the different tools available in a station whereas a rigorous algorithmic procedure based on valid governing equations is adopted in this study. Moreover, in Hsieh et al. (2002), delays caused by maintenance operations within a tool are considered a posteriori whereas, in the analysis proposed here, events that could delay a batch processing are concurrently considered during the calculation of the assignment decision variable. A concept similar to EF is also used by (Noack et al. 2008, 2196) where it is applied to establish the order with which batches assigned to a tool will be processed, that is, it is used for a dispatching strategy rather than an assignment strategy. Similarly, Te Quek et al. (2007, 3) analyse the combined effect of assignment and dispatching on wet stations capacity.

In this study, the simulation algorithm used to support assignment decisions is based on a trial and error approach; scheduling logics and applicable constraints captured in the algorithm faithfully reproduce those observed at a wet station operating in a real semiconductor fabrication facility. The development of the algorithm has been based on a detailed analysis of the tools' manufacturing data. Data analyses have been conducted to characterise the tools' behaviour as the internal scheduling logic was proprietary to the tools' vendor. An accurate representation of the tools' behaviour is achieved as several recipes, bi-directional production flows (Lee, Lee, and Lee 2007, 490) and correct robot transfer movements (Novas and Henning 2012, 192) can be simulated. This also ensures that the scheduling optimisation problem at wet-etch station is not reduced to a flowshop permutation scheduling problem (Bhushan and Karimi 2004, 365; Aguirre, Cafaro, and Méndez 2011, 1825) and the property of parallel processing, which is typical of a batch chamber tool, is fully exploited in the assignment procedure. The assignment strategies investigated in this study adopt the same scheduling algorithm so that the sole effect of the assignment decisions is analysed. The strategies originally developed for a particular wet station characterised by a relatively simple recipe mix have been generalised to be applicable to more complex production scenarios. A recipe describes the order with which a batch will visit the different tanks; an operation establishes the processing times in each tank. Further experiments have also been conducted to investigate both the impact of the a priori exclusion of tools subjected to preventive maintenance from the list of eligible tools at a station and the impact of recipe dedication policies on cycle time performances.

3. WET STATION DESCRIPTION

In the wet station initially considered in this study, four identical tools operate in parallel; these tools present a standard multi-chamber wet etch tool layout (Figure 1). In each tool, the 6 tanks available (T1-T6 in Figure 1) are alternately filled with chemical etchants and deionised water. The chemical etchants are used to etch away the exposed photo-resist from the wafer layers; the etchant action will be terminated by submerging the wafers into the consecutive water tank. As an over-exposure of the wafers to a chemical etchant will most likely damage the wafers, a "Zero Wait" (ZW) constraint is applied to the chemical tanks; this constraint forces a batch to leave the tank once the prescribed processing time is reached and implies that both the associated rinse tank and the robot are available at that time. In order to satisfy the ZW constraint and also to guarantee that a temporary storage is available in case of a sudden chemical tank operational failure, the rinse tank associated with a chemical tank is required to be available and idle while a batch is being processed in the chemical tank. No intermediate buffer is available between any two consecutive tanks; this generates a further scheduling constraint, usually called "No-Intermediate Storage" (NIS). The NIS constraint requires that before a batch can leave a tank, the following tank in its recipe has to be empty. The tanks subjected to the ZW constraint will be considered "unsafe"; on the contrary, the remaining tanks will be considered "safe" tanks since a sensible overexposure time in these tanks will not damage the wafers. The maximum overexposure time in a safe tank varies according to the etchants in the associated chemical tank. Safe tanks can occasionally be used as local storage when the following tanks or the material handling system are not available. Table 1 details the contents and the nature of the tanks at the four tools available at the wet station; in the tanks' columns, the first letter indicates if the tank is filled with a chemical etchant (C) or deionised water (R), the second letter represents a chemical identifier, the third letter refers to the tank classification (e.g., whether safe (S) or unsafe (U)); two different chemical etchants (e.g., a and b) are available in the tools.

Table 2 reports the recipes and the associated number of operations performed at the station.

Value of Real-Time Data for Cycle Time Optimisation at Wet Tools

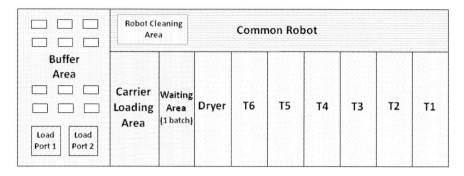

Figure 1. Schematic representation of the wet tools modelled.

Table 1. Details on chemical and rinse tanks operating at the wet tools

# tools	Tanks					
	T1	T2	T3	T4	T5	T6
2	C b - U	R b - S	C a - U	R a - S	C a - U	R a - S
2	C b - U	R b - S	C b - U	R b - U	C a - U	R a - S

Table 2. Recipes and operations performed at the station

Recipe ID	# Op	Recipe		
A1	9	C a	R a	Dryer
A2	6	C b	R b	Dryer

In the tools modelled, an automated material handling system is available. A common robot moves the batches across the tanks, whereas, internal handlers at each tank perform vertical movements so that a batch can be transferred from the robot into the tank. The internal handler at unsafe tanks also transfers batches to the associated rinse tanks. Preemptive robot assignment rules are implemented in the internal scheduler control logic to increase the availability of the dryer; in particular, transfer of batches out of the dryer are prioritised over any other robot movement.

Due to necessary and periodic repouring operations, the chemical tanks present a limited availability; during these operations, a tank is emptied and refilled with the appropriate chemicals whilst the other tanks in the same tool continue processing as usual. Regular preventive maintenance is also performed at the wet tools.

A common buffer upstream of the wet station gathers all the lots that require processing at the station. Here batching is performed so that lots with compatible operations are virtually batched and immediately made available to the assignment system that selects the wet tool that will process the batch. Each operation is characterised by a maximum waiting time (MWT); if a lot in the upstream buffer remains single for the MWT, it will be processed as a single lot.

Two assignment rules are generally considered in the plant where the tools operate. The "Emptiest-Oldest" (EO) strategy assigns batches to the tools with the lowest number of batches currently assigned. In case of equally loaded tools, the tool with the oldest last arrival is preferred. On the contrary, if the Oldest Time Out (OTO) strategy is implemented, the tool with the oldest last batch exiting the tool is preferred. The OTO strategy is adopted in the wet station initially modelled.

Once in a wet tool, the lots are logged in the internal scheduler that generates fully feasible schedules based on the recipe and the robot and tanks' availability. The schedule details the times in and out of each tank. At the prescribed time, the lots are loaded by a mechanical device onto a carrier and moved by the robot following the schedule generated by the internal scheduler. If the first tank to be visited is occupied, the carrier is left in the internal waiting area. Before leaving the tool every batch visits the dryer, then the lots are unloaded from the carrier and return into the internal buffer.

4. SYSTEM MODELLING

In order to investigate the impact of different assignment strategies on cycle time at the wet station investigated, a simulation model was developed. The model presents a modular structure so as to enhance its flexibility; the most relevant modules for the analysis presented here consist of the assignment algorithm and the scheduling algorithm.

The assignment algorithm verifies the inventory status at regular time intervals and based on the lots available creates virtual batches, if possible. When lots ready to be assigned are available, the algorithm identifies the tool to route each lot to based on the assignment strategy applied. Different versions of this algorithm have been developed for simulating the different strategies considered.

The scheduling algorithm generates detailed schedules for the lots to be processed based on deterministic input data. The scheduling logic used is inspired by the behaviour of the internal scheduler of the tools analysed; the control logic applied in the real plant is proprietary to the tools' vendor and cannot be modified without incurring significant costs. Communication with the production engineers involved in this research and analyses of historical data helped develop an adequate understanding of the tools' control logic so that it could be accurately modelled. Historical manufacturing data generated by the internal scheduler were analysed over a 15 day period to gain an understanding of the scheduling logic and validate the production engineers' knowledge regarding the tools' behaviour. In particular, the batch progression within the tools was analysed in order to identify scheduling constraints and pre-emptive rules governing the material handling system. As an interesting outcome of this process, it was found that the internal waiting area that is interposed between the carrier loading area and the tanks (Figure 1) does not constitute a bottleneck for the tool as initially thought. Figure 2 reports a sample of a real batch schedule based on the internal scheduler data; the time scale and the chambers denomination have been modified for confidentiality reasons. Each step line represents a batch with the corresponding schedule; different recipes are performed in the tool so that each batch can visit different tanks as required in its recipe (Table 2). The dashed black lines highlight batches that skip the waiting area while this is occupied by another batch. This provides clear evidence that the internal waiting area cannot be considered as a production flow constraint as it was originally assumed by the production engineers. This assumption had influenced the tools' management, especially in regard to the assignment logic.

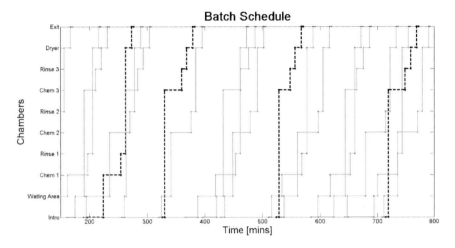

Figure 2. Visualisation of batch schedule.

The tools' control logic described in the previous section has been reproduced in the scheduling algorithm and the following assumptions have been made:

- As different recipes correspond with different tanks sequences, the problem analysed here can be defined as a "non-permutation" schedule (Karuno and Nagamochi 2003, 310). This means that the batches can enter the different tanks in a different order;
- A tank can process one batch at a time;
- Maximum processing times are set for both safe and unsafe tanks, for unsafe tanks the maximum processing time coincides with the minimum processing time whereas for the safe tanks the maximum time is greater than the minimum processing time and is set for both process and efficiency reasons;
- The processing/transfer of batches is generally non-pre-emptive; the only exception is made for transfers out of the dryer that are prioritised;
- The robot can move one batch at a time and cannot be used as a temporary storage for any batch;
- The robot is failure free; the tool is subjected to failures and regular preventive maintenance; repouring operations also limit the tanks availability;
- Transfer times within the tool are dependent on the origin and destination;
- Robot transfers take into account the time required to reach the batch to be moved, transfer the batch to its destination tank, perform cleaning operations and return to the start position of transfers previously scheduled (e.g., for previous batches).
- Unloaded robot trips are considered;
- Setups are needed for both robot and tanks; cleaning operations for the robot arms are required post every transfer involving chemical tanks.

The algorithm that reproduces the control logic is based on a trial and error approach with a backward correction procedure. The algorithm utilises availability matrices for both tanks (Φm) and robot (ΦR) at each tool while it progressively builds the schedules; Φm and ΦR consist of chronologically ordered arrays that indicate the time intervals for which the resource is not available; for the robot, a similar data structure is used to record the position occupied by the robot at the times recorded in ΦR so that accurate travel times can be calculated based on the robot position. Φm also contains information about repouring operations and preventive maintenance that are modelled following the schedules implemented in the plant.

The algorithm works as per the pseudo-code in Figure 3; it considers one batch at a time and completes its schedule before the following batch is considered. Fundamental equations and constraints are used to calculate the start (TS) and finish (TF) time of a batch k in the jth tank of its recipe

$$TS_{kj} \begin{cases} \geq TS_{k(j-1)} + t_{k(j-1)} + tr_{j-1 \to j} & \text{for safe tanks} \\ = TS_{k(j-1)} + t_{k(j-1)} + tr_{j-1 \to j} & \text{for unsafe tanks} \end{cases} \quad (1)$$

$$TF_{kj} \begin{cases} \geq TS_{kj} + t_{kj} & \text{for safe tanks} \\ = TS_{kj} + t_{kj} & \text{for unsafe tanks} \end{cases} \quad (2)$$

where t_{kj} and $tr_{j-1 \to j}$ represents the processing time of batch k in the jth tank and the transportation time from tank j to tank $j+1$ in the recipe, respectively.

```
for i = 1 -> no. batches
        j = first tank to be visited
        while j ≠ dryer
                Calculate TS_j and TF_j
                if robot is available between TF_j-1 and TS_j
                        if tank j is available between TS_j and TF_j
                                Update Φ_j and Φ_R
                                j = j+1
                        else Calculate min TS_j* at which j is available and restart from
                                last safe tank (j = j-x)
                        end
                else Calculate min TF_j-1* at which robot is available and restart from
                        last safe tank (j = j-x)
                end
        end
        i = i+1
end
```

Figure 3. Pseudo-code of the scheduling algorithm.

Considering each tank in the order suggested by the batch's recipe, the algorithm checks the robot availability for a time interval that includes [$TF_{k(j-1)}$; TS_{kj}] and possible additional times related with the robot transfer from its previous position to tank $j-1$ and from tank j to the following position previously scheduled. Further time needed for possible robot cleaning operations required after any transfer can be taken into account. If the robot is available for the entire time interval required, Φ_R is updated by inserting that time interval so that the chronological order is respected. The tank availability is successively checked for the time interval [TS_{kj}; TF_{kj}]. If the tank is available during that interval, Φ_j is updated and the next tank in the recipe is considered. In case either the robot or the tank are not available, the smallest time at which the unavailable resource becomes available for the required time interval is calculated and, based on that, the start and finish times at the immediately previous unsafe tanks are derived using Equations 1 and 2. The iteration restarts from the first safe tank that precedes the tank currently assessed if the over-exposure required by the backward procedure is allowed; otherwise, previous tanks are considered. When the backward procedure is triggered, Φ_R and Φ_m are restored to their previous status at all relevant tanks. It is worth noting that the reverse procedure and the decision of restarting the schedule calculation from the last safe tank available might cause unnecessary occupation of this tank and prolong the cycle time of a batch. However, this is what observed in the real system; moreover, considering both the computational savings and the typical inter-arrival rate of batches to the wet tools modelled, the scheduling logic adopted proves quite efficient. Experiments have been run to analyse the impact of this approach on possible cycle time delays; the results show that this algorithmic step influence the cycle time by less than 5% of the scheduled lots as in most cases the backward procedure forces the schedule calculation to restart from the very first tank in the recipe. Further details on the scheduling logic and the inherent governing equations are provided in (Rotondo, Young, and Geraghty 2015, 264–65). The model has been coded in MATLAB v9®.

In spite being inspired by a specific wet station, the scheduling algorithm described here can be easily adapted to model stations with different configurations as most of the structural parameters are given as inputs to the algorithm. For instance, the tank capacity, the number of tanks, the nature of the tank, whether safe or unsafe, the presence of an internal handler between two adjacent tanks and the number of common robots can be immediately modified. This is proven by the application of the same algorithm to the tools configurations and recipes described in Section 0.

Applications of similar trial-and-error scheduling approaches can also be found in (Yih 1994, 506–9) where the hoist scheduling problem in flexible printed circuit board electroplating lines is analysed. Wet stations and electroplating lines are characterised by similar configurations (i.e., common material handling system, chambers' sequence, etc.) and applicable scheduling constraints (i.e., NIS, processing time tolerances). In (Yih 1994, 509), processing time tolerances are used in the reverse scheduling procedure to avoid schedules' recalculations due to jobs' concurrent requirement of either workstations or hoist. During the schedule calculation, workstation and hoist availability are separately checked in (Yih 1994, 506–9); they are progressively evaluated in (Hindi and Fleszar 2004, 94–96), as implemented in the scheduling algorithm used in this paper. The structural differences between the heuristics developed in (Yih 1994, 506–9; Hindi and Fleszar 2004, 94–96) and the wet stations scheduling algorithm concern the presence of processing time tolerances at all workstations (i.e., these are allowed only at the safe tanks in a wet tool), the possibility of rescheduling more than one job at a time during the reverse procedure, and the uni-directionality of the production flow; bi-directional flow is allowed in complex wet station configurations and is modelled in the scheduling algorithm illustrated here.

4.1. Model Validation

The simulation model has been validated by comparing the average Queuing Time (QT) and Run Time (RT) generated by the algorithm against real observations. Three sets of data, each of which recording 15 days of production, have been used for the model validation. QT is here intended as the time elapsing between the lot's arrival at the buffer upstream the wet station and the time at which the lot enters the first tank of the wet tool to which it is assigned; this implies that the waiting time within the tool is also included in QT. RT is intended as the time interval that starts when a lot enters the first tank and ends when that lot leaves the tool. The sum of QT and RT is defined as CT. The model accuracy proves good as the QT and RT prediction errors are always less than 0.7% and 1.5%, respectively, also when the different tools' configurations are analysed. Also in the validation process, the importance of extracting knowledge from available data to improve the system modelling accuracy is shown. For instance, Figure 4 shows the modelling accuracy improvements as more information on the wet station behaviour was progressively acquired from the data analysis. Three modelling phases are identified in Figure 4; these correspond with the major modelling achievements. The Phase 1 model was mainly based on the production engineers' knowledge. During Phase 2 detailed data analyses were performed in order to refine the modelling of maintenance, repouring and setup operations at each tank; stochastic travel times to the tools from the upstream common buffer and MWT were also introduced in the model. During Phase 3 attention was paid to refining the knowledge of the assignment strategy logic based on both historical data and automation engineers' suggestions. More flexibility was added to the model as virtual capacity limits were introduced in the tools' internal buffer area; this is done in the real system in order to control the batch assignment and avoid congestions at tools.

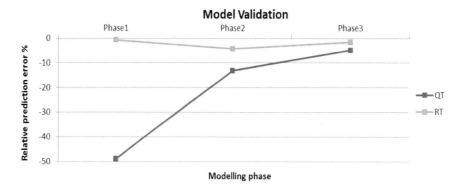

Figure 4. Validation process.

The computational efficiency of the algorithms is also significant as detailed batch schedules are generated in less than 0.01 seconds per batch, on average. For bigger problem sizes, the computational time increases slightly more than linearly; 120 seconds are needed to generate feasible schedules for 6200 lots when the assignment strategy adopted in the plant is applied. For the most computational intensive strategy investigated in this study, the same problem was solved in 130 seconds. The experiments were run on a 2.4 GHz INTEL Core Duo processor.

5. ASSIGNMENT STRATEGIES

Four assignment strategies based on the EF concept were originally developed based on the wet tool configuration described in Section 0. An overview of these strategies is provided in this section; further details on these strategies are available in (Rotondo, Young, and Geraghty 2014, 963–65).

The first strategy (EF strategy) consists of a direct implementation of the EF concept as the tool able to complete a batch at the earliest finish time is selected in the assignment process. Using the scheduling algorithm, the completion time of a batch at the different tools available in the station is calculated by virtually assigning that batch to each tool. If the same completion time is obtained for two or more tools, the tool with the lowest workload since the last shut down for maintenance operations is chosen; this is due to quality reasons and also to ensure a balanced utilisation across the tools. The batches waiting in the upstream buffer are considered in a sequence that is determined by the time at which each virtual batch is formed. For each batch to be assigned the simulation algorithm considers the availability of each tool based on the actual tool's workload, this implies that batches that have been assigned to that tool but are not yet logged into the tool's internal scheduler are also considered in the schedule calculation. This obviously limits the tool's availability in comparison to the availability derivable from the internal scheduler and probably represents one of the most difficult requirements of the algorithm for its implementation in real-time. The control system that currently regulates the batch assignment to the wet tools cannot access data stored in the internal scheduler; as a consequence, information on the tools' current status and their associated batches' schedules are not immediately available. Moreover, if a batch assigned to a tool by the control system has not entered that tool, the internal scheduler ignores its existence and no data regarding that batch's schedule can be found. However, sensors at each tank of a tool can be activated; these sensors would be able to communicate the tanks' status, whether under maintenance or operational and, in this last case, whether occupied by a batch or idle. In case a batch is being processed in a tank, the corresponding sensor would also be able to reveal the time at which the tank will complete the processing and when it will become available for the next batch. Based on the information potentially available the EF strategy has been modified so as to allow a quicker implementation and reduce the computational efforts required. It is worth noting that the EF strategy is still implementable in the system; structural changes could be made to the internal scheduler so as to allow a direct communication with the plant information system or, as an alternative, a parallel scheduling system based on the simulation algorithm could be built to keep a record of all the batches' schedules at the different tools. In this case, the simulated schedules should be periodically validated to make sure that the simulated schedules are aligned with the actual schedules. Three different simplified variants of the EF strategy have been developed; they incorporate progressive information details on the tools' status and, as a consequence, require progressive implementation efforts.

The first variant, "Tank Check" (TC) strategy, requires the least amount of information as it is based on the current status of each single chemical tank excluding the dryer. For each batch to be processed, the algorithm identifies the tanks containing the chemicals required by the batch's recipe. In the station initially analysed, the recipes performed consists of just one chemical bath, the associate rinse bath and the dryer (Table 2). The relevant tanks' status, whether idle or occupied by a batch, is then checked and the time at which the tank will be available is retrieved. In the real system this information would be easily available through the tanks' sensors. The virtual queue at each tank is then considered; this is derived by considering the batches assigned to a particular tool. The plant information system is able to provide this information as it keeps a record of the production route of each batch. In order to consider the effects of the virtual queue on a tool's availability, for each batch in the tool's virtual queue, the tank that will most probably process that batch is identified using the following logic:

- if only one tank with the chemical required by the batch's recipe operates in the tool, the batch is necessarily assigned to that tank;
- if more than one relevant tank is available at the tool, the batch is assigned to the tank that becomes available at the earliest time.

Using this reasonable assumption, information on the actual completion processing time of batches virtually assigned to a tank is not required. Among the tanks that present no virtual queue, the tank with the lowest completion time for the batch currently processed is chosen and the batch is assigned to the corresponding tool. Figure 5 schematises the TC assignment logic.

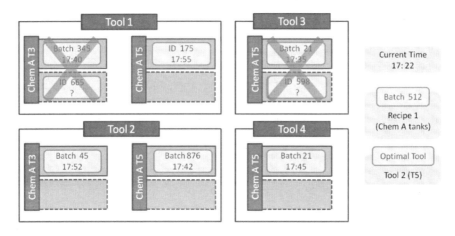

Figure 5. Assignment logic of TC strategy.

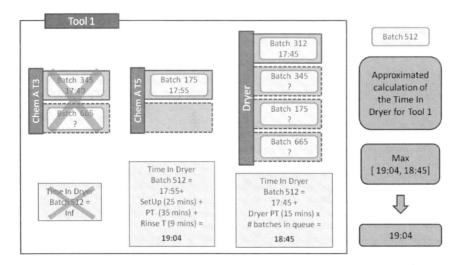

Figure 6. Example of TCD assignment logic.

The second variant, the "Tank & Dryer" (TD) strategy, includes information on the dryer which is not considered in the TC strategy. The dryer represents a critical element of a wet tool as all recipes require a final visit to the dryer (See Table 2). The TD strategy is based on an approximated calculation of the dryer's minimum "Time In" by considering both the tanks' availability and the dryer workload. Firstly, the dryer's "Time In" is calculated from a tank perspective. Following the TC procedure, the chemical tank's "Time In" is calculated for all the available tanks compatible with the batch's recipe; the processing time in the chemical tank, the associated rinse time and the relevant travel times, are then added to the chemical tank's "Time In" so as to obtain an approximated estimation of the dryer's "Time In". In parallel, the dryer's "Time In" is calculated from a dryer's perspective based on the number of batches currently assigned to the corresponding tool. The "Time Out" of the dryer for the currently processed batch would be available through sensors installed within the dryer. Processing times and setup times are then added for each batch previously assigned so that the minimum "Time In" for the batch to be assigned is derived. The assumption that there is no delay between consecutive batches at the dryer is made; this means that no starvation is supposed to be experienced at the dryer. The dryer's "Times In" calculated using both procedures are compared and the maximum of the two is consider an appropriate estimate of the starting processing time in the dryer for the batch to be assigned. The dryer's "Times In" obtained for the different tools are compared and the tool with the lowest resulting time is chosen.

The last variant developed, the "Tank & Dryer with Repour Check" (TDR) strategy, slightly modifies the TD strategy as it incorporates information on the tanks' availability with respect to repouring operations. Indeed, in order to keep the chemical concentrations within the allowed limits, the chemical tanks are subjected to regular repours; the repours are either time- or batch- based and their schedule can be obtained through the tanks' sensors. This variant implements an a priori exclusion of all tanks that are or will be subjected to repours within a sensible time interval. The exclusion time interval is based on considerations on the travel time between the upstream buffer and the tool, and the typical waiting time within the tool. The calculation of the dryer's "Time In" for the tanks not undergoing repouring operations within the set time interval is the same as in the TD strategy; however, the TDR strategy intends to prevent any unexpected delay due to regular maintenance operations.

6. PRELIMINARY FINDINGS

The impact that the assignment strategies described in the previous section have on the CT components has been investigated using the simulation model developed. Real data have been used to investigate the strategies' efficiency. The results obtained are reported in Figure 7. QT and RT are represented as percentages of the CT observed when the current assignment strategy applied in the wet station, that is the OTO strategy, is simulated. As a consequence, for this strategy, the sum of QT (59.2%) and RT (40.8%) is necessarily equal to 100%; however, for all the other strategies the sum of QT and RT is significantly less than 100%. This means that all the strategies developed are effective in generating considerable CT reductions. In particular, QTs increasingly benefit from the applications of strategies that incorporate greater quantity of information. The QT reductions observed progressively increase from 11% for the TC strategy, which requires the least information on the system status, to 28% for the EF strategy, which is based on a detailed representation of the system's status; the QT percentage reductions are calculated with respect to the original QT percentage value, which is approximately 58% of the original CT. A complete knowledge of the tools' status, as required by the EF strategy, represents an ideal situation and generates the best results in terms of CT; however, as relevant information constraints apply to the implementation of the EF strategy in a real plant, the results obtained show that even the introduction of partial information on the system in the assignment process still determines positive effects on the tools' performance and should be encouraged.

Figure 7 also shows that all the strategies investigated do not impact RT; this is quite expectable as, in the wet station modelled, the recipes consist of a single chemical bath with the associated rinse bath and a final visit to the dryer. This implies that there is a very limited possibility of observing an over-exposure in the rinse tank as if the chemical tank is not available the batch is kept in the tool's internal waiting area. Moreover, the reduction of QTs also means that the tools are less congested and the tools' effective capacity increases. It is worth noting that QT includes components that cannot be reduced as they are structural; for example, waiting times for batching operations and travelling times from the buffer upstream of the station to the tools can be considered fixed QT components. QT reductions are mainly due to the extra-capacity created by a more efficient utilisation of the tools.

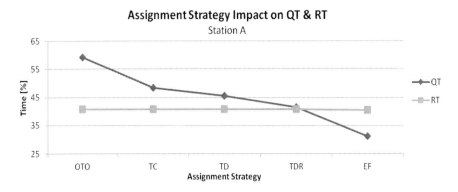

Figure 7. Impact of assignment strategies with different levels of information quality and quantity on CT.

7. ASSIGNMENT STRATEGIES GENERALISATION TO MORE COMPLEX SCENARIOS

The assignment strategies considered in the previous sections were conceived as adaptations of the EF strategy to the information constraints observed in the real plant; in particular, the simplification of the EF assignment logics was inspired by the specific production scenario observed at the wet station modelled. All the recipes performed at that station prescribe to visit one chemical tank, the associate rinse tank and the dryer; moreover, the tools include two identical chemical tanks that are able to satisfy the demand for higher volume operations. When the recipes prescribe to visit more than one chemical tank, the exclusion of tanks that already present a virtual queue, as requested by the TC, TD and TDR strategies, would mean drastically reducing the possibility of making assignment decisions as most tools would be excluded from the remaining decision steps. For this reason, a generalisation of the strategies has been developed so that the TC, TD and TDR strategies could be applied to wet stations with more complex recipe mixes. In particular, two other wet stations, station B and station C, operating in the same semiconductor facility where the original station was observed, have been considered; the previous wet station will be referred to as station A hereinafter. Stations B and C layout is the same as in Figure 1; however, the succession of chemical and rinse tanks differs for the three stations. Table 3 illustrates the wet station and tools' configuration; the same notation as in Table 1 is used in Table 3. Station B configuration suggests that it is possible that a chemical tank is considered safe from a scheduling perspective and that rinse tanks can physically precede the associated chemical tanks; this means that bi-directional flow can exist in a real wet tool. Moreover, in Station B, the two safe chemical tanks are coupled; this means that a batch has to visit them consecutively. This shows that the succession of chemical and rinse tanks is not necessarily alternating as reported in most research studies (Geiger, Kempf, and Uzsoy 1997, 115; Bhushan and Karimi 2004, 364; Zeballos, Castro, and Méndez 2010, 1706). As regards Station C, due to the peculiar recipe mix that determines a non-unidirectional production flow within a tool, a recipe dedication policy is applied so that, at any time, a tool in the station is dedicated to process batches of a specific recipe; the tools' dedication to a specific recipe is rotated at regular time intervals. Due to low capacity requirements, two tanks in one of Station C's tools are left empty. The recipes and the number of corresponding operations performed at the three wet stations are reported in Table 4.

Table 3. Wet station and tools' configuration

Station	# tools	Tanks					
		T1	T2	T3	T4	T5	T6
A	2	C b - U	R b - S	C a - U	R a - S	C a - U	R a - S
	2	C b - U	R b - S	C b - U	R b - U	C a - U	R a - S
B	7	C d - U	R d - S	R e - S	C e1 - S	Ce2 - S	R e - S
C	2	C c - U	R c - S	C c - U	R c - S	C a - U	R a - S
	1	C c - U	R c - S	NA	NA	C a - U	R a - S

Table 4. Recipes and corresponding number of operations at the three wet stations

Station	Recipe ID	#Op	Recipe					
A	A1	9	C a	R a	Dryer			
	A2	6	C b	R b	Dryer			
B	B1	14	C e1	C e2	R e	Dryer		
	B2	7	C d	R d	C e1	C e2	R e	Dryer
	B3	4	C d	R d	Dryer			
C	C1	7	C c	R c	C a	R a	Dryer	
	C2	2	C a	R a	C c	R c	Dryer	
	C3	1	C c	R c	Dryer			
	C4	5	C a	R a	Dryer			

For more complex recipe mixes such as those of Stations B and C, the assignment strategies have been adapted as follows; based on the generalisation approach illustrated here, adaptations to other recipes not similar to those performed in Stations B and C can be derived.

In the TC strategy, the assignment decision is based on the lowest "Time In" for the last chemical tank in the recipe; this "Time In" is calculated with a procedure similar to the calculation of the "Time In" for the dryer for the TC strategy. Figure 8 reports the pseudo-code that illustrates the generalised TC strategy assignment logic.

The earliest time at which the last chemical tank in the recipe becomes available (TimeOut*+virtualQueue_PT(h), in Figure 8) is compared with the cumulative sum of the time at which the previous chemical tanks become available plus the required processing and transportation times to reach the last chemical tank (cumulativePT(t), in Figure 8); the maximum value between these two predictions represents the predicted "Time In" for the last chemical tank for the corresponding tool. The time at which a tank becomes available consists of the "Time Out" of the batch currently processed in that tank if there is no virtual queue (TimeOut*); on the contrary, when a virtual queue exists at that tank a penalty time is applied. The consideration of a penalty time avoids the exclusion of the tank and the associated tool from further decision steps. The penalty time consists of the processing times and setup times of the batches virtually assigned to the tanks. It is worth noting that for ease of representation, terms such as setup times or travel times are not reported in the pseudo-code (Figure 8) but are considered in the assignment algorithm. If more than one tool presents the same minimum "Time In" for the last chemical tank, the tool with the fewest number of batches processed since its last shut down is chosen.

```
for each batch k to be assigned:
    retrieve recipe r required
    retrieve chemical tanks (t's) required by the recipe
    for each t in r:
        for each tool m that can process r:
            cumulativePT(m,0) = Time Log In (k)
            for each t in r:
                if t > 1:
                    cumulativePT(m,t) = cumulativePT(m,t-1)+ RinseT(k,t-1)
                for each tank h with chemical t:
                    retrieve TimeOut*(h) of batch currently processed
                    virtualQueue_PT(h) = 0
                    for each batch v in virtualQueue(h):
                        virtualQueue_PT(h) = virtualQueue_PT(h) + PT(v,t)
                    minTimeInTank(k,h) = max{cumulativePT(t), TimeOut*+virtualQueue_PT(h)}
                cumulativePT(m,t) = min_h(minTimeInTank(k,h))+PT(k,t)
    select tool m* ∋ ` cumulativePT(m*,t_final)=min_m(cumulativePT(m,t_final))
```

Figure 8. Pseudo-code of the generalised TC strategy.

The TD and TDR strategies incorporate the variations made to the TC strategy with respect to the calculation of the earliest "Time In" of the chemical tanks; moreover, for the TDR strategy, following the same logic as the TC strategy, tanks subjected to repours within sensible time intervals are given a penalty time but are not excluded from the consecutive procedure's steps.

8. RESULTS

The efficacy of the strategies' variants described above has been tested on both stations B and C; also for these stations, CT progressively reduces as the strategies applied incorporate more information on the tools' status. This is especially evident at Station B, where both CT components benefit from the implementation of information-based assignment strategies (Figure 9). In this case, RT reductions are possible due to both the relatively high number of operations performed at the station (See Table 4) and the classification of chemical tanks as safe tanks (e.g., T4-T5 in Table 3). Indeed, a better allocation of batches to tools prevents safe tanks from being used as local storage and reduces RT.

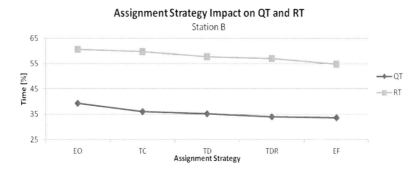

Figure 9. Impact of the assignment strategies' variants on QT and RT at Station B.

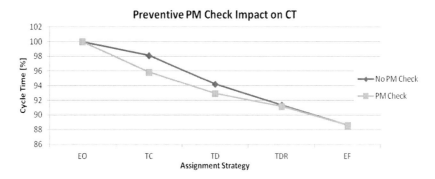

Figure 10. Impact of assignment strategies on CT when different preventive PM check policies are applied.

The impact of information quality on CT performance has been further investigated at Station B; the effects of the a priori exclusions of tools subjected to preventive maintenance (PM) at the moment of the assignment decision or tools with maintenance operations scheduled within a sensible time interval have been compared with the original case when no check for ongoing preventive maintenance operations is performed (Figure 10). The results show that the introduction of this exclusion rule is more effective on the TC strategy, which is the one based on poorer information. As the level of information increases, the impact of the exclusion rule reduces. In the TDR and EF strategy the exclusion rule does not generate CT variations as these strategies are not substantially modified; indeed, repours occur during PM and, as a consequence, tools undergoing maintenance operations are automatically penalised by the TDR logics. Moreover, the EF strategy does not apply any a priori exclusion rule because, by its nature, all the tools are considered for the assignment decision, so effectively, no modification has been made to the EF strategy and the same results are reported in Figure 9.

As regards Station C, interesting results are obtained as the TC and TD strategies generate significant QT reductions that are of the same order of those obtained when the EF strategy is implemented (Figure 11). The limited number of tools available (e.g., three tools operate in Station C) that is further reduced by the recipe dedication policy applied is most probably the reason for this result. As happens in Station A, RT are not affected by the assignment strategy due to both the recipes' structure and the presence of unsafe tanks.

It is worth noting that the application of the TDR strategy has not generated sensible results at this station; as a consequence, the TDR strategy's results have been omitted in Figure 11. This is probably due to the limited number of tools available and the recipe dedication policy applied which further reduces the number of tools available for the assignment decision; moreover, repours are performed at quite high frequency so that the high penalty times assigned to the tools delay the assignment decision and, as a consequence, increases in QT are observed.

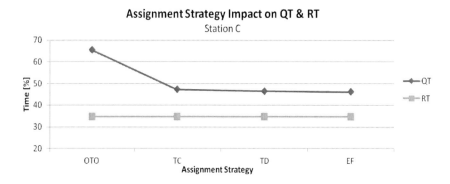

Figure 11. Impact of the assignment strategies' variants on QT and RT at Station C.

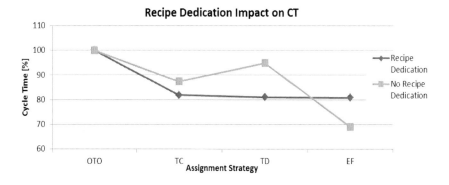

Figure 12. Impact of the assignment strategies on CT under different recipe dedication policies.

The impact of the recipe dedication policy on the efficacy of the alternative assignment strategies has also been analysed; simulation experiments have been conducted to investigate the effect of a "Non dedication" policy on CT for the TC, TD and EF strategies. In these experiments all the three tools at Station C are considered available for processing batches of any recipe. The alternative strategies still generate CT reductions (Figure 12) with respect to the strategy currently adopted (e.g., OTO); however, for the TC and TD strategy, CT increases with respect to the corresponding "Recipe Dedication" scenarios. This is because, in the tool previously dedicated to one recipe (e.g., recipe C1 which constitute 77% of the production volume at Station C) the "non dedication" policy makes the production flow asynchronous and conflicts for the tanks utilisation cause processing delay. In this case, an assignment decision based on a complete knowledge of the tools' status (e.g., EF strategy) can compensate for this issue and generate even more significant CT reductions as the absence of a dedication policy increases the number of tools eligible for the assignment decision. In other words, when the strategy is based on poorer information, sensible constraints that make the production flow synchronous and limit the search domain could facilitate an efficient assignment decision; on the contrary, for strategies based on good quality information, any unnecessary constraint reduces the strategy's efficiency.

Finally, the strategies illustrated here have been assessed by the industrial engineers involved in this study and positive feedback on the results obtained has been reported. The strategies development has been guided by their knowledge on the stations' control system and the possibility to access relevant data. Since significant modifications to both the data management system and the control system are required to implement these strategies, preliminary investigations are currently being carried out internally at the company to assess the applicability of similar assignment concepts to other stations operating in the plant that comprise parallel batch chamber tools. Positive results would support the final decision to allocate resources for the strategies adoption in the plant. In this regard, it is worth noting that, modifications of scheduling related control parameters that were suggested based on experimental results and required no implementation cost have been instigated in the plant. As a result, significant CT reductions, in the order of 10%, have been obtained; the parameters modified included the tools' virtual buffer capacity and MWTs for different operations.

9. DISCUSSION

The underlying concept motivating this work is that by increasing the amount of knowledge about a problem, the solution quality can significantly improve. This concept has been investigated with reference to the scheduling problem at wet etch stations, which is recognized to be a complex scheduling problem due its dynamics and applicable constraints. Reducing CT and RT considerably by using information which should be available in most semiconductor fabs provides fundamental support to the hypothesis that value can be obtained by processing data in real-time and transforming them into information that can support decision making. The value of real-time data and the validity of the general concept has been demonstrated by showing how this could be done by a sequence of changes/refinements of the scheduling algorithms which add more and more information about the problem instance. This could be a valuable approach for other scheduling problems, too. The main limitation of the work presented is that the underlying concept has only be validated for a single problem instance where it worked well. Although it is expectable that the introduction of additional details on a problem instance in the solution algorithm is likely to generate output improvements, there exist potential exceptions to this generic concept. An obvious reference would be to the tool/recipe configuration used for Station C in this chapter; in this case, the TDR strategy, which incorporates more problem specific knowledge, generates higher cycle times than less knowledge-intensive strategies (i.e., TC and TD); the TDR results for Station C are unrealistic (extremely high cycle times probably caused by model blockage) and, hence, not reported in the previous section (see Figure 11 and Figure 12). It is also expectable that there will be alternative tool/recipe configurations not considered here, where counterintuitive experimental results will not fit to the general concept that knowledge on the problem means output quality. The research reported in this chapter stemmed from a collaborative research with one of the leading semiconductor companies operating at the cutting edge of industry; the set of tools/recipes considered for analysis in this chapter was deemed exhaustive by the industrial engineers at the semiconductor facility involved in this research; hence, no further experiment on alternative tool or recipe configurations was conducted. A more theoretical approach to the investigation of factors that could impact the concept validity (i.e., sequence of tanks, recipe structure, operational times, etc.) would bring benefits to the concept validation process and identify potential boundaries to the concept validity. This is subject of current research. Likewise, the concept application to other problem instances is being explored. For instance, the same concept has been applied to a hybrid flow shop scheduling problem whereby different scheduling algorithms have been developed by progressively incorporating additional details on the system modelled. In this case, preliminary results show a less predictable correlation between problem specific knowledge captured in the algorithms and the algorithms' performance. These results will be published in the near future.

Another possible limitation of this research concerns the type of technology considered for analysis. More advanced single wafer wet etching technology has been recently introduced and adopted by companies at the forefront of the semiconductor industry; however, the multi-chamber batch technology is still in use (i.e., for test wafers regeneration). The assignment strategies suggested here are still relevant and will be for at least a decade. Moreover, the assignment problem in single wafer wet processing becomes even more critical since, when this technology is used, the wet station fleet size typically increases; therefore, assignment decisions become more complex and should be driven by dynamic data-based approaches. Hence, the importance of both demonstrating the value of real-time data usage in prescriptive tools and availing of predictive tools able to support trade-off analyses on the cost of data and its potential benefits in the lean manufacturing context emerge.

CONCLUSION

The assignment problem at wet stations presents relevant challenges as the variety of recipes performed and the peculiar scheduling constraints applied generate an asynchronous production flow within the tools that makes their efficient utilisation a difficult task. This study demonstrates that when the assignment decision is based on accurate information on the tools' status, the choice of the tool is made more effectively and, as a result, the tools' capacity increases. This concept has been inspired by the No Free Lunch theorems that state the importance of incorporating problem-specific knowledge into optimisation and search algorithms and also by the application of Data Analytics approaches to manufacturing. Data available in the manufacturing system should be exploited to extract information, refine the industrial practitioners knowledge on the system behaviour and support optimal decisions; for instance, preliminary data analysis during the wet tools scheduling algorithm development have revealed that a chamber of the tools initially considered a bottleneck did not constitute a production flow constraint for the tools. As regards the use of real-time data in decision support tools, the presence of information constraints should never be neglected as access to relevant data required by the solution algorithms may not be possible.

The four strategies illustrated in this study are generalised to wet stations with complex chambers configuration and recipes structure. The strategies consider the effects of a progressive introduction of information quantity and quality into the assignment decision process as an efficient assignment concept is adapted to applicable information constraints. The analysis has been carried out using a scheduling algorithm based on the scheduling logics observed at real wet stations. The scheduling algorithm development has been guided by the tools' internal scheduler characterisation based on historical data analyses. The algorithm generates feasible schedules in short computational times and is integrated with an assignment module that implements the four strategies analysed. The results obtained show that, generally, the queuing times benefit most from the application of information-based assignment strategies as a more efficient workload distribution among the tools available generates extra capacity at no investment cost. This happens especially when the recipe mix performed at the station consists of quite short recipes. When the recipes prescribe to visit more than one chemical tank and several unsafe tanks operate in the tools, run times also reduce when the quantity of information on which the assignment decision is based increases. Further experiments have been run to show that the introduction of further details, such as Preventive Maintenance related information, in the assignment decision process can also improve the results obtained. It has also been observed that the presence of assignment constraints intended to optimise the tools' utilisation by making the production flow synchronous (e.g., recipe dedication policy) has a positive impact on CT performance when the assignment decision is based on partial information on the tools' status; however, when a complete knowledge of the system is possible, this assignment constraint significantly limits the decision's efficiency and should be avoided.

Future work will focus on providing further experimental evidence to the concept that progressive introduction of knowledge about a problem instance in corresponding solution approaches can significantly improve the solution quality. This will be done by extending the experimental plan on the wet etch scheduling problem and using a more systematic approach to the analysis of the most impacting factors on the algorithms efficacy. Application to the same concept will also be investigated within the context of more generic assignment and/or scheduling problems.

ACKNOWLEDGMENTS

The research leading to these results has received funding under the National Development Plan 2008-2013 by the Irish Research Council for Science Engineering and Technology and carried out in collaboration with the Enterprise Ireland & IDA sponsored

Technology Centre, Irish Centre for Manufacturing Research. Ongoing research on this work is being funded from the European Union Seventh Framework Programme (FP7-2012-NMP-ICT-FoF) under grant agreement n° 314364.

REFERENCES

Aguirre, Adrián M., Vanina, G Cafaro. & Carlos, A Méndez. (2011). "Simulation-Based Framework to Automated Wet-Etch Station Scheduling Problems in the Semiconductor Industry." In *Simulation Conference (WSC), Proceedings of the 2011 Winter*, 1816–28.

Aguirre, Adrián M., Carlos, A Méndez. & Pedro, M Castro. (2011). "A Novel Optimization Method to Automated Wet-Etch Station Scheduling in Semiconductor Manufacturing Systems." *Computers & Chemical Engineering*, 35 (12), 2960–72.

Aydt, Heiko., Stephen, John Turner., Wentong, Cai., Malcolm, Yoke Hean Low., Peter, Lendermann. & Boon, Ping Gan. (2008). "Symbiotic Simulation Control in Semiconductor Manufacturing." In *International Conference on Computational Science*, 26–35.

Bhushan, Swarnendu. & Karimi, I. A. (2003). "An MILP Approach to Automated Wet-Etch Station Scheduling." *Industrial & Engineering Chemistry Research*, 42 (7), 1391–99.

Bhushan, Swarnendu. & Karimi, I. A. (2004). "Heuristic Algorithms for Scheduling an Automated Wet-Etch Station." *Computers & Chemical Engineering*, 28 (3), 363–79.

Castro, Pedro M., Adrián, M Aguirre., Luis, J Zeballos. & Carlos, A Méndez. (2011). "Hybrid Mathematical Programming Discrete-Event Simulation Approach for Large-Scale Scheduling Problems." *Industrial & Engineering Chemistry Research*, 50 (18), 10665–80.

Castro, Pedro M., Luis, J Zeballos. & Carlos, A Méndez. (2012). "Hybrid Time Slots Sequencing Model for a Class of Scheduling Problems." *AIChE Journal*, 58 (3), 789–800.

Chae, Bongsug. & David, L. Olson. (2013). "Business Analytics for Supply Chain: A Dynamic-Capabilities Framework." *International Journal of Information Technology & Decision Making*, 12 (01), 9–26. https://doi.org/10.1142/S0219622013500016.

Chae, Bongsug., Chenlung, Yang., David, Olson. & Chwen, Sheu. (2014). "The Impact of Advanced Analytics and Data Accuracy on Operational Performance: A Contingent Resource Based Theory (RBT) Perspective." *Decision Support Systems*, 59 (1), 119–26. https://doi.org/10.1016/j.dss.2013.10.012.

Cheng, Fan-Tien., Chi-An, Kao., Chun-Fang, Chen. & Wen-Huang, Tsai. (2015). "Tutorial on Applying the VM Technology for TFT-LCD Manufacturing." *IEEE Transactions on Semiconductor Manufacturing*, 28 (1), 55–69.

Choudhary, Alok Kumar., Jenny, A Harding. & Manoj, Kumar Tiwari. (2009). "Data Mining in Manufacturing: A Review Based on the Kind of Knowledge." *Journal of Intelligent Manufacturing*, 20 (5), 501.

Emblemsvg, Jan. (2005). "Business Analytics: Getting behind the Numbers." *International Journal of Productivity and Performance Management*, 54 (1), 47–58. https://doi.org/10.1108/17410400510571446.

Gan, Boon Ping., Peter, Lendermann., Kelvin, Paht Te Quek., Bart, Van Der Heijden., Chen, Chong Chin. & Choon, Yap Koh. (2006). "Simulation Analysis on the Impact of Furnace Batch Size Increase in a Deposition Loop." In *Simulation Conference, 2006. WSC 06. Proceedings of the Winter*, 1821–28.

Geiger, Christopher D., Karl, G Kempf. & Reha, Uzsoy. (1997). "A Tabu Search Approach to Scheduling an Automated Wet Etch Station." *Journal of Manufacturing Systems*, *16* (2), 102–16.

Govind, Nirmal. & David, Fronckowiak. (2003). "Process Equipment Modeling: Resident-Entity Based Simulation of Batch Chamber Tools in 300mm Semiconductor Manufacturing." In *Proceedings of the 35th Conference on Winter Simulation: Driving Innovation*, 1398–1405.

Haghighirad, Farzad., Ahmad, Makui. & Behzad, Ashtiani. (2008). *"Chaos in Production Planning"*, *26* (3), 739–50. http://haghighirad.ir/wp-content/uploads/2017/01/chaos-in-production-planning.pdf.

Ham, Myoungsoo. & John, W Fowler. (2008). "Scheduling of Wet Etch and Furnace Operations with next Arrival Control Heuristic." *The International Journal of Advanced Manufacturing Technology*, *38* (9–10), 1006–17.

Harding, J. A., Shahbaz, Srinivas M. & Kusiak, A. (2006). "Data Mining in Manufacturing: A Review." *Journal of Manufacturing Science and Engineering*, *128* (4), 969–76. https://doi.org/10.1115/1.2194554.

Hindi, Khalil S. & Krzysztof, Fleszar. (2004). "A Constraint Propagation Heuristic for the Single-Hoist, Multiple-Products Scheduling Problem." *Computers & Industrial Engineering*, *47* (1), 91–101.

Hsieh, M. D., Amos, Lin., Kenley, Kuo. & Wang, H. L. (2002). "A Decision Support System of Real Time Dispatching in Semiconductor Wafer Fabrication with Shortest Process Time in Wet Bench." In *Semiconductor Manufacturing Technology Workshop*, 2002, 286–88.

Jagadish, H. V., Johannes, Gehrke., Alexandros, Labrinidis., Yannis, Papakonstantinou., Jignesh, M Patel., Raghu, Ramakrishnan. & Cyrus, Shahabi. (2014). "Big Data and Its Technical Challenges." *Communications of the ACM*, *57* (7), 86–94.

Kang, Pilsung., Dongil, Kim., Hyoung, Joo Lee., Seungyong, Doh. & Sungzoon, Cho. (2011). "Virtual Metrology for Run-to-Run Control in Semiconductor Manufacturing." *Expert Systems with Applications*, *38* (3), 2508–22. https://doi.org/10.1016/j.eswa.2010.08.040.

Karimi, I. A., Zerlinda, Y. L. Tan. & Swarnendu, Bhushan. (2003). "An Improved MILP Formulation for Scheduling an Automated Wet-Etch Station." *Computer Aided Chemical Engineering*, *15*, 1181–86.

Karuno, Yoshiyuki. & Hiroshi, Nagamochi. (2003). "A Better Approximation for the Two-Machine Flowshop Scheduling Problem with Time Lags." *Lecture Notes in Computer Science*, 309–18.

Khan, Aftab A., James, R Moyne. & Dawn, M Tilbury. (2007). "An Approach for Factory-Wide Control Utilizing Virtual Metrology." *IEEE Transactions on Semiconductor Manufacturing*, *20* (4), 364–75.

Kusiak, Andrew. (2006). "Data Mining: Manufacturing and Service Applications." *International Journal of Production Research*, *44* (18–19), 4175–91.

Lee, Tae-Eog. (2008). "A Review of Scheduling Theory and Methods for Semiconductor Manufacturing Cluster Tools." In *Proceedings of the 40th Conference on Winter Simulation*, 2127–35.

Lee, Tae-Eog., Hwan-Yong, Lee. & Sang-Jin, Lee. (2007). "Scheduling a Wet Station for Wafer Cleaning with Multiple Job Flows and Multiple Wafer-Handling Robots." *International Journal of Production Research*, 45 (3), 487–507.

Lim, Hwa Kyung., Yongdai, Kim. & Min-Kyoon, Kim. (2017). "Failure Prediction Using Sequential Pattern Mining in the Wire Bonding Process." *IEEE Transactions on Semiconductor Manufacturing*, 30 (3), 285–92. https://doi.org/10.1109/TSM.2017.2721820.

Mauer, John L. & Roland, E. A. Schelasin. (1993). "The Simulation of Integrated Tool Performance in Semiconductor Manufacturing." In *Proceedings of the 25th Conference on Winter Simulation*, 814–18.

Monzon, Shai. & Stepehn, Gray. (2018). *Increasing Product Quality and Yield Using Machine Learning*, no. March. https:// www.intel.com/ content/ dam/ www/ public/ us/ en/ documents/ white-papers/ increase-product-yield-and-quality-with-machine-learning-paper.pdf.

Moyne, James., Jamini, Samantaray. & Michael, Armacost. (2016). "*Manufacturing Advanced Process Control*", 29 (4), 283–91.

Moyne, James. & Brad, Schulze. (2010). "Yield Management Enhanced Advanced Process Control System (YMeAPC) Part I: Description and Case Study of Feedback for Optimized Multiprocess Control." *IEEE Transactions on Semiconductor Manufacturing*, 23 (2), 221–35. https://doi.org/10.1109/TSM.2010.2041294.

Muehlen, Michael zur. & Robert, Shapiro. (2010). "Business Process Analytics." *Handbook on Business Process Management*, 2, 137–57. https://doi.org/10.3905/jai.2007.695270.

Niedermayer, Heiko. & Oliver, Rose. (2003). "A Simulation-Based Analysis of the Cycle Time of Cluster Tools in Semiconductor Manufacturing." In *Proceedings of the 15th European Simulation Symposium*, 349–54.

Noack, Daniel., Boon, Ping Gan., Peter, Lendermann. & Oliver, Rose. (2008). "An Optimization Framework for Waferfab Performance Enhancement." In *Proceedings of the 40th Conference on Winter Simulation*, 2194–2200.

Novas, Juan M. & Gabriela, P Henning. (2012). "A Comprehensive Constraint Programming Approach for the Rolling Horizon-Based Scheduling of Automated Wet-Etch Stations." *Computers & Chemical Engineering*, 42, 189–205.

Oliveira, Marcos Paulo Valadares De., Kevin, McCormack. & Peter, Trkman. (2012). "Business Analytics in Supply Chains - The Contingent Effect of Business Process Maturity." *Expert Systems with Applications*, 39 (5), 5488–98. https://doi.org/10.1016/j.eswa.2011.11.073.

PwC. (2016). "Industry 4.0: Building the Digital Enterprise." *PwC.*, 2016. https://doi.org/10.1080/01969722.2015.1007734.

Quek, Paht Te., Boon, Ping Gan., Song, Lian Tan., Chan, Lai Peng. & Bart, vd Heijden. (2007). "Analysis of the Front-End Wet Strip Efficiency Performance for Productivity." In *Semiconductor Manufacturing, 2007. ISSM 2007. International Symposium On*, 1–4.

Rotondo, Anna., Paul, Young. & John, Geraghty. (2014). "EF-Based Strategies for Productivity Improvements at Wet-Etch Stations." *Production Planning and Control*, 25 (11). https://doi.org/10.1080/09537287.2013.782845.

Rotondo, Anna., Paul, Young. & John, Geraghty. (2015). "Sequencing Optimisation for Makespan Improvement at Wet-Etch Tools." *Computers and Operations Research*, 53. https://doi.org/10.1016/j.cor.2014.04.016.

Scholl, Wolfgang. & Joerg, Domaschke. (2000). "Implementation of Modeling and Simulation in Semiconductor Wafer Fabrication with Time Constraints between Wet Etch and Furnace Operations." *IEEE Transactions on Semiconductor Manufacturing*, 13 (3), 273–77.

Sisselman, Michael E. & Ward, Whitt. (2007). "Value-Based Routing and Preference-Based Routing in Customer Contact Centers." *Production & Operations Management*, 16 (3), 277–91. https://doi.org/10.1111/j.1937-5956.2007.tb00259.x.

Trkman, Peter., Kevin, McCormack., Marcos, Paulo Valadares De Oliveira. & Marcelo, Bronzo Ladeira. (2010). "The Impact of Business Analytics on Supply Chain Performance." *Decision Support Systems*, 49 (3), 318–27. https://doi.org/10.1016/j.dss.2010.03.007.

Wolpert, David H., and William G Macready. 1995. "*No Free Lunch Theorems for Search.*" Technical Report SFI-TR-95-02-010 10: 1–38. https://doi.org/10.1145/1389095.1389254.

Wolpert, David H. & William, G Macready. (1997). "No Free Lunch Theorems for Optimization." *IEEE Transactions on Evolutionary Computation*, 1 (1), 67–82.

Yih, Yuehwern. (1994). "An Algorithm for Hoist Scheduling Problems." *The International Journal of Production Research*, 32 (3), 501–16.

Yugma, Claude., Stéphane, Dauzère-Pérès., Christian, Artigues., Alexandre, Derreumaux. & Olivier, Sibille. (2012). "A Batching and Scheduling Algorithm for the Diffusion Area in Semiconductor Manufacturing." *International Journal of Production Research*, 50 (8), 2118–32.

Zeballos, Luis J., Pedro, M Castro. & Carlos, A Méndez. (2010). "A CP-Based Approach for Scheduling of Automated Wet-Etch Stations." In *Proceedings of the 2nd International Conference on Engineering Optimisation*.

Zeballos, Luis J., Pedro, M Castro. & Carlos, A Méndez. (2010). "Integrated Constraint Programming Scheduling Approach for Automated Wet-Etch Stations in Semiconductor Manufacturing." *Industrial & Engineering Chemistry Research*, 50 (3), 1705–15.

Zhang, Dongsong. & Lina, Zhou. (2004). "Discovering Golden Nuggets: Data Mining in Financial Application." *IEEE Transactions on Systems, Man and Cybernetics Part C: Applications and Reviews*, 34 (4), 513–22. https://doi.org/10.1109/TSMCC.2004.829279.

In: Lean Manufacturing
Editors: F. J. G. Silva and L. Carlos Pinto Ferreira
ISBN: 978-1-53615-725-3
© 2019 Nova Science Publishers, Inc.

Chapter 7

THE EIGHTH WASTE: NON-UTILIZED TALENT

M. Brito[1,*], A. L. Ramos[2], P. Carneiro[3] and M. A. Gonçalves[1]

[1]DME – Departement of Mechanical Engineering, ISEP – School of Engineering,
Polytechnic of Porto, Porto, Portugal
[2]DEGEIT – Department of Economics, Management,
Industrial Engineering and Tourism, University of Aveiro, Aveiro, Portugal
[3]DPS – Department of Production and Systems, University of Minho, Braga, Portugal

ABSTRACT

In a changing economic climate, characterized by great pressure to improve productivity and reduce costs, industries use different management approaches, with lean manufacturing being the most popular in recent years. Lean manufacturing is based on value creation for the customer and the elimination of the waste which occurs during the production process, while improving working conditions.

There are eight types of waste in lean manufacturing. The well-known seven wastes are production process-oriented, while the eighth waste is directly related to management's ability to utilize personnel. This type of manufacturing waste occurs when management in a manufacturing environment fails to ensure that all the employees' potential talent is being well utilized. This talent refers to management's ability to use critical thinking and continuous improvement feedback from employees to improve a lean manufacturing process. When management does not engage with manufacturing employees on topics of continuous improvement and allow employees to produce change for the better, that is considered a manufacturing waste.

Although performance management and the use of the employees' talent are crucial to ensure a competitive advantage, there is a lack of research regarding this type of waste: non-utilized talent. On the other hand, there is a lot of literature addressing process-oriented wastes.

The purpose of this work is to explore this "talent waste" and identify the contributing factors for this type of waste, such as: lack of reward, lack of recognition, lack of motivation, lack of training/knowledge, organizational injustice, undefined goals, etc.

---

[*] Corresponding Author's E-mail: mab@isep.ipp.pt.

A tool is also introduced which takes the form of a questionnaire that will allow managers to evaluate their workers and help them to identify if there is "talent waste" in their companies, as well as the main causes related to this type of waste.

To the authors' knowledge this work is innovative and valuable because it will help companies increase employees' performance, which is very important for organization sustainability.

Keywords: lean manufacturing, wastes, human resources

1. INTRODUCTION

Waste avoidance is a crucial idea of the Lean philosophy, as this approach significantly contributes to maximize value from the customer's perspective. Waste takes place in diverse forms, depending on the types of industry and of working processes. Elimination or reduction of waste to a certain degree requires the ability to identify waste and to make it transparent to the parties involved in the working process (Denzer et al. 2015, 723). Intellect underuse was identified as the eighth waste (Nunes et al. 2015, 892), when management in a manufacturing environment fails to ensure that all the employees' potential talent is being utilized.

In an increasingly competitive global market, more and more organizations must ask the difficult question, "How can we get more out of our employees?" (Westover et al. 2010, 372).

Thus, the factor that sets companies apart is the know-how of its employees, which also forms the company's main asset. With the knowledge and intelligence of its workers, a company is able to effectively manage its own resources, take actions to improve its performance and innovate (Małachowski and Korytkowsk 2016, 165). Thus, Human Resources (HR) must be able to become reliable and be ready to develop new skills, be committed and capable of making changes through teamwork. They are asked to think globally and have a vision for the future. Thus, attention and guidance from HRs are aspects that must be improved (Taba 2018, 65).

A company with multi-skilled employees can function with a lower number of workers. Employees who are single-skilled might sit idle waiting for work to become available. Multi-skilled workers are allocated to work, and do not wait for work to come to them. In a company with a multi-skilled and pro-activity workforce, planning focuses on the needs of the customer, not on the competences of the staff. This leads to a decrease in the number of idle hours. Multi-skilled workers are familiar with reliably learning new skills and adapting to changes. Employee satisfaction improves morale in a business, which leads to better productivity and employee retention rates. An experienced and well-trained multi-skilled staff translates directly into higher productivity, better quality and lower costs (Małachowski and Korytkowsk 2016, 165).

Business owners and managers have tried to find answers to the same questions: "Who is the right person for the job?" and "What can we do to improve morale and productivity in our workplace?" (Westover et al. 2010, 373).

Once these candidates are identified and hired, their skills need to be utilized properly. This may be easier said than done given that it may be impossible to have a worker with great mechanical skills, for example, utilize those skills as a therapist. However, employers can still try to identify particular skills (as appropriate for the job area) that may be helpful to the department or organization. More than anything, these conclusions are based on employee perceptions. Therefore, even organizational efforts to identify and recognize individual skills/talents (even if there is no immediate way to utilize them) may substantially impact improving employee perception of their own skill utilization (Westover et al. 2010, 381).

Businesses require information on labor utilization to optimally use their workers. Information on workforce underemployment is essential because it can enable businesses to improve firm output and profits by using the underemployment information to boost productivity and performance, as well as to reduce turnover and its attendant costs. Likewise, it enables firms to realistically gauge the available labor pool in their labor sheds (Addy et al. 2012, 214).

One of the clearest symptoms of deteriorating condition in an organization is the lack of work satisfaction (Taba 2018, 65). In its most cynical shape, the symptoms are hidden behind layoff, work deceleration, failure, and turnover. The symptoms may also be complaints, poor performance, poor product quality, disciplinary problems, and other issues. On the contrary, high work satisfaction is desirable by managers because it can be linked with a positive result that they expect. High work satisfaction is a sign of a well-run organization and is basically a result of effective behavior management (Taba 2018, 66).

Indeed, it might be necessary and important to track both job satisfaction and underemployment as the two attributes can be used to measure organizational effectiveness. Companies should consider improving job satisfaction as well as reducing underemployment in order to lower turnover and its attendant costs (Addy et al. 2012, 226).

According to Addy et al. (2012, 217-219), underemployment may affect job attitudes, overall psychological well-being, career attitudes, job behaviors and performance, and marital, family, and social relationships, in the following ways:

- Fewer workers believe their jobs adequately suit their education and training, skills, and experience;
- More think they are qualified for a better job;
- More would quit their current jobs for higher income;
- More are prepared to commute longer and farther for a better job;
- More have looked for better jobs in the preceding quarter;
- Fewer are married, more are female, and their median age is a little lower.

Thus, beyond the importance of finding the right person for the right job (Westover et al. 2010, 381) it is also important, according to the literature, to identify whether workers use their capabilities in full and, if not, what the reasons are for not doing so.

In spite of the importance, recognized in the literature, of the eighth waste, which is related to underemployed, the causes and ways of reducing this type of waste have not yet been clarified. The purpose of this study is to explore this "talent waste" and identify the contributing factors for it.

Another of its objectives is the development a tool that allows employers to assess whether this type of waste exists in their companies, i.e., whether or not their employees use their talent in their entirety.

To the authors' knowledge, this work is innovative and valuable as it will help companies increase their employees' performance, which is essential for organization sustainability.

1.1. Background

Lean production began in Japan; as described by Womack et al. (1990, 114-116), it started with the Japanese engineer Eiji Toyoda, who conducted a three-month study of the Ford Rouge plant in Detroit. After studying the factory production system carefully, at what was the largest and most efficient manufacturing complex in the world, he came to the conclusion that mass production would never work in Japan. From this first experiment originated what Toyota came to call the Toyota Production System (TPS), which eventually became Lean production.

In the TPS, the goal was to get things right the first time, and efficiently implement tools aiming at being strongly effective in this process, producing the exact amount using the minimum necessary resources, which included the elimination of waste with an improved production flow with less lead time, lower costs, better quality, greater efficiency in services which meet the customer expectations, and thus reaching the improvement in production efficiency in general. Its main focus was also to ensure the quality of the products being produced, that is, "the result that every organization wants is for their customers to be highly satisfied with the products with guaranteed quality, which leads to profitability in the business" (Santos et al. 2015, 5949).

Ohno, who visited Detroit repeatedly immediately after the war, thought this whole system was rife with muda, the Japanese term for waste in effort, materials and time (Womack et al. 1990, 132).

Lean Management focuses on the elimination of the non-value-added activities or wastes. As presented in Figure 1, there are eight different types of wastes: transportation, excessive inventory, unnecessary movements, over production, over processing, waiting time, quality/defects and intellect underuse (Nunes et al. 2015, 891-892).

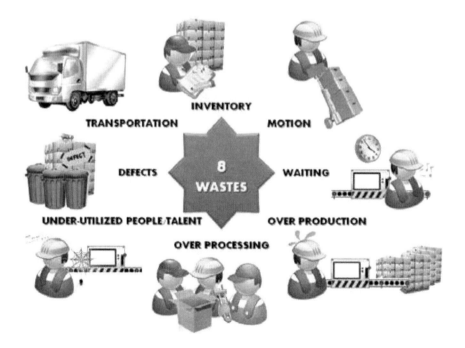

Figure 1. Eight different types of waste (Nunes et al. 2015, 892).

During Lean Manufacturing for a company that seeks Lean production, a mapping for the whole process can be made, assisting in the identification of each step-by-step phase with the information of its effectiveness, adding value to the product with the goal of beating the competition and effectively looking to meet customer needs. The TPS has become a world reference in which many of the organizations apply the tools that exist within the system (Santos et al. 2015, 5949), such as, according to Pil and Fujimoto (2007, 3742-3743):

- Jidoka: the practice of stopping the line when defects are uncovered;
- Total quality control;
- Kaizen: continuous improvement activities;
- Inventory reduction via Kanban;
- Heijunka: leveling of production volume and product mix;
- Reduction of "muda" (non-value adding activities);
- Reduction of "mura" (uneven pace of production);
- Reduction of "muri" (excessive workload);
- Genryo Seisan: production plans based on dealers' order volume;
- Tsukurikomi: on-the-spot inspection by direct workers;
- Poka-yoke: fool-proof prevention of defects;
- Andon: real-time feedback of production troubles;
- Assembly line stop cord;
- 5S: emphasis on cleanliness, order, and discipline on the shop floor;
- Quality control circles;
- Standardized tools for quality improvement;
- Worker involvement in preventive maintenance;
- Reduction of process steps to save equipment;
- Amongst others.

Sugimori et al. (1977, 557) suggests that a vital tenet of TPS is respect for humans, which includes:

- The removal of waste movements by workers;
- Attention to workers' safety;
- Self-display of employees' capabilities by entrusting them with greater responsibility and authority.

As Liker (2004, 145) noted, "The Toyota way preaches that the worker is the most valuable resource – not just a pair of hands taking orders, but an analyst and problem solver". Toyota has a system to assess and formally certify individual skill levels. The skills system, implemented in the early 1990s, consists of four levels, each requiring increasingly greater skill and seniority. The goal of systematically tracking and codifying skill was to broaden employee skills and create systematic training programs. The work-life plans that result give the workers a series of goals to meet, and a reward to more senior workers who may not be promoted to group leader. Workers have ownership and involvement in the course of standardization and are given the tools for such involvement (Pil and Fujimoto 2007, 3750).

In the 1980s, productivity levels were extremely significant and individuals and groups were evaluated based on their performance relative to engineering targets (called the coefficient of production remuneration or CPR). However, this produced a lot of employee dissatisfaction. Data is now pooled across multiple plants, allowing for more systematic comparisons and learning. The importance of productivity in the payment system has been cut to a third of its original level and was replaced by a system that ties a bonus into cost, quality, and safety metrics (Pil and Fujimoto 2007, 3751).

According to Westover et al. (2010, 375) workers enjoying high levels of organizational commitment are more satisfied and motivated in their workplace than those who actively consider other employment.

Job involvement describes how personally involved an employee is in fulfilling his/her work role. Job involvement is a function of personality and organizational climate and is linked with higher levels of organizational effectiveness. Literature demonstrates that job involvement is moderately related to job satisfaction (Westover et al. 2010, 375). Table 1 shows the important outcomes of job satisfaction.

Table 1. Important Outcomes of Job Satisfaction

Variable related with job satisfaction	Direction of relationship
Life Satisfaction	Positive
Job Performance	Positive
Worker Motivation	Positive
Job Involvement	Positive
Organizational Commitment	Positive
Organization. Citizenship Behavior	Positive
Employee Tardiness	Negative
Employee absenteeism	Negative
Withdrawal cognitions	Negative
Employee turnover	Negative
Worker Health	Positive
Perceived Stress	Negative

Source: Westover et al. 2010, 375

Therefore, employers can reduce withdrawal cognitions (and therefore turnover) by focusing on enhancing job satisfaction (Westover et al. 2010, 376).

Westover et al. (2010, 381) shows that job satisfaction, talent use by employer, value congruence, fair pay, and age are each meaningfully related to organizational commitment. Passion and education are both significantly connected to organizational commitment. Unit increases in job satisfaction, value congruence, fair pay, passion, and age lead to rises in organizational commitment. When an employee feels his or her talents are being utilized, their commitment to the organization grows (Westover et al. 2010, 381).

According to Addy et al. (2012, 214) "underemployed persons are workers who believe that their education and training, skills, or experience are not fully used in the jobs they currently hold and that qualify them for higher paying or more satisfying jobs, for which they could quit their current positions".

Some strategies that can be used to enhance labor utilization and job satisfaction could involve on-the-job career development opportunities, such as training and career growth, monetary and non-monetary compensation incentives, and employee-focused management practices. (Addy et al. 2012, 215).

2. METHODOLOGY

This study adopted a quantitative research approach. Seventeen interviews were carried out to selected production workers, managers and executives of a few Portuguese private companies. The questions were specifically designed to identify the contributing factors for talent waste, i.e., why workers fail to use their full talent. This talent refers to management's ability to utilize the critical thinking and continuous improvement feedback from employees to improve a lean manufacturing process.

The answers to the question "Why do you think workers do not use their full talent?" were collected and then validated based on the literature. Since there is not much literature related to the eighth waste, the authors opted to try to understand the relationship that each component (eighth waste reason) has with the worker's performance. The component was validated if a positive relation between them was found.

The objective of this work was not to analyze the differences between the answers and the different respondents but to identify the reasons and put them in a tool, in the form of a questionnaire which will allow managers to evaluate their workers and help them to identify if there is "talent waste" in their companies, as well as its main causes.

3. RESULTS

The respondents (production workers, managers and executives), answered that the eighth waste is related to the lack of one or more than one of the following components: rewards, recognition, justice, evaluation, motivation, goals, self-esteem, knowledge, and resources (Figure 2).

After collecting and analyzing the responses, they were validated in the literature by the relationship that each component (eighth waste reason) has with the worker's performance. The following section explains that relationship.

3.1. Rewards

The reward system may play a role in increasing employees' motivation to work more effectively, increase productivity within the company or compensate for the lack of commitment, if a reward is related to employee's performance. Increased performance typically raises economic, sociological, and psychological rewards (Davis 1981, p. 99).

Additionally, reward systems are important to attract candidates to join the organization, keep them working, and motivate them to work hard. If used effectively, a reward can have an impact on the individual's behavior such as turnover, absenteeism, achievement, and commitment. One of the factors that lead an employee to work in a fun and satisfying way is their commitment to the organization or company where they work. Thus, employees are asked to improve their performance in increasingly heavy and varied tasks. Improved performance is closely related to the level of a worker's organizational commitment to a system of rewards that can meet employee's intrinsic and extrinsic needs (Taba 2018, 67).

Figure 2. Components whose failure leads to the eighth waste.

According to Markova and Ford (2011, 813), organizations ought to critically evaluate all aspects of their reward systems to reflect the uniqueness of their employees.

Incorporating non-monetary rewards in reward systems is required to encourage productivity and creativity of knowledge workers (Markova and Ford 2011, 813).

Intrinsic rewards are categorized as motivational influences, job content, such as achievement, occupation, recognition, probability of development, and responsibility. Herzberg (1959) argues that if these factors are absent from an organization, they do not necessarily lead to dissatisfaction, but if these factors exist, they will be strong motivators and and will lead to good performance.

3.2. Recognition

Bradler et al. (2016, 3085) find that recognition increases ensuing performance substantially, and particularly when recognition is exclusively provided to the best performers.

If one does a good work, he will intrinsically feel pleased with it. If the organization offers a reward to recognize high productivity, giving a salary increase and promotion this will increase employee satisfaction (Taba 2018, 71).

3.3. Justice

The employees need a system of pay and promotion policies that they understand to be fair, unambiguous, and in line with their expectations (Taba 2018, 67).

Equity models assert that satisfaction rests on the perception of how fairly an individual is treated at work. If one sees that another worker is receiving equal or more rewards for doing less work, this will have a negative impact on his/her level of satisfaction. Consequently, an employer's duty, according to this model, is to seek to understand his/her workers' perceptions of fairness and to seek to interact with said employees in a way that helps them to feel treated equitably. (Westover et al. 2010, 374).

If the reward is deemed proper and fair, then the greater the satisfaction will be because employees feel that they receive rewards in accordance with their performance (Davis 1981, p. 99).

3.4. Motivation

Kuvaas and Dysvik (2009, 217) found that the connection between perceived investment in employee development and work effort was mediated by intrinsic motivation. Furthermore, intrinsic motivation was found to moderate the relationship between perceived investment in employee development and organizational citizenship behavior. The form of the moderation demonstrated a positive relationship only for employees with high levels of intrinsic motivation (Kuvaas and Dysvik 2009, 217).

Dinger et al. (2015, 281) found a positive relationship between intrinsic motivation, job satisfaction, and job performance.

3.5. Training and Knowledge

The importance of worker development programs is growing among those organizations, pursuing to get an advantage over their competitors. Training programs are helpful for companies to emphasize the knowledge, expertise and ability of workers (Divyaranjani and Rajasekar 2017, 649).

Training and development programs have a positive correlation with employees' job performance as well as the entire organizational growth since it gives employees the opportunity to expand their learning (Sujatha et al. 2014, 23765).

3.6. Goals

Goals with feedback generally increase worker satisfaction and productivity. On the other hand, monetary incentives further improved worker performance but added no incremental satisfaction gain (Shikdar and Das, 2003, 603).

Payment based on achievements provides an opportunity for employees to get more by being more productive. An effective program also provided employees with a clear achievement target. It allows an employee to monitor his/her performance compared with the goal at any time. Based on psychological research, a simple way to monitor self-achievements is to encourage individuals to compete with themselves and improve on their previous achievement level (Gibson et al. 1997, p. 277).

Achievement was related to the accomplishment of organizational goals which are difficult and challenging. Tasks which are relatively difficult to do turned out to provide motivation for employees. The setting of difficult goals also had the same effect. Meeting objectives which are relatively difficult will boost motivation in a more powerful way than if those objectives are easy to reach (Taba 2018, 71).

3.7. Evaluation Feedback

Performance evaluation is described as comparing the performance of workers and the work, standards, and managing the necessary activities in a systematic way to attain these standards. What makes performance measure a necessity is its focus on the performance of personnel as an objective measure of whether the company is moving in the correct direction or not. This is because the most important problem found in organizations is the difficulty in the determination of how successful the personnel are in the satisfaction of their duties and what their capabilities in their jobs are (Esen et al. 2014, 53).

3.8. Self-Esteem

The effects of personnel management policies, which emphasize job satisfaction, could potentially lead to developments in levels of health, happiness, subjective well-being and workers' self-esteem, all of which are factors that can improve organizational performance (Satuf et al. 2018, 181).

3.9. Resources

Most of the answers revealed that one of the reasons why the workers do not use their full talent is that the organization did not provide them with the necessary resources to develop the work in an optimized way. Examples of these resources are: tools, computer programs, etc.

An interesting aspect found in the literature is that, on average, white-collar employees more often believed that their work was variable and provided opportunities to use one's knowledge and skills than did blue-collar employees. They also felt their work was autonomous, that is, they could influence when and how to do their work. The blue-collar workers in general did not perceive job variety, use of knowledge and skills, development opportunities, and opportunities for decision making in their job as frequently as did the white-collar employees (Seppälä and Klemola 2004, 168).

3.10. Assessment of the Eighth Waste

The objective of this work was also the development of a tool, in the form of a questionnaire to allow managers to evaluate their workers and help them to identify if there is "talent waste" in their companies, as well as the main causes related to it.

The authors decided to make the tool as simple as possible, so that the workers did not waste much time responding and did not have to spend much time to analyze it. Thus, only two questions were raised:

1. Do you use your full talent in your daily work, most of the time?
   Yes ☐   No ☐
2. If the answer to the previous question was no, complete the following sentence: "I do not use my full talent due to the lack of…" (you can choose more than one):
   - Rewards ☐
   - Recognition ☐
   - Justice ☐
   - Motivation ☐
   - Training and Knowledge ☐
   - Goals ☐
   - Evaluation Feedback ☐
   - Self-esteem ☐
   - Resources ☐
   - Other(s) ☐ Identify which: _____

The answers must be anonymous so that the workers feel safe in answering the questions without suffering any kind of retaliation. The goal is to identify the main causes and then implement solutions to combat this waste. At the end this should be a win-win process, for workers and for the organization, i.e., it should enhance employee performance and consequently organizational performance.

CONCLUSION

In a changing economic climate, characterized by pressures to enhance productivity and reduce costs, performance management has a more central role in helping to ensure competitive advantage (Rowland and Hall, 2012, 280).

Despite the fact that performance management and the use of the employees' talent are crucial to ensure a competitive advantage, there is a lack of research regarding this type of waste: non-utilized talent. On the other hand, there is a lot of literature addressing process-oriented wastes. This study explored this "talent waste" and identified the contributing factors for this type of waste, such as: lack of reward, lack of recognition, lack of motivation, lack of knowledge, organizational injustice, undefined goals, lack of evaluation feedback, lack of resources and lack of self-esteem.

In the authors' opinion, the organizations must, first, measure this eighth waste, know the causes and then identify possible solutions or tools to combat it. For this reason, this chapter also presents a tool, in the form of a questionnaire which will allow managers to evaluate their workers and help them to identify if there is "talent waste" in their companies as well as the main causes related to it.

The findings of this study help to raise the awareness of employers, human resource managers, professional and industrial experts of the importance of reducing the eighth waste.

The main limitation of the research is the lack of generalizability of the findings – that it represents data from just seventeen interviews in a small sampling of organizations.

The next steps in this investigation will involve the validation of the components of the eighth waste identified in this study using the questionnaire to a larger number of workers and companies and then the investigation of which tools are the most appropriate to eliminate or reduce each identified cause of the eighth waste.

The authors' of this work believe that a new look at people's development is essential in a culture of lifelong learning. Workers are the essential part of an organization and, when motivated, will use their maximum potential, contributing to the success of the company. And this is where the eighth waste enters!

REFERENCES

Addy A. N., Bonnal, M. & Lira, C. (2012). "Toward a More Comprehensive Measure of Labor Underutilization: The Alabama Case." *Business Economics, 47* (3), 214-227.

Bradler, C., Dur, R., Neckermann, S. & Non, A. (2016). "Employee recognition and performance: A field experiment." *Management Science, 62*(11), 3085-3099.

Davis, K. (1981). *Human Behavior at Work: Organizational Behavior*, 6th ed. McGraw-Hill Inc., New York, NY.

Denzer, M., Muenzl, N., Sonnabend, F. A. & Haghsheno, S. (2015). "Analysis of definitions and quantification of waste in construction." *Proceedings of IGLC 23 - 23rd Annual Conference of the International Group for Lean Construction: Global Knowledge - Global Solutions*, 723-732.

Dinger, M., Thatcher, J. B., Treadway, D., Stepina, L. & Breland, J. (2015). "Does professionalism matter in the IT workforce? An empirical examination of IT professionals." *Journal of the Association of Information Systems, 16*(4), 281-313.

Divyaranjani, R. & Rajasekar, D. (2017). "Measuring the development of workers after effective training in automobile manufacturing companies with reference to Chennai city". *International Journal of Mechanical Engineering and Technology, 8*(6), 649-661.

Esen, H., Hatipoŀlu, T. & Boyaci, A. I. (2014). "A performance evaluation model of a job title using fuzzy approach". *Proceedings of the International Conference on Fuzzy Computation Theory and Applications*, 53-60.

Gibson, J., Ivancevich, J. M. & James, H. (1997). *Organizations*, 8th ed., Binarupa Aksa, Jakarta.

Herzberg, F., Mausner, B. & Snyderman, B. B. (1959). *The Motivation to Work* (2nd ed.). NewYork: John Wiley & Sons.

Kuvaas, B. & Dysvik, A. (2009). "Perceived investment in employee development, intrinsic motivation and work performance". *Human Resource Management Journal, 19*(3), 217-236.

Liker, J. K. (2004). *The Toyota Way: 14 Management Principles from the World's Greatest Manufacturer.* McGraw-Hill: New York.

Małachowski, B. & Korytkowsk, P. (2016). "Competence-based performance model of multi-skilled workers." *Computers & Industrial Engineering, 91,* 165–177.

Markova, G. & Ford, C. (2011). "Is money the panacea? Rewards for knowledge workers." *International Journal of Productivity and Performance Management, 60*(8), 813-823.

Nunes, I. L. (2015). "Integration of Ergonomics and Lean Six Sigma. A Model Proposal." *Procedia Manufacturing, 3,* 890-897.

Pil, F. K. & Fujimoto, T. (2007). "Lean and reflective production: the dynamic nature of production models". *International Journal of Production Research, 45*(16), 3741-3761.

Rowland, C. A. & Hall, R. D. (2012). "Organizational justice and performance: Is appraisal fair?" *Euro Med Journal of Business, 7*(3), 280-293.

Santos, Z. G. D., Vieira, L. & Balbinotti, G. (2015). "Lean Manufacturing and Ergonomic Working Conditions in the Automotive Industry." *Procedia Manufacturing, 3,* 5947-5954.

Satuf, C., Monteiro, S., Pereira, H., (...), Marina Afonso, R. & Loureiro, M. (2018). "The protective effect of job satisfaction in health, happiness, well-being and self-esteem." *International Journal of Occupational Safety and Ergonomics., 24*(2), 181-189.

Seppälä P. & Klemola, S. (2004). "How Do Employees Perceive Their Organization and Job When Companies Adopt Principles of Lean Production?" *Human Factors and Ergonomics in Manufacturing, 14* (2), 157–180.

Sugimori, Y., Kusunoki, K., Cho, F. & Uchikawa, S. (1977). "Toyota production system and kanban system: materialization of just-in-time and respect-for-human system." *International Journal Production Research, 15* (6), 553–564.

Sujatha, S., Vinoth, M., Nyamekye, B. & Bempah Botwe, P. (2014). "Assessing The Impact of Training and Development Programmes on Worker Performance and Business Productivity; A case study of royal bow company ltd. Kumasi, Ghana." *International Journal of Applied Engineering Research, 9*(24), 23765-23770.

Shikdar, A. A. & Das, B. (2003). "The relationship between worker satisfaction and productivity in a repetitive industrial task." *Applied Ergonomics, 34*(6), 603-610.

Taba, M. I. (2018). Mediating effect of work performance and organizational commitment in the relationship between reward system and employees' work satisfaction. *Journal of Management Development, 37* (1), 65-75.

Westover, J. H., Westover, A. R. & Westover, A. A. (2010). "Enhancing long-term worker productivity and performance. The connection of key work domains to job satisfaction and organizational commitment." *International Journal of Productivity and Performance Management, 59* (4), 372-387.

Womack, J., Jones, D. & Roos, D. (1990). *The Machine that Changed the World.* New York: Rawson Associates.

In: Lean Manufacturing
Editors: F. J. G. Silva and L. Carlos Pinto Ferreira
ISBN: 978-1-53615-725-3
© 2019 Nova Science Publishers, Inc.

Chapter 8

QUALITY AND SAFETY CONTINUOUS IMPROVEMENT THROUGH LEAN TOOLS

Gilberto Santos[1], J. C. Sá[2,3]*, J. Oliveira[3], Delfina G. Ramos[4,5] and C. Ferreira[2]

[1]Design School, Polytechnique Institute Cavado Ave, Barcelos, Portugal; IEA - Vigo University, Vigo, Spain
[2]DME – Departement of Mechanical Engineering, ISEP – School of Engineering, Polytechnic of Porto, Porto, Portugal
[3]School of Business Sciences, Polytechnic Institute of Viana do Castelo, Portugal
[4]Technology School, Polytechnic Institute of Cavado and Ave, Barcelos, Portugal
[5]Algoritmi Centre, Minho University, Braga, Portugal

ABSTRACT

The purpose of this chapter is to suggest the use of several lean tools that can be used, indicating the improvements that can be obtained with each of the recommended tools. The main objective is to be a contribution to the organizations, showing how they can detect the wastes in the productive flow through VSM, and at the same time to improve the quality of products through lean tools.

The method presented was a case study. The various lean tools were also collected and described.

This chapter intends to show several lean tools that can be applied in different situations, as well as, the wastes that each can eliminate and the benefits that are obtained from each one. Throughout the production process, with the support of Value Stream Mapping (VSM), it was possible to detect several wastes. A case study is presented as results from a work carried out in a company that is dedicated to the production of mechanical equipment.

This work can be a guide to support organizations that wish to start their lean road. Smart value creation remains today dependent on the maturity of how lean management tools are applied. Lean, innovative technologies, critical knowledge, talent and big data need other dimensions to be sustainable. Quality and safety continuous improvement

---

* Corresponding Author's E-mail: cvs@isep.ipp.pt.

through lean tools, as well as, Industry 4.0 assumes a critical pillar for the new journey toward the future of our organizations.

Keywords: quality, safety, continuous improvement, lean, value stream mapping, case study

1. INTRODUCTION

Increased competition and customer expectations require organizations that make products with high quality to gain powerful competitive advantages in the globalized marketplace (Barbosa et al., 2018). Varying definitions of Lean paradigm can be found in the literature, but all of them share the same principle: cost minimization and waste elimination (Cabral et al., 2012). Thus, Lean production is the philosophy of eliminating waste (Heizer & Render 2004) or the creation of a Lean and balanced flow in a process (Stevenson, 2007). The word Lean was introduced by Krafcik (1988) to describe Toyota's Production System (TPS). In the last decades, Toyota Production System (TPS), just-in-time (JIT), Total Quality Control (TQC) and, later, Lean Production have been implemented by many companies leading them to reinvent their strategic management, management accounting system, performance measurement system and operations management (Chiarini & Vagnoni, 2015). Lean is an ongoing drive toward perfection, sometimes difficult to envision because it is a major paradigm shift (Wilson, 2010). The TPS is the most successful production applications of the Lean concept. TPS has been called 'just-in-time (JIT)', and more recently, 'lean production' (Womack et al., 1990), the common term in the West. Although these practices started in Japan, Lean implementation is now the primary improvement methodology in many countries.

In 1988 Tachii Ohno presented in his book "Toyota Production System: beyond large-scale production" (Ohno, 1988), the seven wastes that are authors for the low productivity of companies. According to this author, the wastes that exist along the production flow are creators of losses and must be removed. The knowledge of their location is crucial for their identification, and this must be performed on Gemba (shop floor) because that is where they occur, and it is necessary to remove (Imai, 1997). Nowadays, companies keep wastes along the production flow and it is more and more important to eliminate them, because they are sources of costs and loss of productivity inside the companies, endangering their future sustainability. Rother & Shook (2003), stated that in all products/services provided to customers there is a determined added value and the challenge is in the visualization of this value stream. These authors, who were already familiar with the Toyota Production System (TPS), developed the Value Stream Mapping (VSM). The purpose of the VSM design was to help company managers and directors see the flow of material and information in their plants. Most recently Sá et al. (2011), developed the Waste Identification Diagram (WID) as an alternative representation to VSM, to more easily identify wastes along the value stream.The utilization of the VSM or WID is essential for any organization. Among other things it allows: flow visualization and visualization of the sources of waste in the value stream. The seven wastes identified by Tachii Ohno (Ohno, 1988) in his major work "Toyota Production System: Beyond Large- Scale Production, continues to be a nightmare for all organizations. Masaaki Imai (Imai, 1997), in his work: "Gemba Kaizen - A Common sense, low cost approach to management" reinforce the need for companies to identify their waste and then eliminate them. This author uses the term "Muda" rather than "waste" to define the term "wastage" and reinforce a need to be identified in Gemba (on the factory floor). That's where they will have to be eliminated! "Muda" is not an acronym, but it is a Japanese word meaning futility, uselessness or wastefulness and is a key concept in the Toyota Production System as one of the three types of variation. Thus, "Muda" is a key removal-of-waste concept (along with Mura and Muri) in lean manufacturing that focuses on the removal of non-value adding steps. Genba (also romanized as Gemba) is a Japanese term meaning "the actual place". The need that companies have to eliminate their desires are not adequate for the final value of the final product, but rather, resources are needed. They bring added value to the final product/service. The principles and practices of Lean began in the late 1980s, but in 1988, through a group of researchers led by James Womack who are doing research at the Massachusetts Institute of Technology (MIT).

Lean principles have long been recognized as a competitive advantage (Pakdil, & Leonard, 2014). Thus, Womack (1990) defined the concept of "Lean" as an internal philosophy of the organization, which sought an elimination of change in the book "The Machine That Changed the World". Later, Womack predicted major changes in the production process for the companies due to the type of orders from the customers and that they would have to adapt to the change. He asserted that the era of standard mass production had ended and that they were now in the era of small quantity and large variety orders. Faced with this new reality, it was necessary to change the production system of the companies to a new production regime based on the detection and elimination of MUDA. Womack, however, found that some companies in the United States, Germany, and England had already embarked on this path due to the fact that there was no growth at the time, that is, they were in a period of economic stagnation. Unfortunately, many companies continued to resist the adoption of this new paradigm of production, since it was sufficient to look at the high stocks that existed in the companies (raw material, product in the process of manufacturing - WIP and finished product) which remained clinging to retrograde concepts. Womack (1990) believed that resistance to change was no more than a psychological issue, since the adoption of lean allowed organizations in the short term to improve their resources and eliminate existing change, thereby enabling them to produce the same quantities but with fewer resources. This situation posed a challenge to the top management that could choose to lay off employees or else bet on innovation and create/launch new products/services to the market and keep all existing workforce.

This second situation would be the best option, as it allowed companies to increase the number of products/services they had to offer to the market and, consequently, to increase the volume of invoicing without increasing labour costs. Another aspect to take into consideration, in order to choose the second situation, was the need not to cause any internal social unrest in the company, because it was imperative that employees accept the changes brought about by the implementation of the Lean, they could put it into practice, and consequently improve the organization's performance. Womack et al. (1990) advocated the idea that top management should install the spirit of teamwork in its employees through a clear focus on Lean tools and techniques, in order to create a culture of polyvalence within the organization's internal structure/Employee turnover. Another idea pointed out by Womack (1990) was that organizations should reinforce to the employees the importance of identifying the problems and their causes, without this being a reason for later reprimand, but a moment of congratulation because the employees of the company have the opportunity for improvement in their company. Later, Womack & Jones (2003), presented the concept of "Lean Thinking", in order to present the key ideas developed by Taiichi Ohno (Ohno,1988) in the Toyota Production System (TPS).

Several authors, such as, Moore & Gibbsons (1997) and McDonald et al. (2002) defined the key areas of Lean Manufacturing as follows:

Flexibility; Elimination of waste; Optimization; Monitoring of processes and Involvement of people. Several authors defined tools/techniques to help organizations implement the Lean culture. Among them, it can be highlighted the work developed by Rother & Shook (2003), with the presentation of Value Stream Mapping (VSM), and Sá et al. (2011) with the presentation of Waste Identification Diagram (VSM), which aimed to design the flow of value of a product. The use of VSM and WID as a tool for the detection of waste as a support for the implementation of the Lean philosophy has reached the most diverse sectors of activity and has contributed to eliminate some outdated concepts.

This chapter aim to demonstrate continuous quality improvement through the use of various lean tools. It also aims to be a contribution to the organizations, showing how they can detect the wastes in the productive flow through VSM or WID, and at the same time to improve the quality of products (Doiro et al., 2017). In addition to the detection of waste, this chapter intends to show several lean tools that can be applied in different situations, as well as the wastes that each can eliminate, and the benefits that are obtained from each one.

2. CASE STUDY

2.1. Presentation of One Company Situation

The company selected for the case study, is a company that is dedicated to the production of mechanical equipment, according Figure 1. Currently the manufacturing process is oriented by production orders, which are issued by the production manager, taking into account the existing orders and the estimate of new orders. The production is subdivided into four main areas of work: reception of materials; conformation (cutting, drilling, bending); welding (longitudinal welding by submerged arc, laying of the bottoms, circular welding of the bottoms by submerged arc, manual assembly of the boiler); painting (pickling, manual painting); assembly, packaging and storage of the final product.

The most outstanding wastes throughout the company were: WIP (work in progress) and high stocks, equipment shutdown and large distances between jobs. When the production managers were questioned about the situations recorded, the response was that the company had always worked in this way and that because of the type of product produced, with all the productive operations associated, there was no other way to do it. The way to show how the productive part of the company could be reorganized was to show clearly to the top management the wastes that exist along the productive flow, as well as the costs that these have for the company. Since the concept of Lean philosophy consists of producing only what the customer needs when he needs it the first step was to know which product had the highest customer demand and the average monthly demand (Womack & Jones, 2003). The information obtained was that the products that had the most output had an average monthly demand in the order of 34 products per month, and the unit cost of this article stood at 300 €.

This information allowed to calculate the takt-time that defines the regularity with which products should go out to the customer. Knowing the monthly demand and considering that a month on average has 22 business days, you get the average daily demand of the customer:

Daily demand = (34 products/month)/(22 working days/month) = 1,55 products/day

That is, the average daily demand is 1.55 products. From this value, we calculate the takt-time that gives us the information on how often the company should have an end product ready to satisfy customer demand. The takt-time value will then be: Takt-time = (8 hours)/(1.55 products/day) = 5.16 hours

With this information, we now know that the company has to have a product ready, of 5.16 hours in 5.16 hours. Gembutsu should be used to collect the information for the value stream mapping process (Imai,1997). It should always be done in Gemba, because that is where the problems happen and there is nothing like going there to see what is really going on - Gembutsu. The most appropriate set of actions. As can be verified through VSM, there is a high level of stock and several workplaces which can be reduced through the implementation of a pull system (Ohno, 1988). However, existing stocks may result from the fact that production is not synchronized with customer demand (Rother &Shook, 2003), but may have emerged because there are a large number of unplanned breakdowns due to malfunctions, which could be reduced through the implementation of the TPM (Total Productive Maintenance) (Borris, 2006). The disorganization and lack of existing standards has been one of the main sources of waste, which can be eliminated through the implementation of 5S (Gapp et al., 2008). There is also a high changeover time, which can be reduced substantially through the SMED (Single Minute Exchange of Die) application (Dillon & Shingo, 1985). It was also analyzed the implementation of the standardized work, with the purpose of optimizing production processes and reducing process variability (Berger, 1997).

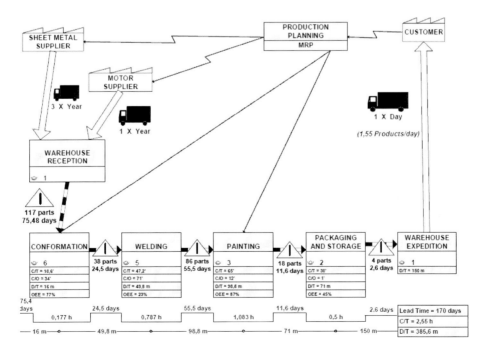

Figure 1. The VSM of the current situation (Oliveira et al., 2017).

The method used was PDCA (Plan-Do-Check-Act). Plan - in the planning of the company, the problems existing was identified through the VSM. It defines the tools that will be implemented in order to reduce waste. Do - the tools are implemented. Check-the impacts of these measures are analysed. Act - if the issues have been resolved, the matter is closed. If not, it needs to go back to the beginning and new tools will be tested.

3. THE USE OF LEAN TOOLS

The following tools will allow the organization to achieve continuous and valuable propositions for their key stakeholders. A lean manufacturing initiative is focused on cost reduction and increases in turnover by systematically and continuously eliminating all non-value-added activities. In a competitive market, lean is the "the solution" to manufacturing industries for survival and success. Lean manufacturing helps organizations to achieve targeted productivity by introduction of easy-to-apply and maintainable techniques and tools. Its focus on waste reduction and elimination enables it to be engrained into organization culture and turns every process into profit.

The first lean tool will be used it will be the Value Stream Mapping (VSM), because it will help us identify wastes along the value stream. Only after we identify our wastes in the organization that we can select the appropriate lean tool to remove them.

3.1. Value Stream Mapping

Value Stream Mapping (VSM) is a method developed by Rother & Shook (2003), which permits an overview of the material flow from the raw material acquisition to the final product expedition. Abdulmalek & Rajgopal (2007) defined VSM as a map to identify waste, improvement opportunities and which use lean tools. In order to apply the VSM methodology, four steps should be followed: 1 - Selection of the product or family of products to use as improvement subject; 2 - Drawing of representation of current state; 3 - Drawing of the future state, without the inefficiencies previously pinpointed. This is referred as value stream design (VSD); 4 - Elaboration of a work plan to achieve the future state.

The use of VSM helps on the identification of waste sources, provides a common language for its analysis and facilitates the understanding of the connections of the material flow. It is also an effective way of registering lead times, setup times and other indicators, in a way which enables the responsible to clearly visualize the system's performance.

3.2. A3 - Problem Solving

Tool A3 (Problem Solving Report), as the name implies, is a A3-size paper sheet develop by Toyota Motor Corporation. This tool makes it possible to identify the problem, determining the causes and suggesting possible solutions for that problem (Mobley et al., 2008). By using this tool efficiently, it enables good management of meetings, always endorsing teamwork on problem solving. It also identifies the team responsibilities for each member of the team (Liker, 2004; Shook, 2008).

This report comprises 7 sections, each section having a specific function: (1) Establish the commercial context and the problem importance and relevance; (2) Describe of the problem conditions; (3) Identify the desired result; (4) Analyze of the situation to establish causality; (5) Propose counter measurements; (6) Define an action plan; and (7) Map the follow-up process (Shook, 2009).

3.3. Daily Kaizen

Daily Kaizen is a team building tool that uses daily meetings to develop an autonomous and dynamic team so it actively improve work processes and the workplace (Monteiro et al., 2015). Dias (2012) argues that this tool enables the entire organization to understand which activities add value. Each department is constituted by its team, being this denominated as "natural team" since they work in the same department or area (Monteiro et al., 2015), and has one leader, i.e., a person responsible for the development of the work in that team Through frequent Kaizen meetings, the teams remain connected and informed, thus providing a high-level control and efficiency. Additionally, it includes a continuous analysis of performance indicators, also allowing its continuous improvement (Monteiro et al., 2015).

This tool comprises a four-level model, that should take place in the following order: (1) create organizational mechanisms, in the team, and (2) in the workstation; then (3) implement normalization; and finally (4) focus on problem solving (This tool consists of a four-level model, in which a certain order is respected: firstly, organizational mechanisms must be created, (1) in the team; as (2) in the workstation; then (3) implement normalization; and finally (4) focus on problem solving.

3.4. 5S

The 5S lean tool was developed in Japan by Sakichi Toyoda, Kishiro Toyoda and Taiichi Ohno in 1960 (Ohno, 1988). This tool aims to achieve a clean and organized workspace in order to maintain an outstanding organizational environment. The method consists on the sequential following of five steps: Seiri (Sort) - Consists on the removal of everything deemed unnecessary. The workplace should only have what is needed to perform the activities; Seiton (Set in order) - There must be a place for everything and everything should be in its place. Quick and visual identification of tools and areas saves time and facilitates processes; Seizo (Shine) - Cleaning the workspace is essential. It reduces the risk of accidents and aids on the inspection of products; Seiketsu (Standardize) - In order to optimize the first three S's, standard must be created and followed; Shitsuke (Sustain) - The last step consists on developing a method to ensure the 5S technique is followed. It requires discipline and focus. Usually, audits are performed to assure the sustainability of the methodology. 5S bring several benefits to a company, being the most relevant one the decrease of waste of time and space. The rewards of applying 5S can be extended to quality, safety and hygiene (Santos et al, 2014).

3.5. Standard Work

Standard work is a lean tool developed by Onho (1988). It is defined as the degree of rules and operational procedures which are formalized and executed. Furthermore, the authors argue teams require autonomy to establish a set of specific rules which facilitate their work. The method aims to eliminate the variation and inconsistency of results by instructing workers to execute manufacturing activities following clearly defined procedures. This goal can be achieved by both defining an optimal procedure and ensuring its performance. There is no room for improvisation. Therefore, the operations are often referred as an inflexible work standard. They are used as a training auxiliary tool as well (Womack & Jones, 2003). Benefits: Variability reduction - The work effort becomes stable and measurable; Cost reduction - By means of waste reduction derived from inefficient work procedures, the system becomes more cost-effective; Quality improvement - If the same operation were to be executed differently depending on the person, the probability of defects would increase; Worker involvement - Standard work shifts the blame for errors from the worker to the system. Thus, people tend to start being more honest about improvement opportunities; Continuous improvement - This tool is essential for continuous improvement, since it facilitates change to improved standards, making it easier, faster and more efficient overall.

3.6. On Point Lesson (OPL)

All corporations should have training sessions for its workers, and they can consist on visual norms since it provides a better understanding and knowledge dissemination of the processes. This can be achieved through One Point Lesson (OPL - One-Point-Lesson, which means step-by-step visual norm). As a visual tool and a step-by-step explanation of a particular task, it allows the collaborator to know how to execute it correctly when performing that task.

This type of document should contain simple, fast and easy-reading language, use diagrams, illustrations and images and must contain only the necessary and indispensable information (The OPL allows quick learning through a timely view of the structure of the equipment, function or method to be used, and still provide knowledge sharing since it demonstrates the best ways to perform a specific work in a practical, direct and dynamic manner (Nieminen, 2016).

3.7. Visual Management

According to Galsworth (2004), visual management is a "self-ordering, self-explaining, self-regulating, and self- improving work environment where what is supposed to happen on time, every time, because of visual devices". It is the basis of several other lean tools, such as 5S and standard work. Hall (1987) defines it as communication with no words nor voice. It consists on the utilization of fast and intuitive means of communication. There are several visual management systems such as informative boards, space delimitations and work instructions. The goal is to empower workers to manage their own work environment, reducing errors and further forms of waste.

3.8. Andon

Implementation of an Andon System helps to control the production, as well as improves the communication of the problems in the production line (Ohno, 1988). Andon, a Japanese word for a visual aid tool, is a production-control device that allows real time information about the production status, giving an alert if there is an imminent problem.

3.9. Heijunka

Product levelling tool, concept in Japanese is Heijunka, distributes products by volume and combines a mix of products uniformly (Dennis, 2006). The tool Heijunka levels the production (Liker & Meier, 2006), allowing workload elimination, Mura. In other words, it provides levelling of the production volume in a continuous, harmonic and efficient flow. Process are previously thought and balanced, so that it is possible to do quick changes in the products in the necessary production amounts (Coleman & Vaghefi, 1994). The Heijunka goal is to not only level the production volume, but also to level other various types of products using the same sequence of products in each production cycle (Matzka, et al., 2012).

Heijunka provides stability onto the production process, converting the uneven clients demand into an uniform and predictable production chain (Grimaud et al., 2014). This tool allows production synchronized with the demand, reduction of the lot production size, and consequently the excess stocks. It increases the variability of the produced products, thus satisfying all the client requirements and maintain the flow value of the system at a constant rate (). The production levelling allows production of a certain amount of several products every day, instead of producing higher lots of product and storage them until it is sell. The goals of Heijunka are reduction of stock levels, adapt flow of production to the market demands which is unstable and unpredictable, improve final product quality, produce smaller lots and minimize inventory.

3.10. Kanbans

Developed by Ohno (1988) on Toyota production lines, Kanbans emerged as a solution to the tendency of factories to overproduce. He looked for a way of reducing this waste by finding a means of delivering only what is necessary when necessary. Kanban can be translated from the Japanese as a card or signal, and is a visual input used in pull systems (Hall, 1987). Arbulu et al. (2003) define Kanban as a lean approach developed in the automotive industry to "pull" materials from the production line in a "just in time" mind-set. The concept of this method consists on promoting the restock of materials only when required, by receiving and sending signals, usually in the form of cards. This process can be either internal or external to the company. In order to assist this practice, it is used in supermarkets and milk-runs. The former is a structure filled with organized product components. The latter is a transport vehicle which provisions the assembly lines with the components they need from the supermarket.

3.11. Jidoka

Jidoka concept, driven by Sakichi, means "automation with a human touch" and consists in the detection of faulty equipment and defects through methods that allow machines to automatically self-manage, regulate and maintain quality control (Audenino, 2012). The main goal of Jidoka is the non-occurrence of any bad quality product (Wilson, 2010).

According to Shingo & Dillon (1989), automation provides the best way to achieve cost reduction in work-labour. Since the equipment automatically detects defects, the operator will not need to constantly supervise it, thus becoming more productive, as it can be working with other machines simultaneously.

3.12. Poka-Yoke

Although the Poka-Yoke concept existed for many years everywhere, and in a variety of ways, it was Shingo & Dillon (1989), who had the idea of designing a tool to achieve zero defects. A Poka-yoke device is any mechanism that prevents an error or defect from occurring by removing the causes of the defects (Shimbun, 1989). This device allows to avoid the production of nonconforming products, and in this way can lead to the elimination of activities that do not add value, such as quality control activity. The purpose of Poka-Yoke is to allow errors to be prevented or immediately detected and corrected (Fisher, 1999).

3.13. Milkrun

Mizusumashi (Japanese term for logistic train) is a lean tool defined by "route programmed for transport of material" (Hugos, 2003), providing a fast, flexible and efficient supply of materials. In the United States, on the milk market, this concept was named "milkrun", and consisted on defining the milkman's milk delivery routes, leaving the bottles full and collecting the empty ones (Gomes, 2012). In predetermined time cycles, and following predefined circuits, the logistics train supplies the production line, collects the materials that will be needed and delivers those that are missing (Coimbra, 2009), based on a principle of "epty box, full box" (Gomes, 2012).

The milkrun tool in a production system requires the use of kanbans, physical cards used to signal when a later process requires more material, and acts as a replenishment order for the upstream process (Coimbra, 2009). The implementation of this tool presents some advantages, including reduced transport costs, reduced pollution emitted since it reduced the number of trucks on the road (Nemoto et al., 2010), improvement of operations and production processes as the insertion of the materials onto the production line becomes more accessible and efficient (Nemoto et al., 2010).

3.14. Overall Equipment Effectiveness (OEE)

The strategic goal of TPM implementations is the reduced occurrence of unexpected machine breakdowns that disrupt production. Overall Equipment Effectiveness (OEE) indicator includes results from all equipment manufacturing into a measurement system that helps manufacturing and operations teams improve equipment performance and, therefore and reduce equipment costs (Ahuja & Khamba, 2008). The OEE key performance result offers a starting-point for developing quantitative variables for relating maintenance measurement to plant strategy. OEE can be used as an indicator of the reliability of the production network. Forming cross-functional teams to solve the root causes/problems can drive the highest improvements and generate real bottom-line benefits. A comparison between the expected and current OEE measures, can deliver the needs for the manufacturing organizations to improve the maintenance policy and affect continuous improvements in the manufacturing systems. According Ahuja & Khamba (2008), OEE offers a measurement tool to evaluate equipment performance and ensure relevant information for productivity improvement. OEE is a productivity improvement process that starts with management consciousness of total productive manufacturing and their commitment to focus the factory work force on training in teamwork and cross-functional equipment problem solving. A key objective of TPM is to eliminate or minimize all losses related to manufacturing system to improve overall production effectiveness (Suzuki, 1994). The OEE measure is fundamental to the formulation and execution of a TPM improvement strategy.

3.15. Single-Minute Exchange of Die (SMED)

Shingo & Dillon (1989) invented the Single-Minute Exchange of Die system for Toyota, helping the companies reducing their changeovers and can achieve quickly improvements in their results as: lead time reduced, lower inventories that will improve quality, productivity, profits and global results. Single Minute Exchange of Dies, is a methodology used to reduce the time machines are down during changeovers. In traditional changeover processes, operations generally start taking place only after the machine has stopped. First, the SMED approach looks to identify the steps that can be performed while the machine is running (external setup operations) and those that can only take place while the machine is stopped (internal setup operations). The strong way to get value with this methodology is to start to transform internal operations to external operations, reducing the down time during changeovers and standardize the process. SMED methodology distinguishes in a changeover two types of operations: Internal Operations (I.E.D.) - Input Exchange of Die that can only be performed with the machine immobilized and External operations (O.E.D. - Output Exchange of Die) that can be performed while the machine is running.

The SMED method has allowed numerous companies to considerably reduce their changeover times. Companies can now go from several hours to a few minutes. Today, when we compare with the past, it is possible to move from hours to less than a minute for a change. The definition of the standards for changeover plays here a key role. The application of SMED is indispensable, since the long changes of a production series are critical problems to guarantee the fluidity of the circulation of products. Improve maintenance quality by establishing periodic inspection and replacement criteria based on provisional cleaning and inspection standards. SMED is also known as a quick changeover. Definition of "changeover": a changeover is the process of preparing a work system, such as a machine, for the manufacture of a (different) product type or product, by loading the required tools, for example.

Quick changeover is a procedure to optimize changeovers so that they require less time. A quick changeover is performed, for example, by separating and reducing the internal and external changeover time, and by means of changeover preparation, quick release systems and changeover carts. Organizations expect from Quick Changeover to make it possible to switch between product variants quickly and flexibly. They can thus respond quickly to changes and produce in small lot sizes, as is required for levelling. The fewer facilities are standing idle, the higher their net production time and the shorter the lead times. This enables us to increase the value added and improve productivity. Quality is ensured by "subsequent approval following changeover". TPM, SMED creates also benefits regarding quality improvements and shorter time-to-market, based on advanced manufacturing processes and mature supply chains operations.

3.16. Yokoten

Toyota has introduced the concept of Yokoten System. It consisted on concise reports of successful problem-solving processes that were implemented at Toyota's various plants (Demeter et al., 2016; Marksberry et al., 2010). Yokoten is a Japanese word that means "sharing best practice" (Baykut, 2011). The Yokoten System was created by Toyota Motor Corporation, and aims information and knowledge to be horizontally dissemination over the entire organization (Baykut, 2011). Yokoten was defined as sharing and propagation of best practices (Jones et al., 2013).

In TPS, the Yokoten is not intended to copy, but to improve what was formerly seen and recorded (Ikonen, 2011). In an organization, such as Toyota, which has multiple industrial units throughout the company, the Yokoten tool helps ensuring that all of these industrial units possess the best performance (Baykut, 2011; Melton, 2005). Marksberry et al., (2010) claims that documentation and standardization of the company's successful experiences are crucial for the Yokoten achievement (Machikita et al., 2016). Implementation of Yokoten, sharing of knowledge along the organization, enables workers from any department to know how to tackle problems efficiently, and focusing in the organizational culture, disseminating knowledge and solutions to eventual problems Howell, 2014).

3.17. Total Productive Maintenance (TPM)

TPM is a production management tool which has been developed in Japan over the past several decades. TPM creates a culture inside the organization which strives to eliminate losses, improving the competitiveness of the organizations. Implementing TPM inside organizations needs support from top management, because this advanced concept requires the commitment, availability and participation of all employees to promote productive maintenance through motivation management or voluntary small group activities.

The failure or success of TPM depends on the determination and commitment of senior management. TPM becomes one of the foundations for operational performance stability and improvements and is a key tool that allows production management to achieve higher levels of efficiency and effectiveness. It also brings to production technologies, efficiency and profitability. Productive Maintenance (TPM) means autonomous, planned, and preventive maintenance of machines and facilities. We want to ensure the best possible use of production facilities through maintenance, care, and inspection measures. We expect from TPM, little or no downtime of facilities due to failures. The targeted elimination of the causes of failures increases plant productivity due to the high technical availability of the operating facilities, improved motivation of associates due to active participation.

With TPM implementation, we maintain a high level of quality in our products by preventing machine based quality losses and a cost optimization due to the reduction of ad-hoc measures and failure and repair costs. TPM is supporting the effort to reach the economic and strategic targets of the organization, focusing especially on achievements in: no loss of quality caused by machine failures, no unforeseen machine breakdown and minimal changeover time. Another goal is a reduction of the maintenance expenditure during the utilization phase and the guarantee of the continuous availability of the equipment's inside the factories. Planned and unplanned downtimes should be minimized systematically and maintenance and repair activities quickly and easily completed, based on standardization of processes.

When implementing TPM strategy with suitable knowledge, support and a correct way, we are increasing the potential of the organizations to achieve the three Zeros of TPM: zero breakdowns, zero defects and zero accidents. TPM is relevant to maximize the effectiveness of the equipment, improving its overall efficiency, helping organizations get benefits with this approach based on a role model comprehensive productive-maintenance system that helps the entire life of the equipment in a good condition to support their business (Mwanzaa & Mbohwaa, 2015). The fundamental concept is to maintain plant equipment with higher standards based on TPM pillars. The organization will see a strong availability in equipment, safety, quality problems and a sustainable productivity, maintaining the business in a safe path (Jadhav et al., 2009). TPM implementation helps organizations have a smart preventive and productive maintenance and involves every level, from top executive to the floor operator. TPM has been a key pillar for helping them increase the productivity and overall equipment effectiveness (Jadhav et al., 2009). In order to handle breakdowns and failures that decrease availability, it is mandatory to analyze the downtime and get to know each machine's reliability and maintenance ability. There are benchmarked key performance indicators to monitor the equipment as Mean Time Between Failures (MTBF) and Mean Time to Repair (MTTR). Machine operators carry out maintenance work independently, including the continuous improvement of the machinery. Associates are actively involved and are jointly responsible for their production facilities.

We define targeted maintenance activities, including the development of control and diagnostic systems, to further increase the reliability of facilities. TPM strategy take future maintenance activities into account as early as the planning and procurement stage, with good accessibility and visualization of maintenance points, for example. TPM is one of the most valued tool with high added value to achieve high standards of efficiency for all organizations. There are eight pillars for the TPM implementation. Each pillar has specific targets to achieve higher levels of performance in each organization (Suzuki, 1984). The establishment of "Jishu-Hozen" Autonomous maintenance system to foster operators that have proficient skill in equipment mechanism to do "Jishu-Hozen" of own equipment. To implement this (Suzuki, 1984), TPM proposes the implementation of the following steps: Initial clean-up, countermeasures for the causes of dirty equipment and improving access to hard-to-clean areas, test maintenance standards, general inspection, autonomous inspection, standardization and all-out autonomous maintenance.

Implementing a planned maintenance system to improve the availability of the equipment has specific goals to achieve to realize efficiency in the maintenance department to eliminate 8 major losses. To support this, it is necessary to implement the daily maintenance, time based maintenance, predictive maintenance, improvement for increasing the service life expectancy, control of replacement parts, failure analysis and prevention of recurrence and blurbification control. Skill training program for operators and a maintenance team, requires specific training to raise the technical skill level of operators and maintenance (Suzuki, 1984). They are relevant to educate operators in taking care of their equipment. Basic maintenance steps, tightening the bolts and nuts, bearing maintenance, maintenance of transmission gears, leakage prevention operation, maintenance of hydraulic and pneumatic equipment and maintenance of electrical control systems.

The pillar, establishment of initial phase management system for products and equipment, is a fundamental pillar for the development and acquisition of technologies with higher levels of maturity, where improvements have been incorporated with information collected by this pillar. With this information, suppliers have conditions to shorten new equipment for development design and manufacturing periods and realize minimum run-up time and stable launching of new products and equipment operation. Higher levels of maturity mean: easy to manufacture, easy to use, maintainability, reliability and review the design. Establishment of Quality Maintenance system is a pillar to realize "Zero-defects" through observing the required equipment maintenance, confirm the standard for quality characteristics and recognize the defect record and assure product quality, investigate the unit process, material and energy utilization, equipment conditions in the production method (Suzuki, 1984). TPM in the office establishes an efficiency achievement system as a pillar to support improvements to the production department like making the operation more efficient, standardizing and innovating production control systems. The eight pillar, establishment system for safety/health and good working environment helps to keep the "zero-accident" level, create a healthy and clean working area, safety measures to protect the operators from equipment accidents, improve the working environment (noise, vibration and dust) and globally make the operation safe (Suzuki, 1984). The best models for TPM implementation in organizations are co-ordinated by a multidisciplinary team, with horizontal organisation and regular meetings. The systematic methodology is based on a practical problem definition, prioritizing activities and always making the improvements based on standards.

It is necessary to check the implementations on the shop floor and results have to be measurable (EFQM User Guide, 2013). The associates involved with TPM activities, are responsible for driving and overseeing TPM activities factory-wide. The following tasks for the planned maintenance are critical for the TPM success: Clear definition of the best sequence of activities, separating the activities/time for each equipment and operator in the complete production line; Work descriptions per equipment, specifying tools, equipment spots and visual instructions for all activities; training the operators in maintenance, change over and standard work activities. Increase the operators' involvement in designing the best sequence of activities and implementation of standards: cleaning, inspection, change over, maintenance activities, process confirmation with analysis of deviations, continuous improvement and standards optimization. These points are relevant to value their operators and create a culture of empowerment for achievement of both organizational and personal targets.

4. TPM AND A CONNECTED INDUSTRY (INDUSTRY 4.0)

The internet of things is already a reality. Industry 4.0, otherwise known as the fourth industrial revolution, integrates people and digitally connected machines with the internet and information technology. The Industry 4.0 describes the networked manufacturing of the future. The result is the "smart factory", which is categorized by versatility, resource efficiency and ergonomic design. Fabrication laboratories can help on development of new business models with new digital technologies (Santos et al., 2018,c; Bravi et al., 2017; Bravi et al., 2018).

Humans, machines, objects and systems are connected via internet and communicate in a dynamic real time optimized. Production data is aggregated in real time for continuous improvement of processes. Associates are key players in this concept, and their work is facilitated to a greater degree than ever by software-based systems. Human-machine collaboration will increase in a normal way and people's health and well-being will be safeguarded through adaptive workplace ergonomics. With this collaboration, we can make processes more efficient, increase utilization and improve the competitiveness of the organizations. The Industry 4.0 applications play today a strong role for the TMP implementation. Mobile IT solutions support the operators and maintenance engineers to increase the speed and quality of all kinds of maintenance. The Industry 4.0 connected industry approach is connecting sensors that observe the equipment functions in a real time, and the maintenance will be done if it is necessary. With this, there are clear benefits: maintenance employees and operators will be informed about the current state of the equipment's and the availability will be increase. The big data available for the TPM teams, helps them make quick decisions in order to maintain higher availability and higher outputs in production lines. In Industry 4.0, people are the key players. Real-time big data will not take away people's power to make decisions or their responsibility, but it will support people by providing relevant information in real time, thus enabling continuous improvement of the production processes. The well-organized management of information is a critical pillar in a global performance of the modern organizations that means the quality of the information can affect results (EFQM, 2012; EFQM, 2013).

To implement successfully quality systems, organizations need to have the adequate information system that provides transparent information that supports the business development in an adequate way.

5. QUALITY CONTINUOUS IMPROVEMENT - LEADING EXCELLENCE WITH TQM AND IMS

The lean production concept identifies extremely efficient and effective production systems that consume fewer resources, creating higher quality (Santos & Milán, 2013; Marques et al., 2018; Santos & Barbosa, 2006) and lower cost as outcomes. Using both practical and project-based perspectives, a key strategy is the elimination of waste (Pettersen, 2009).

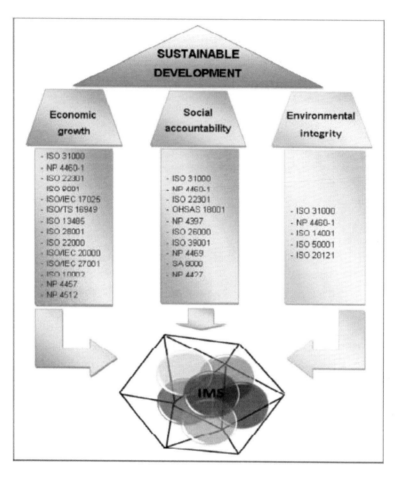

Figure 2. Model of Sustainable Development with IMS - Integrated Management Systems (Rebelo et al., 2016a).

In 1980, the new management practices linked to the TQM (Total Quality Management) philosophy began to emerge in the U.S., in a response from its industrial sector to a global competition where the Japanese companies with its management methodologies, supported by high quality of their products and services, disrupt the US companies' market shares (Sá & Oliveira, 2013). TQM became an interesting concept in the beginning of the 1990s in order to describe how organizations should work to obtain better performance and customer satisfaction. In addition, TQM is often associated with the figures within the field of quality management, for example, Deming and Juran. TQM is perceived as a targeted methodology to adding value to the customers by producing outstanding products and services and by improving their satisfaction (Sá & Oliveira, 2013). TQM can also be seen as a philosophy where the whole organization is involved in continuous improvement processes. The main pillar of TQM is recognizing that associates are the organizations' main assets that means their commitment with the continuous improvement process is imperative (Sá & Oliveira, 2013). EFQM Model is a well-known TQM approach and is a common framework that helps all organizations to improve their businesses. The EFQM Excellence Model offers a holistic view of the organizations, highlighting its strengths and areas for improvement (EFQM, 2013). The EFQM excellence model is adaptable and can be used to assess, to improve an entire organization or just segments, improving their competitiveness of the organizations and deliver value propositions based on their needs, delighting their customers beyond expectations. To achieve the demanding targets regarding quality of services and products the organizations have to understand the basis for managing with agility, and attract new talents to enable them to anticipate and understand their external environment.

According Santos et al. (2011) and Rebelo et al. (2016a), effective quality management within companies, integrated with other management areas like environmental and occupational health and safety, is nowadays assumed to be a strategic way to implement and improve Lean and cleaner production. Thus, the IMS (Integration of Management Systems) take on relevant importance for the implementation of Lean methodologies in organizations (Rebelo et al., 2016b; Ribeiro et al., 2017; Carvalho et al., 2018, Rebelo et al., 2017; Santos et al., 2017). Also, SD (Sustainable Development) and business sustainability can be achieved through a better coordinated management of processes versus associated resources (Santos et al., 2015; Santos et al., 2018,a). A model of SD and Integrated Management Systems is presented in the Figure 2. It is a good way to achieve excellence in business (Santos et al., 2018,b).

6. LEAN SAFETY - INTEGRATING LEAN PRODUCTION WITH SAFETY MANAGEMENT

The new standard ISO 45001 (2018) - Occupational health and safety management systems - Requirements with guidance for use - intends to improve working conditions, management practices, internal and external security communication (for example, safety rules and processes) and ensures compliance with legislation. In addition, its implementation reduces the likelihood of accidents and disruptions in the production process and improves internal climate, image and overall performance.

The implementation of Lean Production Systems is deemed essential for companies that wish to obtain high levels of competitiveness (Tortorella & Fogliatto, 2014). Lean Production adoption entails significant organizational change which requires companies to properly manage the key factors that might influence on the success of the adoption process (Martínez-Jurado et al., 2014). Many health care organizations are already using Lean management techniques to improve quality, patient safety, and cost-effectiveness (Thomas, 2015.) Integrating Lean Production with Safety Management can have positive impacts related to the effects of lean production on quality and productivity. On the other hand, lean production can have a negative impact on working conditions (Delbridge et al., 2000; Saurin & Ferreira, 2010). Recently, a model "Lean-integrated management system for sustainability" was developed by Souza & Alves (2018) to bring to light important themes such as: ergonomics and reuse of materials and resources.

There are tools that workers can easily access and understand, such as 5S, value stream mapping (VSM) and Visual Management to improve the work environment and ergonomics, resulting in direct benefits to the worker (Martínez-Jurado et al., 2014). Nikolou-Walker &

Lavery (2009) emphasize and recommend the importance of considering safety before and when lean manufacturing is implemented. Gonçalves et al. (2019), based on the VSM and WID models, developed the Safety Stream Mapping (SSM) with the goal of integrating safety with lean. The SSM shows along the value stream, activities with high occupational hazards. These activities need to be identified because they cause waiting and non-compliant products (waste), due to the occurrence of occupational accidents and diseases.

The company may benefit from a safe working environment, a safe workforce and a streamlined production process. Through the safe implementation of Lean Manufacturing, non-value-added activities (including accidents and their associated ancillary costs) will be reduced, allowing the company to realize success in terms of improved productivity, efficiency and profitability within a safe working environment. Tools and techniques of Lean Manufacturing (e.g., just-in-time (JIT), total productive maintenance (TPM), pull production, cellular manufacturing, 5S/7S, kaizen, visual control, poka-yoke, and value stream mapping (VSM)) can also facilitate achieving sustainability (Vinodh et al., 2011).

In the literature there are few guidelines on how to combine Lean principles with safety initiatives in the workplace (Mitropoulos et al. 2007). Both conceptual and empirical research in this field link lean production to more worker injuries and worse health. However, these studies tend to look at lean as the key input and worker's health and safety as the main output, implicitly assuming that all lean implementations lead to better productivity. However, many implementations of lean failures and/or are not true for lean principles (Longoni et al., 2013). Approach to safety protection can be simplified, and sometimes eliminated, when a process improvement change removes the hazard or reduces the frequency of dangerous tasks in the workplace (Sammon, 2012). Examples include: i) saving money on Personal Protective Equipment (PPE) if the process improvement changes the procedure or exposure such that the PPE is not needed and ii) getting a production line up and running faster by simultaneously addressing both safety and production needs, both of which are essential to good manufacturing practice. Sammon (2012) quotes the following phrase of Damon Nix of the Georgia Tech Research Institute: "Safety is value-added, and hazards are waste". This idea is very useful to understand the integration of safety and process improvement in the familiar language of a Lean practitioner. When implementing lean production, we can contribute to eliminate the following eight "wastes" which have impact on safety hazards: 1) defects - increased maintenance activities, hazardous material exposure, machine exposure; 2) overproduction - overexertion, extra handling, unnecessary machine interaction; 3) waiting - setups/changeovers - hazardous energy exposure; 4) not using employee ideas - the company misses out on potential safety improvements; 5) transportation - extra handling, slip, trip and fall hazards, exposure to fork lift traffic; 6) inventory - falling loads, traffic congestion, trip hazards, extra handling; 7) motion - overexertion, poor ergonomic design and 8) extra processing - unnecessary machine interaction.

CONCLUSION

Lean Management includes a set of tools that allow companies to receive strong benefits when they implement them properly. The use of lean tools is a simple way and low cost solution to achieve productivity and profitability, using a continuous focus on the elimination of waste through all the organization. Lean Tools are easy to use. They engage all the organization and assures the commitment of all from top to down. They assure the way to empower the collaborators and turn visible all the results of theirs work. With TPM implementation, the organizations collect several benefits, because TPM is a world-class approach, which involves everyone in the organization, working to increase equipment effectiveness. TPM implementation in an organization can ensure higher productivity, better quality, fewer breakdowns, lower costs, reliable deliveries, motivating working environments, enhanced safety, and improved self-confidence of the employees.

The ultimate benefits that can be obtained by implementing TPM are enhanced productivity and profitability of the organizations. TPM aims to increase the availability of existing equipment in a good condition, reducing in that way the need for further capital investment in a new equipment's. To maximize the results, the TPM involves everyone at every level of the organization, from top management to front line employees using overlapping small group activities to achieve the target of zero losses and higher OEEs. With TPM implementation, organizations get high quality products, optimization of costs through stable processes, and short lead time due to flexible production. Aims to create a culture and environment that constantly tries to maximize the effectiveness of the entire production system (increase OEE). TPM is a key pillar in the success of Lean management. The Lean's purpose is to develop critical skills and competencies in organizations. New innovative products and services are coming for the organizations and new critical knowledge is necessary. Critical knowledge is one of the critical success factor to sustain the competitiveness of our organizations. In the world of Industry 4.0, people, machines, and products communicate with each other directly. Manufacturing processes and services are smart and connected across the boundaries of organizations to make manufacturing and services more efficient and flexible. But the success of Industry 4.0 depends largely on the maturity of organizations and knowledge regarding the application of traditional Lean tools such as TPM and others.

This work can be a guide to support organizations that wish to start their lean road. Lean, innovative technologies and big data need other dimensions to be sustainable. Quality and safety continuous improvement through lean tools, assumes a critical pillar for the new journey toward the future of our organizations.

ACKNOWLEDGMENTS

The authors thanks to the company selected, as well as all their collaboratores involved in the performed case study. The authors manifest a full acknowledgement to the Procedia Manufacturing for permission to publish Figure 1 and the Journal of Cleaner Production for permission to publish Figure 2. All figures designed by the authors of this chapter were previously published in the cited journals.

REFERENCES

Abdulmalek, J. and Rajgopal, J. (2007). Analyzing the Benefits of Lean Manufacturing and Value Stream Mapping via Simulation: A Process Sector Case Study. *International Journal of Production Economics*, 107 (1): 223-236.

Ahuja, I. P. S. and Khamba, J. S. (2008). Total productive maintenance: literature review and directions. *International Journal of Quality & Reliability Management*, 25 (7): 709-756.

Arbulu, R., Ballard, G. and Harper, N. (2003). Kanban in Construction. *11th Annual Conference of the International Group for Lean Construction*. Virginia, USA, 1-12.

Audenino, A. (2012). Kaizen and Lean Management Autonomy and Self-Orientation, Potentiality and Reality ... *Proc. 2nd International Conference on Communications, Computing and Control Applications*. 1-6. 10.1109/CCCA.2012.6417921.

Barbosa, L., Oliveira, O., Santos, G.(2018). Proposition for the alignment of the integrated management systems (quality, environmental and safety) with the business strategy. *International Journal for Quality Research*, 12: 925–940

Baykut, M. (2011). *Evaluation of Lean Systems in Rail Maintenance Operations*. Master thesis in Industrial Engineering at the Cleveland State University, October.

Berger, A. (1997). Continuous improvement and kaizen: standardization and organizational designs. *Integrated Manufacturing Systems*, 8 (2): 110-117.

Borris, S. (2006). *Total Productive Maintenance: Proven Strategies and Techniques to Keep Equipment Running at Maximum Efficiency*. New York: McGraw- Hill Education.

Bravi, L., Murmura, F., Santos, G. (2017). Attitudes and Behaviours of Italian 3D Prosumer in the Era of Additive Manufacturing. *Procedia Manufacturing*, 13: 980–986.

Bravi, l., Murmura, F., Santos, G. (2018). Manufacturing Labs: where new digital technologies help improve life Quality. *International Journal for Quality Research*, 12: 957–974.

Bupe, G. M. and Mbohwaa, C. (2015). Design of a total productive maintenance model for effective implementation: Case study of a chemical manufacturing company. *Procedia Manufacturing*, 4: 461-470.

Cabral, I., Grilo, A. and Cruz-Machado, V. (2012). A decision-making model for Lean, Agile, Resilient and Green supply chain management. *International Journal of Production Research*, 50 (17): 4830-4845.

Carvalho, F., Santos, G. and Gonçalves, J. (2018). The disclosure of information on sustainable development on the corporate website of the certified Portuguese organizations. *International Journal for Quality Research*, 12(1): 253-276.

Chiarini, A. and Vagnoni, E. (2015). World-class manufacturing by Fiat. Comparison with Toyota Production System from a Strategic Management, Management Accounting, Operations Management and Performance Measurement dimension. *International Journal of Production Research*, 53(2): 590-606.

Coimbra, E. (2009). *Total Flow Management: Achieving Excellence with Kaizen and Lean Supply Chains*. 1st ed. Kaizen Institute.

Coimbra, E. A. (2005). Introduction to Alternative Logistics. *KAIZEN Forum*, 1-4.

Coleman, B. J. and Vaghefi, M. R. (1994). Heijunka: A Key to the Toyota Production System. *Production and Inventory Management Journal*, 35 (4): 31-35.

Delbridge, R., Lowe, J. and Oliver, N. (2000). Shopfloor responsibilities under lean team working. *Human Relations*, 53 (11): 1459-1479.

Demeter, K., Szász, L. and Rácz, B. (2016). The Impact of Subsidiaries' Internal and External Integration on Operational Performance. *International Journal of Production Economics*, 182: 73-85.

Dennis, P. (2006). *Lean Production Simplified: A Plain-Language Guide to the World's Most Powerful Production System*. Boca Raton, FL: CRC Press.

Dias, T. (2012). *Project Increase Productivity and Reduce Inventory*. Kaizen Institute Consulting Group.

Dillon, A.P. and Shingo, S. (1985). *A Revolution in Manufacturing: The SMED System*. New York: Productivity Press.

Doiro, M., Fernández, F. J., Félix, M. J., Santos, G. (2017). ERP- machining centre integration: a modular kitchen production case study. *Procedia Manufacturing*, 13: 1159–1166

EFQM (2013). *User Guide, Using Lean within the EFQM Framework*.

EFQM-European Foundation for Quality Management (2012). EFQM Excellence Model. *EFQM Information Brochure*.

Galsworth, G. (2004). The value of vision. *Industrial Engineer*, 36(8): 44-49.

Gapp, R., Fisher, R. and Kobayashi, K. (2008). Implementing 5S within a Japanese context: an integrated management system. *Management Decision*, 46 (4): 565-579.

Gomes, F. (2012). *Kanbans and Consignment Implementation Project (CMI / VMI) with Suppliers*. Master's thesis in Industrial Engineering and Management. EE - University of Minho.

Gonçalves, I., Sá, J. C., Santos, G. and Gonçalves M. (2019). Safety Stream Mapping—A New Tool Applied to the Textile Company as a Case Study. *Occupational and Environmental Safety and Health*, 202: 71-79.

Grimaud, F., Dolgui, A. and Korytkowski, P. (2014). Exponential Smoothing for Multi-Product Lot-Sizing With Heijunka and Varying Demand. *Management and Production Engineering Review*, 5 (2): 20-26.

Hall, R.W. (1987). *Attaining Manufacturing Excellence: Just in Time Manufacturing, Total Quality, Total People Involvement*. New York: Dow Jones- Irwin.

Heizer, J. and Render, B. (2004). *Principles of Operations Management*. Upper Saddle River, NJ: Pearson/Prentice Hall.

Howell, V. W. (2014). *Yokoten: Multiplying Lean Success*. May, 29-31.

Hugos, M. (2003). *Essentials of Supply Chain Management*. Edited by John Wiley & Sons.

Ikonen, M. (2011). *Lean Thinking in Software Development: Impacts of Kanban on Projects*. PhD Thesis. Department of Computer Science. University of Helsinki Finland.

Imai, M. (1997). *Gemba Kaizen. A Commonsense, Low-Cost Approach to Management*. New York: McGraw-Hill.

Jadhav, R. M., Alessandro, M. and Sawant, S. H. (2009). Total Productive Maintenance Theoretical Aspect: A Journey towards Manufacturing Excellence. *Journal of Mechanical and Civil Engineering*, 51-59.

Jones, R., Latham, J. and Betta, M. (2013). Creating the Illusion of Employee Empowerment: Lean Production in the International Automobile Industry. *The International Journal of Human Resource Management*, 24 (8): 1629-1645.

Krafcik, J. F. (1988). Triumph of the Lean Production System. *Sloan Management Review*, 30 (1): 41-52.

Liker, J. K. (2004). *The Toyota Way: 14 Management Principles from the World's Greatest Manufacturer*. New York: McGraw-Hill.

Liker, J. K. and Meier, D. (2006). *The Toyota Way Fieldbook: A Practical Guide for Implementing Toyota's 4Ps*. New York: McGraw-Hill.

Longoni, A., Pagell, M., Johnston, D. and Veltri, A. (2013). When does lean hurt? - an exploration of lean practices and worker health and safety outcomes. *International Journal of Production Research*, 51 (11): 3300-3320.

Machikita, T., Tsuji, M. and Yasushi Ueki, Y. (2016). Does Kaizen Create Backward Knowledge Transfer to Southeast Asian Fi Rms? *Journal of Business Research*, 69 (5), 1556-1561.

Marksberry, P., Badurdeen, F., Gregory, B. and Kreafle, K. (2010). Management Directed Kaizen : Toyota's Jishuken Process for Management Development. *Journal of Manufacturing Technology Management*, 21 (6): 670-686.

Marques, C., Lopes, N., Santos, G., Delgado, I and Delgado, P. (2018). Improving operator evaluation skills for defect classification using training strategy supported by attribute agreement analysis. *Measurement*, 119: 129-141.

Martínez-Jurado, P., Moyano-Fuentes, J. and Jerez-Gómez, P. (2014). Human resource management in Lean Production adoption and implementation processes: Success factors in the aeronautics industry. *BRQ Business Research Quarterly*, 17: 47-68.

Matzka, J., Di Mascolo, M., Kai Furmans, F. (2012). Buffer Sizing of a Heijunka Kanban System. *Journal of Intelligent Manufacturing*, 23 (1): 49-60.

McDonald, T., Van Aken, EM. and Rentes, A. F. (2002). Utilising simulation to enhance value stream mapping: a manufacturing case application. *International Journal of Logistics*, 5 (2): 213-32.

Melton, T. (2005). The Benefits of Lean Manufacturing: What Lean Thinking Has to Offer the Process Industries. *Chemical Engineering Research and Design*, 83 (6 A): 662-673.

Mitropoulos, P., Cupido, G. and Namboodiri, M. (2007). Safety as an emergent property of the production system: How lean practices reduce the likelihood of accidents. *Proceedings IGLC*-15. Michigan, USA.

Mobley, R., Keith, R. H. and Wikoff, D. J. (2008). *Maintenance Engineering Handbook*. 7th ed. New York, McGraw-Hill.

Monteiro, M., Pacheco, C., Carvalho, J. and Francisco Paiva, F. (2015). Implementing Lean Office: A Successful Case in Public Sector. *FME Transactions*, 43: 303-310.

Moore, S. M. J. and Gibbons, A. (1997).Is lean manufacture universally relevant? An investigative methodology. *International Journal of Operations & Production Management*, 17 (9): 899-911.

Nemoto, T., Hayashi, K. and Hashimoto, M. (2010). Milk-Run logistics by Japanese automobile manufacturers in Thailand. *Procedia Social and Behavioral Sciences*, 2: 5980-5989.

Nieminen, H. (2016). *Improving Maintenance in High-Volume Manufacturing. Case: Ball Beverage Packaging Europe*. **Master's** Thesis. Lahti University of Applied Sciences, Finland.

Nikolou-Walker, E. and Lavery, K. (2009). A work-based research assessment of the impact of 'lean manufacturing' on health and safety education within an SME. *Research in Post-Compulsory Education*, 14 (4): 441-458.

Ohno, T. (1988). *Toyota Production System: Beyond Large-Scale Production*. New York: Productivity Press.

Oliveira, J., Sá, J. C. and Fernandes, A. (2017). Continuous improvement through Lean Tools: An application in a mechanical company. *Procedia Manufacturing*, 13: 1082-1089.

Pakdil, F. and Leonard, K. M. (2014). Criteria for a lean organisation: development of a lean assessment tool. *International Journal of Production Research*, 52(15): 4587-4607.

Pettersen, J. (2009). Defining Lean Production: Some Conceptual and Practical Issues. *The TQM Journal*, 21 (2): 127-142.

Rebelo, M. F., Santos, G and Silva, R. (2016,a). Model based integration of management systems (MSs) - case study. *The TQM Journal*, 28 (6): 907-932.

Rebelo, M. F., Santos, G. and Silva, R. (2016,b). Integration of management systems: towards a sustained success and development of organizations. *Journal of Cleaner Production*, 127: 96-111.

Rebelo, M., Santos, G. and Silva, R. (2017). The integration of standardized Management Systems: managing Business Risk. *International Journal of Quality & Reliability Management*, 34 (3): 395-405.

Ribeiro, F., Santos, G., Rebelo, M. and Silva, R. (2017). "Integrated Management Systems: trends for Portugal in the 2025 horizon." *Procédia Manufacturing*, 13: 1191-1198.

Rother, M. and Shook, J. (2003). Learning to See: Value Stream Mapping to Add Value and Eliminate Muda. Massachusetts: *Lean Enterprise Institute*, Massachusetts.

Sá, J. C., Carvalho, J. D. and Sousa, R.M. (2011). Waste Identification Diagrams". *CLME'2011-IIICEM – 6º Congresso Luso-Moçambicano de Engenharia – 3º Congresso de Engenharia de Moçambique*, Moçambique, Maputo, 207–208.

Sá, J. C., Oliveira, J. (2013). Generating Value with TQM and Six Sigma. *IRF 2013 - 4th International Conference on Integrity, Reliability and Failure*. Funchal, Portugal.

Sammon, T. (2012). How Will Combining Safety and Lean in Process Improvement Efforts Save Money?. *Georgia Manufacturing Extension Partnership*. [retrieved July 2018]. URL: http://gamep.org/wp-content/uploads/2012/08/How-Will-Combining-Safety-and-Lean-in-Process-Improvement-Efforts-Save-Money. pdf.

Santos, D., Rebelo, M. and Santos, G. (2017). The Integration of certified Management Systems. Case Study - Organizations located at the district of Braga, Portugal. *Procedia Manufacturing*, 13: 964-971.

Santos, G. and Barbosa, J. (2006). QUALIFOUND - a modular tool developed for Quality Improvement in Foundries. *Journal of Manufacturing Technology Management*, 17 (3): 351-362.

Santos, G. and Milán, A. L. (2013). Motivation and benefits of implementation and certification according ISO 9001 - The Portuguese experience. *International Journal for Quality Research*, 7 (1): 71-86.

Santos, G., Afonseca, J., Lopes, N., Félix, M. J. and Murmura, F. (2018,b). Critical success factors in the management of ideas as an essential component of innovation and business excellence.*International Journal of Quality and Service Sciences*, 10 (3): 214-232.

Santos, G., Bravi, L. and Murmura, F. (2018,a). SA 8000 as a Tool for a Sustainable Development Strategy. *The Corporate Social Responsibility and Environmental Management Journal*, 25: 95-105.

Santos, G., Mendes, F. and Barbosa, J. (2011). Certification and integration of management systems: the experience of Portuguese small and medium enterprises. *Journal of Cleaner Production* 19 (17-18): 1965-1974.

Santos, G., Murmura, F. and Bravi, L. (2018,c). Fabrication laboratories: The development of new business models with new digital technologies. *Journal of Manufacturing Technology Management*, 29 (8): 1332-1357.

Santos, G., Rebelo, M., Barros, S., Silva, R., Pereira, M., Ramos, G., Lopes, N. (2014). Developments regarding the integration of the Occupational Safety and Health with Quality and Environment Management Systems, Chapter of the book *Occupational Safety and Health – Public Health in the 21st Century*, edited by Ilias Kavouras and Marie-Cecile G. Chalbot. New York: Nova Science Publishers – Chapter 6: 113-146.

Santos, G., Rebelo, M., Lopes, N., Alves, M. R. and Silva, R. (2015). Implementing and certifying ISO 14001 in Portugal: motives, difficulties and benefits after ISO 9001 certification. *Total Quality Management & Business Excellence*, 27 (11-12): 1211-1223.

Shimbun, N. K. (1989). *Poka-Yoke: Improving Product Quality by Preventing Defects*. Portland, Oregon: CRC Press.

Shingo, S., Andrew, P. and Dillon, A. P. (1989). *A Study of the Toyota Production System: From an Industrial Engineering Viewpoint*. Boca Raton (FL): CRC Press.

Shook, J. (2008). Managing to Learn: Using the A3 Management Process to Solve Problems, Gain Agreement, Mentor and Lead. *Lean Enterprises Inst Inc*; Pap/Chrt edition, June.

Shook, J. (2009). Toyota's Secret. *MIT Sloan Management Review*, 50(4): 30-33.

Souza, J. and Alves, J. (2018). Lean-integrated management system: A model for sustainability improvement. *Journal of Cleaner Production*, 172: 2667-2682.

Stevenson, J. (2007). *Operations Management*. 9th ed. Boston (MA): McGraw Hill-Irwin.

Suzuki, T. (1984). *TPM in Process Industries*. New York: Productivity Press.

Tortorella, G. L. and Fogliatto, F. S. (2014). Method for assessing human resources management practices and organisational learning factors in a company under lean manufacturing implementation. *International Journal of Production Research*, 52 (15): 4623-4645.

Vinodh, S., Arvind, K. R. and Somanaathan, M. (2011). Tools and techniques for enabling sustainability through lean initiatives. *Clean Technol. Environ. Policy*, 13: 469-479.

Wilson, L. (2010). *How to Implement Lean Manufacturing*. New York: Mc-Graw Hill.

Womack, J. P. and Jones, D. T. (2003). *Lean Thinking - Banish Waste and Create Wealth in Your Corporation*. New York: Free Press.

Womack, J. P., Jones, D. T. and Roos, D. (1990). *The Machine that Changed the World: The Story of Lean Production*. New York: Macmillan Publishing Company.

In: Lean Manufacturing
Editors: F. J. G. Silva and L. Carlos Pinto Ferreira
ISBN: 978-1-53615-725-3
© 2019 Nova Science Publishers, Inc.

Chapter 9

LEAN MANUFACTURING APPLIED TO THE PRODUCTION AND ASSEMBLY LINES OF COMPLEX AUTOMOTIVE PARTS

Conceição Rosa[1], Francisco J. G. Silva[1,*], Luís Pinto Ferreira[1] and J. C. Sá[2]

[1]DME – Departement of Mechanical Engineering,
ISEP – School of Engineering, Polytechnic of Porto, Porto, Portugal
[2]Polytechnic Institute of Viana do Castelo,
School of Business Sciences, Viana do Castelo, Portugal

ABSTRACT

Lean Manufacturing tools have known a wide implementation in the automotive industry, since the flexibility and agility required to this industrial sector have grown in a extremely significant fashion. Despite the strong automation associated with body-in-white work and automotive assembly, many component manufacturers still deal with very handcrafted processes, where being competitive requires applying Lean tools capable of meeting the demands of small series, high quality and relatively short lead times.

This chapter intends to describe the application of Lean tools in a mechanical cable production industry, where the multiplicity of models aggregated to each production line, as well as the need to save time on the dedicated lines, required a careful analysis and the application of several Lean tools in these lines, in order to eliminate wastes of time and to recover the competitiveness of the product. Some Lean tools were used, such as Value Stream Mapping, Visual Management, 5S, Standard Work, SMED and the PDCA cycle, which, together or in a single way, allowed productivity gains of more than 40% in any of the lines, being that all these lines had quite different characteristics and needs. It has been proven that, when properly applied, these tools have extreme usefulness, leading to positively surprising results.

---

[*] Corresponding Author's E-mail: fgs@isep.ipp.pt.

Keywords: lean tools, automotive components industry, value stream mapping, SMED, PDCA, visual management, standard work, assembly lines, workstation, productivity

1. INTRODUCTION TO AUTOMOTIVE INDUSTRY

The automotive industry currently generates a great impact on world economy, presenting itself as a sector which is completely globalized, subject to constant development, and at the forefront of technology. It is one of the most important industrial activities worldwide, thus playing a fundamental role in the economy of the most industrialized countries, and already occupies a relevant position in developing economies. It is the true "industry of all industries" (Rosa et al. 2018, 555-556) (Costa, Silva, and Campilho 2017, 4043-4044).

1.1. The Global Context of the Automotive Industry

Dating back to the end of the 19th century, the origin of the automotive industry was essentially based on craft production; each car was always unique and built to suit the customer's preferences. It was, however, short-lived due to its lengthy manufacturing process, the small quantities of items produced and its elevated cost. The problems experienced by the industry were surpassed by Henry Ford in 1913 when he conceived his first assembly line, with a continuous work flow (Rosa, Silva, and Ferreira 2017, 1036). By using this technique, designated as "mass production," Ford drastically reduced costs, which subsequently led to a drop in car prices. However, supply to the customer was limited to the type of car produced in huge quantities by a standardized assembly line (Kocakülâh, Brown, and Thomson 2008, 17).

Lean production has made it possible to associate the benefits of craft production, in which the customer's specifications are produced exactly with those of mass production, where great quantities are manufactured. This allows for the production of a great number of goods which present a wide variety (Singh et al. 2010, 158).

The timeline in Figure 1 consists of a representation of some of the most significant milestones in the automotive industry over the years. It also presents the development of different production systems, from the period of craft production to the present day (Clarke 2005, 71-113).

Mass production flourished and besides the "Big Three" in the North American automotive industry - Ford, General Motors and Chrysler - the system was adopted by practically all companies in the automotive industry on a worldwide scale. The year in which mass production reached its peak was also the year when the three North American companies began to decline. The most significant cause was that many other companies - Renault, Fiat, Mercedes - produced on a similar scale but offered a much greater diversity in products. Subsequently, a stagnation period occurred in mass production, both in North America as well as in Europe. This would have continued indefinitely if another new industrial philosophy had not emerged in Japan: the Toyota Production System (TPS) (Pyzdek, Keller, and Dekker 2003, 210) (Goicoechea and Fenollera 2012, 621).

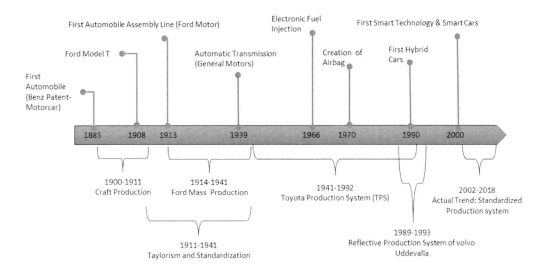

Figure 1. The principal milestones (above) and the principal production systems (below).

TPS provided the grounding for Lean thinking, the principles which underpin most of the production systems in the automotive industry. Replicated over time by many companies (Neves et al. 2018, 697), this consists of a culture which includes training and practice. It makes use of methods and tools to eliminate waste, motivate teams, enhance equipment and increase productivity. Its main objective resides in generating the continuous improvement of processes and people (Rosa, Silva, and Ferreira 2017, 1036-1037) (Correia et al. 2018, 664).

1.2. The Importance of the Automotive Sector in World Economy

The automotive industry is today a driver of world economy, and contributes significantly to each country's economic activity, influencing different sectors which range from textile production to the metalwork industry. It deals with a wide range of technologies, skills and organizational processes, with a view to the development and production of components, modules and systems, and is set within the logic of a product which is complex, global and integrated (Machado and Moniz 2007, 3-4) (ATKearney 2013, 3).

The impact of this industry on the economy is clearly demonstrated by business figures and the great number of job positions generated. According to data from the International Organization of Motor Vehicle Manufacturers (OICA), in the year 2017, this sector was responsible for the existence of 12.5 million job positions worldwide, as well as the production/assembly of a massive number of vehicles (over 95 million) (OICA 2017). Since 2010, there has been an upward trend in the sector, which is demonstrated by the figures presented in Figure 2.

Variation in automotive production figures is quite common, since countless parties are involved in this industry. As such, various factors must be considered, such as the consumers' power of acquisition, or even the price of petrol, which influences transportation. The global financial crisis in 2008 caused an unprecedented downturn in the market. Automotive manufacture has, however, gradually recovered since the second trimester of 2009, which is chiefly due to the direct support provided by governments, as well as greater demand (OECD

2011). Europe, the United States and Japan/Korea have been the main producers of motor vehicles on a worldwide scale. In the last few years, however, one has seen some changes to this regard, and China now occupies the position of the greatest car manufacturer in a global context (Figure 3) (ACEA 2018, 17-20).

The growth of the Asian market, and especially that of China, can be seen in Figure 4, where sales have almost quadrupled in the last ten years, reaching the impressive figure of approximately 25 million vehicles in 2017 (Scotiabank 2018).

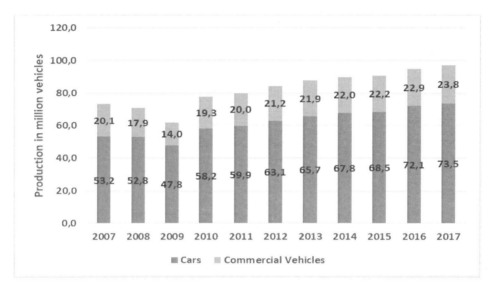

Figure 2. Vehicles produced worldwide annually, from 2007 to 2017.

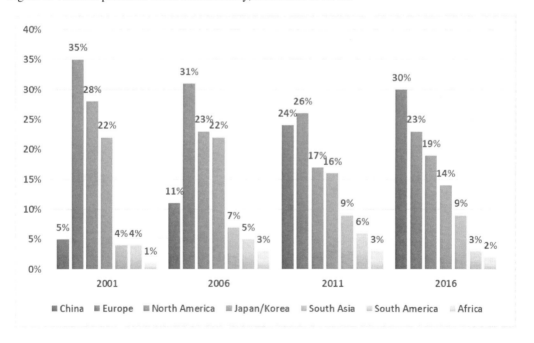

Figure 3. Percentage of participation in motor vehicle manufacture, on a world scale (ACEA 2018, 17-20).

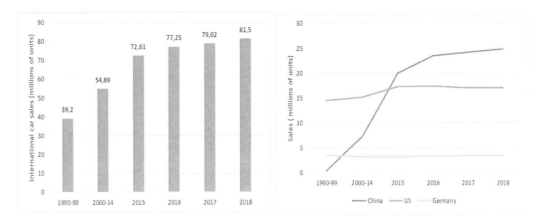

Figure 4. Vehicles sold worldwide from 1990 to 2018 (left), and vehicles sold in the three most significant countries worldwide from 1990 to 2018 (right) (Scotiabank 2018).

The automotive industry plays a crucial role in generating employment, with a job multiplier potential of between five and seven, whereas this figure is about three in other industries. The automotive industry in Europe is responsible for 12.6 million jobs; 2.5 million of these are direct job positions and 10.1 million are indirect (ACEA 2018, 13). Another way of measuring the importance of the automotive industry in economies is through its percentage of contribution to the GDP. On a worldwide scale, the sector contributes with approximately 3% to the GDP, with an even greater participation in the case of emerging markets (Blatt 2017, 14).

2. THE MANUFACTURE OF COMPONENTS FOR THE AUTOMOTIVE INDUSTRY

The automobile is manufactured by a global industry, with a hierarchically flexible but structured value chain, and is a significant part of everyday life in communities. Cars today contain a high number of components obtained from a highly competitive market and from a wide range of companies which supply automotive parts. This includes a great spectrum of sectors, from electronics, chemicals and plastics, to metal and rubber materials, amongst others (INTELI 2005) (Moniz 2008, 18).

All over the world, large corporate groups have triggered growth by acquiring other companies, and have thus managed to compete on a worldwide scale. This global growth, which was successfully achieved through the acquisitions and fusions mentioned previously, has led to an effort to consolidate the value chain. By resorting to strategies such as modularization, outsourcing, as well as the sharing of platforms and components, it has been possible to establish efficient scale economies in assembly. Furthermore, it has also become fundamental to share responsibilities with suppliers regarding the areas of design, development and manufacture (INTELI 2005).

By intervening within a network logic - OEMs (Original Equipment Manufacturers), suppliers, customers and support structures - the sector is of decisive importance in world economy as a complex system of value (Veloso, Henry, and Roth 2000, 2-3). These organizations exercise different roles in the automotive industry supply chain. The position assumed by each of these is determined by each company's participation in the final product, the level of technology used, the complexity of the product supplied, the complexity of production and its activities as a supply company (Lampón and Lago-Peñas 2013, 2-3).

The supply chain is divided into four distinct groups (Lesková and Kovácová 2012, 96-98) (Goicoechea and Fenollera 2012, 624):

- OEMs (Original Equipment Manufacturers): are responsible for most of the aspects related to the design of the car, the production of engines and transmissions, as well as the assembly of these on their own premises. These companies are big employers and, as such, they possess great negotiation power along the entire supply chain.
- TIER 1 (First tier suppliers): are mostly suppliers who are globally positioned and possess a production or assembly capacity which is very similar to that of the OEMs. They are the direct suppliers of components, sub-modules, modules and systems for assembly lines linked to car brands. They play an increasingly significant role in the areas of design, production and foreign investment.
- TIER 2: are TIER 1's key suppliers and do not supply the OEMs directly. Besides producing simple components, they can supply TIER 1 with low integration level sub-modules. The location of their production or assembly premises is usually in close proximity to those of TIER 1. They may possess a global or regional scope.
- TIER 3: are local companies which manufacture parts or components for those in TIER 2 and, occasionally, for those in TIER 1. They simply comply with the quality and quantity criteria for parts or components made of plastic, metal, aluminium, and so on. All of these suppliers are usually assessed by the OEMs, which evaluate prices, quality, delivery deadlines and operational performance.

In order to attain high levels of efficiency, the component industry is subjected to pressure; it must thus adapt to the OEMs or higher-level suppliers' forms of organization and methodologies of production. This pressure is revealed both in the cost and final prices, as well as in the general quality of the products themselves. This pressure is also extended to subcontracted companies (Araújo et al. 2017, 1539) (Sabbagha et al. 2016, 70).

The OEMs' business strategies impact greatly on supply companies (especially those of TIER 1). As OEMs displace their activities to emerging countries, suppliers tend to do the same due to the need to comply with delivery deadlines and the associated costs of logistics. In the case of suppliers who have business dealings with several OEMs, one of the ways of minimizing costs resides in the selection of a central location. This is of great benefit since it enables them to serve their customers more rapidly and efficiently (Lampón and Lago-Peñas 2013, 5-6)

The automotive industry has been subjected to stringent requirements with regard to: product quality, productivity, competitiveness and steady improvement. Standardization fosters a reduction in costs for suppliers and customers, enhances market transparency, helps to create new businesses and maintain those already in existence. It is a way of assuring customers that the products/services are of suitable quality and safety, and that they respect the environment.

With the purpose of providing a guarantee for their products, OEMs periodically carry out audits of their suppliers. These audits help to detect discrepancies and failures in the requirements demanded of the products and, at times, of the process itself. This control reduces reorganization costs, in the case of failures, and minimizes the risk of a loss in the product characteristics demanded by the consumers. In this context, motor vehicle manufacturers require that suppliers comply with the stringent technical specifications, which are established by the quality management norms for suppliers of the automotive sector, known as IATF 16949:2016.

Through this specification, the norms for quality systems in the automotive sector have been replaced and standardized. Also identified are the requirements which must be met by the quality systems in the areas of design/development, manufacture, installation and service, related to any product in the automotive sector. The level of competition in the motor vehicle sector has led to the certification of suppliers, which now constitutes a customer requirement and is, therefore, no longer optional (Goicoechea and Fenollera 2012, 621-623)(AICEP 2016, 12-13).

3. COMMAND CABLES FOR CAR VEHICLES

Command cables are used as the mechanical means of movement transmission between two or more systems (Pinto and Silva 2017, 519). The fact that one can open a car door by activating a device is a good example of this. The movement of the device will activate a cable, which will transmit this mechanical impulse to the car's locking system, thus opening it.

Regardless of its makeup and position in the vehicle, the purpose of this device is to transmit a trigger force to the receptor system, which will activate the system, as illustrated in Figure 5.

Figure 5. The cable as a mechanical means of transmitting movement.

On request, two different cable actions are activated: the pull system and the push/pull system.

- The Pull System - The cables transmit the force as a form of traction to a mechanism. These are used in accelerators, trunk doors, hand brakes, fuel lids, clutches, cruise control, etc.

- The Push/Pull System - The cables transmit the force as a form of traction and compression to a mechanism. It is used in systems which require a two-way force, as is the case of the open/close system in air-conditioning.

When compared to other systems, the use of cables as a means of movement transmission constitutes a great advantage, since they can be adapted and shaped to fit different geometries. This feature allows for their adaptation to the layouts of various drive systems (Hi-Lex 2006), which are indispensable in motor vehicles, as is illustrated in Figure 6.

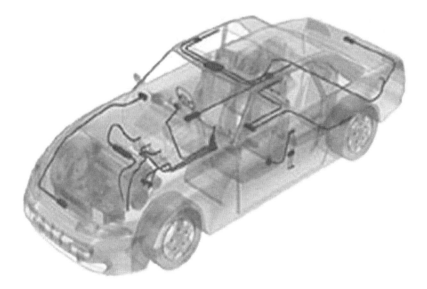

Figure 6. Systems using command cables.

The systems used to command and control movement by means of cables in vehicles include: clutch cables, brake cables, accelerator cables, adjustment of window-opening cables, seat command cables, fuel lid release cable, trunk release cable, door opener cable, etc.

3.1. The Makeup of Command Cables

The basic composition of command cables consists of (Figure 7): inner cable/metal cable (1), inner end piece/cable end piece (2), outer casing/conduit (3) and the outer end piece/conduit end piece (4). However, various components of different characteristics are usually integrated, depending on the function, position, force, number of requests, and so forth (Moreira et al. 2017, 1386-1387).

Figure 7. Command cable.

Table 1. Command cable components

	Type	Name of Part	Shape of Part	Functions
1	Inner cable/Metal cable	Single-Twisted Core		Transmits power and displacement
		Multi-Twisted Core		
		Coat Core		
2	Outer Casing/Conduit	Flat wire		Guides and protects the inner cable
		Armored		
		Braided		
3	Inner cable end fitting/metal cable end fitting	T-end		Positions the metal cable in the mechanism
		Nipple end		
		U-fitting		
		I-end		
		Screw end		
4	Outer casing end fitting/conduit end fitting	Casing cap		Positions the conduit in the mechanism
		Screw cap		
5	Other components	Guide, bushing		Adjusts the cable to the layout
		Bend pipe		Adjusts the cable to the layout
		Protector		Prevents noise
		Clamp (bracket)		Adjusts the cable to the layout
		Damper		Prevents noise
		Boots, seal cover		Prevents the entry of waste into the conduit
		Grommets		Prevents the entry of waste into the conduit

Table 1 presents the components which often constitute the makeup of command cables (Hi-Lex 2006) (CMAcable 2015).

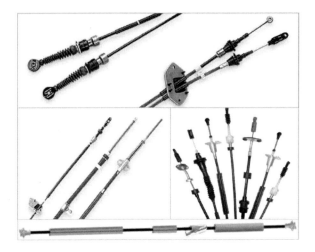

Figure 8. Examples of mechanical command cables.

4. PROBLEMS AND WASTE

Taichii Ohno and Shigeo Shingo developed the Toyota Production System (TPS) after the Second World War in order to cope with market impositions; there was little demand, yet diversity was also required (Holweg 2007, 422). The TPS system, designated as Lean manufacturing or Lean production after the publication of the seminal study by Womack & Jones, led to the Lean Thinking philosophy of the 90s. The latter consists of a production system which seeks constant improvement through the elimination of waste (Correia et al. 2018, 664).

In the different approaches used to identify waste, the purpose is to reach a condition where production capacity meets what is required by the customer (Sundar, Balaji, and Kumar 2014, 1875). In other words, companies must be in possession of the processes, materials, staff and technology to produce the right quantity of the product and/or service required by the customer at the right time (Rosa, Silva, and Ferreira 2017, 1036)

In an organization, one can identify three types of activities: value adding activities - those which, in the customer's perspective, make the product more valuable; non-value adding activities - those which do not represent any value for the end customer and are unnecessary, thus constituting pure waste which must therefore be eliminated in the short term; and non-value adding activities which are necessary - those which are of no value to the end customer but which are, however, required for the current production process. In this last case, they constitute a necessary waste and imply radical changes to the process if they are eliminated in the short term (Duquer and Cadavid 2007, 72).

When Taiichi Ohno and Shigeo Shingo developed the TPS, they identified the seven types of waste which add no value to the product as being: surplus production, waiting time, transport, over processing, excessive stock, unnecessary movement and defects (Singh and

Singh 2015, 82). An eighth type of waste was later added, which is related to the improper utilization of human potential (Kulkarni, Kshire, and Chandratre 2014, 431).

Table 2. Eight types of waste, causes and solutions

Waste	Description	Causes	Measures/solutions
Surplus production	When the quantity produced exceeds the quantity ordered	Production of large batches Errors in demand forecasts Weak production planning Market instability	Pull production systems One-piece flow Heijunka box
Waiting time	Time during which people, equipment, products or information are on hold, without the execution of any activities.	Balancing of the line Machine action cycle Breakdowns Material shortage Lack of autonomy Production delays Operator's limited flexibility	Standardization of operations SMED Pull production systems Heijunka Box
Transport	The unnecessary movement of raw materials, unfinished goods and end products between production operations.	Layout inadequate for production requirements Use of intermediate storage zones between operations and production stages	Production in cells Kankan
Over processing	Existence of operations which are unnecessary or incorrect in relation to product design.	Badly defined production method No staff training Inadequate machines	Standardization of operations One-piece flow Training of operators
Excessive stock	The existence of surplus raw material, WIP, as well as finished products.	Weak production planning Weak balancing of production processes Unstable market	One-piece flow Heijunka Box Production in small batches Visual management techniques SMED Pull production systems
Movement	Unnecessary movements for the execution of operations.	Unstable market Badly defined work methods No staff training	5S Standardization of operations One-piece flow Staff training
Defects	Products which do not meet the customer's requirements.	Human error No standardization in inspection No Poka-yoke	Problem reduction methodologies Standardization of operations TPM Automation Poka-Yoke
Improper utilization of human potential	Not listening to workers and not making use of each person's inherent skills.	Bad management of human resources Insufficient knowledge of workers' skills	Promotion of team work Problem-solving methodology Involvement of staff in the entire process

Sources: Melton 2005, 665-666; Kulkarni, Kshire, and Chandratre 2014, 430-431; Andrés-López, González-Requena, and Sanz-Lobera 2015, 26-29.

Once the different types of waste have been identified, one must subsequently devise strategies which aim to eliminate these, thus optimizing production (Barbosa et al. 2017, 1240).

Table 2 presents the eight forms of waste, as well as the causes and the solutions which must be implemented to eliminate them.

5. BEST LEAN TOOLS DEVOTED TO COMPONENT PRODUCTION FOR THE AUTOMOTIVE INDUSTRY

Lean philosophy provides several support tools: Kaizen, VSM, Visual management, 5S, Standard Work, SMED, PDCA cycle (Plan - Do - Check - Act) are used by companies to assist in the identification of opportunities for waste reduction, and in the improvement of the efficiency of production processes (Kumar and Abuthakeer 2013, 1032) (Rocha, Ferreira, and Silva 2018, 632).

A successful Lean company should not, however, rely solely on the tools which comprise the Lean production system; it must also promote its own principles (Spear 2004, 3). Lean thinking implies altering the management of operations, and the entire change begins with a mindset: it is essential for everyone to participate in this process, which must begin at senior management level. The top level management domain at a company must become conscious and participatory; it must seek to develop a keener sense of what requires transformation in the current situation. This sense of change will gradually extend to mid-level management, until it reaches all of the staff (Rosa, Silva, and Ferreira 2017, 1037).

"Tools and techniques are not secret weapons to transform a company. The continuous success experienced by Toyota in the implementation of these tools ensues from a deeper company philosophy, based on understanding people and human motivation. Its success essentially anchors on its ability to promote leadership, teams and culture in order to create strategies, build relations with suppliers and maintain a learning organization" (Liker 2004a, 18).

This is a culture which includes training and practice; it draws on methods and tools in order to eliminate waste, motivate workers, enhance equipment and heighten productivity (Natasya et al. 2013, 1293) (Godina et al. 2018, 730).

5.1. Kaizen

The practice of Kaizen is endless and aims to eliminate everything which could be considered waste. It relies on good sense, making use of economical solutions which are supported by the workers' motivation and creativity (Singh and Singh 2009, 53-55).

Kaizen is not only based on the continuous improvements developed and implemented by experts and managers. It involves everyone, since it relies on the extensive knowledge, skills and experience of those who deal directly with the processes at hand (Halevi 2001, 197-198).

Kaizen events identify the root cause of the problems; they are action-oriented, driven by data and facts, and challenge collaborators to think outside the comfort zone, within a safe environment which is conducive to creativity and productivity (Liker 2005, 263).

5.2. VSM

VSM mirrors the path followed by a product. Based on the visual representation of all the processes involved in the flow of materials and information, this technique facilitates the detection of waste and enables one to devise a value flow, from the supplier to the customer (Rother and Shook 1999). It is a tool which allows for an approach to the improvement of processes in a systematic manner (Azizi and Manoharan 2015, 154) with the purpose of finding different types of waste, and eliminating these by means of Lean methods and techniques (Rohani and Zahraee 2015, 7).

In brief, VSM provides a picture of the overall status, and helps to define the path which should be followed in the direction of excellence (Sundar, Balaji, and Satheesh Kumar 2014, 1876-1877).

5.3. Visual Management

Visual management constitutes a Lean concept which emphasizes the placing of crucial information, in a clear and structured manner, in areas used by workers. Its objective is to enhance the efficiency and effectiveness of operations, making things visible, logical and intuitive (Liker 2004a, 172).

Regarding Gemba, the visual signs may appear as markings painted on the floor or on walls, Kanban cards, Heijunka boxes, tool shadowboards, light signals, clothing of different colours, etc. Visual systems play a fundamental role in many of the other Lean tools, examples of which are SMED (Single Minute Exchange of Die) and Standard Work, amongst others (Liker 2004a, 167).

Visual management involves the fully visible display of information on how to execute the activities required, the means to do so (tools, equipment, etc.), what the standards are, as well as the performance indicators and objectives. This is undertaken by resorting to production boards, parameters, signaling the need for assistance and the sharing of knowledge and good practices (lessons learnt). The great advantage of creating routines, which value visual management, resides in the fact that they will be understood and carried out by everyone; this will occur much more easily than in the case of written procedures. Furthermore, this system assists staff in achieving enhanced management and the control of processes: it prevents errors, waste of time and provides those involved with greater autonomy (Feld 2001, 87-90).

5.4. 5S

The 5S method is one of the most commonly used Lean tools, and is used to organize, tidy and clean workplaces (Choomlucksana, Ongsaranakorn and Phrompong 2015, 105). Originally developed in Japan, the system is based on five words, all of which begin with the letter "S" and correspond to the five stages which define the methodology. The meaning of each of these is presented in Table 3 (Miroslava et al. 2016, 331).

Table 3. 5S means

S	KeyWord	Description
Seiri	Sort/classify/clear	Remove what is unnecessary.
Seiton	Organize/tidy	"A place for each object and each object in its place," with clear identification provided.
Seiso	Clean	Keep the workplace clean, eliminate all sources of dirt.
Seiketsu	Standardize/conform	Procedures and rules to maintain the first three stages. It basically consists of the creation of cleaning and inspection routines for all workers, thus ensuring that workplaces are kept perfectly clean. Procedimentos e regras para manter as três primeiras etapas. No fundo, consiste em criar rotinas de limpeza e inspeção para todos os trabalhadores, de forma a manter os locais de trabalho em condições perfeitas
Shitsuke	Discipline	Acquire the habit of respecting and correctly using the previously developed standard procedures and controls. Monitoring of this stage includes the assessing of compliance with the procedures established, and holding evaluation meetings.

Source: Gitlow 2009, 30; Miroslava et al. 2016, 331

The key to success resides in focusing on a change in people's attitudes and behavior. The staff must be made aware of the importance of the concepts, and the way in which these should be used to generate a pleasant, safe and productive environment.

5.5. Standard Work

The standardization of processes and activities requires the documentation of the operation methods involved, ensuring that everyone follows the same procedures, uses the same tools in the same way, and knows what to do when confronted with various situations. This concept is employed to guarantee stability of the process, so that operations are always undertaken in an established sequence and in the same manner, within a specific time period and producing the least amount of waste, thus achieving greater productivity and better quality (Antoniolli et al. 2017, 1121-1122).

Standardized work encompasses information relating to quality, safety, maintenance and work patterns. Work process standards (workstation files) must contain information relating to cycle time, intermediate stock levels and the work sequence, with a clear definition of the start and end of the process. This does not mean that standardization imposes a rigid form of working; instead, it points to the best, easiest, safest and most efficient ways found until that moment in time. Whenever improvements occur, these must be introduced so as to achieve even better standards (Kulkarni, Kshire, and Chandratre 2014, 431) (Ar and Ashraf 2012, 1734). Standardized work supports continuous improvement; namely, only on the basis of standardization can one know if an alteration has actually produced an improvement (Antoniolli et al. 2017, 1112).

Through the implementation of this methodology, one has seen significant gains in terms of productivity, in the reduction of failures and operation times. One should also underscore its importance as a learning tool, as well as its valuable contribution to the training and discipline of workers, thus enabling improved collaborators' skills and promoting the staff's multi-functionality (Martin and Bell 2011, 46).

5.6. SMED

The change of products, tools or adjustments carried out during the process is designated as setup or changeover. The period during which production is interrupted to change or adjust tools is known as setup time, and corresponds to the pause between the last good item manufactured and the first good item produced in the next batch (Costa et al. 2013, 951).

Shigeo Shingo developed the methodology called Single Minute Exchange of Die (SMED) with the objective of reducing setup time. It can be translated as a rapid tool change within a one-digit-minute time period. The proposal is that setups should be undertaken within ten minutes, which can be achieved if there is a rational approach to the tasks carried out by the machine operator (Sousa et al. 2018, 612-613) .

The appearance of the SMED methodology in the 1960s was driven by a great need to reduce the setup time of a press. Due to a lack of resources, another machine could not be acquired to meet the requirements. Shigeo Shingo then dedicated some time to the observation of the setups undertaken on the press and realized that most of the activities carried out during setup could be executed in the periods before or after machine downtime. He thus discovered an aspect which is fundamental to SMED - the definition of internal and external activities, which can be translated as (Ram et al. 2015, 39-40) (Ulutas 2011, 1195):

- Internal activities - those which must be carried out when the machine/equipment is stopped, for instance when tools are changed on a press or on the mold of an injection machine;
- External activities - those which can be undertaken before the machine stops and after setup ends, for example when fetching tools or putting away documents related to previous production.

In brief, the SMED methodology anchors on six stages to reduce setup time (Figure 9) (Staudter 2009, 305) (Sugai, Mcintosh, and Olívio 2007, 324-325) (Godina et al. 2018a, 730-731) (Kumar and Bajaj 2015, 35-36).

The activities which are designed to reduce setup time range from the very simple - such as a change in the location of tool-keeping - to the implementation of sophisticated devices to prepare and change matrixes (Rosa et al. 2017, 1035). These actions are usually very well received by staff and are implemented with enthusiasm, since many result from the ideas and experience of the workers themselves, who are on the ground and deal with problems daily (King 2009, 8) (Esa, Rahman, and Jamaludin 2015, 220). Knowledge of the technical aspects of equipment and tools, work organization (who does what and when), as well as knowledge of the method (how to do) constitute essential requirements for the successful implementation of SMED methodology. It is also important to foster teamwork so that the gains inherent to the methodology might prosper (Ribeiro et al. 2011, 47-48). The reduction of setup time through SMED produces many benefits for the companies using it. These lead to the reduction of batch sizes, less stock, improvement in quality, reduction of lead time, increased flexibility and greater productivity (Costa et al. 2013, 951).

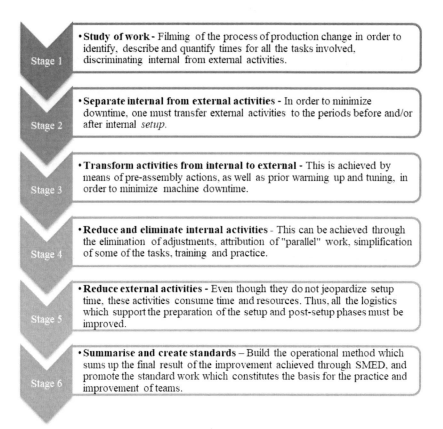

Figure 9. SMED methodology stages.

In the context of an increasingly competitive world, which changes by the second, it is paramount for companies to be always ready to respond quickly to shifts in the market and to the "normal" changes undertaken and demanded by their customers (Rosa et al. 2017, 1035) (Simões and Tenera 2010, 297). Productivity, quality and flexibility are the main pillars of the automotive industry (Costa et al. 2018, 3041).

5.7. PDCA

The PDCA cycle, also known as the Deming cycle, is a development method which focuses on continuous improvement. It can be applied to each process in the company, or to the system as a whole, with the purpose of preventing the stagnation of processes and achieving a state of perfection. It consists of four phases: Plan, Do, Check and Act (Sokovic, Pavletic, and Pipan 2010, 477).

Each phase comprises several steps, which must be implemented before the following one can be executed.

Figure 10 presents the eight compulsory steps required for the method to generate positive results and ensure durability (Jorgenca 2012) (Slack N., Chambers S., and Johnston 2001).

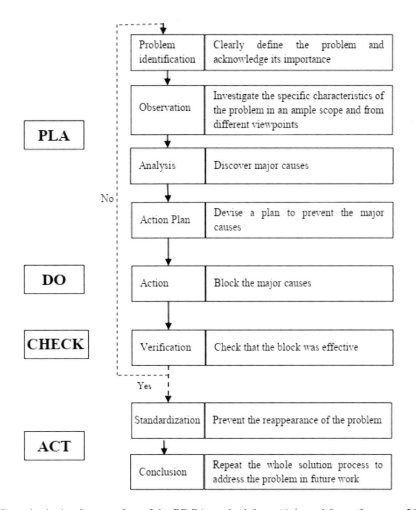

Figure 10. Steps in the implementation of the PDCA methodology (Adapted from (Jorgenca 2012)).

This is the time to act upon the entire organizational system, stimulating behaviours and emotions which will sustain the motivation of those who were greatly responsible for the whole project. The creation of mechanisms to ensure the preservation of a good work environment will determine the continuity of success, with greater productivity and the elimination of waste (Demeter 2011, 37) (Liker 2004, 209-210) (Womack, Jones, and Roos 1990, 55).

6. CASE STUDY

6.1. Definition of the Product

This chapter deals with a study undertaken on three assembly lines, which produce mechanism driver cables. The assembly lines designated as line A and line B manufacture driver cables to open doors; assembly line C produces cables for seat adjustment.

The function of the door-opening cables is to unlock the door when activated by the latch on the vehicle. In the case of the seat cables, once these are activated, their purpose is to move the position of the seat backrest, thus allowing the driver to adjust it to his physical condition and providing the comfort required for travel, especially long distance trips.

6.2. Definition of the lines

The assembly lines consist of a set of equipment, which is organized in workstations. The equipment is arranged sequentially, and is interconnected by means of a continuous system of material transportation. The various operations which occur at each of the workstations on the assembly line are carried out manually, with the aid of tools or small pieces of equipment.

The design of the assembly line meets the conditions which determine success. It is flexible enough to accommodate a variety of projects and enables the optimized use of resources. It is in line with the concept of one-piece-flow with zero inventories in the process, and all the stations are equipped with Poka-Yoke systems to prevent errors and, consequently, the production of scrap.

In Figure 11, Figure 13 and Figure 15 are represented the production lines dealt in this work, while in Figure 12, Figure 14 and Figure 16 are represented the flowcharts of each of the lines addressed in this study. Assembly line A consists of four workstations (station 1, station 2, station 3, station 4) and line B for three workstations (station 1, station 2, station 3) on each of these lines only the assembly of a single reference is processed. The components necessary for the assembly of the final product are supplied at the edge of the line, and in cases where it is necessary to facilitate the reach of the components during the operations, its decanting is done for small boxes attached to the lines. In order to make the production situation visible in real time, the quantity of work carried out, compliant and non-compliant, as well as all anomalies that have resulted in line stops and quality problems, are recorded at each working hour.

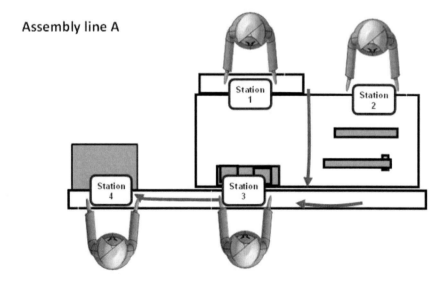

Figure 11. Layout for line A.

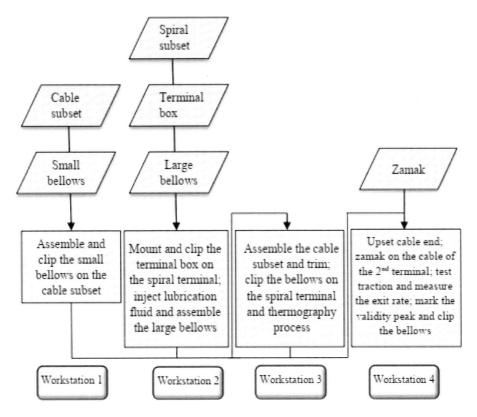

Figure 12. Flowchart for line A.

Assembly line B

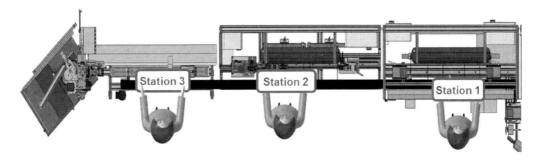

Figure 13. Layout for line B (Rosa et al. 2018, 557).

The assembly line C consists of eight workstations (station 1, station 2, station 3, station 4, station 5, station 6, station 7, station 8), in which 15 references are produced. There are stations that accumulate several operations, some of which are specific to certain references. Taking as an example the stations 3, where it is possible to clip sections, place labels, mount clips and clip the terminal. References with a section, do not clip the sections, just place labels and mount clamps, while references with two sections clip the sections but do not mount clamps. In this assembly line there are jobs that are not used by all references, namely jobs 2, 6, 7 and 8, while all references carry out operations in jobs 1, 3, 4 and 5.

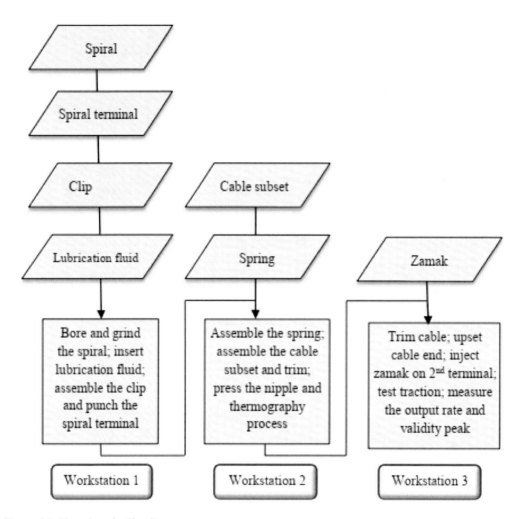

Figure 14. Flowchart for line B.

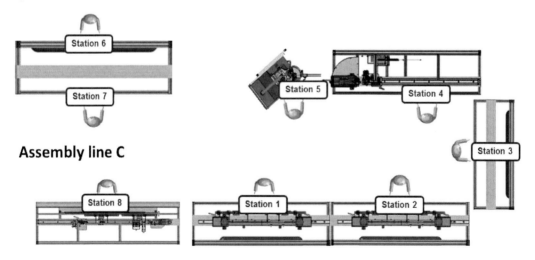

Figure 15. Layout for line C (C Rosa et al. 2017, 1036).

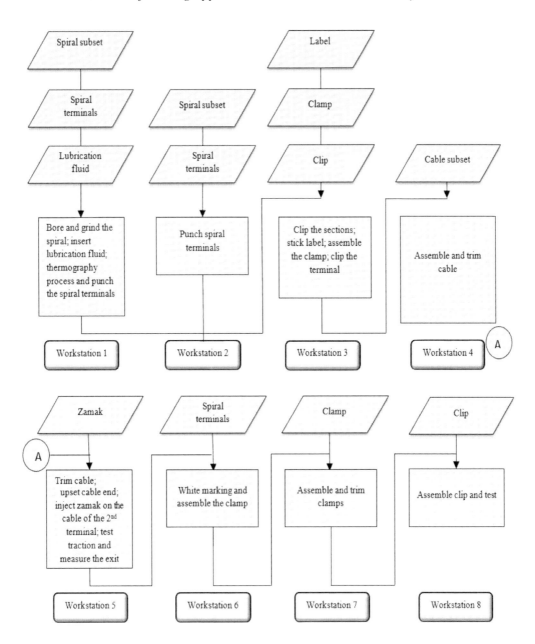

Figure 16. Flowchart for line C.

Although the basic makeup of the different cables is similar, there are specifications which lead to different compositions. This means that different components and processes are used in their assembly. The processes which occur on practically all lines are: the boring and grinding of the spiral; the punching of terminals on the spiral; the insertion and trimming of the cable; the injection of zamak on the metal cable of the 2nd terminal; thermography process; the injection of lubrication fluid and the verification of the specific characteristics of the cable (output rate and resistance to traction).

Presents the processes which are most commonly used in the assembly lines of command cables.

Table 4. Processes used in command cable assembly lines

	Process	Purpose	Equipment	Product
1	Boring and grinding of the spiral	Eliminates the burr after cutting		Before After
2	Punching of the spiral terminals	Attaches terminals to the spiral		Before After
3	Introduction of the metal cable on the spiral and trimming	Inserts the cable in the spiral and corrects its length to ensure that the exit rate will meet specifications after the 2nd terminal is injected with zamak		Before After
4	Injection of the 2nd cable terminal with zamak	Develops a part to allow the cable to connect with the driver systems		Before After
5	Thermography	Engraves registration data on the cable to enable its identification and tracking (date of manufacture, reference, etc.)		Before After
6	Injection of lubrication fluid	Lubricates the spiral interior to ease sliding of the cable		Subgroup with lubrication fluid
7	Test bench (usually positioned next to the machine used to inject Zamak on the 2nd terminal)	Checks specific cable characteristics (output rate and resistance to traction), thus ensuring that they meet customer requirements. After verification, markings are executed (date of manufacture, reference, etc.), or simply the validity peak		Guaranteed output rate and resistance to traction

6.3. Problems and Wastes Identification

In the process of assembling a product, the low productivity is worrying, as it affects the results of the company and can condition the customer satisfaction. Associated with low productivity, there are several problems that cause waste. In the assembly lines of cables to drive mechanisms, the following problems are highlighted as causes of inefficiency:

- Lack of reliability/robustness of the equipment: this leads to failure of operations, resulting in downtime and, sometimes, generation of defective products. Technologically upgrading the equipment is a way of getting around the problem, without resorting to the acquisition of new equipment. In the case under study, equipment has design deficiencies that are not easy to fill with an economic upgrade and short in time, causing the situation to drag on indefinitely.
- Excessive manual operations: the existence of manual operations in the process makes it time consuming, leading as well that these tasks are monotonous and repetitive, as in the case of assembly and manipulation of components. These short-time and repetitive operations are tiresome to the operator. In this case, its realization automatically or semi-automatically, in addition to making the process faster, improves working conditions as well as decreases the probability of error. In the present case, there is still an excessive number of manual operations, due to the defective design of the industrialization process, and the relatively low cost of labor.
- High setup times, especially in tuning operations: in a logic of monetization of the assembly lines, when the weekly orders are of small volume, each assembly line is used to produce several references, which induces a higher number of setups, leading to a greater waste of time. In these cases, the application of the SMED methodology is a priority. Some of the assembly lines of the company where this study was developed allow the realization of more than a dozen different products from the same family, but which require preparation of the line for each reference.
- Lack of polyvalence: the specialization of operators in a single workstation affects productivity, especially if the production process is mostly manual. The lack of technical knowledge of the equipment and the processes inherent in the different jobs makes it difficult to replace the operators in case of absenteeism, and also in the case where the physical fatigue inherent in the repetitive activities is counteracted by employee turnover between jobs. The turnover of workers in the company under study makes it difficult to train and specialize them.
- Lack of definition and standardization of the working method: in the absence of definition and standardization of the working method, operators will hardly perform the task using the best method, and taking the same way, which compromises the speed of execution and the quality of the work and product. In the absence of standardized methods, the training to be provided to operators is an arduous task, thus compromising the versatility of the team. The company under study has made an additional effort to improve this situation but it is not optimized yet.
- Existence of operations that do not add value to the product: there are operations that do not add value to the product and, consequently, the customer does not pay. Therefore, all efforts must be concentrated on eliminating or at least reducing them. In assembly lines, the following are examples of these operations: transportation of component housings, component decantings, transfer of components supplied in boxes to the existing brackets for the boxes in the line, unnecessary movements due to inefficiencies in the design of the equipment, extra operations to fill process inefficiencies, need of passing components between stations, quality control operations and production control. As many other companies, the company where the

study was carried out still has deficiencies at this level, and studies are under way to overcome this situation.
- Lack of ergonomics in the workstations: jobs require unnecessary physical effort, due to the layout they present. Small layout changes lead to reduced operator fatigue and, consequently, increased productivity. There are ergonomic problems to be solved, due to the lack of maturity of the industrialization team at this level.
- Product design problems: Product design cannot be dissociated from the assembly process. Small changes in product design, which do not affect its functionality and are insignificant to the customer, facilitate the assembly process, with positive results in process stability. In the automotive industry, there is negotiating power between component manufacturers and OEMs to address these issues, since they are duly explained and justified.
- Unbalance of the workstations: balancing an assembly line is the key to eliminating/reducing waiting times. The existence of a discrepancy between the times of occupation of the workstations increases the cycle time of the process, which is directly associated with low productivity. This requires a detailed study of the work cycles at each station, and the homogeneous distribution of tasks across the different workstations.

These wastes will be dissected forward, taking into account the operations performed on the line.

6.4. Methodology

The methodology adopted in solving problems associated with assembly lines comprises three stages: diagnosis, action and verification. In the diagnosis stage, the problems are identified and their root cause. Through Kaizen events, analyzes of products, materials, defects, flows, wastes, times and methods of operation are performed so that, using the VSM of the current state, the current process is analyzed, identifying critical points and potentialities. In the action stage, Lean tools are implemented: Visual Management, 5S, Standard Work and SMED. In the verification stage, the results obtained and benefits of the implemented actions will be evaluated and discussed. The systematic monitoring of the three stages is done using the PDCA cycle.

6.5. Applying Lean Tools

VSM is an excellent tool as a starting point in the process of optimizing an assembly line, since, in a systematic way, it assists in the identification of wastes and allows to idealize the value flow. The VSM diagram of the production line studied in this paper can be seen in Figure 17.

Based on the VSM diagram of the current state, the Kaizen group outlines an action plan that will focus on the critical points identified, with the goal of eliminating/reducing waste and, in this way, achieving the future state projected from the perspective of an optimized process. This group, made up of people from different departments, identifies the root cause of the problems and, based on the extensive knowledge and experience of those who work directly with the processes, presents economic and effective solutions, most of which rely on the experience of the operators.

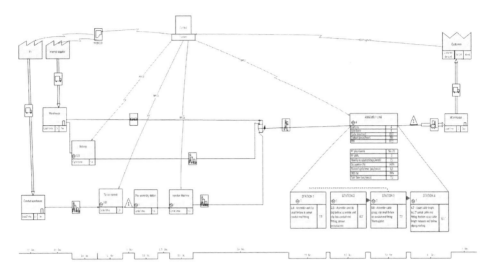

Figure 17. VSM diagram of the production line under study (Rosa, Silva, and Ferreira 2017, 1038).

Figure 18. Example of an action managed through a PDCA cycle.

Table 5. Solutions implemented on assembly lines for inefficiencies of the process

Detected problem: Inefficiencies of the process - Robustness and reliability of the process		
Description	Implemented solutions	Applied Lean tools
Cable diverts the clamp, forcing the operator to reintroduce it	To ensure orientation of the cable, magnets and a funnel-shaped part were placed	Kaizen
After cutting, the cable surplus is trapped in the gripper that pulls it	A breath of air is applied and a program for the tweezers was performed to make an open/close movement, in order to shake the tip of the cable	Kaizen
The pressed clip is misaligned with the terminal	The cavity where the spring is positioned is widened, which allows to avoid torsion of the spiral during punching of the clip, and also to facilitate positioning of the subassembly in the device	Kaizen
The clip becomes damaged when pressing, if incorrectly positioned	A poka-yoke system was installed. The clip only fits if it is in the correct position	Kaizen
Zamak runner breaks inside the mold, which forces the operator to use the compressed air gun for its removal	Widening of the runner channel, so that it only breaks when the cutter is activated	Kaizen
Extraction of the cable in the zamak machine often fails	Extractor stroke has been increased to ensure correct extraction of the cable in the zamak machine	Kaizen

Proposals to improve processes resulting from VSM analysis of the current state need to be listed and followed in order to ensure their implementation and effectiveness. The use of the PDCA methodology allows the implementation and follow-up of the defined actions, as well as guaranteeing their registration to future memory. An example of an action managed by a PDCA cycle can be seen in Figure 18.

Table 6. Solutions implemented on assembly lines for excessive manual operations

Detected problem: Excessive manual operations		
Description	Implemented solutions	Applied Lean tools
Assembly of the bellows in the cable subassembly	Implementation of a device to insert the small bellows into the cable subassembly, and connect it to the zamak terminal	Kaizen
Assembly of the terminal box on the spiral subassembly	Implementation of a new automatic machine to inject mass into the cable and connect the terminal box with the spiral subassembly	Kaizen

Table 7. Solutions implemented on assembly lines for high setup times

Detected problem: High setup time		
Description	Implemented solutions	Applied Lean tools
The setup tools are not organized and the identification of the reference to which they belong is not intuitive, which makes the search process slow	Organization of setup tools: • Placement of the tool corresponding to each reference in individual boxes and properly identified; • Placing the blocks of the thermo-grooving code on a plate with small slots, with magnets that allow the block to be housed in the position identified for that purpose; • Placing a chute to accommodate the label rollers in the position identified for each reference. Identification of tools: • Creation of a color code; • Based on the combination of 6 base colors, 21 codes were created to identify the tools corresponding to the different references; • Elaboration of a map with the picture of the equipment, highlighting the respective setup tools, which facilitates the work during the setup; • Construction of a table where, through the color code, the association between the setup tool and the reference to which it belongs is made; • Identification of the toolboxes with the corresponding color code assigned to each reference; • Elaboration of the set-up sheets, with the detailed procedures; • Training sessions to operators have been created.	Kaizen 5S Visual management Standard work SMED
The tools used require the use of screwdriver with the need to rotate several times, in order to perform the tightening of the parts	Replacement by quick-tightening tools	Kaizen SMED
Air Hoses with Threading Connections	Replacement by quick-release couplings, fixed to a metal plate, which allows to organize the pipes and to fix the points of connection	Kaizen SMED
Unavailability of the necessary keys to support setup	A set of necessary keys was placed on each workstation	Kaizen SMED
Quick wear parts were only available in the parts warehouse	A support cart was made available to support the setup, equipped with quick-wear parts and some support tools	Kaizen SMED
The preparation for the next production is done with the line stopped. The concept of external setup was not practiced	The preparation of the materials for the next production will no longer be an internal setup task, guaranteed by the operators, and will be assumed by the team leader as an external setup task	Kaizen SMED

Table 8. Solutions implemented on assembly lines for lack of versatility

Detected problem: Lack of versatility		
Description	Implemented solutions	Applied Lean tools
Low performance of operators in some jobs due to the need to make turnover to reduce fatigue.	Elaboration of detailed worksheets Ongoing training of the operators.	Standard Work

Table 9. Solutions implemented on assembly lines for operations that do not add value to the product - Detected problem: Operations that do not add value to the product

Description	Implemented solutions	Applied Lean tools
The debarking / grinding operation needs to be performed due to the cutting process	This operation started to be performed off-line using an automatic process	Kaizen
Manual removal of the runner and sprue	The operation was carried out by a device coupled to the zamak injection machine	Kaizen
Transport of the components to the assembly line, and move to the place defined to put on the empty boxes	Placement of a rack next to the line, with capacity for supply and recovery the components' boxes, avoiding the displacement of the operator to the place where the empty boxes were put on	Kaizen
Need to perform decantings to supply the components to the assembly line	Create appropriate brackets on the line for components to be made available in the supplier boxes, and allow only subsets in the appropriate boxes to be delivered to the existing brackets	Kaizen
Need to conduct visual inspections to ensure the quality of the provided subassemblies	Do not accept subsets that do not meet quality standards	Kaizen
Unsuitable working methods and at the discretion of each worker	Detail worksheet, mainly for tasks that require greater expertise and with particular attention to those referring to the bottleneck	Standard Work
To register the quantity of final product produced in the system and print the label to be placed on the packaging, the operator must move approximately 15 meters to the printer and wait for the availability of the registration system	Implementation of a system that allows the label to be printed on a printer placed on the assembly line	Kaizen

Table 10. Solutions implemented on assembly lines for lack of ergonomics in the workplace

Detected problem: Lack of ergonomics in the workplace		
Description	Implemented solutions	Applied Lean tools
Pick up and transport packaging of the final product from the place of packaging to the pallet	Due to the lack of space to move the pallet next to the place where the packaging is made, the line was rotated 180º, bringing the workstation closer to the pallet	Kaizen
The location of the components is outside the range of the hand	Replace components box bracket	Kaizen

VSM, Kaizen and PDCA are Lean tools used in the diagnosis and follow-up stage of problems. In its resolution, solutions are implemented based on 5S, Visual Management, Standard Work and SMED, Lean tools that stand out for the simplicity in the application and excellent results that they provide. The use of improved processes working on the equipment itself is also often used, since the lack of reliability and robustness of the equipment negatively affect its performance. Table 5 presents the Lean tools implemented in the assembly lines under study.

Table 11. Solutions implemented on assembly lines for problems in product design

Detected problem: Problems in product design		
Description	Implemented solutions	Applied Lean tools
The introduction of the cable into the spiral terminal fails and the operator, to overcome the difficulty, uses an improper method to perform the operation	Elimination of edges inside the terminal, changing the part design	Kaizen

Table 12. Solutions implemented on assembly lines for assembly line balancing

Detected problem: Assembly line balancing		
Description	Implemented solutions	Applied Lean tools
There is a mismatch in the occupation time of several workstations included in the assembly line	Redistributing the tasks by the stations, in order to obtain a balance between the times spent in each position	Kaizen

6.6. Results and Critical Analysis

Table 13, Table 14 and Table 15 show the results achieved in the three assembly lines where the previously presented solutions were implemented.

Table 13. Results obtained in assembly line A (Rosa, Silva, and Ferreira 2017)

	Before the improvements	After improvements	Variation (%)
Number of Workers	4	4	0
Occupied surface (m2)	25	25	0
Output (pcs/hour)	350	493	+41
Cycle time (sec/pc)	10,3	7,3	-29
PPH (pieces/person/hour)	87,5	123,3	+41
Balance assembly line (%)	85.8	88,4	+ 3

Table 14. Occupancy rate of the production line A and time needed to satisfy the customer demand

Year		Year 1	Year 2	Year 3	Year 4
Annual demand		1 750 000	2 100 000	2 640 000	2 300 000
Time required to fulfill the request (week hours)	Before improvements	106,5	127,8	160,7	140,0
	After improvements	75,5	90,6	113,9	99,2
	Variation	- 29	- 29	- 29	- 29
Occupancy rate (considered 3 shifts)	Before improvements	94,7	113,5	142,8	124,4
	After improvements	67,1	80,5	101,3	88,1
	Variation	- 29	- 29	-29	-29

From the results obtained in line A (Table 13), the increase in output was 41%, with an hourly production of 350 pieces to 493. The increase of 83 pieces per hour was due to improvements in equipment and the remaining 60 pieces provided from the disposal/reduction of waste.

Table 15. Results corresponding to production line B (Rosa et al. 2018, 561)

Number of Workers		3	3	0
Output (parts/hour)		229	327	+ 43
Cycle time (seg)		15,7	11,0	- 30
PPH (parts/person/hour)		76	109	+ 43
Balance assembly line (%)		90,4	97,8	+ 7,4
Time required to fulfill customer's demand (weekly hours)		44	31	- 30
Occupancy rate	(1st shift)	100%	83%	- 17
	(2nd shift)	17%	-----	- 17

Table 16. Results corresponding to production line C (Rosa et al. 2017, 1041)

Reference	Setup time Before improvements (min)	Setup time After improvements (min)	Variation (%)
reference 1 => reference 2	20	10	-50,0
reference 2 => reference 3	20	10	-50,0
reference 3 => reference 4	30	10	-66,6
reference 4 => reference 5	20	10	-50,0
reference 5 => reference 6	20	10	-50,0
reference 6 => reference 7	20	10	-50,0
reference 7 => reference 8	30	10	-66,6
reference 8 => reference 9	20	10	-50,0
reference 9 => reference 10	20	10	-66,6
reference 10 => reference 11	30	10	-66,6
reference 11 => reference 12	30	10	-66,6
reference 12 => reference 13	20	10	-50,0
reference 13 => reference 14	20	10	-50,0
reference 14 => reference 15	20	10	-50,0
reference 15=> reference 1	30	10	-66,6
Total	360	150	-58,3

The benefits of increased output are evident in the analysis of the ability to satisfy customer orders. Prior to the improvements, the assembly line was unable to meet customer demand in any of the four years with the highest volume. After improvements, only in year three, which corresponds to the peak of contracted quantities, it presents limitations on the capacity to respond, since the load factor is slightly higher than 100% (101.3%, see Table 14). However, this situation can be easily overcome by choosing to work with two shifts over the weekend.

In line B, it was intended to increase the output to 324 parts/hour. Expectations were exceeded, because the work done resulted in 327 parts per hour, which translates into a gain of 43% compared to the initial value of 229 parts per hour. Line balancing also increased positively, with 90.4% initially, reaching 97.8% at the end (see Table 15), which shows that the workload is conveniently distributed across the various stations, canceling out the hypothesis of the so-called "wait" waste. The time required to satisfy the client's weekly orders is now 31 hours per week, instead of the initial 44 hours (see Table 15), which eliminated the need to work on a second shift.

Table 16 shows the results obtained in line C, whose intervention was exclusively directed to the improvement of setup times. The results presented assume that during the course of the week all the references attributed to the line are produced and that there are tuning technicians available to ensure that the setup in all workstations of the line (total of 8) is carried out at the time imposed by the bottleneck station. It should be noted that after the improvements made to the assembly line and the training provided to the workers, the setup of each position was carried out by its worker, except for one station that needs to exchange a mould, in which case the worker is assisted by the technical tuner worker.

As shown in Table 16, the time that the line is unavailable weekly due to setups has reduced by at least 58.3%. At least the line started to have an additional 210 minutes of availability, which in reality corresponds to a longer time, since it was difficult to simultaneously have assistance of the number of technicians enough to guarantee the setup in the workstation corresponding to the bottleneck time.

CONCLUSION

This chapter aimed to present the approach used in the application of Lean tools in assembly and production lines of automotive components, specifically cables to drive mechanisms, and the improvements achieved in the production process by this way. Given the economic landscape that world is facing, the great challenge is to reduce costs and increase competitiveness without investment. In this context, it is crucial that companies have a solid foundation in Lean and are enthusiastic about continuous improvement of processes and people.

Lean tools and techniques such as: Value Stream Mapping, Kaizen, Visual Management, 5S, Standard Work, PDCA cycle and SMED were used to improve the efficiency of production processes aiming waste reduction opportunities.

In this context, VSM is a powerful tool that provides a global view of the value chain and the production processes involved. Sources of waste and critical points are easily identified by this means. Based on the analysis of the current state and based on Lean concepts, improvements are identified and can be implemented in order to lead to the future state. The systematized monitoring of the problem solving evolution is guaranteed using the PDCA methodology.

This approach is reflected in the case studies of the three lines presented in this work. In the case of line A and B, the objective was to increase the output and in the case of line C the reduction of setup time, with a common objective being to increase productivity.

In the case of lines A and B, actions were taken to improve equipment and eliminate/reduce waste associated with supply problems, operator movements, reliability and robustness of equipment, job balancing, definition and standardization of the work methods, among others, thus achieving a productivity increase of 41% in line A, rising the production in 143 parts per hour, and in the case of line B there was a productivity upsurge of 43%, with an increase of 104 pieces per hour.

In the case of line C, which aimed to reduce setup time by applying the SMED methodology, along with other Lean (5S, visual management and standard work) tools, several improvement actions were implemented regarding the organization and identification of the tools, the use of fastening and fastening tools, reorganization of internal and external tasks, detailed setups, visual aids and specific training was also given to operators, thus achieving a reduction of the minimum weekly setup time of 58.3%, corresponding to 210 min. This gain is higher in practice, given that the number of technicians available hardly was the three that would be required to perform the setup in the time corresponding to the bottleneck station, which implied longer delays than theoretically expected. In addition to the gain in availability achieved in the line, it should also be noted that the setups are now assured in the various workstations by the operators themselves, without the intervention of the tuning technicians, except in the change of references that require the exchange of zamak mold, in which the tuning technician maintained his intervention, acting in parallel with the operator.

With this type of approach, companies, in addition to improving the speed of response to customer orders, also increase the availability to receive new projects on existing lines. Lean tools have a preponderant role in the results obtained, as they lead to a systematic approach that allows to find viable solutions of low cost that result in a significant increase of the productivity, never ceasing of betting on the people's involvement, that despite the natural resistance to the initial change, when properly involved, present suggestions and adhere to the improvements implemented.

SUGGESTIONS FOR FUTURE WORK

The work regarding continuous improvement can never be given as concluded. The continuity of this analysis in the remaining lines is mandatory, as it allows the survey of existing problems or inefficiencies and, through a careful study, to find viable and low cost solutions that will lead to a significant increase in the reliability and efficiency of the processes, resulting in increased productivity to ensure success.

REFERENCES

ACEA. *The Automobile Industry Poket Guide 2017/2018*. Brussels: ACEA - *European Automotive Manufacturers Association*, 2017. Accessed April 28, 2018. https://www.acea.be/uploads/publications/ACEA_Pocket_Guide_2017-2018.pdf.

AICEP. Indústria Automóvel e Componentes [Components and Automotive Industry]. Lisboa: *Portugal Global*. 2016. Accessed April 29, 2018. http:// portugalglobal.pt/ PT/ RevistaPortugalglobal/2016/Documents/Portugalglobal_n87.pdf.

Andrés-López, E., I. González-Requena, and A. Sanz-Lobera. 2015. "Lean Service : Reassessment of Lean Manufacturing for Service Activities." *Procedia Engineering* 132: 23–30.

Antoniolli, L., P. Guariente, T. Pereira, L. P. Ferreira, and F. J. G. Silva. 2017. "Standardization and Optimization of an Automotive Components." *Procedia Manufacturing* 13: 1120–27.

Ar, Rahani, and Muhammad Ashraf. 2012. "Production Flow Analysis through Value Stream Mapping: A Lean Manufacturing Process Case Study." *Engineering Procedia* 41: 1727–34.

Araújo, W F S, F J G Silva, R D S G Campilho, and J A Matos, 2015. "Manufacturing Cushions and Suspension Mats for Vehicle Seats: A Novel Cell Concept." *The International Journal of Advanced Manufacturing Technology* 90: 1539–45.

ATKearney. "The Contribution of the Automobile Industry to Technology and Value Creation," 2013. Accessed May 02, 2018. https:// www.atkearney.com/ documents/ 10192/ 2426917/ The+ Contribution+ of+ the+ Automobile+ Industry+ to+ Technology+and+Value+Creation.pdf/8a5f53b4-4bd2-42cc-8e2e-82a0872aa429.

Azizi, Amir, and Thulasi a/p Manoharan. 2015. "Designing a Future Value Stream Mapping to Reduce Lead Time Using SMED-A Case Study." *Procedia Manufacturing* 2: 153–58.

Barbosa, B, M T Pereira, F J G Silva, and R D S G Campilho. 2017. "Solving Quality Problems in Tyre Production Preparation Process: A Practical Approach." *Procedia Manufacturing* 11: 1239–46.

Blatt, Benjamin. 2017. "Research of the German Automotive Industry." *Journal of Social Sciences* 1: 14–24.

Choomlucksana, Juthamas, Monsiri Ongsaranakorn, and Phrompong Suksabai. 2015. "Improving the Productivity of Sheet Metal Stamping Subassembly Area Using the Application of Lean Manufacturing Principles." *Procedia Manufacturing* 2: 102–7.

Clarke, Constanze, 2005. *Automotive Production Systems and Standardisation*. Germany: Physica Verlag (Springer).

CMACABLE. *Cable Manufacturing & Assembly Company*, 2015. Accessed July 12, 20*18*, http://www.cmacable.com/engineering/catalog/#book/Cover.

Correia, Damásio, F J G Silva, R M Gouveia, Teresa Pereira, and Luís Pinto Ferreira. 2018. "Improving Manual Assembly Lines Devoted to Complex Electronic Devices by Applying Lean Tools." *Procedia Manufacturing* 17: 663-71.

Costa, Eric, Sara Bragança, Rui Sousa, and Anabela Alves. 2013. "Benefits from a SMED Application in a Punching Machine." *World Academy of Science, Engineering and Technology* 7: 951–57.

Costa, M J R, R M Gouveia, F J G Silva, and R D S G Campilho. 2018. "How to Solve Quality Problems by Advanced Fully-Automated Manufacturing Systems." *The International Journal of Advanced Manufacturing Technology* 94: 3041–63.

Costa, R J S, F J G Silva, and R D S G Campilho. 2017. "A Novel Concept of Agile Assembly Machine for Sets Applied in the Automotive Industry." *The International Journal of Advanced Manufacturing Technology* 91: 4043–54.

Costa, Eric Simão Macieira da, Rui M. Sousa, Sara Bragança, and Anabela Carvalho Alves. 2013. "An Industrial Application of the SMED Methodology and Other Lean Production Tools." *4th International Conference on Integrity, Reliability and Failure of Mechanical Systems* 1: 1–8.

Demeter, Krisztina. 2011. "Factors Influencing Employee Perceptions in Lean Transformations." *International Journal of. Production Economics* 131: 30–43.

Duquer, Diego F. M., and Leonardo R. Cadavid. 2007. "Lean Manufacturing Measurement: The Realationship Between Lean Activities and Lean Metrics." *Estudios Gerenciales* 23: 69–83.

Efrari. 2018. *Cabos Flexíveis* [*Flexible Wire-Ropes*] Accessed July 10, 2018, http:// www.efrari.com.br.

Esa, Mashitah Mohamed, Nor Azian Abdul Rahman, and Maizurah Jamaludin. 2015. "Reducing High Setup Time in Assembly Line: A Case Study of Automotive Manufacturing Company in Malaysia." *Procedia - Social and Behavioral Sciences* 211: 215–20.

Feld, William M., 2001. *Lean Manufacturing: Tools, Techniques and How to Use Them*. New York: CRC Press.

Ficosa. 2018. *Systems for Doors and Seats*. Accessed July 12, 2018, https:// www.ficosa.com/products/systems-for-Doors-and-Seats/.

Gitlow, Howard S. *A Guide to Lean Six Sigma Management Skills*. New York: CRC Press. 2009.

Godina, Radu, Carina Pimentel, F J G Silva, and João C O Matias. 2018. "Improvement of the Statistical Process Control Certainty in an Automotive Manufacturing Unit." *Procedia Manufacturing* 17: 729-36.

Goicoechea, I, and M. Fenollera. 2012 "Quality Management in the Automotive Industry." *DAAAM International* 51: 619–32.

Halevi, Gideon. *Handbook of Production Management Methods*. Woburn: Butterworth Heinemann. 2001.

Hi-Lex, Corporation. 2006. *Automobiles and Industrial Equipment-Control Cable System*. Accessed July 10, 2018, http://www.hi-Lex.co.jp/e/cable/report/index.html.

Holweg, Matthias. 2007 "The Genealogy of Lean Production." *Journal of Operations Management* 25: 420–37.

INTELI, 2005. *Diagnóstico da Indústria automóvel em Portugal* [*Diagnosis of the Automotive Industry in Portugal*]. Accessed March 09, 2017, http:// www.inteli.pt/uploads/documentos/documento_1375704640_1255.pdf.

IATF, 2016. "IATF 16949 - Quality management system for organizations in the automotive industry." *International Automotive Task Force*. Pontoise, France.

Jorgenca, Jorge H. M. 2012. "Melhorias de Processos Segundo o PDCA - Parte I" ["Processes Improvements According to PDCA – Part 1"]. Accessed July, 18 2018, http://jorgenca.blogspot.pt/2012/04/melhorias-de-Processos-Segundo-O-PDCA.

King, P. L. 2009 "SMED in the Process Industries." *Industrial Engineer* 41: 30–35.

Kocakülâh, Mehmet C, Jason F Brown, and Joshua W Thomson. 2008 "LEAN Manufacturing Principles and Their Application." *Cost Management* 22: 16–27.

Kulkarni, Prathamesh P, Sagar S Kshire, and Kailas V Chandratre. 2014 "Productivity Improvement through Lean Deployment & Work Study Methods." *International Journal of Research in Engineering and Technology* 3: 429–34.

Kumar, B. Suresh, and S. Syath Abuthakeer. 2013 "Implementation of Lean Tools and Techniques in an Automotive Industry." *Journal of Applied Sciences* 53: 1689–99.

Kumar, Vipan, and Amit Bajaj. 2015. "The Implementation of Single Minute Exchange of Die with 5'S in Machining Processes for Reduction of Setup Time." *International Journal on Recent Technologies in Mechanical and Electrical Engineering* 2: 32–39.

Lampón, Jesús F., and Santiago Lago-Peñas. 2013. "Factors behind International Relocation and Changes in Production Geography in the European Automobile Components Industry." *MPRA* 29: 1-22. Accessed July, 1 2018. https:// mpra.ub.uni-muenchen.de/ 45659/8/MPRA_paper_45659.pdf.

Lesková, Andrea, and Lúvica Kovácová. 2012. *Automotive Supply Chain Outline.* 7: 96–104. Accessed June 26, 2018, http://pernerscontacts.upce.cz/26_2012/Leskova.pdf.

Liker, Jeffrey. 2004. *The Toyota Way.* New York: McGraw-Hill.

Machado, Tiago, and António Moniz. 2007. "Models and Practices in the Motor Vehicle Industry - Contrasting Cases from the Portuguese Experience." *MPRA* 23, 1–21. Accessed July 4, 2018, https://core.ac.uk/download/pdf/7306212.pdf.

Martin, Timothy D., and Jeffrey T. Bell. 2011. *New Horizons in Standardized Work.* New York: CRC Press.

Melton, T. 2005. "The Benefits of Lean Manufacturing -What Lean Thinking Has to Offer the Process Industries." *Icheme* 83: 662–73.

Miroslava, M, Vanessa Prajová, Boris Yakimovich, Alexander Korshunov, and Ivan Tyurin. 2016. "Standardization - One of the Tools of Continuous Improvement." *Procedia Engineering* 149: 329–32.

Moniz, António. 2008. "Competitivity in the Portuguese Automotive Sector and Innovative Forms of Employment Management." *MPRA* 22: 1–24. Accessed July 8, 2018, https://mpra.ub.uni-muenchen.de/6970/1/MPRA_paper_6970.pdf.

Moreira, B M D N, Ronny M Gouveia, F J G Silva, and R D S G Campilho. 2017. "A Novel Concept Of Production And Assembly Processes Integration." *Procedia Manufacturing* 11: 1385–95.

Natasya, Amelia, Abdul Wahab, Muriati Mukhtar, and Riza Sulaiman. 2013. "A Conceptual Model of Lean Manufacturing Dimensions." *Procedia Technology* 11: 1292–98.

Neves, P, F J G Silva, L P Ferreira, T Pereira, A Gouveia, and C Pimentel. 2018. "Implementing Lean Tools in the Manufacturing Process of Small Textile Products." *Procedia Manufacturing 17: 696-704.*

OECD. 2011. "Recent Developments in the Automobile Industry." *OECD Economics Departement Policy Notes* 7: 1–11.

OICA. 2017. *Production Statistics. International Organization of Motor Vehicle Manufacturers.* Accessed July, 9 2018. http:// www.oicfa.net/ category/ production-statistics/2017-Statistics/.

Pinto, Helder, and F J G Silva. 2017. "Optimisation of Die Casting Process in Zamak Alloys." *Procedia Manufacturing* 11: 517–25.

Pyzdek, Thomas, Paul A Keller, and Marcel Dekker. 2003. *Quality Engineering Handbook.* New York: Marcel Dekker, Inc.

Ram, Khushee, Sanjeev Kumar, D P Singh, Mechanical Engineering, and Khushee Ram. 2015. "Industrial Benefits from a SMED Methodology on High Speed Press in a Punching Machine : A Review." *Pelagia Research Library* 6: 38–41.

Ribeiro, Domingos, Fernando Braga, Rui Sousa, and S. Carmo Silva. 2011. "An Application of the SMED Methodology in an Electric Power Controls Company." *Romanian Review Precision Mechanics, Optics and Mechatronics* 3: 115–22.

Rocha, Hugo Tiago, Luís Pinto Ferreira, and F J G Silva. 2018. "Analysis and Improvement of Processes in the Jewelry Industry." *Procedia Manufacturing 17: 631-9.*

Rohani, Jafri Mohd, and Seyed Mojib Zahraee. 2015. "Production Line Analysis via Value Stream Mapping: A Lean Manufacturing Process of Color Industry." *Procedia Manufacturing* 2: 6–10.

Rosa, Conceição, F J G Silva, L Pinto Ferreira, and R Campilho. 2017. "SMED Methodology: Reduction of Setup for Steel Wire-Rope Assembly Lines." *Procedia Manufacturing* 13. 1034–42.

Rosa, Conceição, F J G Silva, and Luís Pinto Ferreira. 2017. "Improving the Quality and Productivity of Steel Wire-Rope Assembly Lines for the Automotive Industry." *Procedia Manufacturing* 11: 1035–42.

Rosa, Conceição, F J G Silva, Luís Pinto Ferreira, Teresa Pereira, and Ronny Gouveia. 2018. "Establishing Standard Methodologies To Improve The Production Rate Of Assembly Lines Used For Low Added-Value Products." *Procedia Manufacturing 17:555-62.*

Rother, Mike, and John Shook. 1999. *Learning to See*. USA: Brookline, Massachusetts.

Sabbagha, Omar, Mohd Nizam, Ab Rahman, Wan Rosmanira, and Wan Mohd. 2016. "Impact of Quality Management Systems and After-Sales Key Performance Indicators on Automotive Industry: A Literature Review." *Procedia - Social and Behavioral Sciences* 224: 68–75.

Scotiabank. 2018. *Global Economics/Global Auto Report*. Accessed July 6, 2018, https://www.scotiabank.com/ content/ dam/ scotiabank/ sub-Brands/ scotiabank-Economics/english/documents/global-Auto-report/GAR_2018-07-06.pdf.

Simões, Andreia, and Alexandra Tenera. 2010. "Improving Setup Time in a Press Line - Application of the SMED Methodology." *IFAC* 43: 297–302.

Singh, Bhim, S. K. Garg, S. K. Sharma, and S. K. Grewal. 2010. "Lean Implementation and Its Benefits to Production Industry." *International Journal of Lean Six Sigma* 1: 157–68.

Singh, Jagdeep, and Harwinder Singh. 2009. "Kaizen Philosophy : A Review of Literature." *The Icfai University Journal of Operations Management* 8: 51–73.

Singh, jagdeep, and Harwinder Singh 2015. "Continuous Improvement Philosophy - Literature Review and Directions." *Benchmarking: An International Journal* 22: 75–119.

Slack N., Chambers S., and R. Johnston. 2001. *Operations Management*. England, Person Education Limited.

Sokovic, M, D. Pavletic, and K. Kern Pipan. 2010. "Quality Improvement Methodologies - PDCA Cycle, RADAR Matrix, DMAIC and DFSS." *Journal of Archievements in Materials and Manufacturing Engineering* 43: 476–83.

Sousa, E, F J G Silva, L P Ferreira, M T Pereira, R Gouveia, and R P Silva. 2018. "Applying SMED Methodology in Cork Stoppers Production." *Procedia Manufacturing 17: 611-22.*

Steven C, J. 2004. "Learning to Lead at Toyota Learning to Lead at Toyota." *Harvard Business Review:*1-10.

Staudter, Christian. 2009. *Design for Six Sigma + Lean Toolset*. Frankfurt: Springer.

Sugai, Miguel, Richard Mcintosh, and Novaski Olívio. 2007. "Metodologia de Shigeo Shingo (SMED): Análise Crítica e Estudo de Caso" ["Shigeo Shingo Methodology (SMED): Critical Analysis and Case Study"]. *Gest. Prod* 14: 323–350.

Sundar, R., A. N. Balaji, and R. M. Satheesh Kumar. 2014. "A Review on Lean Manufacturing Implementation Techniques." *Procedia Engineering* 97: 1875–85.

Ulutas, Berna. 2011. "An Application of SMED Methodology." *World Academy of Science, Engineering & Technology* 5: 100–103.
Veloso, Francisco, Chris Henry, and Richard Roth. 2000. "Can Small Firms Leverage Global Competition? Evidence From the Portuguese and Brazi Automotive Supplier Industries." *4th International Conference on Technology Policy and Innovation*, 1–16.
Womack, James p., Daniel T. Jones, and Daniel Roos. 1990. *The Machine That Change The World*. New York: Rawson Associates.

In: Lean Manufacturing
Editors: F. J. G. Silva and L. Carlos Pinto Ferreira

ISBN: 978-1-53615-725-3
© 2019 Nova Science Publishers, Inc.

Chapter 10

SMED APPLIED TO COMPOSED CORK STOPPERS

Eduardo Sousa[1], F. J. G. Silva[1,], Carina M. O. Pimentel[2] and Luís Pinto Ferreira[1]*

[1]DME – Departement of Mechanical Engineering, ISEP – School of Engineering, Polytechnic of Porto, Porto, Portugal
[2]DEGEIT – Department of Economics, Management, Industrial Engineering and Tourism, University of Aveiro, Aveiro, Portugal

ABSTRACT

A growing focus among organizations is their flexibility in production. The market pressure for products with large variety in small batches leads organizations to maximize their resources in order to be able to adapt their processes to constant changes. One of the most known barriers in processes flexibility is the equipment changeover time. To reduce this time and thus increase the processes flexibility and the organization competitiveness in the market, the Lean methodology can be applied, being Single Minute of Exchange of Die (SMED) methodology the most used Lean tool in this context.

This chapter intends to present a practical case with mixed Lean tools application, being mainly focused on SMED study and implementation in cork industry equipment, leading to reduce the changeover time. There are several publications about SMED application in the most varied industrial areas, being this work an innovation in the cork industry. The equipment under study performs the connection between cork stoppers to capsules, which is done by glueing them with hot-melt glue.

In order to start this kind of studies, it is important to know the initial equipment condition and for this the equipment data need to be initially collected. With these data, it is possible to identify the improvement opportunities. In order to recognize which activities really add value to the product, the Value Stream Mapping (VSM) technique is usually used. The SMED is applied as a mean to reduce the downtime caused by changeover activities. To follow the development of the SMED project, another valuable lean tool is the A3 report. Lastly, in a way to improve the monitoring of possible deviations during production, the Overall Equipment Efficiency (OEE) KPI can be implemented as an overall equipment efficiency indicator.

---

[*] Corresponding Author's E-mail: fgs@isep.ipp.pt.

The aforementioned lean tools were successfully applied in the composed cork stopper manufacturing process, resulting in a 43% changeover time decrease from 66:56 min to 37:59 min through the application of SMED and, at the end of this study it is possible to prove that lean tools and principles are a powerful approach to get solid profits without large investments, as well as their important contribution to increase the processes flexibility.

Keywords: SMED, VSM, Lean, Manufacturing, Cork stoppers, Quick changeover

1. INTRODUCTION

In general, the market pressure leads organizations to improve productivity and quality while reducing costs (Costa et al., 2018, p. 3041). The global competitiveness and technological development give customers the opportunity to choose customized products with unique details to meet their specifications (Costa, Silva and Campilho, 2017, p. 4043). In the cork industry this paradigm is also verified, being the high differentiation of the product and small lots have become important issues. As demand for these types of products increases, organizations must be flexible in terms of production (Moreira et al., 2018, p. 624). This requirement is increasingly necessary to the organization's competitiveness because with the frequent changes only the most flexible organizations will be able to succeed over time (Araújo et al., 2017, p. 1539). One of the ways to make the production more flexible is by applying Lean methodologies in order to reduce waste (Womack, Jones and Ross, 1990, p. 13).

Lean methodology has become a widely acceptable and adoptable for a best manufacturing practice across countries and industries (Rosa, Silva and Ferreira, 2017, p. 1036). Due to its global lead in cost, quality, flexibility and quick respond, this methodology was applied in multiple industries (Neves et al., 2018, p. 697; Martins et al., 2018, p. 648; Rocha, Ferreira and Silva, 2018, p. 641). The main goals of this methodology is the product cost reduction and productivity improvements by eliminating wastes or nonvalue-added activities (Rosa et al., 2018, p. 556). The most important TPS tools to remove the different kinds of production waste are the Value Stream Map (VSM), cellular manufacturing, Total Productive Maintenance (TPM), and Single Minute Exchange of Dies (SMED). A great quantity of literature has addressed the first four tools, while information on SMED is a bit scarce (Reza et al. 2016, 1239). According to (Nash and Polin, 2008, p. 39), the VSM should allow the identification of all activities, whether or not they add value; from the raw material to the customer.

The A3 methodology is frequently used to solve problems. This methodology was developed by Toyota Motor to show the essential information about the problems and that it should be perceptible in a short space of time and outlined in an A3 sheet, which must include the following points: Background, Current Condition, Targets to improve, Gap Analysis, Planned Milestones, New Confirmed State and Learned lessons (Bassuk and Washington, 2013, p. 2). The demand for diversity has enforced the adoption of the Lean Production, which outcomes in the production of smaller lots. This shows an association between lot size and setup times (Rosa et al., 2017, p. 1036; Correia et al., 2018, p. 664). If the time required to carry out the setup is too long or unwieldy, there is a propensity to the production of larger lots. Besides limiting the operating flexibility and making it more difficult to meet the customer's needs; this process also generates other types of waste. To respond to the increasingly competitive market demands, setups must be performed quickly (Sundar, Balajib and Kumarc, 2014, p. 1878; Moreira et al., 2017, p. 1386; Nunes and Silva, 2013, p. 330).

The SMED methodology was developed by Shigeo Shingo as a way of responding to the emerging Japanese automobile industry need for reducing batch size (Ulutas, 2011, p. 1194). This methodology reports the issue of time reduction in the preparation, exchange, equipment tuning, and which tools are associated with these exchanges (Pinto, 2014, p. 71). Changeover time is defined as the time needed to set up a production system to run a different product with all the requirements. This is a typical example of waste, since changeover is a non-added value activity that incurs hidden costs (Goubergen and Landeghem, 2002, p. 206; Costa et al., 2018, p. 564). SMED application outcomes in a higher productivity, improved quality, less stock, reduced lead-time, greater flexibility and smaller lot sizes (Rosa et al., 2017, p. 1036).

The OEE calculation data is used to understand the impact of the equipment improvements; this indicator was presented by Nakajima (1988, p. 1; Antoniolli et al., 2017, p. 1121), in the Total Quality Management context, as a key indicator of equipment performance. One of the main reasons for the application of OEE, among researchers and practitioners, is because it is a simple, yet comprehensive measure of internal efficiency (Hedman, Subramaniyan and Almström, 2016, p. 129).

The present study was developed with a focus on an equipment of the cork industry, aimed at reducing its setup time. The study's main objective was to design the VSM of the sector, apply the SMED methodology accompanied by an A3 model and the OEE (Overall Equipment Effectiveness) calculation of the equipment. This type of investigation is an innovation in this kind of industry since there is a lack of published works on this sector.

2. CORK

2.1. Cork as a Natural Material

Cork it's a light, elastic material, practically impermeable to liquids and gases, being also a thermal and electrical insulator and an acoustic and vibration absorber; whose applications have been known since ancient times, mainly as a floating artifact and as a sealant, whose market, since the beginning of the 20th century, has expanded enormously, especially in the development of various cork based agglomerates. It is also innocuous, resistant to rot, and can be compressed with practically no lateral expansion. Macroscopically, cork is made up of alveolate cells layers, whose cell membranes have a certain impermeability degree and are full of a gas, considered similar to air, which occupies around 90% of its volume. It has an average density of around 200 kg/m^3, and low thermal conductivity. Cork also has remarkable chemical and biological stability and good resistance to fire (Mestre and Gil, 2011, p. 53).

With some centuries of history, cork is one of the most appreciated natural products worldwide, being considered the Portuguese economy jewel. The cork industry, in addition to its strong connection with the wine industry, has also a wide expansion of its applicability to other types of products, being therefore a reference product in the world market.

The cork industry is considered a "Portuguese" industry, since about 49.6% of the world production of this raw material (Figure 1) is of Portuguese origin, as can be seen in Figure 2 (Corticeira Amorim, 2017).

This is due in part to the fact that Portugal has a high cork oak area, with approximately 34% of the world area (APCOR, 2017, p. 12).

In Portugal, 84% of the area occupied by cork oak is located in south Portugal, after the Tejo River, with a growth of 0.8% between 2005 and 2010, in this area. Cork Oak is also the second dominant forest species in the Portuguese forest (APCOR, 2017, p. 12).

2.2. Cork Stoppers Industry

In Portugal, there are 7 thousand companies linked to the cork industry, making up 2% of all companies in the country. Of these companies, the largest part, about 72%, is dedicated to the cork stoppers production for the wine industry (APCOR, 2017, p. 20).

According to INE (National Statistics Institute), in 2015, Portugal reached 899.3 million euros in exports related to the cork industry, which are led by cork stoppers exports, representing 644.4 million euros (APCOR, 2017, p. 22).

The main export destination for cork stoppers is the USA, followed by France, Italy and Spain, since they are strong wine tradition countries.

Figure 1. Cork oak forest.

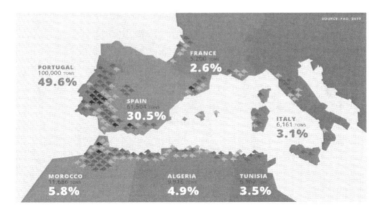

Figure 2. Annual cork production by country (Corticeira Amorim 2017).

In the following figure (Figure 3) it is possible to observe the evolution of cork stoppers exports in relation to the various types of products.

In this industry, according to GEP (Strategy and Planning Office) and MTSSS (Ministry of Labor, Solidarity and Social Security) data, between 2011 and 2014, the number of companies increased by 12% (APCOR, 2017, p. 16).

In the following figure (Figure 4) can be verified a sector crisis between 2008 and 2009, that has been recovering from year to year. By 2014, there were 670 companies, employing about 9,000 workers, and producing about 40 million corks per day (APCOR, 2017, p. 16).

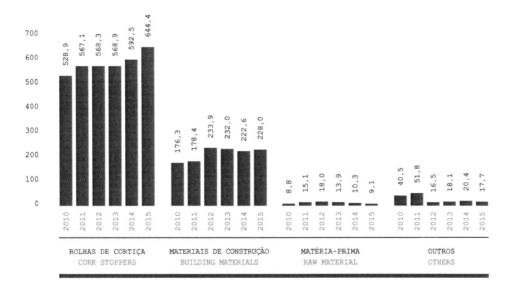

Figure 3. Evolution of the cork products sales.

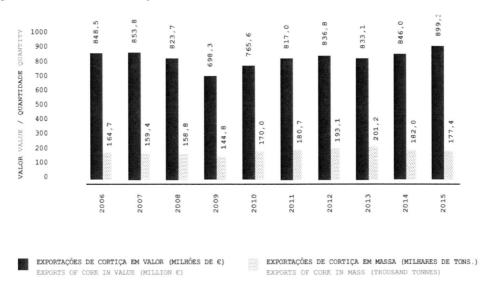

Figure 4. Evolution of the cork Portuguese exports.

2.3. Types of Cork Stoppers

In the production unit in which this study was approached, there are three types of stoppers, with two different finishing types.

The stoppers can be classified as follows (Pereira, 2007, p. 263):

- Natural Cork - Cork from natural cork, in which it does not undergo any kind of conformation or finishing process at the porosity level (Figure 5);
- Colmated Cork Stopper - This stopper is a natural cork of greater porosity that is subjected to a finishing operation at the level of the pores, in which is tried to improve its visual aspect (Figure 6);
- Micro Agglomerated Cork Stopper- This cork is obtained through extrusion or molding processes, in which unused cork is used in the production of natural stoppers (Figure 7).

Figure 5. Natural cork stoppers.

Figure 6. Colmated cork stoppers.

Figure 7. Micro agglomerated corks.

2.4. Composed Stoppers

The stoppers denominated by bartops are composed by a cork stopper and a top. The cork stopper can be classified into two categories: natural cork stoppers (natural or colmated) and agglomerated cork stoppers.

Each type of corks previously mentioned has a finish, they can be chamfered or rounded (Figure 8).

In addition to the different types of corks produced, appropriate to each customer, there is one more element that differentiates the final product, the top.

They are added to the finished cork stopper. This top can be made of wood, plastic, metal or even glass (Figure 9).

Figure 8. Different types of cork stopper finishes.

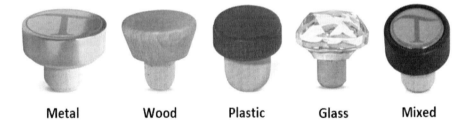

Figure 9. Different top types.

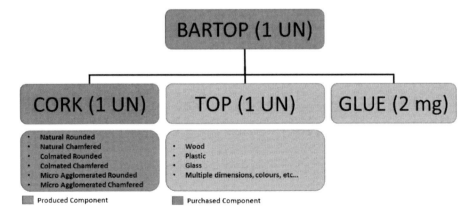

Figure 10. Product tree.

This is the main factor that justifies the enormous heterogeneity of the bartop product. The top value can surpass the stopper value, making it the main element of added value for the customer.

In most orders, wooden and plastic tops are used, both are purchased from external suppliers. In addition to the material, the capsules also differentiate through their dimensions, color, relief, embossing, among other characteristics.

As is noticeable, there is a large range of products, in Figure 10 the product tree is depicted.

3. COMPOSED STOPPERS MANUFACTURING PROCESS

In order to understand the composed stoppers manufacturing process, it is necessary to describe the entire process, from the bottle to the cork oak. The cork life cycle begins with the extraction of the cork oak bark, being this operation called stripping, which takes place during the most active phase of cork growth. This phase takes place between mid-May or early June, until the middle or end of August (APCOR, 2018).

It takes about 25 years for a cork oak to start producing cork and to be profitable. After this period, its exploitation will last on average about 150 years. With this first stripping, called "Desboia", a very irregular cork with a high hardness is extracted, which becomes almost impossible to work with it. This is usually referred to as virgin cork, which will be used in other applications than corks (such as decks, insulation, etc.), since they haven't the necessary quality to be used for corks (APCOR, 2018).

Nine years later, the second stripping is performed, in which a material with a more regular structure is obtained, less hard, but not yet adequate to manufacture corks, being denominated by secondary cork (APCOR, 2018).

It is only with the third stripping and in the followings, that it is obtained a cork with the proper properties for the cork stopper production, since it already has regularity in its structure with smooth coast and belly. From this moment forward, and in nine years cycles, cork oak will provide cork for the cork stopper manufacture for about 150 years and will be the target of 15 stripping cycles throughout its life (Costa and Pereira, 2004, p. 53).

The stripping operation is a traditional hand craft practice, which becomes quite arduous given the climatic conditions, the land and the ants. It is a specialized art that requires a lot of skill, a certain hardness but also delicacy. The worker need to have a well-trained hand to give the right ax that cuts the cork but not the tree. The ax is tailor-made for stripping, has a wooden handle with a wedge to lift the cork without ever touching the trunk (Green Cork Project, 2017) (Figure 11).

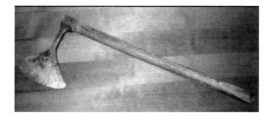

Figure 11. Stripping ax.

According to (APCOR, 2018), the process of stripping is then carried out in six stages:

1. Open: it is to strike the cork in the vertical direction, choosing the deepest crevice of the "enguiado" (the grooves of the bark). At the same time, the edge of the ax is twisted to separate the plank from the interlock.
2. Separate: in this step, the board is separated with the introduction of the ax's edge between the belly of the board and the interlocking. Then, a twisting motion of the ax is executed between the trunk and the cork to be separated.
3. Divide: with a horizontal cut, the size of the corkboard to be exited is defined, and the one that is in the tree. During tracing, the sequelae left in the interlock are frequent, and sometimes these mutilations end up altering the trunk geometry and should be minimized to the maximum.
4. Extract: The board is carefully removed from the tree so as not to break. The larger the boards are, the greater is their commercial value. It is the dexterity and ability of the workers that allows to obtain the planks in whole. Once the first board is removed, these operations are repeated to release the entire trunk.
5. Remove: after the extraction of the boards, some fragments of cork are adhered to the base of the trunk. To remove the possible parasites that exist in the chocks of the cork tree, the workers gives some blows with the eye of the ax.
6. Marking: finally, the tree is marked, using the last digit of the year in which the extraction was performed.

In the following figure (Figure 12) it is possible to observe in a schematic way all the steps of the descaling process.

After the stripping process, the cork boards are stacked either in the forest, or in warehouses inside the factory premises. There they remain exposed to the open air, sun and rain (Figure 13) for a period of not less than six months (CIPR, 2013, p. 18)

Figure 12. Stripping process stages (APCOR 2018).

During this stage, piles of cork boards are formed following very strict rules (defined by the International Code of Cork Stopper Practices - CIPR), so that cork can stabilize (Fernandes 2004, p. 20).

From the plank to the cork stopper, the cork goes through a set of steps that differ in the type of cork that is intended to be produced. Natural cork stoppers are manufactured by drilling from a single piece of cork, and the technical stoppers are produced from a body formed by agglomerate of cork granules, which can also be applied to the tops, discs of natural cork.

In this study, as the stoppers in the process are mostly natural stoppers, it will be this manufacturing process that will be approached.

After the rest period, the raw boards are subjected to a water boiling operation. This is characterized as being a treatment in tanks with boiling water, for about 1 hour (Figure 14). When subjected to this treatment, cork expands, especially in the radial direction, and there is also a decrease in porosity, which solubilizes a small fraction of cork extractives, particularly tannins. Another purpose of this boiling operation is also to improve the mechanical properties, facilitating subsequent cutting operations, either for the production of corks, discs or granules (Costa and Pereira, 2004, p. 53).

Figure 13. Example of cork stacked during rest period.

Figure 14. Planks boiling process.

After boiling, the cork is stabilized (Figure 15). It is only after this period, of two to three weeks, that the boards are selected. The stabilization attends to flatten the boards and allows them to rest. This is the only way that cork gets the necessary consistency for its transformation into corks. The stabilization allows the cork to reach the ideal moisture content for its processing, which is about 14% (APCOR, 2018).

After the stabilization time, the selection of planks and slicing operation takes place. In this operation, boards are cut in order to obtain cork pieces of a more homogeneous caliber and quality, which are classified in caliber and quality classes (Fernandes, 2005, p. 15) (Figure 16).

This operation is carried out by very specialized workers, the trawlers, who select, board by board if it is suitable for the production of corks, visually appreciating its appearance in the belly, the cross-sections and radial and the coast. At this stage, if any cork that does not have the technological capability to manufacture stoppers or disks, whether due to insufficient caliber or quality, is separated, calling this waste separation. This operation has a high degree of subjectivity of the operator and depends on the industrial unit where it is inserted, with a tendency for the tracing and grading of the boards to be more rigorous than in a single unit.

Figure 15. Stabilization of cork after boiling.

Figure 16. Selection of planks and slicing operation.

In the tracing process, the tracer observes the cork board and determines the areas that need to be separated, using straight tracing knives in wooden pews. To determine the cut lines, a few rules are followed (Pereira, 2007, p. 271):

1. Take advantage of natural cracks in the coast to spoil the board as little as possible;
2. Separate parts of cork that are not usable (due to insufficiency of thickness or severity of defects) and will constitute scrap;
3. In view of the minimum standard dimensions of boards for routing and the consequent production of corks, the thickness criterion may be sacrificed;
4. Pay attention to the need to obtain plank dimensions with an easy handling in the routing;
5. Separate the pieces, which are pieces of cork without dimension for plank or utility for the industrial processing, and destined, like the refuse, to the granulating industry. According to (CIPR, 2013, p. 5), the pieces correspond to pieces with a surface of less than 400 cm^2.

Once drawn, the board is trimmed, straightening one or more sides of the boards by faceting with a knife, to facilitate classification, improve the section's appearance and allows a better routing. At this stage, the wastes are the bits, the scraps, and the trimming or trimming chips (Figure 17).

All operations described above (Discard, Rest, Cook and Trace) refer to the macro task of preparing the cork. Most companies in the cork sector, have their manufacturing units in Montijo and Ponte de Sor areas, where they store the cork and make the tracing.

The following steps are usually carried out in the units installed in the area of Santa Maria da Feira, where the pallets are transported with planks ready to be used in the production of stoppers. In these units, the first operation is the routing (Figure 18), which consists of cutting transverse strips of cork (traces), perpendicular to the axial direction of the cork, on a disk saw (radisher) with a height depending on the length of the corks (Costa and Pereira, 2004, p. 54).

The traces are drilled by drilling in the cross-section with a hollow cutting cylinder with an inner diameter greater than 1 mm which is the desired diameter for the cork, whereby the axial axis of the cork corresponds to the axial axis of the cork plank, by boring (Figure 19).

The mass utilization of the drawn planks for the manufacture of corks is small, representing less than 25% of the initial mass of cork prepared. By-products, which are intended for grinding, consist mainly of cuttings and the punching traces (Figure 20) (Pereira, 2007, p. 276).

Figure 17. Clipping from the slicing operation.

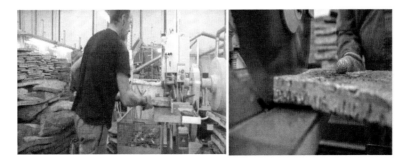

Figure 18. Slicing operation.

Figure 19. Punching process.

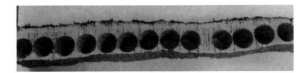

Figure 20. Punching by-products.

Already with the stopper in cylindrical format, these go to a greenhouse where they stagnate for 24 hours, to lower their humidity, and in order to be possible to rectify it. This operation consists of a mechanical operation of abrasion/polishing of the tops and surfaces (cork body), to obtain the desired dimensions (length and diameter, usually 29 mm x 21 mm).

At this point, the stoppers are ready to be subjected to a first electronic choice, which consists of separating the finished stoppers into different visual classes, and the determination of each class is made by automatic control of the surface of the stoppers. During this phase, in addition to defining the qualities, defective corks are also eliminated. The correct determination of the classes is the main factor that influences the profit margin of the final product.

After the stoppers are separated by visual classes, these go to another mechanical finish. Since the stoppers are size 29 mm x 21 mm, then it may be necessary to rectify the stopper to a lower gauge and perform a finishing operation.

The finish consists of a mechanical abrasion operation (similar to the grinding), where the chamfered or rounded finish of the cork is carried out.

After the rectification, the stoppers are washed, this can be made using hydrogen peroxide or peracetic acid. This bath is used to clean and disinfect corks, but other methods such as microwaves or ozone can be used, this operation gives the cork a more uniform appearance and a lighter color.

After the washing/disinfection, the stoppers are processed in an equipment called INNOCORK®, whose main purpose is the optimization of the sensorial performance of the stoppers. The drying is then carried out, in which the moisture content is stabilized, thus optimizing the performance of the stopper as a seal, while reducing microbiological contamination.

Eventually, the stoppers may be subjected to the sealing process, giving rise to the sealed stoppers. The sealing consists of covering the pores on the surface of the corks (lenticels) with a mixture of cork powder resulting from the rectification of natural corks. For the fixing of the powder in the pores (lenticels) is used a glue based on natural resin and a glue with water. This process aims to improve the visual appearance of the stopper and its performance.

Afterwards, if the stoppers are approved, they proceed to a second electronic choice, in which an improvement of the visual class of the product is performed. The stoppers are then treated superficially, and thereafter glued to a capsule, in the encapsulation process. This capsule can be made of plastic, wood, porcelain, metal, glass or other materials.

When the production is complete, the stoppers are packed in plastic bags filled with SO_2 (sulfur dioxide), a gas that inhibits microbiological development. Finally, they will be transported to the bottler of wines or spirits.

3.1. Work-Flow

The cork preparation process and the stoppers production has been explained previously (section 3), and to better understand the productive flow of the sector where the project is inserted, Figure 21 briefly describes each of the involved processes.

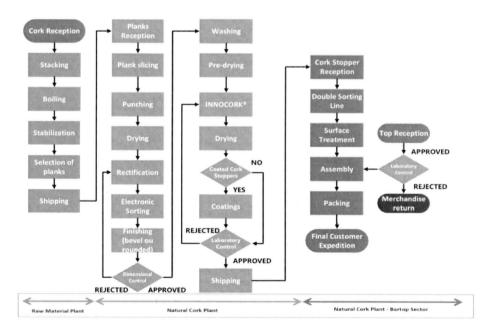

Figure 21. Work-flow diagram.

- Reception of corks: it is, along with the reception of the capsules, the activities that initiate the productive flow of the sector of the bartops. The stoppers are supplied by the natural corks unit and are already delivered by the laboratory after had been pre-selected.
- Receipt of capsules: consists of receiving the non-cork material, the capsules. Both wood and plastic, are delivered by external suppliers, according to the needs of production. The capsules are still under quality control, and only after approval do they proceed to the encapsulation.
- DSL (double sorting line): consists of a double line of choice, first with a manual choice on a mechanical carpet, to remove critical defects, and secondly an electronic choice to improve the customer's visual class, by selection equipment by artificial vision (Figure 22).
- Surface treatment: the visual class of the cork plots is approved, they undergo a surface treatment process to allow adequate product functionality (lubrication and waterproofing of the cork surface) (Figure 23).
- Capping/Assembly: completing the treatment, which lasts one hour, the corks go to the encapsulation. In this operation, a capsule is added to the stopper through the glue-hot-melt process, forming the final product, the bartop. These capsules can be made of materials such as wood, plastic, glass or metal, and are always purchased from outside suppliers. Since the stoppers are already attached to the capsules, they undergo a final control operation (on the equipment mat), in which the operator removes any defects before proceeding to the bag, which goes directly to the customer (Figure 46).
- Packing and shipping: finally, having the bags with the desired quantity (the machine counts itself), these are sealed with SO2 (in order to guarantee the preservation of the bartops during transportation), they are packed in cardboard boxes, placed on pallets and shipped (Figure 24).

Figure 22. DSL process.

Figure 23. Surface treatment process.

Figure 24. Sealing and packiging process.

3.2. Equipment

The equipment that will be the focus of this study is the capsular machine (Figure 25).

In this equipment, the operation of joining two products (top and stopper) is carried out. These can vary greatly from order to order, making tool changes a constant activity.

In the equipment under study, the capsules are fed with the operator to deposit the capsules in a tops container (filling container), and these, through a vibrator, are oriented with the top, top down, that is, whenever it changes tops there are adjustments that need to be made (Figure 26).

In the case of corks, they are supplied through a cart which is fixed to a support, where the stoppers fall into a cork stopper container, and are then guided to the tubes by a vibrator (Figure 27).

After the stoppers are in the tubes that guide them to the machine, they pass in a guidance system. This system detects the position of the stopper and rotate it so that it is not glued upside down on the top (Figure 28).

SMED Applied to Composed Cork Stoppers 241

Figure 25. Assembly/capping machine.

Figure 26. Assembly machine top supply system.

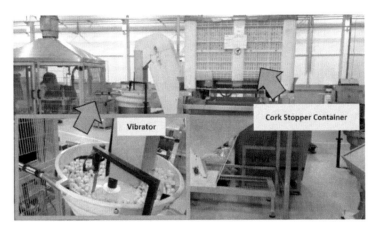

Figure 27. Assembly machine cork stopper supply system.

Figure 28. Cork stopper orientation system.

After the stoppers are in the correct position, they are "pushed" through an upward cylinder, where they travel through an inverted "U" circuit, until they are placed on the turntable (Figure 29).

The stoppers are then placed on the turntable (one by one) and, during rotation, the glue is applied to the stopper, and at a certain point, the top and the stopper plate met, gluing stopper with the capsule (Figure 30).

After joining, the bartops are left on a rug that moves them to the final visual inspection table, passing through a sensor that detects slopping corks (Figure 31). The operator removes any defective products, and then the assembly process ends.

Figure 29. Stopper route to the turntable.

Figure 30. Top and stopper junction point.

Figure 31. Final assembly control system.

4. THE PROBLEM AND THE CHALLENGE

As previously mentioned, the bartop sector start with the tops and stoppers reception and ends with the assembly of these two products. The assembly takes place in the capping production area, which is focused on detailing, identifying and analyzing the problems.

Since there may be a nearly infinite combination of stoppers and tops, there are frequent tool changes in the joining process. As in some cases, there are changes between very different gauges, this type of tool changes ends up taking up a lot of productive time.

To characterize the initial situation of the equipment, the tool changes data were collected and grouped, referring to a month of production- October 2017. The total data set has a relatively small size, and there is a large heterogeneity in the registers, not only at the numerical level (daily production, changeover times, etc.), but also from writing point of view (record of causes). In order to simplify the current condition analysis, the tool change time data were organized in the graph of Figure 32.

After analyzing the data, it's possible to visualize in Table 1 the time spent, stoppers quantity and monetary value lost in tool changes during the month of October 2017.

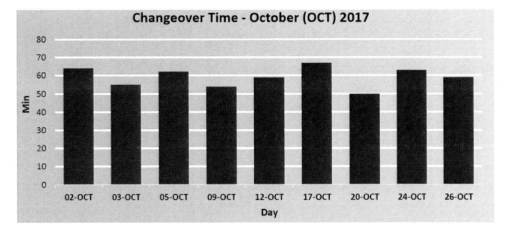

Figure 32. Changeover time - October 2017.

Table 1. Minutes, amount of stoppers and monetary value lost in tool changes

Month	Minutes	Quantity of corks	€ Value
October 2017	533	62,000	4,500

These data shows that there's a huge improvement opportunity regarding tools change. In the next chapter, it will be presented some ideas to try to achieve a changeover time reduction and, consequently, a reduction of all the waste that exists.

4.1. Wastes Identification Trough VSM Tool

The material and information flows were monitored from the beginning of the process to enable the VSM draw, since the stoppers production until the final product (bartop) is ready to be shipped. As far as the flow of information was concerned, it was necessary to schedule a meeting with all those involved in the process.

The VSM presented in Figure 33was based on a wooden top bartop with a cork stopper for white drinks, because these ones have a longer delivery lead time and the stopper suffers a surface treatment also with a longer lead time. The batch size was 52 thousand bartops, since it's the typical quantity per order according to previous year history.

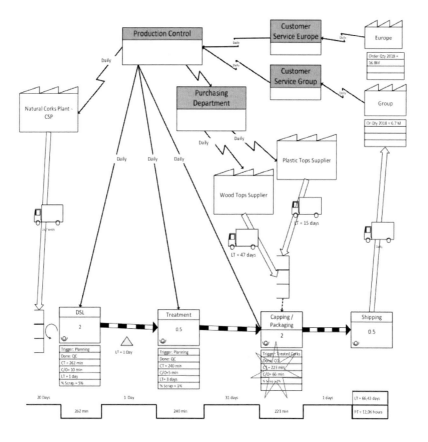

Figure 33. VSM of current state.

The production process starts with the customer order. The data related to this order is sent to the production control, daily. Then, the production control places the corks requests to the Natural Corks Plant and the tops requests to the purchasing department. These tops only enter in the production flow in the capping operation.

Concerning the corks, Natural Corks Plant takes about 20 days to deliver the request. Upon the cork reception, they undergo a visual cataloguing process called DSL (Double Sorting Line); with the visual quality ready, the corks go through the surface treatment process. Upon the end of this process, the stoppers are moved to the capping operation.

All of this process is continuous, making only stock in the final gauge before capping since there are costumers that the lead time required does not allow to make MTO (made-to-order), and consequently, there is a need to have stoppers finished in stock. The Capping operation is the process that determines the rhythm of the productive flow. Observing the VSM, it is possible to verify that the total lead time is 66,43 days and the processing time is 12,08 hours. This data reveals the need to eliminate or minimize activities that do not add value to the process.

In a future state, the change in the VSM will be in the changeover time that will influence the lead time. With the projected goal of reducing the changeover time by 50%, the change in lead time will be reduced from 66,4 days to 65,8.

4.2. A3 Report

To start and follow the development of the present project, the following A3 model was made (see Figure 34).

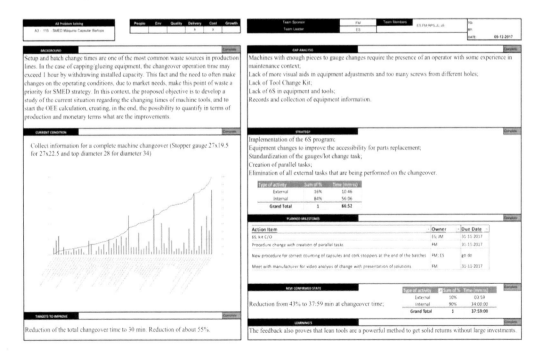

Figure 34. SMED project A3 model.

4.3. Internal and External Tasks

In order to identify improvement opportunities, it was necessary to understand the internal processes of the sector in greater detail and also, all the activities related to the capping machine setup process. The first step of SMED was to interview the workers, which was very important, in order to realize what was done correctly or not in the changeover operation, and the existing improvement possibilities, as well as recording the changeover situation. After it was complete, a meeting was held with the workers, in order to classify tasks as internal or external (step 1 of SMED methodology). From this analysis, it was found that 84% of the tasks are internal and 16% are external; they take respectively 56:06 minutes and 10:46 minutes, totalizing the changeover process in 66:52 minutes (see Table 2).

Table 2 shows the distribution of the tool changeover times, observed between External and Internal Activities. This data makes it possible to conclude that there are some aspects in the changing line process with the need of improvement. The enormous amount of time spent in activities that could possibly be external can be explored and worked.

Table 2. Distribution of changeover time in external and internal activities

Type of activity	Percentage (%)	Time (mm:ss)
External	16	10:46
Internal	84	56:06
Total	100	66:52

4.4. Equipment Improvements

In the equipment changes defined, it was planned to move the internal tasks to external, and at the same time reduce the internal ones. Some of these modifications will be described below.

Figure 35. Implemented improvement in tube sensors.

Whenever it was intended to make a stopper diameter change, the tubes had also to be replaced. To remove these tubes, it was necessary to unscrew the sensors and divide the tubes into parts. So, the solution to this, then implemented, was to fix the sensors externally to the tubes. With this modification the worker will no longer have to unscrew, fix and adjust the sensors, as well as no longer need to divide the tubes, making it possible to pre-prepare the tube kit to be used (see Figure 35).

Another improvement identified, was in the three shafts height regulation. Initially, the operators needed to adjust the shaft, and with the help of the calliper, measure the height of one shaft and adjust the other two with the characteristics of the first. To improve this system, it was created a part that only needs manual adjustment by turning a handle. This process puts the three pins at the same height (see Figure 36), with only one movement. With this improvement, the use of the calliper, as well as the key used, was eliminated, by the rotation of a handle.

Figure 36. Implemented improvement in shaft height regulation.

Figure 37. Implemented improvement in the part of the cork stopper.

In the part of the cork stopper orientation, were also identified improvements needs. Each time the tool is changed, the part of the cork guide has also to be changed, but the fixing screw of this part was so poorly accessed that workers had to go under and unscrew the cylinder in order to remove this part. The implemented improvement was to modify the insert, so that the screw was easily accessible, in order to ease the exchange of this part, and it was no longer needed to unscrew the cylinder (see Figure 37).

In the same section of the machine; the guides that lead the cork to the orientation part have been modified. Previously, to change the guides, the worker had to use a wrench to loosen these and then re-tighten them. Since the quick grip that these guides had, only served for the worker to open them. The improvement made was to replace these fasteners by fastening, improving access to the worker and eliminating one more key (see Figure 38).

Figure 38. Implemented improvement in the guides that lead the cork to the orientation part.

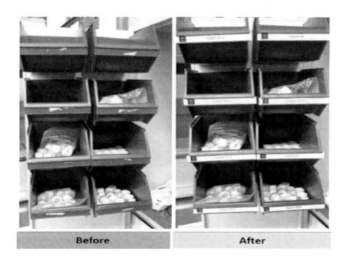

Figure 39. Improvement implemented in the organization of the parts.

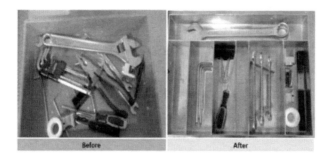

Figure 40. Changeover Kit.

Furthermore, besides the changes made to certain components of the equipment, which led to a reduction in the number of keys needed and the normalization of screw size, the 6S methodology was also used. Of the tasks identified in the video analysis, 08:55 min, that is, 12% of the tool changeover time was associated to the search for tools and/or poor organization of these. In order to reduce this unproductive time, were also implemented other improvements:

- Identification of each set of cups with the corresponding diameter (see Figure 39);
- Creation of a Changeover Kit (see Figure 40);

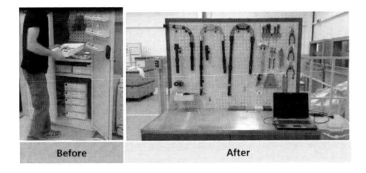

Figure 41. Improvement implemented in the material preparation space for the change.

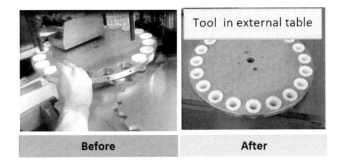

Figure 42. Improvement implemented in "cup" change operation.

Improvement of the preparation space of the parts before the change (see Figure 41). We sought the improvement of the preparation zone for changing tools so that, the operator had more space and also the tools were organized.

The activity of changing the stoppers "cups" (a piece of composite material that holds the stoppers before the join to the top) is considered internal, and it is not possible to pass it to external, however, it was found a solution to shorten its duration. Prior to this implementation, the worker changed these parts one by one using a small key. As an improvement, a new "dish" (the wheel-shaped part where the cups are fixed) was purchased, so that the worker could externally prepare the cups dish for the next stopper gauge. Just by unscrewing two screws this exchange is performed, instead of changing cup by cup (see Figure 42).

5. MEASURING THE BENEFITS

To conclude the SMED project, it was recorded a video of the changeovers with all improvements implemented, so that it was possible to test if the main goal of the project was reached and perhaps make more ideas for future improvements. In Table 3, it is possible to verify the new division into internal and external tasks. The difference in the total duration of the tool change is noticeable, decreasing from 66:52 min to 37:59 min, resulting in a reduction of 28:53 min. The time spent on external tasks dropped from 10:46 min to 03:59 min, but the most significant decrease was on the time spent on internal tasks, that went down from 56:06 min to 34:00 min. In terms of the distribution of tasks between internal and external, before the SMED application it was 84% and 16%, respectively, and after improvements implementation, began to be divided into 90% and 10%. This internal percentage increase is due to the reduction of setup stages, from 66 to 47, and also contributed to the fact that some external activities were eliminated.

Table 3. Distribution of changeover time in external and internal activities before and after improvements

Type of activity	Minutes			After Improvements		
	Percentage (%)	Time (mm:ss)	Number of setup stages	Percentage (%)	Time (mm:ss)	Number of setup stages
External	16%	10:46	16	10%	03:59	11
Internal	84%	56:06	50	90%	34:00	36
Total	100%	66:52	66	100%	37:59	47

Through the Table 4, it is possible to understand the financial impact that the implemented improvements have brought to the production process. With the reduction of 28:53 minutes in the changeover time (around 43%) and with an average of 10 changes done monthly, it is possible to quantify a saving of around 2340 € per month and per machine.

Table 4. Savings related to implemented improvements

Implemented improvements	Time (mm:ss)
Changeover Time Before Improvements	66:52 min
Changeover Time After Improvements	37:59 min
Monthly Average Changeovers	10
Financial savings in each changeover	234 €
Financial savings per month	2340€

6. START OF OEE CALCULATION

The OEE is a simple metric that includes three contributive factors of a given equipment efficiency. The OEE calculation of the capping machine was a proposed procedure as a complement to the improvement work that would be developed, in order to monitoring the evolution of the machine performance over time. To calculate the OEE, it is necessary to analyze the production records, as well as the equipment stops. By performing this analysis it was possible to obtain the values presented in Table 5.

The calculated OEE values show some points that must be improved in the equipment operation, and that the main cause of the low efficiency is mainly the high number of micro-stops (duration less than 5 minutes) and the low working speed, which contribute to the fact that the lowest percentage value is the first performance. In fact, by observing the capsule machine operation, it is obvious that the recommended production rates frequently are not met, and the equipment is stopped for several reasons throughout the working shift.

Other source of the great difference between availability and performance indexes, is the way that the registers are carried out by the workers. Many stops influencing the machine performance are certainly more than 5 minutes, and their causes should be recorded.

Regarding the values calculated for the two shifts, it is verified that the availability index is higher in the second shift, while the performance index is higher in the first shift. This is likely, due to the fact that the responsible worker for the records in the first shift allows a better classification of stops and micro-stops, being more rigorous and providing more detailed information on the recorded stops.

Table 5. OEE calculation for shift 1 and 2

Parameter	Availability	Performance	Quality	OEE
Shift 1 (%)	93%	83%	99%	76%
Shift 2 (%)	97%	57%	99%	55%

CONCLUSION

Starting with the initial situation, in which the capping machine changeovers were a process with space for improvements and several organizational and normalization difficulties, the challenge was to accomplish a 55% reduction in the changeover time and also to start the OEE calculation for the capping equipment.

This practical experience of applying the SMED methodology allowed to realize that this method is strongly related to the understanding of lean concepts, which are indispensable for creating work conditions that allow workers to work more efficiently. An example of this efficiency was the activities identified as internal, that with only the installation of two holders, were eliminated from the setup, not requiring the worker to make more adjustments.

The improvements implemented had a positive impact on the process, and the main goals were almost achieved. The 43% reduction in the average changeover time is significant; there is three identical equipment's, but with different parts, so there is a future need to normalize the various equipment tools to create a certain stability in the setup process. All the improvements implemented represent a small step if it is considered that are there still many hypotheses to be explored. The search for improvement should be permanent and not limited to the duration of a project. This thought should be followed, not only because it will help the maintenance of the daily improvements, but also because there are more improvements to be developed and applied in the future.

Conclusively, it is important to highlight that a good workplace environment, trust in the employees and cooperation among all, are fundamental to the success of any project; this fact is even clearer in a project of this kind, that requires an involvement by all and perspective a change of how the work is done. An enthusiastic workforce is a powerful tool, and, in this project, in particular, this was a great advantage since many of the ideas were born trough this involvement. This dynamic must be preserved and will surely continue to bring benefits to both the company and all employees.

REFERENCES

Antoniolli, I., Guariente, P. Pereira, T., Ferreira., L. P., Silva, & F. J. G. (2017). "Standardization and Optimization of an Automotive Components Production Line." *Procedia Manufacturing* 13, 1120-1127. doi: 10.1016/j.promfg.2017.09.173.

APCOR. (2017). *APCOR Directory*. Santa Maria de Lamas: Portuguese Cork Association.

APCOR. (2018). *Cork Harvesting*. http://www.apcor.pt/en/cork/processing/cork-harvesting/.

Araújo, W. F. S., Silva, F. J. G., Campilho, R. D. S. G., Matos, & J. A. (2017). "Manufacturing cushions and suspension mats for vehicle seats: a novel cell concept". *The International Journal of Advanced Manufacturing Technology* 90, 1539-1545. doi:10.1007/s00170-016-9475-6.

Bassuk, J. A. & Washington, I. M. (2013). "The A3 problem solving report: A 10-Step scientific method to execute performance improvements in an academic research vivarium." *PLoS ONE*, 8 (10), 1-9. doi:10.1371/journal.pone.0076833.

CIPR. (2013). *Liège - Código Internacional de Práticas Rolheiras [International Code of Cork Stopper Practices]*. 1-90.

Correia, D., Silva, F. J. G., Gouveia, R. M., Pereira, T., & Ferreira, L. P. (2018). "Improving manual assembly lines devoted to complex electronic devices by applying Lean tools." *Procedia Manufacturing* 17, 663-671. doi: 10.1016/j.promfg.2018.10.115.

Corticeira, Amorim. (2017). *Floresta com Futuro [Forest with Future]*. http://www.amorim.com/a-cortica/localizacao-do-montado/.

Costa, A. & Pereira, H. (2004). "Caracterização e Análise de Rendimento da Operação de Traçamento na Preparação de Pranchas de Cortiça para a Produção de Rolhas." ["Characterization and Analysis of the Tracing Operation Yield in the Preparation of Cork Boards for the Production of Cork Stoppers"]. *Silva Lusitana*, 12 (1), 51-66. ISSN: 0870-6352.

Costa, C., Silva, F. J. G., Gouveia, R. M., & Martinho, R. P. (2018). "Development of Hydraulic Clamping Tools for the Machining of Complex Shape Mechanical Components." *Procedia Manufacturing* 17, 563-570. 10.1016/j.promfg.2018.10.097.

Costa, M. J. R., Gouveia, R. M., Silva, F. J. G., & Campilho, R. D. S. G. (2018). "How to solve quality problems by advanced fully-automated manufacturing systems." *International Journal of Advanced Manufacturing Technology*, 94 (9-12), 3041-3063. doi: 10.1007/s00170-017-0158-8.

Costa, R. J. S., Silva, F. J. G., & Campilho, R. D. S. G. (2017). "A novel concept of agile assembly machine for sets applied in the automotive industry." *International Journal of Advanced Manufacturing Technology*, 91, 4043-4054. doi: 10.1007/s00170-017-0109-4.

Fernandes, P. (2004). *Influência do período de estabilização da cortiça e da cozedura na largura dos anéis de crescimento, no coeficiente de porosidade da cortiça e em algumas características tecnológicas das rolhas de cortiça natural [Influence of the stabilization period of cork and firing on the width of the growth rings, on the porosity coefficient of cork and on some technological characteristics of natural cork stoppers]*. Graduation Thesis. Lisbon: Instituto Superior de Agronomia.

Fernandes, R. M. O. S. (2005). *Estudo da influência do calibre e da qualidade das pranchas de cortiça delgada no rendimento do processo fabril de produção de discos de cortiça natural [Study of the influence of the caliber and the quality of the thin cork boards in the yield of the manufacturing process of natural cork disks*]. Graduation thesis. Lisbon: Instituto Superior de Agronomia.

Goubergen, D. V. & Landeghem, H. V. (2002). "Rules for integrating fast changeover capabilities into new equipment design." *Robotics and Computer Integrated Manufacturing*, *18* (3-4), 205-214. doi:10.1016/S0736-5845(02)00011-X.

Green Cork Project. (2017). *O descortiçamento e a cortiça*. [*Discarding and cork*] (in Portuguese)." http://www.greencork.org/a-floresta-a-cortica-e-a-rolha/o-descorticamento-e-a-cortica/.

Hedman, R., Subramaniyan, M. & Almstrms, P. (2016). "Analysis of critical factors for automatic measurement of OEE". *Procedia CIRP*, *57*, 128 - 133. doi:10.1016/j.procir.2016.11.023.

Martins, M., Godina, R., Pimentel, C., Silva, F. J. G., & Matias, J. C. O. (2018). "A Practical Study of the Application of SMED to Electron-beam Machining in Automotive Industry." *Procedia Manufacturing* 17, 647-654. doi: 10.1016/j.promfg.2018.10.113.

Mestre, A. & Gil, L. (2011). "Cork for Sustainable Product design." *Materials Science & Technology*, *23*, 52-63.

Moreira, A., Silva, F. J. G., Correia, A. I., Pereira, T., Ferreira, L. P., & de Almeida, F. (2018). "Quality Reduction and Quality Improvements in the Printing Industry." *Procedia Manufacturing* 17, 623-630. doi: 10.1016/j.promfg.2018.10.107.

Moreira, B. M. D. N., Gouveia, R. M., Silva, F. J. G. & Campilho, R. D. S. G. (2017). "A Novel Concept of Production and Assembly Processes Integration". *Procedia Manufacturing*, *11*, 1385-1395. doi: 10.1016/j.promfg.2017.07.268.

Nakajima, S. (1988). *Introduction to TPM: Total Productive Maintenance*. Cambridge: Productivity Press Inc. ISBN: 0-915299-23-2.

Nash, M. A. & Poling, S. R. (2008). *Mapping the total value stream - A comprehensive guide for production and transactional processes*. New York: Taylor & Francis Group. ISBN: 978-1-56327-359-9.

Neves, P., Silva, F. J. G., Ferreira, L. P., Pereira, T., Gouveia, R. M., & Pimentel, C. (2018). "Implementing Lean Tools in the Manufacturing Process of Trimming Products." *Procedia Manufacturing* 17, 696-704. doi: 10.1016/j.promfg.2018.10.119.

Nunes, P. M. S. & Silva, F. J. G. (2013). "Increasing Flexibility and Productivity in Small Assembly Operations: A Case Study." In: *Advances in Sustainable and Competitive Manufacturing Systems, Lecture Notes in Mechanical Engineering,* Azevedo, A. (Editor), Springer International Publishing Switzerland. doi: 10.1007/978-3-319-00557-7.

Pereira, H. (2007). *Cork: Biology, Production and Uses*. Amsterdam: Elsevier. ISBN: 9780080476865.

Pinto, J. P. (2014). *Pensamento Lean: A filosofia das organizações vencedoras* [*Lean Thinking: The philosophy of winning organizations*]. Lisbon: Lidel.

Reza, J. R. D., Alcaraz, J. L. G., Loya, V. M., Fernández, J. B., Macías, E. J., & Sosa, L.A. (2016). "The effect of SMED on benefits gained in maquiladora industry." *Sustainability*, *8* (12), 1237-1255. doi:10.3390/su8121237.

Rocha, H. T., Ferreira, L. P., & Silva, F. J. G., (2018). "Analysis and Improvement of Processes in Jewelry Industry." *Procedia Manufacturing* 17, 640-646. doi: 10.1016/j.promfg.2018.10.110.

Rosa, C., Silva, F. J. G., Ferreira, L. P., & Campilho, R. (2017). "SMED methodology: The reduction of setup times for Steel Wire-Rope assembly lines in the automotive industry". *Procedia Manufacturing* 13, 1034-1042. doi: 10.1016/j.promfg.2017.09.110.

Rosa, C., Silva, F. J. G., & Ferreira, L. P. (2017). "Improving the quality and productivity of steel wire-rope assembly lines for the automotive industry". *Procedia Manufacturing* 11, 1035-1042. doi: 10.1016/j.promfg.2017.07.214.

Rosa, C., Silva, F. J. G., Ferreira, L. P., Pereira, T., & Gouveia, R. (2018). "Establishing Standard Methodologies to Improve the Production Rate of Assembly Lines Used for Low Added-Value Products." *Procedia Manufacturing* 17, 555-562. doi: 10.1016/j.promfg.2018.10.096.

Shah, R. & Ward, P. T. (2003). "Lean manufacturing: context, practice bundles, and performance." *Journal of Operations Management, 21*, 129-149. doi: 10.1016/S0272-6963(02) 00108-0.

Sundar, R., Balajib, A. N. & Kumarc, R. M. S. (2017). "A review on lean manufacturing implementation techniques." *Procedia Engineering, 97*, 1875 - 1885. doi: 10.1016/j.proeng. 2014.12.341.

Ulutas, B. (2011). "An application of SMED Methodology." *Engineering and Technology International Journal of Industrial and Manufacturing Engineering, 5*, 1194-1197. doi:10.1999/1307-6892/14919.

Womack, J. P., Jones, D. T. & Ross, D. (1990). *The Machine That Changed the World*. New York: Macmillan Publishing Company. ISBN: 0-89256-350-8.

In: Lean Manufacturing
Editors: F. J. G. Silva and L. Carlos Pinto Ferreira
ISBN: 978-1-53615-725-3
© 2019 Nova Science Publishers, Inc.

Chapter 11

LEAN PRODUCTION IN THE PORTUGUESE TEXTILE AND CLOTHING INDUSTRY: THE EXTENT OF ITS IMPLEMENTATION AND ROLE

Laura Costa Maia[*], *Anabela Carvalho Alves and Celina Pinto Leão*
R&D Centro ALGORITMI, Department of Production and Systems,
School of Engineering, University of Minho, Guimarães, Portugal

ABSTRACT

This chapter aims to explore Lean Production (LP) awareness, the extent of its implementation and role in the Portuguese Textile and Clothing Industry (TCI). For this purpose, a questionnaire, applied in two periods (2011 and 2015), and a set of interviews carried out in 2018 with managers of selected companies, were used for data collection. The study covered a representative sample of TCI companies that accepted the challenge, reflecting the existing pattern and extent of LP implementation and the role within the Portuguese economy. The questionnaire was developed to understand if TCI companies know what LP model is, if implemented, and the evolution over the last years. It was also intended to understand the level of implementation of the LP model in this industry. In 2015, a low percentage of companies had considered implementing LP (approximately 18%). Conversely, the majority of companies considered implementing this model in the near future (approximately 56%). The companies that implemented LP reported difficulties in the implementation and pointed out some measures taken to overcome them. Through face-to-face interviews, the authors wanted to confirm the current condition of LP model implementation and if it was, in fact, implemented, understanding the way how companies have been implementing LP model. The interviewed companies were the ones that responded positively to the 2015 questionnaire. At the same time, it was intended to identify current difficulties and if they were the same as reported in 2015. With these figures, the authors wanted to discourse about current requirements and needs to implement LP in the TCI. Also, attended to the arrival of the Industry 4.0, that facilitates the process optimization, what is the role of LP on this context and to enhance the importance and why companies should implement LP.

---

[*] Corresponding Author's E-mail: lauracostamaia@gmail.com.

Keywords: lean production, lean thinking, textile industry and clothing industry, 4.0 industry

1. INTRODUCTION

The work organization model Lean Production (LP) is currently considered a strategy of competitiveness and sustainability for companies, as it reduces costs and increases productivity through waste elimination, i.e., non-added value activities. At the same time, it responds to external customers' demands, by promoting on-time delivery of high-quality products, and to internal customers' demands, by respecting them and their working environment (Maia, Alves, and Leão 2013).

Lean Production was first introduced by the Toyota Production System (Monden 1998; Ohno 1988), which, in turbulent times, helped the Toyota company to outperform compared with other surviving companies all around the world. This performance was supported by the key idea of "doing more with less", where less means less effort, less space, less investment in tools, less engineering hours to develop new products and less stock (Womack et al., 1990). By putting this idea into action, LP increased productivity without requiring large investments and financial efforts on the part of companies.

This method is particularly relevant in times of recession, as showed in Lean Enterprise Institute's 2009 study conducted in United States' companies from almost a decade. Also, the U.S. Census Bureau's 2013 survey, concluded that Lean helps companies to grow, becoming more profitable, productive, and innovative. Moreover, the pursuit of Lean Production must continue, as there is still a wide variation between the most and least productive players within industries. Also, the process of simplifying, consolidating, and removing inefficiencies from operations is extending to new areas, such as resource productivity (Manyika et al., 2012) and their principles and practices enable employees to improve productivity and to achieve continuous improvement (Donofrio and Whitefoot 2015). Recent surveys show that LP is widely implemented in all kinds of goods and services companies around the world (Bhamu and Sangwan 2014; Jasti and Kodali 2014; Alves et al. 2014).

The Portuguese Textile and Clothing Industry (TCI) had been suffering from low productivity, countless waste, and external threats, and so needing a business strategy formulation and new production models (ATP, 2017). This situation creates motivation for a change to make some improvements in this sector. To do this, a study was carried out on this specific subject. This study involved several phases and materials. Firstly, a survey to understand the level of knowledge of TCI companies about LP, if they had this model of production implemented, and the level of LP model implementation. Three years later, a follow-up study was undertaken within a sample from the group of companies that answered the previous questionnaire and showed interested in implementing LP (nine from 18 respondents, around 56%). These interviews enabled to know if the companies implemented LP or not and the most important reasons behind this decision.

So, the main objective of this chapter is to present the main results of the 2015 questionnaire and the findings of the follow-up study. Also, the authors wanted to understand how Portuguese TCI companies see LP in 2015 and if their interest was genuine to the point that they implement or try to implement LP.

Furthermore, some behavioral changes have been made, allowing some openness to different work organization methods. However, a great work of disclosure and awareness is required before going on the field and achieving the necessary competency to face competition and open markets. For these reasons, increasing the application of LP tools is important to achieve optimal performance of the organization, by reducing waste and improving processes' efficiency. Changes are taking place but with a slow step, as evidenced by the authors' study carried out in 2011 and 2015. By the end of this period only one more company had implemented LP, with a proportional increase of, approximately, 6% to 15%.

This chapter is organized into five sections, after the introduction, where the objectives are presented, second section presented the context study. A brief review of the LP model is conducted in section three. The fourth section highlights the research methodology used to collect data, and the instruments used. The results are analysed and discussed in section five. The sixth section presents the final remarks.

2. CONTEXT STUDY

This study was developed in the context of the Portuguese textile and clothing Industry (TCI). TCI is largely represented in the Portuguese industry and always had an important role in the national economy. This industry comprises two large sectors: 1) the textile sector (fibre production, spinning, weaving, knitting, and finishing); and 2) the clothing sector (manufacturing of clothing and accessories). TCI is mainly constituted by small and medium-sized enterprises (SME).

For the Portuguese Institute of Support to Small and Medium-sized Enterprises and Innovation (IAPMEI), a micro-company has less than ten workers, small company less than 50 workers and medium company less than 250 workers. Also, SMEs are companies with a turnover of less than 50 M€ (IAPMEI 2016). SMEs are the most important source of jobs and a niche of business and innovation, particularly in the European economy. There are about 21 million SMEs in Europe, which represent 99.8% of all enterprises in Europe, employing 87 million workers and accounting for about 58% of the total gross value added of the non-financial market segment.

In Portugal, the huge importance of SMEs in the national economy is also reflected in a significant contribution of these companies in the gross domestic product. In Portugal, SMEs represent 67% of the total value created in the economy. The dynamism shown by Portuguese SMEs, as well as their strong contribution to employment and growth, plays a vital role in the future of the Portuguese economy (Teixeira 2016).

In 2011, as described in Maia, Alves, and Leão (2013), Portuguese TCI became more dynamic and competitive, increased its investments in technology and became increasingly modern, changing its business and production strategies. The objective of these strategies was to increase their performance, developed a culture of quality and innovation, respond quickly to customers and dominate the distribution channels. This evolution is mandatory to respond to the delocalisation and closure of companies and maintain TCI as one of the Portuguese's most important manufacturing industries.

According to an interview with João Costa (past president of the Textile and Clothing Association of Portugal – ATP) conducted on March 2015, the TCI exportation had increased

30% since 2009, representing around 6% of national exports (Silva 2015). In recent years, the value chain has changed from large series and products of low added value to companies seeking to be increasingly innovative and creative, supported by technology centres, and focusing on external dynamic promotion (60 fairs in 35 countries in 2014) (Silva 2015). This growth continues, namely, in January of 2018, which Portugal exportations grew in textiles, despite fell in clothing (INE 2018). France and Italy are the main drivers of the great growth of Portuguese textile and clothing exports, which increased by 12.2% in April compared to the same month in last year and 2.5% in the first four months of 2017 compared to the same month of 2017 (Neves 2018). Based on these figures and on the importance of this industry, investments that do not require both capital investment and financial outlay are still worth.

Based on their TCI expertise, the authors believe that much more could be done at the organisational level. TCI is an industry that still face many challenges, with many wastes and confusion due, for example, to disorganisation, too long waiting and setup times and difficulty in understanding the importance of organisation. These problems in this industry often arise because this is a traditional industry that includes, normally, many familiar companies, which entails a resistance to change (Maia et al. 2015).

Additionally, companies in this industry have a great proportion of unskilled human resources and low willingness to change. Workers do not understand why they have to change after many years working in the same way. Therefore, they do not realise that, after 20 or 30 years using the same work methods, they could be not using the most adequate methods and have to change to cope with market changes. It is not always easy to change the way of work, even when it would benefit all (Silva 2015).

3. LITERATURE REVIEW

Lean Production (LP) is an organisation model focused on the customer, aimed to eliminate waste (activities that add no value to the products) and promote on-time delivery of quality products, materials, and information. The term "Lean Production" appeared first in the book "The Machine That Changed the World" (Womack, Jones, and Roos 1990) to describe the Toyota Production System (TPS) (Monden 1998; Ohno 1988). The TPS was developed in the Toyota Company in the 50's by (Ohno 1988). Afterwards, the Toyota principles (Liker 2004) inspired other companies to change and improve their production systems, which triggered the search for Womack's book. This book has been influencing the management thinking since its publication (Samuel, Found, and Williams 2015).

The LP has evolved into a philosophy of thinking, the Lean Thinking (Womack and Jones 1996), whose basic principles are: 1) value, 2) value stream, 3) continuous flow, 4) pull system and 5) pursue perfection. Following these principles, companies are able to create value and wealth. These principles require dedication by all people, but the last one, pursue perfection (principle 5), is the principle that requires the strongest and most continuous commitment by people in order to improve all the processes and activities in the company. This improvement is not only related to the improvement of processes and operations, as referred, but also, and more importantly, to the improvement of the worker's conditions and behaviours.

These changes, as mentioned before, are implicit in the key idea of LP: "doing more with less," where less means less occupied space, fewer transports, lower inventories, and most importantly, less human effort and fewer natural resources. More occupied space with broken machines or piles of inventory waiting to be processed, constant and irregular movement of trucks, stressful people working hard in some periods but without work in other periods, among others, are symptoms of an inefficient production system that is incapable of regularly responding to the demand. Normally, this means the presence of *muda, muri*, and *mura*, the 3M, which are three Japanese words that mean waste, physical strain or overburden and variability, respectively.

Waste or *muda* are non-value activities performed in the company that add no value to the products, from the customers' viewpoint, who are not willing to pay for it. For this reason, these activities must be identified and eliminated. According to Ohno (1988), there are seven types of waste: overproduction, excess of inventory/stocks, defects, waiting time, motion, handling and movements, and inappropriate processing or over-processing.

Muri or overburden is related to the physical strain of the system with a particular influence in the way operators work. Workers are sometimes overloaded with activities or are stressful due to work, at times feeling pressured to meet operational results and to maintain machines working continuously. This situation contradicts the LP principles that, from the very beginning, intend to be a respect-for-human system (Sugimori et al. 1977). In Lean companies, operators are intended to assume an active and participating role in the companies' life, thus assuming a critical thinking and problem-solving attitude to pursue a continuous improvement (Alves, Dinis-Carvalho, and Sousa 2012).

Mura is the variability in the system as a condition to improve quality, effective planning and safety, and to help to prevent work musculoskeletal disorders (WMSD). This regularity is achieved through standardisation, and it is related to the idea that "all work shall be highly specified as to content, sequence, timing, and outcome" (Spear and Bowen 1999). When a standard is created, everyone knows what to do and how to do it. The operations must be followed exactly as they are defined, and there is no margin for improvisation and, consequently, for accidents (Ungan 2006).

Nowadays, an increasing number of companies are implementing lean to improve processes, to increase productivity and to reduce costs by eliminating waste. This implementation is happening in companies of all kinds of goods and services, as reported by international surveys (Silva et al. 2010; Jasti and Kodali 2016; Bhamu and Sangwan 2014; Jasti and Kodali 2015; Panizzolo et al. 2012) and more focused studies (Alves et al. 2011; Alves et al. 2014a). Some reports, namely, one by the United States' National Academy of Engineering, recommended that LP should be adopted to improve productivity and achieve continuous improvement, as well as to be prepared to meet the challenges of the 21st century (Donofrio and Whitefoot 2015).

Currently, Lean Thinking is multidisciplinary, and its principles are being translated from operations to education, in areas such as services, office work, healthcare, supply chain, logistics, among others (Alves et al. 2014b). Synergies between Lean and other subjects like ergonomics (Saurin and Ferreira 2009; Tortorella et al. 2017; Vieira et al. 2012; Queta, Alves, and Costa 2014; Vicente et al. 2016; Oliveira et al. 2017; Arezes, Dinis-Carvalho, and Alves 2015; Eira et al. 2015), project management (Tenera and Pinto 2014; Carroll 2008; Pullan, Bhasi, and Madhu 2013; Bauch 2004), and sustainability (many times referred as Lean-Green) (Abreu, Alves, and Moreira 2017; Abreu, Alves, and Moreira 2016; Alves et al. 2016;

Pampanelli, Found, and Bernardes 2014; Garza-Reyes 2015; Li and Found 2016; Larson and Greenwood 2004; Pojasek 2008; Prasad, Khanduja, and Sharma 2016) are known.

However, the implementation of Lean is not an easy or simple task because it involves changing the mind-set, the culture of a company (Womack and Jones 1994; Yamamoto and Bellgran 2010; Bhasin and Burcher 2006; Bhasin 2012). It demands a new paradigm, and congruence between planning and acting, which is failing in most organisations (Flumerfelt et al. 2014). To achieve this, it is important to establish a methodology to guide this implementation. Each industry or organisation has its own work pattern and requirements, and, sometimes, a personalised or focused methodology is better than a general one. Moreover, companies' stakeholders may choose from a wide array of Lean tools and principles that can be used to remove inefficiencies. The problem lies in determining which tool or tools are more suitable and effective in preventing failures. So, methodologies for Lean implementation are essential for its success (Karim and Arif-Uz-Zaman 2013). These are the main reasons why many methodologies of Lean implementation are found in the literature (Maia, Alves, and Leão 2011).

Nevertheless, many times, no methodology is used when implementing Lean because they are still unknown to companies, particularly in the case of SMEs (Achanga et al. 2006; Cowger 2016; Hu et al. 2015). In other cases, the chosen methodology may not be appropriate for that particular industry as it have been developed within a different context (Hodge et al. 2011). Nevertheless, and although Lean tools differ from application to application, the goal is always incremental and breakthrough improvement because Lean is considered a philosophy and a mind-set (Bhasin and Burcher 2006; Yamamoto and Bellgran 2010) of continuous improvement and is not a "flavour of the month" or an endpoint, as remembered by Cowger (2016). Also, learning how to expose and fix problems is the way to create sustainable advantages that will remain for years to come.

It is this learning capability and the system-thinking and sustainability competencies pulled by the Lean Thinking philosophy (Alves, Dinis-Carvalho, and Sousa 2012; Flumerfelt et al. 2015; Alves, Flumerfelt, S., & Kahlen 2017) that will make the difference in the new context introduced by Industry 4.0 concept (Kagermann, Wahister, and Helbig 2013). In contrary of what some people said that Industry 4.0 will replace Lean, Industry 4.0 will have a positive impact by enabling some lean tools' implementation as some authors had been discussing (Kolberg, Knobloch, and Zühlke 2017; Kolberg and Zühlke 2015; Netland 2015; Pereira et al. 2016; Costa et al. 2017; Chen et al. 2013; Wong, Wong, and Ali 2009).

This was already pointed out by Womack and Jones (2005a) that advocated information technology advances would lift the technical barriers for managing providers' logistics and connecting consumers and providers, allowing to have cost effective solutions, lean solutions (Womack and Jones 2005b). A better connection between producers and consumers will reduce the overproduction, considered by Ohno (1988) as the worst waste because is the cause of all the others. Furthermore, takes from the planet resources that will miss to the future generations (Brundtland 1987).

Beyond the cost effective solutions, they need to be waste-free, eco-efficient, and integrated in a system that optimizes the whole, not just part. To achieve this, people need to be global thinkers but, at the same time, to customize local solutions. This line of thinking is needed to all industries and services. Lean provides this and Industry 4.0 provides the digital technology leading to a better control of the entire textile fabrication process (Weisenberger

2017; Chen and Xing 2015). Also, new solutions in the textile industry need to be developed to monitor and conserve the natural resources (Hussain and Wahab 2018).

4. RESEARCH METHODOLOGY

The research methodology adopted to address the objectives defined for the study included two main instruments that were developed and implemented in different periods: 1) a survey based on a questionnaire in 2015, and 2) a follow-up interview in 2018. The follow-up interview was only performed in companies chosen from the questionnaire results which criteria for their selection was their interest to implement Lean Production.

The questionnaire, entitled "Implementation of the Lean Production model in Textile and Clothing Industry – LPmodTCI", was used as the preferred instrument for obtaining data. Its objective was to investigate whether the LP model was known and/or if it was implemented in the Portuguese TCI, also checking the extent of LP implementation contribution to the improvement of business productivity and company sustainability. There is also a second instrument, a follow up interview with selected TCI companies. To this end, the results of the questionnaire applied previously were used and companies that meet the selection criteria were chosen.

To respond to these objectives, the main research questions were:

1. Which production model is implemented in your company? Are you satisfied with it?
2. What is your knowledge about Lean Production model and/or Lean Production related concepts and tools?
3. Is Lean Production model implemented in your company? If yes, how was the LP implementation process? If no, should the company implement it?
4. What is your level of satisfaction with LP and its main understanding?
5. Did the companies interested in implementing Lean in 2015, really implement it? If not, why?
6. Which difficulties were encountered in the LP implementation?

The questionnaire enables to obtain information to answer the first four research questions. The development of a questionnaire involves processes and steps that need to be well structured, founded and organised to prevent any loss of information in data analysis and to achieve the defined objectives. Since the questionnaire was used for collecting data, it was important to define the target population and to achieve the minimum statistical data set necessary to ensure results' consistency and reliability (Saunders, Lewis, and Thornhill 2008).

Additionally, the interviews capture the perspectives of TCI participants in the last two research questions. Structured interviews were chosen to be used and built on the basis of the last part of the previously applied questionnaire (more detailed in the following sections). Figure presents the phases and methods of the research methodology used.

4.1. Questionnaire Design and Implementation

This section presents the questionnaire design and implementation, namely, target population, design, validation and distribution methodology, objectives and follow-up.

4.1.1. Target Population

The population targeted by this study were companies registered in the database of the Portuguese Association of Textile Engineers and Technicians (Associação Portuguesa de Engenheiros e Técnicos Têxteis - APETT). Since the target areas were companies that produced mainly goods of discrete production, business activity related to dyes and chemicals companies, trade companies and other activities, were excluded for being outside the focus area of this study. That leaves the companies that produced mainly goods of discrete production. The questionnaire was sent to the group of 488 TCI companies, all located in the north of Portugal.

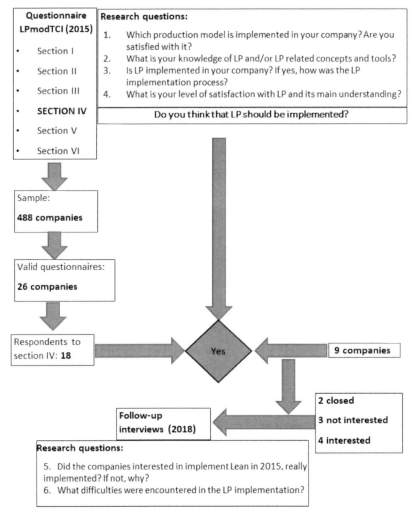

Figure 1. Research methodology phases and methods.

After the period of three months, the survey was kept open for an additional month, and after a new email sent remembering the importance of the companies' participation. The response rate to the questionnaires was 5.3% (26 completed questionnaires were received and considered valid for analysis, of 488 companies). Data results from the completed and submitted questionnaires were analysed with the help of Microsoft Excel® and SPSS®.

Analysing the profile of the respondents, it is possible to realise that about 16% were general managers and about 12% CEOs. Some designers, approximately, 8% also answered to the questionnaire. One official accounting technician filled one questionnaire. Of the respondents, 42% work less than 10 years in the company and 15% more than 30.

Most of the respondents' companies had between 1 and 100 workers (77%), thus being included in the definition of micro, small and medium companies. The majority of the companies were medium-sized (48%), followed by small sized (40%). Two of the three companies with the highest turnover were large. The third one was a medium-sized company that may have these results due to the specificity of its goods (technical textiles). This situation was expected because it follows the national pattern of this industry. On average, the ratio between the total number of workers in company and the number of workers that worked in the manufacturing area (Shop floor) was about 85%, which shows that this sector still has a strong component of human work labour. It is important to notice that results included fewer large companies and one company with a yearly turnover higher than 40M€.

The worker's average age was around 31 and 40 years old (54%), thus following the national and TCI industry pattern (PORDATA 2015).

The activity sectors of the companies that accepted the challenge and answered to the LPmodTCI questionnaire are summarised in Table 6.

It should be noted that this distribution follows that of the companies on the APETT database, used for the questionnaire. As it is possible to see in Table 6 the clothing sector differed substantially from the others TCI sectors, accounting for around 77%. The category "others" included embroidery, sports clothing, carpentry, printed fabrics.

The respondents' companies were mainly national (92%), with 8% multinational companies. Almost every company mentioned that did not have foreign participation in their capital, which corroborates the percentage of national companies. Regarding the main market of the respondents' is the "national market" with 23% and "other markets" 79%. These seem to follow the country's pattern because TCI exportations increased between those years.

Table 6. Percentages of TCI companies by sector of activity

Sector of activity	% Companies
Clothing	76,6
Knitwear	10,0
Dyeing/Stamping/Finishing	6,6
Home Textiles	13,3
Fabrics	3,3
Yarn	0,0
Technical Textiles	3,3
Socks	3,3
Accessories	0,0
Others	13,3

Afterwards, the dimension of the target, for the follow-up interviews, was smaller and according to the results previously obtained. That is, according to questionnaires results, the target were the companies that showed enthusiasm to implement LP, verifying what has been done since then.

4.1.2. Design, Validation, and Distribution Methodology

The LPmodTCI questionnaire for the diagnosis of LP awareness and level of implementation included three main parts: (1) characterisation of the respondent and respondent's company, (2) identification of the company's production model and (3) identification of the LP model implemented, divided into six sections (described in more detail in the following section).

Once developed, the questionnaire was validated by a group of TCI companies, composed of company's directors and textile engineers and technicians. The questionnaire was completed with the assistance and presence of the authors, who observed the respondents' reactions and collected their views and potential difficulties in answering a particular question. This phase was important to determine whether there was a need to modify the language used in the questions or the questions' sequence, to minimise or eliminate deviations from the objective. Also, a semantic analysis was performed to avoid misunderstandings (Saunders, Lewis, and Thornhill 2008). The reviews were positive, and no major adjustments were required.

In 2015, the main results obtained in the previous 2011 questionnaire and the 2015 questionnaire link were included in the cover letter. The 2015 questionnaire link was also identified in an article published in a technical textile journal that focused on the analysis of the obtained results (Maia, Alves, and Leão 2014). This self-administrated questionnaire was available online between September 2014 and January 2015. In December, a second message was sent, reinforcing the importance to participate in the study.

4.1.3. Objectives of the Main Parts and Sections

As previously mentioned, the LPmodTCI questionnaire was divided into three main parts, each one divided into two sections, totalling 47 questions distributed as described in Table 2. The first part was related to the characterisation of the respondent's company. The second part was related to the identification of the production model set up in the company. The third part, to be answered by companies that had implemented LP, was aimed to know why and how these companies implemented lean, what difficulties did they felt during this implementation, what LP principles and tools were implemented, what was the level of satisfaction with this model and what benefits had resulted from this implementation.

Table 2. Parts and sections of the questionnaire LPmodTCI

Parts	Sections	Section title	N° of questions
1	I	Data of the person responsible for completing the questionnaire.	4
	II	General information about the company.	13
2	III	Identification of the production model of the company.	4
	IV	Knowledge of the Lean Production model.	8
3	V	Lean Production implementation process.	12
	VI	Satisfaction with Lean Production model.	6
		Total number of questions	47

The types of questions used to collect the data were both open and closed. Since the questionnaire was self-administrated, whenever possible, closed questions were chosen. In the closed questions, the respondent was asked to select only one option or more, according to the purpose and specificity of the question. The open questions were used for identifying the respondent and characterising the company, allowing respondents to identify more easily different methodologies and concepts and to express their view. Closed questions were used on the third part (respondents from companies that implemented LP), to measure their satisfaction concerning the LP implementation, by using a five-point scale.

4.2. Interviews Preparation and Implementation

According to the results of the LPmodTCI questionnaire obtained in 2015, some companies showed willingness to implement LP. So, in 2018 to find out if these companies that responded positively indeed implemented LP, the interviews were prepared. Also, it was important to discover the difficulties confronted by those that implement and to ask about the reasoning behind for not yet been implemented. The authors contacted by telephone the respondents of 2015, to schedule a meeting to do the follow-up interview.

Based on this fact the follow-up interview conducted in 2018 were based on some points: if the company had already implemented LP; if the company still want to implement lean; if the company did not implement what were the reasons; if the company had already implemented, the difficulties they faced and if they know the relationship between lean and industry 4.0.

From the 26 respondents in 2015, nine companies admitted that they would like to implement LP. From the contacts established with them in 2018, some situations happened in the last three years. Consequently, two of these companies closed and others three were not interested in being interviewed (one because the person who filled the questionnaire no longer works in the company and another because they had no available time). Nevertheless, they confirmed the interest in implementing Lean. According to this, only four were interviewed. The Table 3 presents the companies that accepted to be interviewed.

Table 3. Companies' characterisation

Company	1	2	3	4
Product type	Knitwear; Clothing; Home Textiles).	Clothing	Clothing; Underwear exterior and interior	Knitwear; Clothing; Workwear.
Multinational	No	No	No	No
Foreign capital	No	No	No	No
Main markets	E.U.; U.S.A	E.U.	France, Italy, U.S.A.	Scandinavia
National market (%)	3,5	15	0	3
Company size	Medium	Small	Medium	Medium
Business volume (Yearly)	- 7 M€	- 7 M€	- 7 M€	- 7 M€

5. RESULTS, ANALYSIS, AND DISCUSSION

This section presents the results, analysis, and discussion of the questionnaire applied in 2015 and the follow-up study based on face-to face interviews in 2018. It starts with the TCI companies' characterisation and then analyses the responses given to the main research questions: 1) production models' identification and level of satisfaction; 2) knowledge of LP and related concepts and tools; 3) LP implementation process; 4) satisfaction level and understanding of LP. The results reflected the answers obtained in 2015. The interviews were focused in the responses provided to the questionnaire section IV, question 4.7 and subsequent. The section ends with a discussion and interpretation of the obtained results, accompanied, whenever possible, by a comparison with other studies.

5.1. Questionnaire Results and Discussion

This section is divided according to the sections of the questionnaire. The first subsection enables the companies' production models identification and the level of satisfaction among them with the models. In the second subsection are presented the results of the knowledge of LP and related concepts followed by the LP implementation process in subsection four. Last subsection presents the results of satisfaction level and understanding of LP.

5.1.1. Companies' Production Models

Before looking at this issue of companies' production models, it should be identified the strategies used to face the competition. It was important to introduce this question because this could indicate an interest by the companies in implementing new and efficient management strategies. So, ten sentences, each corresponding to a different strategy, were included in a multiple-choice question, for respondents to identify the strategies adopted by their companies. An "Other" option was included to obtain different responses. The obtained results are presented in Figure .

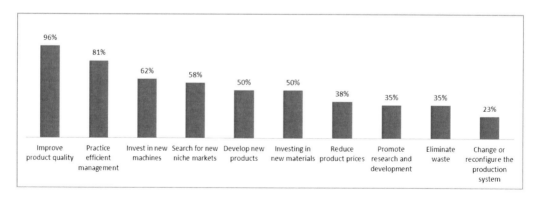

Figure 2. Strategies adopted by the company to face the competition.

"To improve the products' quality" was the most identified strategy (around 96%), followed by "Practice an efficient management" (around 81%). "Eliminate waste" and "Change or reconfigure the production system" were the strategies less chosen what could indicate that companies did not relate those with practicing an efficient management. These two are compulsory when implementing LP (Alves et al. 2015).

Beyond the strategies presented, the respondents filled the "Other" category with "Training" and one respondent mentioned that "There was no competition in Portugal".

In order to understand the companies' knowledge on the different existing production models, the designations of 17 production models were listed: 1) Social-Technical System; 2) Anthropocentric Production Systems; 3) Techno-centric Manufacturing Systems; 4) Virtual Enterprises; 5) Lean Production (in Portuguese); 6) Kanban System; 7) Taylor/Ford Production Model; 8) Agile Production; 9) One-piece-flow; 10) Modular Production; 11) Lean Production (in English); 12) Kaizen System; 13) Autonomous or Semi-autonomous Work Groups; 14) Toyota Production System (TPS); 15) Quick Response; 16) Just-in-Time System (JIT); 17) Non-Stock Production or production without stocks (Carmo-Silva, Alves, and Moreira 2006; Alves 2007).

It should be noted that some of these designations are intentionally associated with the same production model, e.g., Techno-centric Manufacturing System and Taylor/Ford Production Model. Also, some production models listed were repeated: Lean Production (one in English and the other in Portuguese: "Produção Ligeira" or "Produção Magra"). These repetitions were included to understand how these models were known in the industrial environment.

The first objective was to identify if the companies know or unknown the production models referred. The Figure summarises the results obtained with regard to those unknown production models.

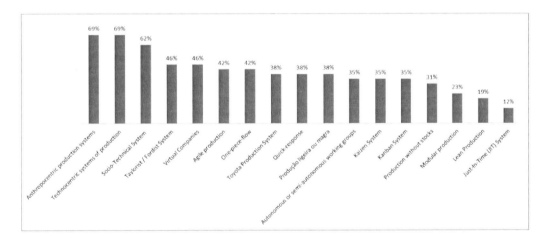

Figure 3. Models that were unknown or never heard of by the companies.

The Anthropocentric Production Systems and Techno-centric Manufacturing Systems were the most frequently unknown models (around 69% each), followed by Social-Technical System, with 62%. This behaviour was expected since the designation of these three models is more used by sociologists (Graça 2002a; Graça 2002b; Kovács and Moniz 1994). The least identified models were the JIT system, Lean Production and modular production system. The modular production this is the common designation, mainly, in textile US industry for Toyota Sewing System (Black and Chen 1995). It seems that these three models sounds familiar to

TCI companies, probably because they appeared in the 80s as a new work organisation method (Moniz 1989).

Lean Production in Portuguese and in English showed different values, with the term in Portuguese ("Produção ligeira ou magra") almost doubling as compared to the English term ("Lean Production"), 38% and 19%, respectively. This finding could indicate that the lack of knowledge of the LP model terminology in Portuguese does not necessarily correspond to the lack of knowledge of the term in English. This reflects the results of the previous questionnaires applied in 2011 (Maia, Alves, and Leão 2016), in the sense that more respondents seemed more familiar with the LP model.

Based on the same production models, respondents identified the ones that were closest to the model adopted by their company. The obtained results are presented in Figure .

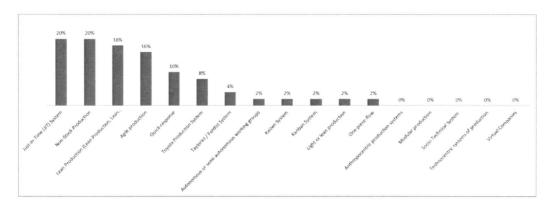

Figure 4. Production model closest to the model adopted by the company.

As expected, and according to the results obtained previously, the production model identified as the closest to the one adopted by the company was the JIT and Non-Stock Production with 20%, followed by LP with 18%. Agile Production with 16% appears immediately next. Inversely, they do not relate LP with the Portuguese designation (2%) nor with kaizen, kanban, working groups and one-piece-flow that are tools for LP implementation. Therefore, it seems that only few respondents distinguish well these designations or their associations. This could indicate a total unawareness about what is LP.

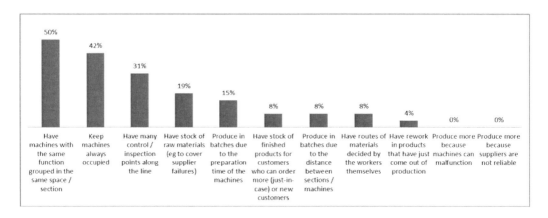

Figure 5. Solutions identified as the ones that companies adopt to face some problems.

Then the solutions adopted by the companies to overcome problems that arise very frequently was identified based on a set of eleven different solutions (Figure).

It is possible to verify that solutions such as "have machines with the same function together...", "keep machines always busy" or "having points of control...", thus revealing a traditional behaviour by the companies. Also, selecting the solution "having materials routes decided by the workers themselves," one of the measures advocated by an LP model, reveals an openness by the company to engage the worker in the decision process. Notice that "Other" alternative category not received any response.

The last question on the section III allowed the identification of companies' satisfaction with the production model adopted. Almost 67% of the respondents mentioned being satisfied with the production models adopted by their companies to meet company needs and employees' needs. The percentage of companies that were unsatisfied with the production model was 33%.

Some of the reasons identified by the companies regarding production model satisfaction were the following: "there is always something to improve"; "there are many inefficiencies to solve"; "we need a better way to organise the work we do and a quick response to our orders."

5.1.2. Awareness of LP and Related Concepts and Tools

The section IV of the questionnaire was related to the awareness of LP as a production model and related concepts and tools. To achieve this awareness, respondents were asked to select one or more concepts from a list of thirteen concepts related to LP. After this, it was mentioned that these concepts were associated with lean and respondents were asked if they recognised that. Then, respondents were asked about known tools and if they can relate them to the LP. In this case, a definition of LP was made in a way to provide uniformity of knowledge. Finally, this section finished by asking if respondents' companies implemented LP or if they would like to implement LP in their company in the near future.

Concerning the level of lean-related concepts that companies had, the most known lean-related concept was waste (73%), followed by Value (69%), JIT and Quality in the origin (65%), perfection and value chain (62%) and continuous flow (50%) (Figure).

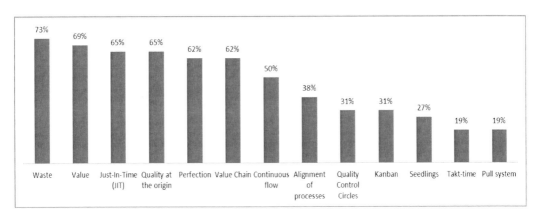

Figure 6. Lean-related concepts identified as the most known by TCI companies.

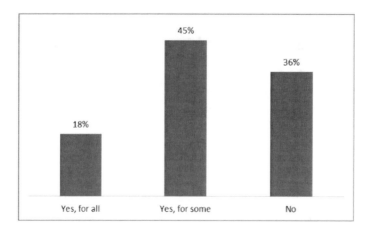

Figure 7. Knowledge of the association between the previous concepts and lean production.

When asked if they knew that these concepts were related to LP just around 20% of the respondents answered that all concepts were related to LP. Almost 50% (45%) identified only some of these concepts as LP key concepts (Figure).

More than half of the respondents (59%) referred knowing LP as a model of production focused on the client, whose main objective is to eliminate activities that add no value (waste) to the products. Also, they knew that LP allows providing quality products timely delivery at low cost. This results, show that while the general concept and definition of LP is relatively clear among the companies' managers, it was more difficult to relate LP to the concepts at least to more than one. This is indicative of the need for a greater knowledge about this issue.

These results where enhanced when respondents identified the tools that are related with LP. Some tools were presented with different designations, like levelling production and Heijunka (in Japanese) or overall equipment efficiency (OEE) written in English and in Portuguese. The same happened with production in small batches and one-piece-flow, which are almost identical concepts, and Six-Sigma and DMAIC, which have the same source (Figure).

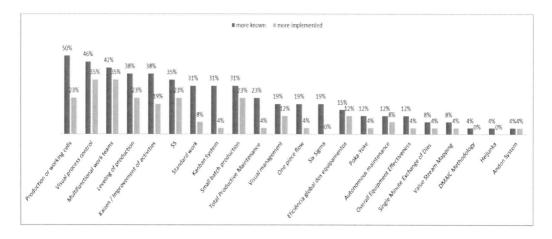

Figure 8. Lean Production tools more known and more implemented by TCI companies.

The most known selected tools included production or working cells (50%), visual control of the processes (46%), multifunctional work teams (42%), Kaizen and levelling production (38%). It was surprising that the same tool but with different designation, have different percentages which could indicate, once more, the unawareness of what the lean tools are and illiteracy on this subject. When asked if they were aware of the association only 22% said "Yes for all" and 47% "Yes for some."

There is a difference between tools knowledge and tools implementation, as Figure 3 shows, the results for the question: "What tools have you implemented?" multifunctional work teams presented the highest percentage (35%), in parallel with visual control process. On the other hand, no reply was obtained for the Six-Sigma, DMAIC and Heijunka tools (0%).

To identify if the companies have LP implemented or are awaiting to implement, two final questions were asked: 1) Is LP model implemented in your company?; 2) Should be implemented? The results showed that a large percentage of the respondents' companies (82%) did not have LP implemented. Nevertheless, the interest in implementing this model was almost 56%, as show in Figure , and a significant percentage does not know if they want to implement LP (28%).

At this point, to those respondents that do not implement Lean, the questionnaire proceeds with an open answer allowing the free expression of the participants by giving their opinions relating this issue. Only one comment was received, from a respondent with training in LP: "The production direction and I had training in lean manufacturing some years ago. We already had production control and processes, so we only changed what made sense in our reality. The implementation of work cells instead of production lines that we previously did made much difference in productivity."

The uncertainty by some respondents might result from lack of knowledge regarding LP. The authors believe that LP should be more widely rolled out across the TCI. One way of doing this is by publishing LP model in Portuguese journals and to make available to the community (Maia, Alves, and Leão 2012; Maia, Alves, and Leão 2014).

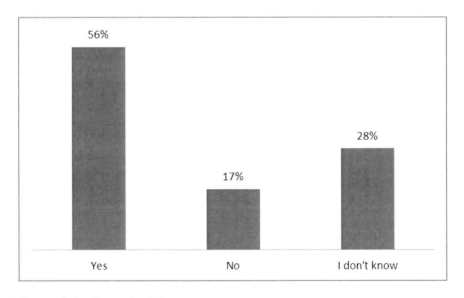

Figure 9. Interest in implementing LP.

Respondents agreed that LP could be implemented in some areas or that it would work in some situations. However, with reservations regarding the LP implementation in environments with a high diversity of models, small quantities and long production times. They believe that some concepts could be applied and are applied (in a specific company) but not in a systematic way. Once more, this could indicate their unawareness about LP, as many authors advocate regardless of the industry: goods or services companies (Alves et al. 2014b; Amaro, Alves, and Sousa, n.d.; Jasti and Kodali 2016; Jasti and Kodali 2015; Bhamu and Sangwan 2014).

From the 18 companies that answered to this question (almost 70% of the companies that fully answered to the questionnaire) nine respondents answered positively, mentioning that LP was a model to be implemented. For these respondents the questionnaire ended. The interviews referred in the research methodology were performed to these companies. More details are given in section 5.2 of this chapter.

5.1.3. LP Implementation and Implementation Process

The section V of the questionnaire was only answered by the companies that have positively answered to the question regarding Lean implementation. From the 26 companies that returned valid questionnaires, just 15% answered positively. However, of these only three respondents completed the section V of the questionnaire. The reasons for LP implementation, how the company implemented LP, the known and used methodologies for LP implementation, the involvement or not of the employers and if they had training, and the difficulties felt during LP implementation were the central themes of this section.

The first question of this section V was to know the reasons for the Lean implementation in the company. Four reasons were provided to the respondents to select: 1) The top management of the company has defined this implementation as a strategic objective to be achieved; 2) The company was not satisfied with the previous model; 3) Other companies (e.g., a client company) implemented this model and influenced the decision; 4) Someone from the company heard (e.g., a training course) about this model and thought that this company could apply. An "Other" category was also provided. The most selected reason was the first and the last two reasons were never selected.

Regarding how the respondents implemented Lean, five options were made available: 1) The company appealed only to experience and collaborators; 2) The company had hired a consulting firm; 3) The company used a team of employees and people from a consulting firm; 4) The company used an implementation methodology; 5) Other. Two of the respondents selected the second option, that is, a consulting firm was to implement LP, and the third indicated that had outsourced a service from an industrial director ("other" category). Using an implementation methodology has not been chosen by any respondent, although many of these methodologies are known (Maia, Alves, and Leão, 2011) it seems they were not of companies aware.

To better understand the companies' awareness or not, six methodologies were presented: 1) Hoshin Kanri 2) Strategy deployment (in English); 3) *Desdobramento da estratégia* (in Portuguese); 4) Policy management; 5) Toyota Production System methodology; 6) Kaizen Institute methodology. It should be noted that the second, third and fourth options are the same but in different languages. It was interesting to observe that all the identified methodologies were selected by the respondents. However, when asked if they knew that these methodologies were related with lean, each respondent answered differently, that is, one answered "Yes, for all", the other "Yes, for some", and the third "No".

Referring the methodology used to the implementation, eight options were presented: 1) the company did not use any methodology, 2) Hoshin Kanri, 3) Strategy development, 4) *Desdobramento da estratégia*, 5) Policy management, 6) Toyota Production System methodology, 7) kaizen Institute methodology, and 8) Methodology developed by the company itself (different the others identified). The last option, methodology developed by the company, was the most selected. This seemingly contradictory with the previously results that identify no methodology was chosen to implement Lean.

In a LP implementation, workers involvement and training is considered one of the key factors (Achanga et al. 2006; Sohal 1996; Motwani 2003; Farhana and Amir 2009; Netland 2016; Salonitis and Tsinopoulos 2016; Alhuraish, Robledo, and Kobi 2017). For this reason, it was included in the questionnaire three questions related with the LP implementation communication and training to the shop-floor and others employees. Two of the companies mentioned that workers from the shop floor were informed and the remaining company mentioned that they were not informed. Similar answers were obtained regarding training, i.e., two companies had provided training, and the other no training.

It was also important to know the difficulties which companies have arisen during the LP implementation process. For this, the three remaining questions of section V were related with these. Results indicate that the main difficult identified during LP implementation were some resistance from the production employees. It is important to notice that the LP implementation was only applied to production areas. Others difficulties were identified, to know: the benefits' quantification, the lack of knowledge of the lean principles and tools and, the lack of communication. This is in according what have been found in the literature (Silva et al. 2010; Cowger 2016; Motwani 2003; Keitany and Riwo-Abudho 2014; Pearce, Pons, and Neitzert 2018).

To rate the difficulties that the companies encountered in the LP implementation, a 5-point Likert scale was used from 1 meant "very difficult" to 5 "very easy." The three companies selected options from 1 to 3, so indicating that there is a relatively difficulty with the implementation of Lean. The justifications given for such difficulties were mainly concerned to the 1) "resistance to change by some employees", that the 2) "technicians did not dominate the processes that had enormous variability", and that 3) "the process is still in transition and is not easy at all, but it is not impossible".

5.1.4. Satisfaction Level and Understanding of LP

Finally, section VI was related with the satisfaction with the Lean model in general. The companies were unanimous in considering that the LP implementation was important. To rate the degree of satisfaction with LP, a 5-points Likert scale was used from 1 meant "not at all satisfied" to 5 "very satisfied." The three companies indicated that they were satisfied with the LP model (4, "satisfied"). The justifications given were relating to the "quality and productivity increase"; and the "production costs and final product improvements". One mentioned that the LP implementation is "considered being at 40% of the way".

The greatest benefits with LP implementation were: an increase of quality products, stocks, costs and lead time reduction, new products' development in less time, an increase of production flexibility of new and different products, and, profits, and productivity and employees' satisfaction. This is according with the literature reviewed (Resende et al. 2014;

Buehlmann and Fricke 2016; Fullerton and McWatters 2001; Melton 2005; Das, Venkatadri, and Pandey 2014; Alves et al. 2011).

The final question of the questionnaire was related with respondents' understanding of LP. To obtain the results six options were presented: 1) A production model aimed to achieve continuous improvement; 2) A set of tools and techniques to improve operations; 3) A production model aimed to reduce waste; 4) A system to innovate, develop new products and improve relations between suppliers and customers; 5) A philosophy of integrated management; 6) A way of life.

The most selected option was "A set of tools and techniques to improve operations", but all were selected, except the option 5: "an integrated management philosophy". Once more, this showed the unawareness of the companies in seeing LP as a philosophy as showed by the literature (Samuel, Found, and Williams 2015; Yamamoto and Bellgran 2010; Bhasin and Burcher 2006; Amaro, Alves, and Sousa, n.d.).

5.2. Follow-Up Interviews Results and Discussion

In this section are presented the results, analysis, and discussion of the follow-up interviews conducted in the four companies that accepted to participate in the study. After the results analysed, it was possible to conclude that neither of these four interviewed companies has implemented LP (since 2015). But all of them have this as a main goal for the company. The reasons presented by the companies were lack of time, the resistance of the shop floor employees to the change, not having in the company a person with sufficient knowledge about Lean principles or about their tools, the fact that is a long process and it seems, that a large amount of capital is required, in their opinion.

To employ a consultant to perform the implementation can be expensive, so companies are waiting for their economic availability, or appeal to the knowledge of their human resources, even though they recognise that they need help and support. All the interviewees showed interest to learn more about Lean, and seemed aware of the advantages that its implementation could bring to the success and future of the company.

Finally, when asked about the arrival of industry 4.0 and the key role that Lean can play to facilitate and to attain the objectives, three of the companies considered that could be very useful to implement Lean to face the changes. They considered that a company more organized, easily accommodate any necessary changes. One of the companies admit that they do not know anything about Industry 4.0.

5.3. Discussion and Limitations

The present study was designed to investigate LP awareness, extension, and implementation in the Portuguese TCI. Data collection was performed through two instruments that meets the objectives of the study: an online questionnaire and a follow-up interview. Key findings of the questionnaire intended to answer the defined set of research questions that led to the research: How many TCI know the meaning of LP and/or related concepts and tools? How many have a LP model implemented? What is the stakeholders' knowledge on the production model implemented and the degree of satisfaction?

This set of questions enables the knowledge that the respondents have regarding the production model of their company and their level of satisfaction with the model (section 5.1). In general, the respondents has little or no knowledge of production models. This can be explained by the many different designations used by companies for the same production model, which is quite common in Industrial Engineering. Thus, it is perfectly understandable that they did not recognise some terms. Most respondents seem to be satisfied with their company's production model. Nevertheless, they point out many similar needs in both periods' comments: a better organisation, continuous improvement and waste elimination.

Also, the knowledge of the LP model and related concepts and tools were understood. Respondents do not know the meaning of lean, reinforcing that LP do not have a long tradition in Portugal. In the 80s and 90s, new methodologies such as JIT production, Six Sigma, PDCA, Quality Control and Quality Control Circles (Kovács & Moniz, 1994) appeared, but LP as a philosophy is recent. The respondents worked in SMEs (only three were from in large companies), which could indicate, once more, the lack of knowledge of this model. As Cowger (2016) referred, there is a lot of confusion around the lean paradigm, as there are many terms for the same tool, a lot of experts, books, and so on. Thus, SMEs do not have the knowledge nor the time to evaluate what to believe/implement and cannot afford to expend their limited resources and time to pursuing something that they do not really believe in.

Figure 4 (section 5.1.1.) revealed that respondents did not seem to know the relation between waste elimination and practising an efficient management, which lean promotes as a philosophy of integrated management. Additionally, they did not seem to know the importance of an appropriate production system for eliminating many wastes and for sustaining the competitiveness of a company (Alves et al. 2015).

These same reasons could be behind the most important finding of this questionnaire, which was the small number of respondents that considered that their companies have implemented LP. Only four of 22 respondents responded affirmatively to this question. Nevertheless, just three respondents answered to the remaining questions of the questionnaire about the LP implementation process and satisfaction with that. Nevertheless, they referred some difficulties in the implementation process. Answers about the implementation process were very different, with resistance to change being an inhibitor.

Respondents were satisfied with LP implementation, but it seems that they were not totally aware of the importance of LP as a philosophy of integrated management or as a business philosophy (Bhasin and Burcher 2006; Bhasin 2012; James P Womack and Jones 1994). Three years after this questionnaire the situation, mainly, in four of the companies do not seem to improve concerning their knowledge about LP, according the results of the interviews.

CONCLUSION

As final remarks, and according to the survey, it is important to highlight that LP is not the most familiar production model in Portuguese TCI companies. Most of them do not know or understand LP as an integrated management philosophy. Only a small percentage of respondents considered that their companies have implemented LP, and admitted that LP was an added value for the company, despite the resistance by workers and the difficulties of implementation. These findings are a clear indication of the much work that needs to be done in Portuguese TCI to ensure success.

Although some Portuguese TCI companies are currently conscious of the fact that only by organising companies and changing mentalities they will survive and face the competition, they still have much to evolve. Accordingly, LP and lean thinking principles can only help if they are adequately implemented and a simple and focused methodology can be used. This methodology should prepare and train people to change the TCI companies' way of work, to eliminate wastes, create value and satisfy the costumers. A methodology with this intention has already been created (Maia et al. 2017). Using this methodology, companies could see that other strategies; production and organisation models are available.

More awareness and training, particularly in this TCI sector, should also be promoted and continued. Strategies like reconfiguring production systems, redefining layouts, eliminating processes, studying new ways to do particular tasks, providing training, improving the flow of production, replacing some obsolete machinery or tools, organising production, and performing preventive maintenance are some solutions that could be adopted by companies to improve the production flow and achieve the Lean Thinking third principle. Also, technologies that better align the consumers and producer's needs, such as Industry 4.0 technologies, pulling from the first what they really need, will reduce the overproduction.

Unfamiliarity about Lean and its transversally and universality exists. It is important to know that Lean is a journey for life, without ending, and that its principles are universal and applicable to all situations, as only true lean companies are sustainable. So, there is a need to divulge Lean more as an integrated management, as well as production models and their tools, for a better knowledge of the production system and increased chances for people to select the right system. Additionally, the authors considered important to create a glossary to standardise the terminology. Following this study, the authors intend to develop a deeper study using correlation factors between some of the variables to understand better their relation.

REFERENCES

Abreu, M.F., A.C. Alves, and F. Moreira. 2017. "Lean-Green Models for Eco-Efficient and Sustainable Production." *Energy* 137. doi:10.1016/j.energy.2017.04.016.

Abreu, M, A. C Alves, and F Moreira. 2016. "Comparing Lean-Green Models for Eco-Efficient Production." In: *Proceedings of 29th International Conference on Efficiency, Cost, Optimisation, Simulation and Environmental Impact on Energy Systems*. Portoroz.

Achanga, Pius, Shehab Esam, Rajkumar Roy, and Geoff Nelder. 2006. "Critical Success Factors for Lean Implementation within SMEs." *Journal of Manufacturing Technology Management* 17 (4): 460–71.

Alhuraish, Ibrahim, Christian Robledo, and Abdessamad Kobi. 2017. "A Comparative Exploration of Lean Manufacturing and Six Sigma in Terms of Their Critical Success Factors." *Journal of Cleaner Production* 164 (October). Elsevier Ltd: 325–37. doi:10.1016/j.jclepro.2017.06.146.

Alves, A.C., Flumerfelt, S., Kahlen, F.-J. 2017. *Lean Education: An Overview of Current Issues*. Springer International Publishing Switzland 2017. doi:http://doi.org/10.1007/978-3-319-45830-4.

Alves, A.C., Dinis-Carvalho, J., and Sousa, R.M. 2012. "Lean Production as Promoter of Thinkers to Achieve Companies' Agility." *Learning Organization* 19 (3): 219–37. doi:10.1108/09696471211219930.

Alves, A.C., Dinis-Carvalho, J., Sousa, R.M., Moreira, F. and Lima, R.M. 2011. "Benefits of Lean Management: Results from Some Industrial Cases in Portugal." In *6º Congresso Luso-Moçambicano de Engenharia (CLME2011)*, ch. 9, Edições INEGI. http://hdl.handle.net/1822/18873.

Alves, A.C., Rui M. Sousa, Dinis-Carvalho, J. and F. Moreira. 2015. "Production Systems Redesign in a Lean Context: A Matter of Sustainability." *FME Transactions* 43 (4). Belgrade University: 344–52. doi:10.5937/fmet1504344A.

Alves, A.C., Sousa, R.M., Dinis-Carvalho, J., Lima, R.M., Moreira, F., Leão, C.P., Maia, L.C., Mesquita, D., and Fernandes, S.. 2014. "Final Year Lean Projects: Advantages for Companies, Students and Academia." *Project Approaches in Engineering Education*, 1–10. http://hdl.handle.net/1822/30172.

Alves, A.C., Kahlen, F.-J., Flumerfelt, S., and Siriban-Manalang, A.-B.. 2014b. "Lean Production Multidisciplinary: From Operations To Education." In *7th International Conference on Production Research - Americas*. doi:10.13140/2.1.1524.0005.

Alves, A.C., Moreira, F., Abreu, F. and Colombo, C.. 2016. "Sustainability, Lean and Eco-Efficiency Symbioses." Edited by Marta Peris-Ortiz, João J. Ferreira, Luís Farinha, and Nuno O. Fernandes. *Multiple Helix Ecosystems for Sustainable Competitiveness - Innovation, Technology, and Knowledge Management. Innovation, Technology, and Knowledge Management*. Cham: Springer International Publishing. doi:10.1007/978-3-319-29677-7.

Amaro, P., Alves, A.C., & Sousa, R. (2019). "Lean Thinking: a transversal and global management philosophy to achieve sustainability benefits." In *Lean Engineering for Global Development*. Springer (In Press).

Arezes, P.M., Dinis-Carvalho, J. and A.C. Alves. 2015. "Workplace Ergonomics in Lean Production Environments: A Literature Review." *Work* 52 (1): 57–70. doi:10.3233/WOR-141941.

ATP - Associação Têxtil e Vestuário de Portugal. 2017. *Fashion from Portugal - Directory 2017*. Retrieved from http:// www.atp.pt/ fotos/ editor2/ 2017/ Diretorio% 20ATP% 202017.pdf, accessed in January 2019.

Bauch, Christoph. 2004. *Lean Product Development: Making Waste Transparent*. Thesis: MIT Sociotechnical Systems Research Center (SSRC).

Bhamu, Jaiprakash, and Kuldip Singh Sangwan. 2014. "Lean Manufacturing: Literature Review and Research Issues." *International Journal of Operations & Production Management* 34 (7): 876–940.

Bhasin, Sanjay. 2012. "An Appropriate Change Strategy for Lean Success." *Management Decision* 50 (3): 439–58. doi:10.1108/00251741211216223.

Bhasin, Sanjay, and Peter Burcher. 2006. "Lean Viewed as a Philosophy." *Journal of Manufacturing Technology Management* 17 (1): 56–72. doi: 10. 1108/ 17410380610639506.

Black, J.T., and J.C. Chen. 1995. "The Role of Decouplers in JIT-Pull Apparel Cells." *International Journal of Clothing Science and Technology* 7 (1).

Brundtland, Gro H. 1987. "Our Common Future: Report of the World Commission on Environment and Development." *United Nations Commission* 4 (1): 300. doi:10.1080/07488008808408783.

Buehlmann, Urs, and Christian F. Fricke. 2016. "Benefits of Lean Transformation Efforts in Small- and Medium-Sized Enterprises." *Production & Manufacturing Research* 4 (1): 114–32. doi:10.1080/21693277.2016.1212679.

Carmo-Silva, S., A.C. Alves, and F. Moreira. 2006. "Linking Production Paradigms and Organizational Approaches to Production Systems." *Intelligent Production Machines and Systems - 2nd I*PROMS Virtual International Conference* 3-14 July 2006. doi:10.1016/B978-008045157-2/50090-0.

Carroll, B. 2008. *Lean Performance ERP Project Management: Implementing the Virtual Lean Enterprise*. Auerbach Publications Taylor & Francis Group. doi:10.1017/CBO9781107415324.004.

Chen, James C., Chen Huan Cheng, Potsang B. Huang, Kung Jen Wang, Chien Jung Huang, and Ti Chen Ting. 2013. *Warehouse Management with Lean and RFID Application: A Case Study. International Journal of Advanced Manufacturing Technology* 69 (1–4). doi:10.1007/s00170-013-5016-8.

Chen, Z., and M. Xing. 2015. "Upgrading of Textile Manufacturing Based on Industry 4.0." In *5th International Conference on Advanced Design and Manufacturing Engineering (ICADME2015)*, 2143–46.

Costa, F., M.D.S. Carvalho, J.M. Fernandes, A.C. Alves, and P. Silva. 2017. "Improving Visibility Using RFID – the Case of a Company in the Automotive Sector." *Procedia Manufacturing* 13. doi:10.1016/j.promfg.2017.09.048.

Cowger, Gary. 2016. "Half Measures Gets Less than Half Results." *Mechanical Engineering The Magazine of ASME* 138 (1): 30–35.

Das, Biman, Uday Venkatadri, and Pankajkumar Pandey. 2014. "Applying Lean Manufacturing System to Improving Productivity of Airconditioning Coil Manufacturing." *International Journal of Advanced Manufacturing Technology* 71 (1–4). doi:10.1007/s00170-013-5407-x.

Donofrio, Nicholas M, and Kate S Whitefoot. 2015. *Making Value for America: Embracing the Future of Manufacturing, Technology, and Work*. National Academy of Engineering.

Eira, R., L.C. Maia, A.C. Alves, and C.L. Leão. 2015. "Ergonomic Intervention in a Portuguese Textile Company to Achieve Lean Principles." In *SHO2015: International Symposium on Occupational Safety and Hygiene*, edited by P. Arezes, JS Baptista, MP Barroso, P Carneiro, P Cordeiro, N Costa, R Melo, AS Miguel, and G Perestrelo, 100–102. Guimarães, Portugal: Portuguese SOC Occupational Safety & Hygiene.

Farhana, F, and A Amir. 2009. "Lean Production Practice: The Differences and Similarities in Performance between the Companies of Bangladesh and Other Countries of the World." *Asian Journal of Business Management* 1 (1): 32–36. http://www.airitilibrary.com/searchdetail.aspx?DocIDs=20418752-200907-201008160043-201008160043-32-36.

Flumerfelt, S., F.-J. Kahlen, A. Alves, J. Calvo-Amodio, and C. Hoyle. 2014. "Systems Competency for Engineering Practice." In *ASME International Mechanical Engineering Congress and Exposition, Proceedings (IMECE)*. Vol. 11. doi:10.1115/IMECE2014-40142.

Flumerfelt, Shannon, Franz-Josef Kahlen, Anabela C. Alves, and Anna Bella Siriban-Manalang. 2015. *Lean Engineering Education: Driving Content and Competency Mastery*. ASME Press.

Found, Samuel, D.P. and Williams J.S. 2015. *How Did the Publication of the Book The Machine That Changed The World Change Management Thinking? Exploring 25 Years of Lean Literature*. 35(10): 1386–1407. doi:10.1108/02683940010305270.

Fullerton, Rosemary R., and Cheryl S. McWatters. 2001. "Production Performance Benefits from JIT Implementation." *Journal of Operations Management* 19 (1): 81–96. doi:10.1016/S0272-6963(00)00051-6.

Garza-Reyes, Jose Arturo. 2015. "Lean and Green – a Systematic Review of the State of the Art Literature." *Journal of Cleaner Production* 102. Elsevier Ltd: 18–29. doi:10.1016/j.jclepro.2015.04.064.

Graça, L. 2002a. *Novas Formas de Organização do Trabalho I*. [*New forms of work organisation. Part one*]., Retrieved from: http:// www.ensp.unl.pt/ luis.graca/ textos164.html.

Graça, L. 2002b. *O Caso Da Fábrica de Automóveis Da Volvo Em Uddevalla (Suécia)*. [*Volvo Automobile Assembly Plant at Uddevalla, Sweden. Part One.*], Retrieved from: https://www.ensp.unl.pt/luis.graca/textos44.html.

Hodge, George L., Kelly Goforth Ross, Jeff a. Joines, and Kristin Thoney. 2011. "Adapting Lean Manufacturing Principles to the Textile Industry." *Production Planning & Control* 22 (3): 237–47. doi:10.1080/09537287.2010.498577.

Hu, Qing, Mason, Robert, Williams, Sharon J, and Found.Pauline 2015. "Lean Implementation within SMEs: A Literature Review." *Journal of Manufacturing Technology Management* 26 (7). Emerald: 980–1012. doi:10.1108/JMTM-02-2014-0013.

Hussain, Tanveer, and Wahab, Abdul. 2018. "A Critical Review of the Current Water Conservation Practices in Textile Wet Processing." *Journal of Cleaner Production*, July. doi:10.1016/j.jclepro.2018.07.051.

IAPMEI. 2016. *Definição de PME*. [*SME definition*]. Retrieved from https:// www.iapmei.pt/ getattachment/ produtos- e- servicos/ qualificacao- certificacao/ certificacao- pme/ recomendacao-da-comissao-2003-361-ce.pdf.aspx, accessed in january 2019.

INE. 2018. *Exportações de Janeiro crescem nos têxteis e descem No vestuário*. [*January exports grow in textiles and fall into clothes*]. Retrieved from: https://jornal-t.pt/noticia/exportacoes-de-janeiro-crescem-nos-texteis-e-descem-no-vestuario/

Jasti, Naga Vamsi Krishna, and Rambabu Kodali. 2015. "Lean Production: Literature Review and Trends." *International Journal of Production Research* 53 (3): 867–85. doi:10.1080/00207543.2014.937508.

Jasti, Naga Vamsi Krishna . 2016. "An Empirical Study for Implementation of Lean Principles in Indian Manufacturing Industry." *Benchmarking: An International Journal* 23 (1): 183–207.

Kagermann, H., Wahister W., and Helbig, J.. 2013. *Recommendations for Implementing the Strategic Initiative INDUSTRIE 4.0: Securing the Future of German Manufacturing Industry*. Retrieved from https:// www. din. de/ blob/ 76902/ e8cac883f42bf28536e7e816 5993f1fd/recommendations-for-implementing-industry-4-0-data.pdf

Karim, Azharul, and Arif-Uz-Zaman, Kazi. 2013. "A Methodology for Effective Implementation of Lean Strategies and Its Performance Evaluation in Manufacturing

Organizations." *Business Process Management Journal.* Vol. 19. doi:10.1108/14637151311294912.

Keitany, P., and Riwo-Abudho, M.. 2014. "Effects of Lean Production on Organizational Performance: A Case Study of Flour Producing Company in Kenya." *European Journal of Logistics Purchasing and Supply Chain Management* 2 (2): 1–14.

Kolberg, Dennis, Knobloch, Joshua, and Zühlke, Detlef. 2017. "Towards a Lean Automation Interface for Workstations." *International Journal of Production Research* 55 (10): 2845–56. doi:10.1080/00207543.2016.1223384.

Kolberg, Dennis, and Zühlke, Detlef. 2015. "Lean Automation Enabled by Industry 4.0 Technologies." *IFAC-PapersOnLine* 48 (3). Elsevier: 1870–75.

Kovács, I., and Moniz, A.B.. 1994. "Trends for the Development of Anthropocentric Production Systems in Small Less Industrialised Countries: The Case of Portugal." *MPRA Paper No. 6551.*

Larson, Tim, and Greenwood, Rob. 2004. "Perfect Complements: Synergies between Lean Production and Eco-Sustainability Initiatives." *Environmental Quality Management* 13 (4): 27–36. doi:10.1002/tqem.20013.

Li, Ai Qiang, and Found, Pauline. 2016. "Lean and Green Supply Chain for the Product-Services System (PSS): The Literature Review and A Conceptual Framework." In *Procedia CIRP.* Vol. 47. doi:10.1016/j.procir.2016.03.057.

Liker, Jeffrey K. 2004. *The Toyota Way: 14 Management Principles from the World's Greatest Manufacturer.* McGraw-Hill Education.

Maia, L.C., Alves, A.C., and Leão, C.P. 2012. "Implementar o modelo de Produção Lean na ITV: Porquê e Como?" ["Lean Production model implementation in Textile and Clothing Industry: Why and How?"]. *Nova Têxtil* 99: 18–23. http:// repositorium.sdum.uminho.pt/handle/1822/20082.

Maia, L.C., Alves, A.C., and Leão, C.P.. 2013. "Sustainable Work Environment with Lean Production in Textile and Clothing Industry." *International Journal of Industrial Engineering and Management* 4 (3).

Maia, L.C., Alves, A.C., Leão, C.P., and Eira, R.. 2017. "Validation of a Methodology to Implement Lean Production in Textile and Clothing Industry." In *ASME International Mechanical Engineering Congress and Exposition, Proceedings (IMECE).* Vol. 2. doi:10.1115/IMECE2017-71464.

Maia, L.C., Eira, R., Alves, A.C., and P. Leão, C.. 2015. "Perceptions and Understandings on the Need of Change: Viewpoints across Management Levels." *Advances in Intelligent Systems and Computing.* Vol. 354. doi:10.1007/978-3-319-16528-8_23.

Maia, Laura C, Alves, Anabela C, and P Leão, Celina. 2016. "Lean Production Awareness and Implementation in Portuguese Textile and Clothing Companies." In *Proceedings of the Regional Helix 2016, International Conference on Triple Helix Dynamics; Topic: 9. Industry-Academia Interactions and Sustainability.* Castelo Branco, Portugal.

Maia, L. C., Alves, A.C., and P. Leão, Celina. 2014. "Implementar o Modelo de Produção Lean na ITV para promover Sistemas Eco-Eficientes." ["Lean Production model implementation to promote eco-efficient systems"]. *Nova Têxtil* 101: 3–10. http://hdl.handle.net/1822/36861.

Maia, L. C., Alves, A. C., and Leão, C. P. 2011. "Metodologias para implementar Lean Production: Uma Revisão Crítica de Literatura." ["Methodologies to implement Lean Production: a critical literature review"]. In *6º Congresso Luso-Moçambicano de*

Engenharia (CLME2011): A Engenharia No Combate à Pobreza, Pelo Desenvolvimento e Competitividade. ch. 9, Edições INEGI, http://hdl.handle.net/1822/18874.

Manyika, James, Jeff Sinclair, Richard Dobbs, Gernot Strube, Louis Rassey, Jan Mischke, Jaana Remes, et al. 2012. *Manufacturing the Future: The next Era of Global Growth and Innovation*. McKinsey Global Institute.

Melton, T. 2005. "The Benefits of Lean Manufacturing." *Chemical Engineering Research and Design* 83 (6): 662–73. doi:10.1205/cherd.04351.

Monden, Y. 1998. *Toyota Production System: An Integrated Approach to Just-in-Time*. Industrial Engineering and Management Press, Insti.

Monforte, Priscila Morcelli, Ualison Rébula Oliveira, and Henrique Martins Rocha. 2015. "Failure mapping process: An applied study in a shipyard facility." *Brazilian Journal of Operations & Production Management* 12 (1): 124. doi: 10.14488/ BJOPM. 2015. v12. n1.a12.

Moniz, A. 1989. "Modernização da Indústria Portuguesa: Análise de um inquérito sociológico." ["Modernization of Portuguese Industry: Analysis of a sociological survey"]. *MRMA Paper 6968*, University Library of Munich, Germany.

Motwani, Jaideep. 2003. "A Business Process Change Framework for Examining Lean Manufacturing: A Case Study." *Industrial Management & Data Systems* 103 (5): 339–46. doi:10.1108/02635570310477398.

Netland, Torbjørn H. 2015. "Industry 4.0: Where It Leave Lean?" *Special Feature Industry 4.0* April: 22–23. http://www.leanmj.com/.

Nasser, Ahdmad. 2016. "Critical Success Factors for Implementing Lean Production: The Effect of Contingencies." *International Journal of Production Research* 54 (8): 2433–48. doi:10.1080/00207543.2015.1096976.

Neves, R. 2018. "Italianos e Franceses dão brilho às exportações Portuguesas de têxteis e vestuário." ["Italian and French Brings Brightness to Portuguese Exports of Textiles and Clothing"]. *Jornal de Negócios*. https:// www.jornaldenegocios.pt/ empresas/ industria/ detalhe/ franceses- e- italianos- dao- brilho- as- exportacoes- portuguesas- de- texteis- e- vestuario.

Ohno, Taichii. 1988. *Toyota Production System: Beyond Large-Scale Production*. Portland: Productivity Press.

Oliveira, B., Alves A.C., P. Carneiro, and Ferreira A.C.. 2017. "Integration of Ergonomics and Lean Production to Improve Productivity and Working Conditions." Edited by G.P. (eds.) Arezes, P., Baptista, J.S., Barroso, M.P., Carneiro, P., Cordeiro, P., Costa, N., Melo, R., Miguel, A.S., Perestrelo. *Occupational Safety and Hygiene - SHO2017*. doi:107-109.

Pampanelli, Andrea Brasco, Found, Pauline, and Bernardes, Andrea Moura. 2014. "A Lean & Green Model for a Production Cell." *Journal of Cleaner Production* 85. Elsevier Ltd: 19–30. doi:10.1016/j.jclepro.2013.06.014.

Panizzolo, Roberto, Garengo, Patrizia, Sharma, Milind Kumar, and Gore, Amol. 2012. "Lean Manufacturing in Developing Countries: Evidence from Indian SMEs." *Production Planning & Control: The Management of Operations* 23 (10–11): 769–88.

Pearce, Antony, Pons, Dirk, and Neitzert, Thomas. 2018. "Implementing Lean - Outcomes from SME Case Studies." *Operations Research Perspectives* 5: 94–104. doi:10.1016/ j.orp.2018.02.002.

Pereira, A., M.F. Abreu, D. Silva, A.C. Alves, J.A. Oliveira, I. Lopes, and M.C. Figueiredo. 2016. "Reconfigurable Standardized Work in a Lean Company - A Case Study." In *Procedia CIRP*, 52:239–44. doi:10.1016/j.procir.2016.07.019.

Pojasek, Robert B. 2008. "Framing Your Lean-to-Green Effort." *Environmental Quality Management*, 85–93. doi:10.1002/tqem.

PORDATA. 2015. *Tabela População Activa: Total e Por Grupo Etário*. [Active population: total and by age group]. Retrieved from: https:// www.pordata.pt/ Portugal/ População+activa+total+e+por+grupo+etário+-29, accessed in January 2019.

Prasad, Suresh, Khanduja, Dinesh, and Surrender, K Sharma. 2016. "An Empirical Study on Applicability of Lean and Green Practices in the Foundry Industry." *Journal of Manufacturing Technology Management* 27 (3). Emerald: 408–26. doi:10.1108/JMTM-08-2015-0058.

Pullan, Thankachan Thomas, Bhasi, M., and Madhu, G. 2013. "Decision Support Tool for Lean Product and Process Development." *Production Planning & Control* 24 (6). doi:10.1080/09537287.2011.633374.

Queta, Vanessa, Alves, Anabela Carvalho, and Costa, Nelson. 2014. "Project of Ergonomic Shelves for Supermarkets in a Lean Work Environment." In *International Symposium on Occupational Safety and Hygiene*.

Resende, Vitor, Alves, Anabela C., Batista, Alexandre, and Silva, Ângela. 2014. "Financial and Human Benefits of Lean Production in the Plastic Injection Industry: An Action Research Study." *International Journal of Industrial Engineering and Management* 5 (2): 61–75. http://www.scopus.com/inward/record.url?eid=2-s2.0-84909986334&partnerID=tZOtx3y1.

Salonitis, Konstantinos and Tsinopoulos, Christos. 2016. "Drivers and Barriers of Lean Implementation in the Greek Manufacturing Sector." Edited by Michel Leseure. *Procedia CIRP* 57 (3): 189–94. doi:10.1016/j.procir.2016.11.033.

Saunders, Mark, Lewis, Philip, and Thornhill, Adrian. 2008. *Research Methods for Business Students. Research Methods for Business Students*. doi:10.1007/s13398-014-0173-7.2.

Saurin, Tarcisio Abreu, and Ferreira, Fabricio, Cléber . 2009. "The Impacts of Lean Production on Working Conditions: A Case Study of a Harvester Assembly Line in Brazil." *International Journal of Industrial Ergonomics* 39 (2): 403–12. doi:10.1016/j.ergon.2008.08.003.

Silva, Cristóvão, Tantardini, Marco, Staudacher, A.P., and Salviano K.. 2010. "Lean Production Implementation: A Survey in Portugal and a Comparison of Results with Italian, UK and USA Companies." In *Proceedings of 17th International Annual EurOMA Conference -Managing Operations in Service Economics,* (Eds.) R. Sousa, C. Portela, S.S. Pinto, H. Correia, Universidade Católica Portuguesa, 6-9 June, Porto, Portugal, 1–10.

Silva, José Paulo. 2015. "João Costa: Têxtil e Vestuário é Sector Promissor e Com Futuro." ["João Costa: Textiles and Clothing is a Sector Promising and With Future"]. *Correio Do Minho*. Retrieved from: https://correiodominho.pt/noticias/joao-costa-textil-e-vestuario-e-sector-promissor-e-com-futuro/85083.

Sohal, Amrik S. 1996. "Developing a Lean Production Organization: An Australian Case Study." *International Journal of Operations & Production Management* 16 (1): 91.

Spear, S., and Bowen, H.K. 1999." Decoding the DNA of the Toyota Production System." *Harvard Business Review* 77 (5): 96–106. doi:http://search.ebscohost.com/login.aspx?direct=true&db=buh&AN=2216294&site=ehost-live.

Sugimori, Y., Kusunoki, K., Cho, F., and Uchikawa, S.. 1977. "Toyota Production System and Kanban System Materialization of Just-in-Time and Respect-for-Human System." *International Journal of Production Research* 15 (6): 553–64. doi: 10.1080/00207547708943149.

Teixeira, Vasco. 2016. *PME Portuguesas no radar da Moody´s. [Portuguese SMEs on Moody's Radar.]* Retrieved from: https://correiodominho.pt/cronicas/pme-portuguesas-no-radar-da-moody-rsquo-s/7584.

Tenera, Alexandra, and Carneiro Pinto, Luis. 2014. "A Lean Six Sigma (LSS) Project Management Improvement Model." *Procedia - Social and Behavioral Sciences* 119 (October): 912–20. doi:10.1016/j.sbspro.2014.03.102.

Tortorella, Guilherme Luz, Garcia, Lizandra, Vergara Lupi, and Pereira Ferreira, Evelise. 2017. "Lean Manufacturing Implementation: An Assessment Method with Regards to Socio-Technical and Ergonomics Practices Adoption." *The International Journal of Advanced Manufacturing Technology* 89 (9–12): 3407–3418. doi:10.1007/s00170-016-9227-7.

Ungan, Mustafa. 2006. "Standardization through Process Documentation." *Business Process Management Journal* 12 (2): 135–48. doi:10.1108/14637150610657495.

Vicente, S., Alves A.C., Carvalho S., and Costa N.. 2016. "Improving Safety and Health in a Lean Logistic Project: A Case Study in an Automotive Electronic Components Company." In *SHO2015: International Symposium on Occupational Safety and Hygiene*, edited by P. Arezes, JS Baptista, MP Barroso, P Carneiro, P Cordeiro, N Costa, R Melo, AS Miguel, and G Perestrelo. Portuguese SOC occupational safety & hygiene.

Vieira, Leandro, Balbinotti, Giles, Varasquin, Adriano, and Gontijo Leila. 2012. "Ergonomics and Kaizen as Strategies for Competitiveness: A Theoretical and Practical in an Automotive Industry." *Work* 41: 1756–62. doi:10.3233/WOR-2012-0381-1756.

Weisenberger, S. 2017. "Sewing Digital Transformation into the Fabric of the Textiles Industry." *Techonology and IIOT*. http://www.industryweek.com/iot-and-digitization-manufacturing.

Womack, J., and Daniel T. Jones. 2005b. *Lean Solutions: How Companies and Customers Can Create Value and Wealth Together.* New York, USA.: Siman & Schuster.

Womack, J.P., and Daniel T. Jones. 2005a. "Lean Consumption." *Harvard Business Review*. www.hbr.org.

Womack, James P., and Daniel T. Jones. 1996. *Lean Thinking: Banish Waste and Create Wealth in Your Corporation.* Edited by Free Press. New York: Free Press.

Womack, James P., Daniel T. Jones, and Daniel Roos. 1990. *The Machine That Changed the World: The Story of Lean Production.* New York: Rawson Associates.

Womack, James P., and Daniel T. Jones. 1994. "From Lean Production to the Lean Enterprise." *Harvard Business Review* 72 (2): 93–103. doi:http://search.ebscohost.com/login.aspx?direct=true&db=buh&AN=9405100922&site=ehost-live.

Wong, Yu Cheng, Kuan Yew Wong, and Anwar Ali. 2009. "A Study on Lean Manufacturing Implementation in the Malaysian Electrical and Electronics Industry." *European Journal of Scientific Research* 38 (4): 521–35.

Yamamoto, Yuji, and Monica Bellgran. 2010. "Fundamental Mindset That Drives Improvements towards Lean Production." *Assembly Automation* 30 (2): 124–30. doi:10.1108/01445151011029754.

In: Lean Manufacturing
Editors: F. J. G. Silva and L. Carlos Pinto Ferreira
ISBN: 978-1-53615-725-3
© 2019 Nova Science Publishers, Inc.

Chapter 12

LEAN MANUFACTURING APPLIED TO A COMPLEX ELECTRONIC ASSEMBLY LINE

F. J. G. Silva[*], *Andresa Baptista, Gustavo Pinto and Damásio Correia*

Mechanical Engineering Department, School of Engineering - Polytechnic of Porto (ISEP), Porto, Portugal

ABSTRACT

Manual assembly lines are usually deeply studied before implementation. Nevertheless, several problems upsurge when the product needs to be slightly changed. This is very common in complex electric and electronic devices usually produced in small batches, where the customers are demanding more and more features and the product needs to be continuously updated. However, these updates sometimes create huge difficulties for the previously installed assembly line, generating as well, line unbalancing and wastes of time regarding the initial situation. In this chapter, a deep study of an adjusted assembly line of electronic devices was carried out using Value Stream Mapping (VSM) method to fully understand and document the different tasks and operations. The Lean Line Balancing (LLB) was also applied in order to reduce the line bottleneck by balancing the line, becoming the Cycle Time (CT) lower than Takt Time (TT) in each workstation and the workers are not overburdened with their task. Standardized processes and standardized work were also applied. During the line layout development stage, assembly fixtures, wastes reductions and visual management techniques were applied as well, different concepts were generated and proposed and, finally, the best solutions were selected. Throughout the study, many benefits for the studied manual assembly line were found, which can be considered as a strong motivation to apply Lean Manufacturing (LM) tools for better line efficiency and production rate. The implementation of Lean methodologies endorsed an increase in the productivity of approximately 10% maintaining the same workforce, only modifying workstations and work methodologies. This chapter describes the importance of Lean manufacturing concept in modern industries.

---

[*] Corresponding Author's E-mail: fgs@isep.ipp.pt.

Keywords: lean production, production systems, assembly line, line balancing, VSM, lean line design, poka-yoke, quality tools

1. INTRODUCTION

Market liberalization and global business competitiveness have challenged companies to focus on reducing production costs (Dombrowski et al. 2016, 607; Costa et al. 2018, 3041). Flexibility in the production of a company is, nowadays, a mandatory paradigm (Nunes and Silva 2013, 330; Araújo et al. 2017, 1539; Magalhães et al. 2019, 26). It is important to consider that the market continuous change leads the companies to adopt new strategies in order to develop novel products in a flexible way, which is a challenge for the production lines that follow the Lean Principles (Costa et al. 2017, 4044; Neves et al. 2018, 697). Therefore, some assembly systems are now developed as manual assembly lines with the objective of producing differentiated products always starting from a common product structure (Bortolini et al. 2016, 927; Moreira et al. 2017, 1386). In order to determine a combination of resources such as energy, construction processes, production lines, services, and supply chains, there are methods such as mathematical models, statistical analysis and Lean tools to improve efficiency and productivity (Rohani and Zahraee 2015, 7; Rocha et al. 2018, 641). The competitive global markets, short product life cycles, and increased customer expectations have forced organizations to recognize the vital importance of investing and focusing in relentlessly work on eliminating waste from the manufacturing process (Rosa et al. 2018. 556; Rosa et al. 2017, 1035).

This work was developed in an assembly line of security and safety systems cameras with a large number of components and some different versions. The production line studied produces final assembling of one of the most complex video units requested nowadays by the market, creating an additional demand in all manufacturing process to fulfil the product design requirements. Being these video cameras a make-to-order product and classified as low manufacturing volume, the techniques used in the line to assemble these units are based essentially on a manual process. A relevant aspect of this camera related to the robustness design is that the components used are in large quantity mechanical parts. The operation of this final assembly line acquires a particular character since the video-cameras are assembled in different workstations and each assembly step in each workstation is made up of few operations. The productive capacity of this assembly line depends on the number of operators established. In order to determine the operators' movements through the assembly line, a Standard Balance is used, which is a diagram that defines the ideal product movements to guarantee the greatest number of units produced. This type of product has many process sequences to ensure a more efficient assembly and also to ensure that the product is fail-safe, thus ensuring a high-quality level. The manufacturing conditions on the line are full of procedures as checklists, setups and process confirmations in different workstations. All these aspects require special attention in the work developed since it diverges from the typical scenarios found in projects under this scope.

The main goals of this study can be summarized as:

- Optimize the productivity of a manual assembly line by applying operations' analysis following the Lean production principles;
- Design of the current VSM through the collection of data and identifying the path the material and information flow follows;
- Analyze the current layout intending to reduce the total area used, which would reduce the walking distance of the operators between workstations and with these results productivity could also be improved;
- Investigate the balance losses in the current assembly line with the aim of improving the cycle time and the tack time;
- Study the current assembly fixtures with the purpose of identifying the non-value-add handlings and also with a view to improve the changeover time;
- Optimize the assembly process sequence taking into consideration the reduction/elimination of tasks which do not add value to the product.

This work is divided into seven main sections. This first one consists of an introduction to the subject and the presentation of the research objectives. In section 2 the literature review is presented to support the Lean tools used to study the present problem. Section 3 presents the methodology to be followed. Section 4 focused on the description of the current state line, namely the problems' description. Section 5 shows the development of the case study with the improvements implemented. Section 6 presents the results and discussion dedicate to the Lean Line Design and Production Process Optimization applied to the developed work. Finally, in section 7, the main conclusions of this study are presented.

2. LITERATURE REVIEW

To respond to specific problems of the automobile industry the Toyota® Company had the need to develop a management tool, denominated by Toyota Production System, nowadays usually known as Lean Manufacturing (LM) to become more competitive and agile in the market (Sousa et al. 2018, 612; Fernando et al. 2007, 70). This management philosophy smoothes the workflows to improve production processes. The implementation of LM brings, in the long term, the creation of value focused on the needs of the client, society, and economy. The value created arises by reducing costs, improving deadlines, improving product quality, eliminating wastes, among others, which is reflected in increased productivity (Womack and Jones 2003; Nagi et al. 2017, 594; Wilson 2010). The Lean philosophy embraces several methods and tools to support its implementation. The following figure describes a summary of some Lean tools that exist in Lean manufacturing, however, just some of them will be covered in detail in the literature review (Martins et al. 2018, 648; Correia et al. 2018, 664).

Figure 1. Lean tools.

2.1. Value Stream Mapping (VSM)

The first book that marked the origin of the studies on Value Stream Mapping was edited by Mike Rother and John Shock under the name "In learning to see" (Rother and Shook 1999), describing in detail its application in the industry. Value Stream Mapping is a tool that allows to understand the flow of material and information as a product goes through the value stream. It is also a special type of flowchart that uses symbols known as "the language of the Lean" to represent the flow of inventory and information leading to improve them using other Lean tools. Figure 2 shows a simple example of how the flow mapping of a single process is represented by this methodology.

This tool also identifies factors that can generate waste in the process, such as the time spent on activities that do not generate value, the cycle time and setup time, leading to a clear identification between the activities that generate real value which the customer is willing to pay, and the time wastes that should not be charged in terms of cost to the customer (Venkataraman et al. 2014, 1188).

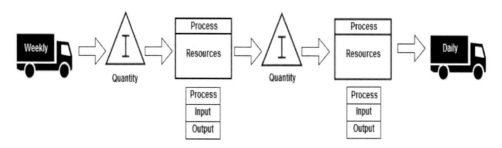

Figure 2. A simple value stream mapping.

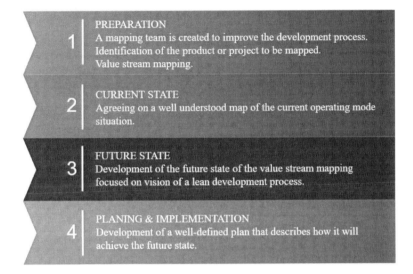

Figure 3. The value stream mapping process implementation.

After identifying the process factors that do not generate value, it is necessary to know all elements that integrate the value stream map and that will create a more transparent visualization and presentation of the interrelationships between the processes. The elements are known as: metrics and process steps, customer requirements, inventory, information and physical flows, total lead time and takt time, and material flows from the suppliers (Venkataraman et al. 2014, 1188).

The visualization of the current situation which is created in this way provides an excellent basis for detecting weaknesses (i.e., bottlenecks) and identifying opportunities for improvement (Sousa et al. 2018, 614).

Figure 3 represents the recommended process for using the value stream mapping (Locher 2008).

2.2. Lean Line Design (LLD)

The Lean Line Design (LLD) theory is a tool that allows optimizing the layout of an existing or a new production line aiming at improving productivity. This tool also allows to design an assembly in a sequential manner, with the elimination of wastes, maintaining its flexibility (Koppal et al. 2013, 75).

To simplify the worker's movement on the Lean line between workstations, the line is usually drawn in a U-shape layout, called Lean line. This approach ensures the reduction of routes in the change between workstations (Bosch Group 2015).

LLD method allows to implement principles of manufacturing using factors such as:

- Continuous flow (one-piece flow) and small lot sizes lead to short delivery times;
- For different customer demands, able to guarantee high and constant productivity;
- Ensures a balanced workload for workers with flexible allocation of work content;
- The concept of simplified line and a low degree of automation translates into a low investment and the reduction of necessary spaces.

Toward the implementation, the following step-by-step approach can be applied.

2.3. Balancing Line

Line balancing is a technique that helps to minimize any possible imbalances between workloads assigned to each operator, in order to achieve the optimal/necessary execution rate. Thus, it is needed to analyze the design process of the line, the layout and cycle time of each of the workstations (Rosa et al. 2017, 1037). These considerations allow, in assembly line balancing, the assignment of tasks to a more adequate set of workstations (Lam et al. 2016, 438; Sabuncuoglu 2009, 288). The cycle time is defined by the time interval between the beginning of the first and the end of the last task. Moreover, the cycle time can also be considered the time between the accomplishment of two equal products in the same production line.

With this technique, it is intended to achieve a similar cycle time in each workstation, however, given the variety of products and assembly operations, this practice is very difficult to achieve (ElMaraghy and ElMaraghy 2016, 5).

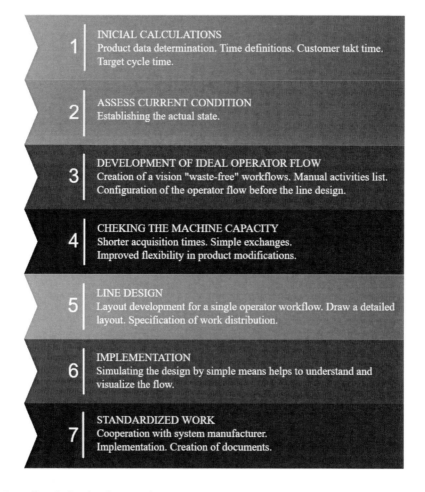

Figure 4. Lean line design implementations steps.

2.4. Single Minute Exchange of Die (Quick Changeover)

For the processes to be waste-free, it is essential to reduce the setup times. This allows a greater number of tool changes in a shorter time interval, resulting in greater flexibility of production (Sousa et al. 2018, 612; Rosa et al. 2017, 1035). The benefits of the single-minute exchange of die (SMED) can be classified by:

- Setup less than ten minutes;
- Greater flexibility in the production system;
- Higher productivity due to larger production time available.

It is recommended to use the available resources in a synchronized way, dividing the preparation tasks into internal and external tasks, allowing the external tasks to be performed with the equipment still in production, which can translate into increased productivity and shorter delivery deadlines (Bosch Group, 2003).

The advantages for operators are as follows:

Figure 5. Single minute exchange of die implementation.

Easier: clear changes in processes cause fewer problems and better distribution of workload.

Faster: quick setup allows more available time for another job.

Safer: better organization translates into less effort during transitions, and standardizing and planning the work improves the operator safety.

In order to accelerate the changeover (CO) process, the following factors should be considered: activities planning, activities optimization and organization optimization. The implementation comprises eight steps that can be seen in Figure 5.

The reduction of internal tasks in CO operations can be achieved in an extremely easy way through the use of modular clamping systems, which will allow the tools to be clamped always in the same position without operator concerns. The dimensions of the base of the tools should be standard so that they easily fit into the housing of the machine designed for this purpose (Costa et al. 2018, 569; Castro et al. 2017, 1440). Poka-Yoke systems can also be considered to avoid positioning errors. When there are significant differences in the tools, the use of adjustable fastening systems can be considered, but with the loss of CO time and concerns that there are no errors in the adjustment.

2.5. DMAIC (Define, Measure, Analyse, Improve and Control)

DMAIC is a data-driven quality strategy used to improve processes. It is an integral part of a Six Sigma initiative, but in general, can be implemented as a standalone quality improvement procedure or as part of other process improvement initiatives, such as Lean. The DMAIC cycle approach encompasses a number of quality activities, methods and tools, associated with each phase of the cycle. Some of these tools and techniques used are for example: Project Charter, Gantt Diagram, SIPOC (Supplier, Input, Process, Output, Customer), Brainstorming, Pareto Diagram, Ishikawa Diagram, 5 W, LLD, VSM, PDCA, A3 Sheet, among others. At the beginning of the cycle (Define), the problem definition and all aspects relevant to the project are performed. During the second step (Measure), the current situation is measured, translating the problem into a measurable form. The next step is the analysis of the collected data, in order to determine the root causes of the problems (Analyze). The objective of the improvement phase is to eliminate the previously identified causes (Improve) and in the last step, the control and monitoring of the process must be carried out, preventing a reversion of this to the initial state (Control) (Costa et al. 2017, 1105).

In order to implement this tool, it is necessary to know some parameters and answer the following questions in each step, see Figure 6.

3. METHODOLOGY

The operators in the assembly line carry out a variety of tasks to produce the final products. There are many tasks in the different stages of the work process which includes tightening the components using fasteners, using mount fixtures, operating machines, testing the assembled product and building the package.

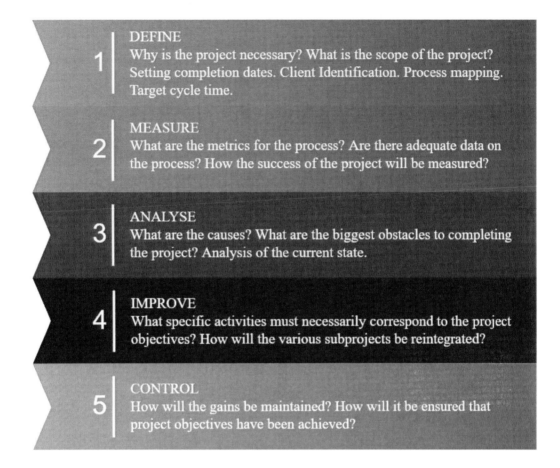

Figure 6. DMAIC implementation.

The procedure involves analyzing the process flow with the help of Lean tools regarding the main goal of reducing wastes. Essentially, it is necessary to: collect process data about the different tasks and compile them; realize what is value and waste and get improvement opportunities.

A process line under an intensive Lean environment in which workers can be rotated through all workstations within the cell to finish the product was considered for the study. The line will have to be monitored to observe operator flow, process sequence, fixtures used and ergonomic aspects. Layout details including tool reach, parts and distance between each workstation were considered. This industrial study was carried out in parallel with a new industrialization project for a new model that will be integrated into the same production line. Taking this project into account, it was necessary to evaluate the assembly sequence and specify the necessary tools and workstations. In order to have efficient industrialization of this new model, it was important to consider how it will be possible to combine tools and probably the same workstations, in order to design the layout and the appropriate assembly process that contemplates this new reality.

Table 1 presents the application of the DMAIC methodology of this project.

Table 1. DMAIC adapted methodology

Phases	Description	Tools
Define the Problem (D)	Identify and define the problems to characterize the initial situation of the working conditions and to identify new improvement opportunities in the assembly line considered in this work.	VSM 5W LLD
Measure/Map the Process (M)	Collection of information from the process; Creation of a process map in the initial situation; Taking pictures from the tasks and tools.	VSM
Analyze the Process (A)	Identification of waste, value and activities without added value; Identify, organize, and prioritize tasks; Select and verify the causes of the problems; Define possible additional information.	Ishikawa Diagram
Improve (I)	Enumeration of all possible solutions; Classification of solutions and selection of the best; Development of performance metrics; Application of 5S methodology to the assembly line; Communication and implementation of the solution; Measurement of results; Determination of the necessary follow-up time.	Brainstorming PDCA A3 Sheet
Assess/Control the Process (C)	The purpose of this step is to sustain the gains; Evaluation of the modifications' results; Lessons learned; Monitor the improvements to ensure continued and sustainable success.	Planning Standardization

4. CASE STUDY – CURRENT STATE

This study deals with a complex manual electronic devices assembly line composed of 16 workstations and 5 test tables at the end of the line to assess and measure all products in a period of 12 hours. The product complexity requires a rigid schedule and rigorous respect of the takt time. For these reasons, the line balancing, the fixtures used for assembly and process sequence properly architected are a core activity and needs to be efficient and effective. Although, the company where this study was implemented already has established a culture way of production based on the LM philosophy, in which the organization work methodologies are focused on a Lean mindset environment. For these reasons, it was a remarkable challenge and more difficult to find production wastes. However, after a thorough observation of this assembly line, some problems were detected to be analyzed, as the area occupied, the time needed for the operator to change workstations, the production capacity and the flexibility of the line to get new products. In order to overcome this drawback and to eliminate wastes, a new ergonomic incorporated layout and a new approach to fixtures design needed to be developed. Change in work methodologies and redesign of the layout lead to cycle time reduction and removal of non-value-added activities.

4.1. Applying Value Stream Mapping (VSM) to Define the Current State

The VSM analysis permits identifying specific root causes bringing problems across the value stream. The VSM method is only an analytical method and does not remove the manufacturing issues or root causes by itself. Therefore, it is a relevant tool to identify problems that other Lean methods could fix. Take this in consideration, the aim of using VSM approach is to help a better visualization of what kind of problems need to be analyzed, to work deeper on improvements solutions with the intent of eliminating wastes and increase productivity.

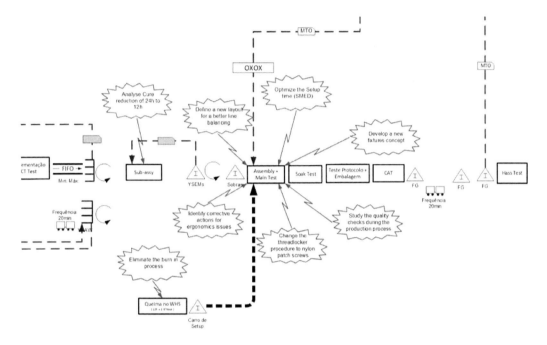

Figure 7. Detailed VSM of the final assembly flow.

Table 2. Identified issues, its causes and potential solutions, resulting from the 5W methodology

Identified Issue	Proposals to improve
Sub-assemblies curing time with a duration of 24 h	Reduce the curing time for 12 h, to be possible a cure supermarket optimization, generate an increase of the storage capacity.
Inappropriate line layout	Design a new layout for better balancing and ergonomic aspects.
Setup times too long	Optimize the setup times.
Low production capacity	Develop a new fixtures concept; Eliminate unnecessary assembly operations which do not add value to the product.
Ergonomics	Identify all the critical ergonomic issues, to possibly create solutions to avoid operator fatigue.
Inadequate quality confirmation during the mount process	Quality checklist revision, to assure the adequate evaluation take into consideration what is a real need to be check by the operators.

Another relevant aspect to notice is that the VSM presented in this work has much information omitted, due to the restrict confidentiality rules from the company where this work is implemented.

Due to the variety and complexity of electronic devices assembling, the VSM chart illustrated in Figure 7 represents the flow value stream chain of the product version with the high demand of the production line studied in this work. The aim of this technique is mainly focusing on the final assembling line of this kind of products, taking into account the identification of improvement opportunities related to reducing cycle time outcome for a better line balancing and workspace optimization.

Table 2 presents all identified problems, their initial causes identified by the method five-whys and possible solutions identified by brainstorming discussions with the operators and team leaders of the company.

4.2. Using Lean Line Design (LLD) to Describe the Current State

Lean line design is one of the best systematic approaches to streamline processes and eliminate waste. The goal is to redesign production and logistics to improve overall efficiency and flexibility, as well as reduce the proportion of investment, the required space and shorten the output time. This method can be used to combine two assembly lines to reduce the area occupied or just to optimize a complex assembly line to achieve better overall performance. In this case, the LLD method was applied to an optimization of the production line, improving the time of output of the different versions of a product family, but all the work carried out taking into account that the desired version has the highest forecast for this line, not considering versions already defined as end of life (EOL) products, that is, product variants whose quantities planned to produce in this line are very low and a deadline is already established to stop the production of these models.

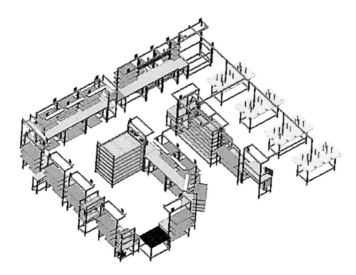

Figure 8. Assembly line 3D model.

The Lean Line Design method is complemented by different techniques such as line balancing, layout design, 5S, walking flow and ergonomics, those can be applied to achieve the expected results. Therefore, it is necessary to have the proper knowledge of each tool to have a clear vision of what will be necessary to redesign a new assembly line. Thus, the reason for applying this tool was to eliminate some ergonomic problems that were identified as critical, decrease target cycle time, reduce the area occupied, improve walking flow, match all workstations and equipment in the same space. Figure 8 shows the 3D model of the main assembly area.

4.2.1. Current State

The actual state was established by recording the current data and times for the line that needs to be redesigned as a Lean line in the future step. The actual state promotes a better understanding of the work steps and their interaction. In order to have a complete understanding of the productive organic of an assembly line, it is important to design several options in terms of the number of operators available to work on the line, because this information will give a better functional notion about the decisions that will have to be taken to redesign the line in the future. Thus, the data collection was based on an assembly line scenario with a flow operation contemplating the number of operators of 6 and 8, as these are the most common realities in this assembly line, otherwise, it would be an exhaustive data collection if one had to look at all the options.

Most of the workstations involved were analyzed, excepting workstations 6, 13, 14 and 16, because these ones were considered as optimized.

Before going further into the study of identifying possible improvement solutions, it is time to analyze the production tasks, their integral movements, and the layout configuration. To assist in the process of analyzing this problem, we use a very useful quality tool in order to analyze all possible causes that may influence line balancing and efficient layout. Thus, in Figure 9 below is the analysis of the actual state using the Ishikawa Diagram.

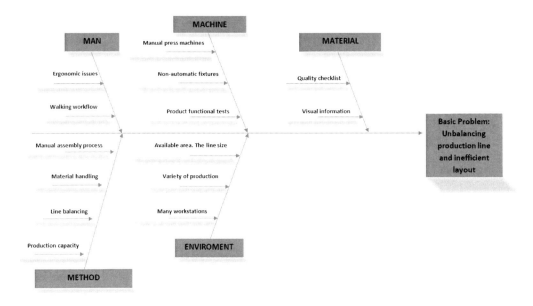

Figure 9. Ishikawa diagram showing issues for unbalancing line.

All these possible causes are closely related to another one and have a considerable impact on the assembly line performance, as well as production cost. The possible causes detected for the problem can upsurge from different types and the most critical will be analyzed and mention on further.

4.2.1.1. Man

In terms of causes related to the operator, one identified some ergonomic issues present in the line as the necessary time and human physical effort to handle and assemble some parts. One also detected the walking workflow, which considering both current situations (6 or 8 operators in the production line), there is a lot of time walking through the stations and some cross paths between each operator.

4.2.1.2. Machine

In this production line, four manual press machines are identified, which have a certain impact in relation to the occupied area and for some assembly operations, this equipment is not so close to the workstation as it is needed. Because it is a low volume production assembly line, most of the tools and fixtures used in the process are fully manual utilization, increasing the time required to perform the operations. Another factor that has a strong impact online output is the functional tests needed to verify product performance. The reason for this is because, for example, testing one unit in the final test (Soak test) takes 12 hours, and although each test table is capable of having 12 units, it is only possible to test one unit at a time (12 units each 12 hours).

4.2.1.3. Method

The sequence of the assembly process is too manual, this type of approach origins having a high cycle time per workstation and, workstations 5, 9 and 12 (Figure 11), have the longest established time for building one part of the assembly. The assembly line balancing standard is not determining to have an ideal distributing of the total manufacturing workload uniformly among all the workstations present along the assembly line. Therefore, in both cases registered, in the current state, the line has a 7% unbalancing. The overall performance of the production system is greatly affected by this distribution of work.

4.2.1.4. Environment

Regarding the area available for this assembly line, it is visible in Figure 13 that one is facing a large assembly line, which is the largest line in the plant. The reason for this is due to the complexity of the product manufactured. The layout is divided into two areas, one that is the main area and the other is a secondary area.

In summary, after analyzing the current state of the assembly line studied, one can understand that it is necessary to redesign the layout to ensure a better production performance of the line. In this way, it must: combine the two assembly areas in just one. This will relieve the space for other production lines, if necessary; decrease the time it takes for operators to move through the workstations to have more time in value-added activities; improve workload distribution across workstations uniformly and achieve a better line balancing; arrange the assembly tasks ideally to achieve greater production capacity on the assembly line.

4.2.1.5. Layout Analysis

The studied assembly line is categorized as a line that produces each model in batches, being the different models of the same product family, having thus, a very similar architecture. Basically, whenever there is a need to produce a different model it is necessary to set up the line, which consists of replacing the raw material in each workstation and changing the assembly fixtures needed. It is more economical to use one assembly line to produce multiple products versions in batches than to create a separate line for each model. The current layout is designed in a U-shaped configuration, where the occupied space is divided into two areas, the main area that has the dimensions of 143 m^2 and a second secondary area with the dimensions of 16 m^2. The location of this assembly line at the factory only allows having material supermarkets on two sides that are close to the corridor side, which allows simple and easy delivery operation.

Figure 10 shows an overview of the entire area occupied by the studied assembly line, so that the distance between the main and the second area is visible, as well as the existence of another assembly line in the middle of both. Different sections of the layout were highlighted and respectively identified with captions.

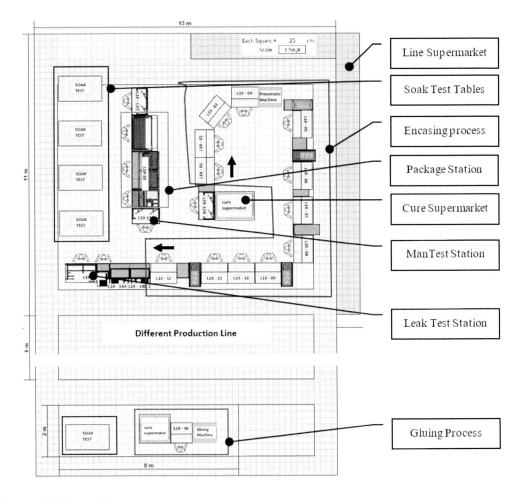

Figure 10. Current line layout.

4.2.1.6. Line Balancing Process Analysis

The line balancing process is performed first by plotting the operator line balancing chart. To plot the operator line balancing, Customer Takt Time, Planned Cycle Time and Operator Cycle Time are calculated. The Customer Takt Time (CTT) is based on the entire call-offs for one line/production unit. It forms the basis for the line design and the calculation of the target cycle time. For calculating CTT it was considered one shift of eight hours and the parameter customer demand was defined taking into account the maximum line capacity, simulating a reality of six and eight operators on the assembly line. The reason for this approach is because the version of the product being studied is a production make to stock. Thus, the CTT value can be given by equation (1):

$$\text{Costumer Takt Time (CTT)} = \frac{\text{Available work time per shift}}{\text{Customer demand rate per shift}} \; [\text{s/pcs}] \tag{1}$$

The values obtained for the CTT can be observed in Table 3.

Table 3. Costumer Takt time for six and eight operators

Simulation	Your available work time per shift [s]	Customer demand [pcs]	CTT [s/pcs]
6 Operators	29100	24	1212,5
8 Operators		32	909,4

With CTT result obtained from the previous equation, it is time to calculate the Target Cycle Time (TCT). It is necessary considering the performance and quality losses of the line (OEE – Overall Equipment Effectiveness). Changeover losses are scheduled losses and therefore considered separately. The equation to calculate TCT can be given by equation (2):

$$\text{Target Cycle Time (TCT)} = \text{CTT} \times \frac{\text{OEE (\%)}}{100\%} \; [\text{s/pcs}] \tag{2}$$

The values obtained for the TCT can be observed in Table 4.

Table 4. Target Cycle Time for 6 and 8 operators

Simulation	OEE [%]	CTT [s/pcs]	TCT [s/pcs]
6 Operators	94,24	1212,5	1143
8 Operators		909,4	857

The starting point for the design of a Lean line is to plan the flow of operators. To do this, one gathers all repeating manual tasks/work steps, as well as the time required for them, and arranges them into a sensible order. This creates a "waste-free" sequence of manual, cyclically recurring activities. Then, one gathers all non-manual, automated process steps and determines the machine capacity. One calibrates manual and automated steps so that they lie on the target cycle time.

Process flow would help to know the assembly sequence carried out in the line. Figure 11a shows the process flow currently carried out in the most produced unit on this assembly line. A stack diagram is prepared for all manual operations in the assembly line, which are put in a stack in a scaled manner, completely independent from the physical workstation, as shown in Figure 11a. The time study of each station is carried out and arranged in a stack diagram, as shown in Figure 11b.

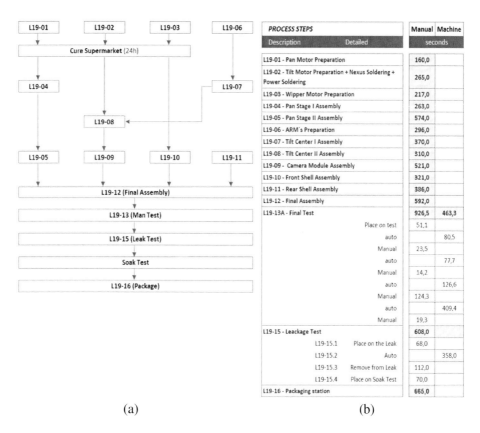

(a) (b)

Figure 11. (a) Process flow diagram; (b) Cycle time for each workstation.

Table 5. Flow assembly sequence for eight operators

Operator	Flow Assembly Sequence			Walking [s]	Operator Time [s]
Operator 1	L19-01	L19-05		58	**792,0**
Operator 2	L19-02	L19-03	L19-04	26	**771,0**
Operator 3	L19-07	L19-08		9	**689,0**
Operator 4	L19-06	L19-09		28	**845,0**
Operator 5	L19-10	L19-11		7	**714,0**
Operator 6	L19-12	L19-5.1		14	**674,0**
Operator 7	L19-13	L19-5.3	L19-5.4	10	**655,3**
Operator 8	L19-16			33	**698,0**
			Total =	185	

After the information obtained through process flow diagram to build a single unit and known the cycle time for each station as shown in Figure 9, it is time to start showing the stack workflow for each operator, considering the scenarios of six and eight operators.

In Table 5 is shown the actual state flow assembly sequence with the condition of having eight working operators in the line.

The current condition of the layout with the cycle times of all operations is plotted in the operators' balance chart as shown in Figure 12.

The operator line balancing chart plotted for the current condition has a bottleneck loop which cycle time is close to the Customer Takt Time, which is the loop made by the operator 4. The cycle times of all operations are less than the Targetted Cycle Time. This line is unbalanced by 7% and the production capacity is 32 pcs/working day (wd). In Figure 13 is demonstrated in the actual state layout the workflow for each operator. The walking time for each operator gives a total of 185 seconds moving across stations.

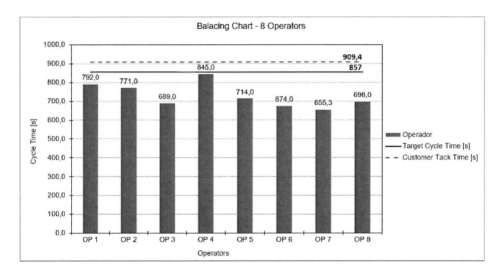

Figure 12. Current line balance chart for eight operators.

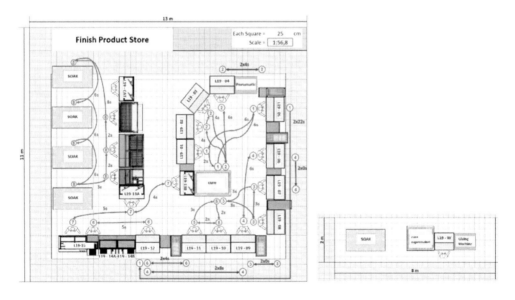

Figure 13. Current flow diagram layout for 8 operators.

Table 6. Flow assembly sequence for six operators

Operator	Flow Assembly Sequence				Walking [s]	Operator Time [s]
Operator 1	L19-03	L19-04	L19-05		61	**1115,0**
Operator 2	L19-02	L19-06	L19-07		19	**950,0**
Operator 3	L19-01	L19-08	L19-09		31	**1022,0**
Operator 4	L19-10	L19-12			15	**928,0**
Operator 5	L19-11	L19-13			13	**862,3**
Operator 6	L19-15.1	L19-15.3	L19-15.4	L19-16	39	**954,0**
		Total =			178	

Table 6 shows the same reasoning considering just 6 operators.

The layout's current condition with the cycle times of all operations is plotted in the operator's balance chart, as shown in Figure 14.

The operator line balancing chart plotted for the current condition has a bottleneck loop which cycle time is closer to the Customer Takt Time, which corresponds to the loop made by the operator 1. The cycle times of all operations are less than the Targetted Cycle Time. This line is unbalanced by 7% and the production capacity is 24 pcs/wd. Figure 15 shows the workflow for each operator of the current state layout. In this scenario, it is possible to see a huge crossover between loops of different operators and the distance to walk between stations is longer compared to the current scenario of eight operators. The walking time of each operator gives a total of 178 seconds passing through the stations.

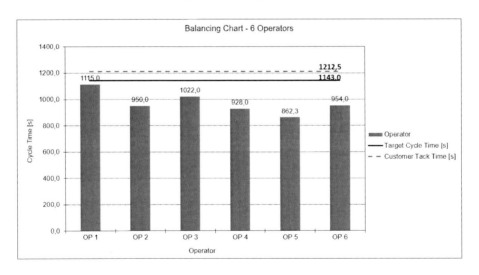

Figure 14. Current line balance chart for 6 operators.

4.2.2. Production Process Optimization

In order to improve productivity and efficiency in each workstation on this complex manual assembly line, a deep knowledge of the product design is mandatory, in a way to understand where the production wastes are and how it could be possible to optimize this assembly line. After the operation process has been analyzed, the operator value-added and/or non-value-added activities were identified for every workstation. At each workstation, all operations processing cycle times were collected. The following topics are summarizing some wastes reductions. It is intended that the optimization of the production process be adjusted in such a way that there are no constraints on the operator taking into account the specificities of the product. The objectives of the production process optimization are:

- Create an ideal workspace on the assembly line that allows workers to perform their duties without obstruction or delays;
- Minimize the production cost;
- Maximize yield and/or efficiency;
- Improve ergonomics aspects.

The production process optimization topic is complemented by different approaches such as SMED, 5S, TPM, Poka-Yoke, Assembly Fixtures, Quality Aspects, and Ergonomics.

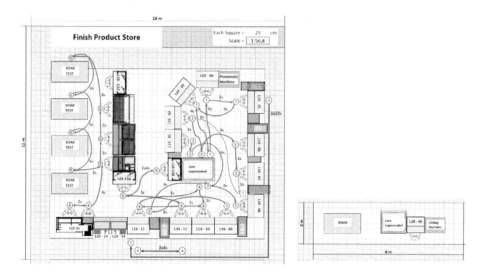

Figure 15. Current flow diagram for 6 operators.

Regarding the process optimization in this assembly line, the actual state was established by observing and recording the current production scenario. The current state promotes a better understanding of manufacturing processes and the interaction between them. In order to have a better understanding of the production process, it is important to detail all the assembly tools used in this line, in all the workstations and to understand their functionality to perform a quality assembly. Another relevant observation in the presentation of the current scenario is to identify processes that do not add value to the product, thus allowing the possibility of reducing or eliminating the current need, such as completing a quality checklist and placing thread locker adhesive in all the screws.

4.2.2.1. Assembly Fixtures

After observation, registration and analysis of all fixtures used for assembly of two models of one product in a complex manual production line, it was possible to deepen the knowledge about the process and the improvements needed to develop a new approach concept to redesign the mounting fixtures. The complexity criteria are intended to be used by engineers in manufacturing engineering for identification of potential quality issues in the development of assembly solutions.

4.2.2.2. Workstation 01 – Pan Motor Preparation

In this station, it was necessary to consider the following aspects: (1) the position of the pinion and the position of the tilt motor related to the mounting plate, which is fulfilled with the fixture in Figure 16a, (2) press fit two bearing into the pan motor plate: this operation is done by one adaptor used on a manual press equipment, as depicted in Figure 16b.

4.2.2.3. Workstation 02 – Tilt Motor Preparation

In this station, it was necessary to consider the following aspects: (1) the position of the pinion in the tilt motor related to the mounting plate: this is fulfilled with the fixture verified in Figure 16a, (2) press fit one bearing into tilt motor plate. This operation is done by one adaptor used on manual press equipment, as shown in Figure 17b and 17c.

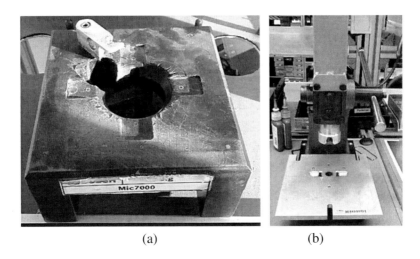

(a)　　　　　　　　　　(b)

Figure 16. (a) The assembly fixture; (b) the manual Arbor.

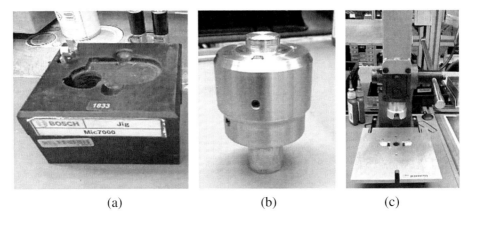

(a)　　　　　　　　(b)　　　　　　　　(c)

Figure 17. (a) The assembly fixture; (b) The adaptor for press fit the bearing; (c) the manual Arbor.

4.2.2.4. Workstation 03 – Wiper Motor Preparation

In this station, it was necessary to consider the following aspects: (1) ensure that the part is properly mounted. The design only permits to attach the parts in one way. The fixture is illustrated in Figure 18, (2) plug in the motor to check if it works and set up a correct orientation.

4.2.2.5. Workstation 04 – Pan Stage I Assembly

In this station, it was necessary to consider the following aspects: (1) press fit one gear onto a shaft, a bearing into Pan body and at last, press fit a leap seal into the Pan body. For each press fit task, it is necessary to set up the equipment. The manual press and the adaptors used are illustrated in Figure 19, (2) another relevant factor in this task is the enormous force required to perform the mechanical interference between the gear and the shaft. Due to the use of a hand press, the force applied is made by an operator, (3) after making the press fit, it is time to fix the Pan body to the fixture, seen in Figure 20a, and tighten the pan base with the help of a special tool, as is shown in Figure 20b.

Figure 18. Fixture for mount wiper assy.

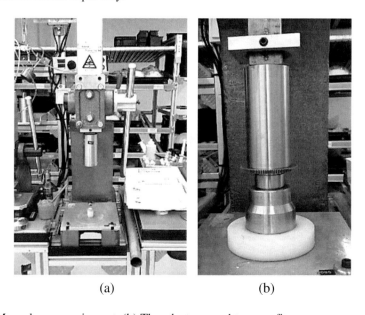

(a)   (b)

Figure 19. (a) Manual press equipment; (b) The adaptors used to press fit.

Lean Manufacturing Applied to a Complex Electronic Assembly Line 307

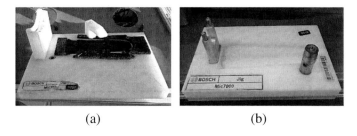

(a) (b)

Figure 20. (a) Fixture to mount Pan assy stage 1; (b) Special tool to tight Pan base.

4.2.2.6. Workstation 05 – Pan Stage II Assembly

In this station, it was necessary to consider the following aspects: (1) quality confirmation by a fixture of the previous assembly, (2) press fit one ferrite onto a plastic part, this operation is done by hand, (3) the rest of the assembly process performed on this workstation only requires an accessory, with the intention of ensuring some stability during assembly.

4.2.2.7. Workstation 07 – Tilt Center I Assembly

In this station, it was necessary to consider the following aspects: (1) press fit one gear onto a shaft, two bearings into Tilt centre holes and at last press fit two leap seals into the same place of the bearings. For each press fit task, it is necessary to set up the equipment. The manual press and the adaptors used are illustrated in Figure 21, (2) after performing the press fitting operations, it is necessary to do a quality inspection related to the position of the tilt gear, and per last it is needing to tighten the arms onto the tilt centre. In the Figure 22 is shown the fixtures.

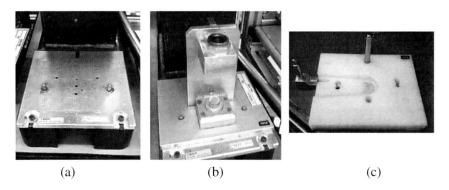

(a) (b) (c)

Figure 21. (a) Manual press equipment; (b) The fixture to press fit bearings and leap seals; (c) Press fit tilt gear.

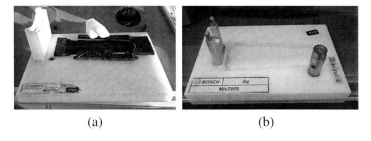

(a) (b)

Figure 22. (a) Fixture used to tight the arms; (b) Quality fixture to check the tilt gear position.

4.2.2.8. Workstation 08 – Tilt Center II Assembly

In this station, it is only needed one fixture to maintain the previous assembly in vertical position, (Figure 23).

4.2.2.9. Workstation 09 – Camera Module Assembly

In this station, it was necessary to mount the camera module, a fixture able to guaranty a proper orientation of the different parts is needed. In Figure 24 is shown the fixture in use.

4.2.2.10. Workstation 10 – Front Shell Assembly

In this station, it was necessary to consider the following aspects: (1) the position and the way which the optical window is tightened is critical, (2) it is necessary to press fit one bushing, assuring the proper position. This operation is shown in Figure 25a. The fixture illustrated in Figure 25b is to be used in the remaining operations.

Figure 23. Fixture to maintain the previous assembly in a vertical position.

Figure 24. Fixture to mount camera module.

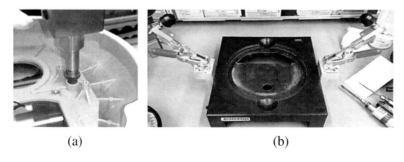

(a)   (b)

Figure 25. (a) Manual press fit of one bushing; (b) Fixture to assembly front shell.

4.2.2.11. Workstation 11 – Rear Shell Assembly

In this station, it was required to get this assembly well done, it is necessary to place the rear shell in three different positions. In Figure 26 are shown two fixtures used in this workstation.

4.2.2.12. Workstation 12 – Final Assembly

Considering this station, it was important consider the following aspects: (1) the fixture used needs to support all the previous assemblies, to be easier for the operator to handle them, (2) before starting to close the unit, it is necessary to perform a torque confirmation in one type of screws.

4.2.2.13. Workstation 15 – Leak Test

In this station, it was necessary to consider the following aspects: (1) a special cap is needed which would need to be tightened on the unit for attaching the air tube. This adaptor is a spare due to wearing. The adaptor is shown in Figure 27a, (2) after the leak test, the unit is placed in the fixture shown in Figure 27b for tightening the yoke caps parts.

4.3. Results and Discussion about the Current State

Table 7 and Figure 28 show the classification of the mounting accessories used on the assembly line, with the time values resulting from the use of the assembly tools. Through the analysis of the graph shown in Figure 27, it is evident that the most consuming time activity is the need to make fixation adjustments and the second one is the time required to perform the assembly operations.

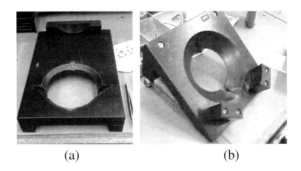

(a) (b)

Figure 26. (a) Principal fixture for rear shell; (b) Fixture use to add inclination.

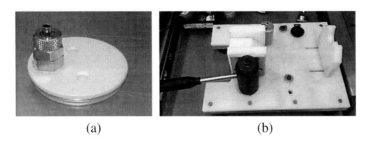

(a) (b)

Figure 27. (a) Special cap to plug the air tube; (b) Fixture use to fix the unit to tight yoke caps.

Table 7. Fixtures classification percentage

Classification	Time [min]	[%]
Fixtures Adjustments	3,7	65
Operation	1,6	28
Quality Checking	0,4	7

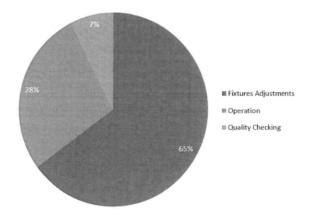

Figure 28. Mounting jigs - current state.

5. CASE STUDY – IMPROVING SOLUTIONS

In this section, the current state of the line layout, its operator unbalance, and the efficiency of the assembly process will be analyzed. In terms of the Lean line design, what matters it to get a clear view of the actual constraints of the current layout and what impact it has on the line balancing. Another important point will be to understand the assemblies and other activities of the process, in order to identify the wastes and then the opportunities for improvement, to obtain Leanness in the assembly process.

5.1. Applying Lean Line Design (LLD)

5.1.1. Improvement Solutions

To achieve a well-balanced assembly line, it is important to eliminate factors that cause losses and wastes. Thus, in order to implement the improvements identified through the analysis of the current state described in LLD, it is important that the organization have the same understanding of the improvement objectives and where these improvements will lead to.

Thus, a plan to carry out the improvements is needed. One of the managing techniques for continuous improvement is the PDCA cycle, which stands for Plan, Do, Check, and Act. **Plan**: In this step, the targets to achieve are defined and quantified, to measure the effectiveness of the improvements; **Do**: Implement the action plan. In this step, all the changes are made, and the new data is collected; **Check**: It is time to check the data collected in the previous step. Therefore, it will be necessary to analyze data trends and other deviation occurrences; **Act**: Apply any needed corrective measures to the new implementation. If something in the new process is not running according to plan and can be easily fixed, it should be done in this step.

One way to apply the PDCA concept is the A3 sheet technique that is an excellent visual method, which will be useful to help a group of people with a collaborative engagement to solve the issue of production capacity, the occupied area, and ergonomics problems. This kind of approach drives the group to address the wastes and identify all the activities which may be eliminated or modified, with the intent to improve the assembly line.

In Figure 29 the details regarding the implementation of this A3 process are structured. It is important to have in this document an explicit description of the project, the current situation and the target that the group needs to achieve, with a clear understanding of what performance indicators the project will influence. The next step in this method is to determine all the activities that must be done with an appropriate schedule. Finally, it is necessary to specify the metrics necessary in the different areas to be worked, in order to quantify the gains and benefits of implementing this project.

In order to be able to explore different layout realities in an efficient and fast way, a methodology not common in this practice of studying new layout configurations was used. Therefore, the method used to select the best layout option with the aim of ensuring the best balance in the production line was an Excel sheet designed to have an easier understanding of the configuration that will give the required results. Thus, to be able to architect this new way of studying layouts, it was necessary to gather some structural information of the production line, such as the total area that is occupied, the correct dimensions of the assembly line, the respective work elements that constitute it and also obtain the 3D model of the line to snapshot each work element. Next, an Excel scaled template was created with the appropriate area and pointing out all the work elements separately, as shown in Figure 30.

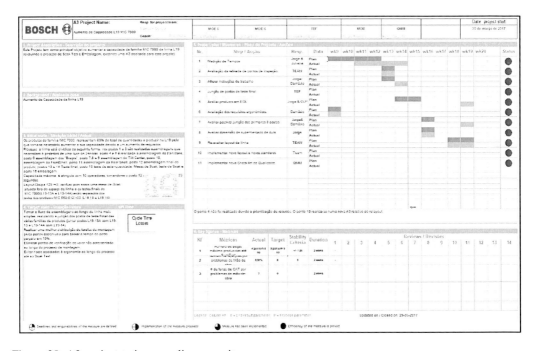

Figure 29. A3 project to increase line capacity.

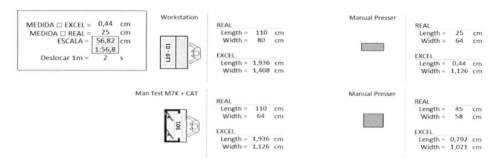

Figure 30. Example of the procedure used to create a scale process.

Typically, the traditional method used in workshops to explore different layout options is to have the assembly line printed with a well-known scale and cut out all the elements of the line individually and play with them for different realities. This type of approach makes the layouts creation and selection process more time-consuming and less flexible.

With the Excel sheet, it is possible to calculate the best balance with respect to the input data, which can be manually manipulated for further modifications and construct different scenarios. It is possible to modify all input data in a matter of minutes, making visualization of different options and their results in a short time, and bringing great flexibility for the evaluation process.

The data input to the model developed in Excel file for this work includes: production type (mixed model), operation names and times (seconds), work zones of operations; product model (important for mixed model balancing), number of stations, number of operators at each station, desired takt time, and precedence of operations.

Balance losses at an assembly line are inevitable. There is a very small possibility to achieve and preserve a perfect balance of workload at a flow line production system due to the dynamic characteristics of this type of system. Also, it is not possible to distribute the process precisely equally among stations in this system. Therefore, it is clear that at flow line production systems there are always balance losses. However, by understanding the sources of these losses, balance losses at assembly lines can be minimized.

For capacity-oriented balancing, the following should be considered: a number of stations should be minimized for a given cycle time and cycle time should be minimized for a given number of stations, and flow time and waiting time of pieces should be minimized.

With the knowledge of the assembly operations and their operator times, the design of an assembly system includes the balancing of the desired assembly line considering the number of operators and workstations expected in the line and arranging the configuration of the equipment in each station according to the results of the balancing process. Theoretically, balancing and planning an assembly line layout are two dissociated procedures that do not influence each other while designing a production system from scratch.

Alternative layout options for the new configuration of the assembly line are created taking into consideration the capacity, line balancing, occupied area, and the investments cost. Therefore, while planning the layout, stations with high replacement costs are tried to be kept at their existing locations. Sub-assembly stations are planned after the pre-assembly stations are lined.

5.1.1.1. First Scenario

At the first scenario (Figure 31), it was taken into consideration the following concerns: (1) reduce the flow walk time from the packaging operator across the Soak tests, (2) the cure sub-assemblies go to the workstations used by the point of use provider (POUP) operator, (3) combine the workstations 2 and 3, (4) combine all the workstations in the same area, with the objective of reducing occupied area, (5) fewer costs in changing the current layout.

5.1.1.2. Second Scenario

At the second scenario (Figure 32), it was taken into consideration the following concerns: (1) combine all the workstations in the same area, with the objective of reducing occupied area, (2) fewer costs in changing the current layout.

5.1.1.3. Third Scenario

The third scenario (Figure 33) had many similarities with the first option designed. Nevertheless, one of the differences is the proximity of the soak tests tables to the packaging operator.

5.1.1.4. Fourth Scenario

With the second scenario (Figure 34), it was taken into consideration the following possibilities: (1) reduce the flow walk time from the packaging operator across the Soak tests, (2) reduce the flow walk time across stations at all the entire line, (3) the cure sub-assemblies go to the workstations were used by the POUP operator, (4) combine all the workstations in the same area, with the objective of reducing occupied area, (5) better distribution of the operations through the workstations.

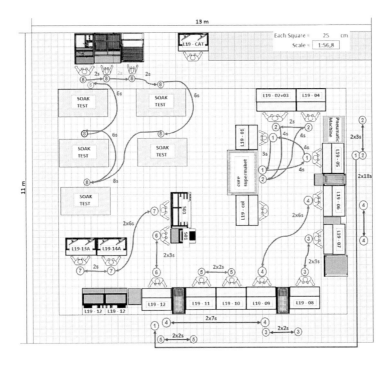

Figure 31. First option for flow diagram for 8 operators.

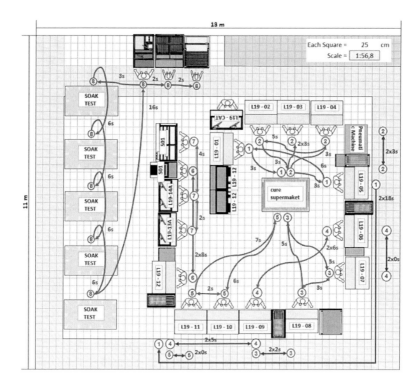

Figure 32. Second option for flow diagram for 8 operators.

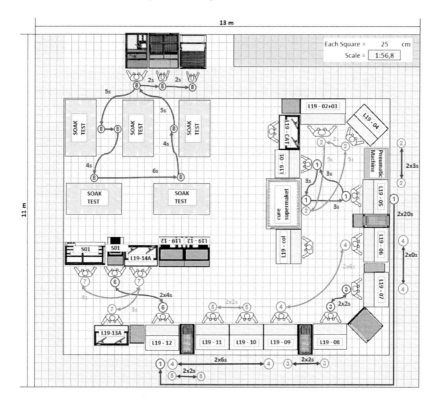

Figure 33. Third option for flow diagram for 8 operators.

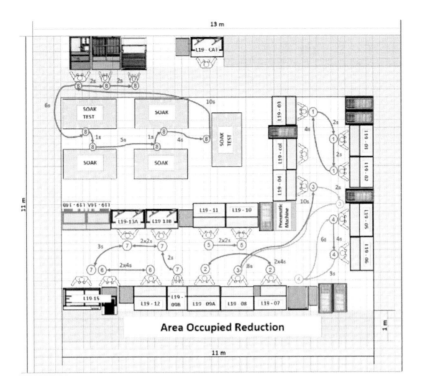

Figure 34. Fourth option for flow diagram for 8 operators.

In summary, four different layout options were designed with a focus on getting the desired assembly line. For this purpose, the strategic location of the workstations, the loops for each operator to perform the tasks of assembly on the line, an organization of work tools, fixture and material handling setup were considered during the design of a new layout configuration. In this way, to achieve a Leaner assembly line, the sources of wastes in the production process are determined.

Therefore, multiple alternative concepts of the layout were developed to select the best approach to be implemented. In Table 8 are registered the different properties took into account to have the most exact decision about which layout option creates more improvements to the line. The evaluation of layout option concepts implies and involves both comparison and decision making. Thus, for better determination, the assess was determined by a quantitive decision matrix method. The decision matrix method is a useful technique for decision-making in management. It is particularly powerful where someone has several alternatives to choose from, and many different factors to consider.

Typically, the approach used to evaluate which layout option will be selected is a qualitative evaluation method. This type of procedure determines a decision based on feelings and emotions, which gives rise to a probability of choosing the wrong option. There are pros and cons in this method but what the matrix does is collect all the emotions and thoughts around the decision and put them into a single vision that can force you to evaluate them in the most objective way possible.

Therefore, to reduce the possibility of a poor decision in the layout selection, it was select a quantitative decision matrix method for this analysis. Due to the reason this method is measurable, it is helpful to clarify which is the smarter decision for the business.

Properties Collection of Layouts Purpose

On the Table 9 are specified the four layout options presented in previous figures. All the relevant factors needed to consider are in the column's headings. The factors took into consideration were area occupied, ease of implementation, the capacity required, unbalancing, etc.

Attributes Table

The next step is scoring each option for each of the factors determined in the previous table. Score each option from 0 (less important) to 5 (more important). Note that is acceptable having the same rate for different factors. The most relevant factor in this analysis is the required capacity and the less significant is the ease of implementation (costs, manpower and time to implement). Table 9 shows the correlation between each property and is needing to determine the importance index (ω_i):

- Compare the properties to each other in terms of importance;
- If property 1 is more important than property 2 then, in the comparison between these two properties, value 1 is assigned to prop1 and value 0 to prop2, for instance;
- The ωi value will be obtained by summing all values 1 to be divided by the number of possible combinations of properties;
- The possible combinations are $N \times (N - 1) / 2$ (N: number of properties).

Table 8. Properties data collect for different Layouts options

LAYOUT	Area Occupied [m²]	Ease of Implementation [€] (1)	Capacity Required [pcs/wd]	Walking [m]	Unbalancing [%]	Target Cycle Time [s]
Option 1	93,8	2000	32	76,5	7	857
Option 2	93,8	3000	32	98	7	857
Option 3	93,8	2500	32	75,5	7	857
Option 4	82,8	6000	35	50,5	2	784

(1) Ease of Changing from current layout, take into account the Cost (Investments), Manpower, Time

Table 9. Attributes table for each property

		Possible combinations										N	5
													10
		1	2	3	4	5	6	7	8	9	10		
		1-2´	1-3´	1-4´	1-5´	2-3´	2-4´	2-5´	3-4´	3-5´	4-5´	Σ	ω
Capacity Required		60	70	70	80							280	0,280
Area occupied		40				55	55	70				220	0,220
Operator flow			30			45			50	60		185	0,185
Balancing				30			45		50		60	185	0,185
Ease implementation					20			30		40	40	130	0,130
											Σ	1000	1

+ Important 1 - Capacity Required - Important 5 - Ease implementation
2- Area occupied
3- Operator flow
4- Balancing

Matrix Selection Method

The performance index (γ) is an algebraic formula that expresses a compromise between two characteristics or properties. In its simplest form, a performance index is usually a multiplication between these two variables, having on one side the importance index (ω) of each layout option and the other variable is the weighted index (β), regarding each option, as well. Thus, for each property determined for different represented layout options, it is necessary to calculate the performance index, thus, after obtaining all performance index results independently, the final performance index is the sum of all. Table 10 shows all the calculated values to obtain the option that has the highest performance index, which is the most adequate for the implementation.

The performance index (γ) is obtained by the following equation:

$$\gamma = \sum \omega_i * \beta_i \tag{3}$$

where,
ω – Importance index related to each layout option.
β - Weighted index related to each layout option.

5.2. Production Process Optimization

Before proceeding to identify possible improvement solutions, it is time to analyze the current state of the assembly tools and the previously identified process tasks, which are the completion of the quality checklist and the application of glue to the screws. To assist in the process of analyzing these problems, one used a very useful quality tool to analyze all the possible causes that may influence an unfavorable design, use of inadequate fixation systems and the unnecessary activities of operations that do not add value to the mounting process. Thus, in Figure 35 below is the analysis of the current state using the Ishikawa Diagram.

Table 10. Matrix selection method

Layout	Capacity Required ↑		Area Occupied ↓		Operator Flow ↓		Balancing ↓		Ease Implementation ↓		Perf. γ
	ω 1-	0,28	ω 2-	0,22	ω 3-	0,19	ω 3-	0,19	ω 3-	0,13	
Op.1	32,0	25,6	93,8	19,4	76,5	12,2	7,00	5,29	2000	13,0	75,52
	91,4		88,3		66,0		28,6		100,0		
Op.2	32,0	25,6	93,8	19,4	98,0	9,5	7,00	5,29	3000	8,7	68,50
	91,4		88,3		51,5		28,6		66,67		
Op.4	32,0	25,6	93,8	19,4	75,5	12,4	7,00	5,29	2500	10,4	73,08
	91,4		88,3		66,9		28,6		80,00		
Op. 5	35,0	28,0	82,8	22,0	50,5	18,5	2,00	18,5	6000	4,3	91,33
	100		100		100		100		33,33		

All these possible causes are closely related to one another and have a considerable impact on the assembly line performance, as well as production cost. The possible causes detected for the problem can be from different types and the most critical will be analyzed and mention further.

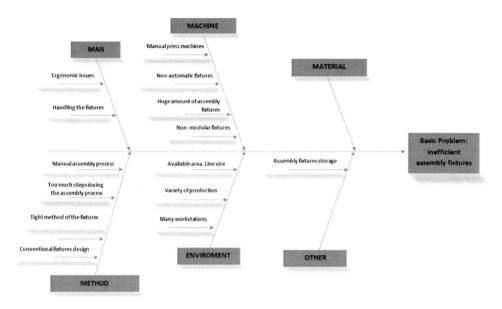

Figure 35. Ishikawa diagram showing issues for inefficient assembly fixtures.

5.2.1. Man

Some ergonomic issues present in the line, as the tools needed to press fit some components of the product. Another subject that could have an impact is the handling of the fixtures. The operator needs to use a lot of fixtures to execute the assembly desired, and the probability of having errors is higher.

5.2.2. Machine

Six manual press machines exist in this production line, whose have a certain impact in relation to the occupied area, the time needed to press fit the components and also the accuracy of those operations. Due to a low volume production system, most of the tools and fixtures used in the process are fully manual utilization.

5.2.3. Method

In this assembly line, the mount process required to build this kind of complex electronic products are identified as very manual sequences. Thus, the current assembly tools used in this line is all of the manual handlings, having no operator automatisms that allow faster operation. One of the consequences of this type of assembly is the influence on the cycle time of each workstation. So, it is clear that the handling of the assembly tools has a significant time percentage related to the cycle time. When observing the assembly process of this product, it was identified the need to use different assembly tools for each workstation, which in some cases, requires set up of support tools, in order to guarantee the correct construction of the unit. In some workstations, the assembly fixtures need to be tight with screws to the work table. The reason for this is to verify that in certain assemblies of the product it is necessary to apply forces and torques to ensure the correct assembly of the product. Taking this into account, it is important that the assembly equipment is really stable. Therefore, during the changeover process of the production line, this operation of fixing the assembly fixtures to the work table requires a certain time, being one of the factors that contribute to the elevated changeover time of this assembly line.

5.2.4. Environment

Taking into consideration that in this assembly line, some different versions are manufactured, which belong to one product family. So, due to this reality of manufacturing, the number of fixtures needed for the building of those different versions is high. Another aspect is the number of workstations in this assembly line. This aspect impacts the number of fixtures required for each workstation to execute properly the attachment of the parts.

5.3. Improvements Solutions

In order to implement new solutions for issues identified through the analysis of the current state described, it is important to apply a structured method.

Therefore, implementation planning is necessary. Thus, in this case, PDCA cycle and managing techniques for continuous improvement were also used. **Plan:** Design just one assembly fixture assuring the possibility of mounting two assemblies from two product versions. With this in mind, one of the objectives is reducing the cost of fixtures manufacturing and eliminate the chance of duplicate fixtures, reducing the need to occupy space; **Do:** Implement the action plan. In this step, all the changes are made, and the new data is collected; **Check:** It is time to check the functionality of the new assembly fixtures and the reduction of some process tasks determined in the previous step. Therefore, it will be necessary to analyze data trends and other deviations occurrences; **Act:** Apply any needed corrective measures to the improvements done. If something in the new process is not running according to plan and can be easily fixed, it should be done in this step.

5.3.1. Assembly Fixtures Improvement

In order to overcome this drawback, the new design concept for the fixtures was based on modular fixtures systems, which consist of using interlocking standardized elements such as dowel pins, clamps, and connections. Flexibility is achieved through the interchangeability of different modules, which enables a modular fixture to be easily disassembled and re-assembled. The great benefits of this design are: reducing the setup time (SMED improvements); be easier to maintain, having spares to change when needed; reducing costs of new fixtures with the same configurations, only needing to design some elements for the new models and a better 5S implementation.

5.3.1.1. Workstation 01 – Pan Motor Preparation

In this station, the assembly fixture development took into consideration some features in order to fulfil the following aspects: (1) just one assembly fixture design has been assured, that is, two or more fixtures have been combined into one, (2) the only changeover for this jig is two adaptors illustrated as item 01 in Figure 36, which the only need, done in an easier way, is placing those adaptors on the square hole of the main fixture, (3) the requirement for the pinion position is guaranteed by the item 04. This is a mobile part of the fixture, which could rotate and moves axially, (4) for a future new product version, under normal condition, it will only be necessary to design and buy the item 03 and 01, reducing tool costs.

5.3.1.2. Workstation 02 – Tilt Motor Preparation

In this station, the assembly fixture development took in consideration some concerns in order to fulfil the following aspects: (1) the design concept follows the same approach as the previous workstation. In Figure 37 can be seen fixture used in this workstation, (2) the only difference between the previous workstation is the need for flipping the part identified as item 03 when changing over for another product.

For press fit tasks, these two workstations share the same manual press arbor. Taking this into account, the tools designed to accurately press the bearings fulfilled the following aspects: use of only one plate to ensure the correct orientation and position of all different pieces and the placement of the used press adapter in the equipment is made by a quick indexing plunger shown in item 03 of Figure 38. With this action, the changeover is easier and can reduce operation time.

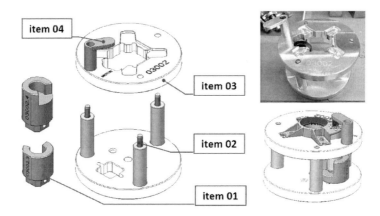

Figure 36. New assembly fixture to workstation 01.

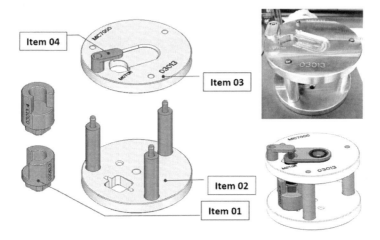

Figure 37. New assembly fixture to workstation 02.

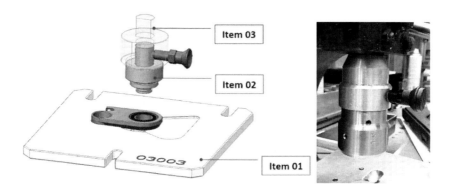

Figure 38. New press tools to workstation 01 and 02.

5.3.1.3. Workstation 03 – Wiper Motor Preparation

In this station, the same design architecture already described in the previous workstation was assumed.

5.3.1.4. Workstation 04 – Pan Stage I Assembly

This workstation was one of the most changed assembly stations in this line, having a huge impact on reducing changeover time and improving ergonomic aspects. Thus, the new changes in this workstation are: (1) the automatic press shown in Figure 39, improvement action with the greatest impact on product quality and assembly. With this investment, it was possible to eliminate three manual press machines, (2) the tools designed for this machine considered the need to press different product versions. Thus, the used adapters placement in the machine is done by a rapid indexing approach.

Figure 39. Press pneumatic machine for workstation 04.

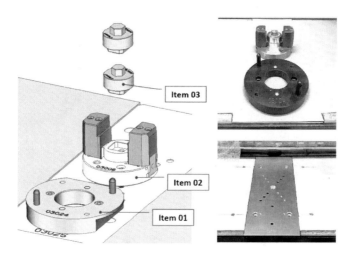

Figure 40. New assembly fixtures to workstation 04.

In spite of the press tasks performed on the press machine, it is necessary to complete the remaining assembly operations on the workbench, and for this, some other tools have been redesigned with the intention of improving the remaining assembly process. Thus, the new changes in this station are: due to the human effort required to make this assembly, the equipment used needs to be really stable. Thus, an adapted workbench was designed with an aluminium plate machined in the middle. In Figure 40 it is possible to see how easy it is to attach to this worktable all the mounting fixtures for this workstation. Other improvements were the tools defined as items 01 and 02 in Figure 40. The accessory identified as item 02 is to fix the product assembly firmly, and for this, the clamps work with springs to lightly tighten the product. The only step required to configure this fixture is the change of item 03, which is also a spare part due to the fatigue of the tool.

5.3.1.5. Workstation 05 – Pan Stage II Assembly

In this workstation, the optimization of the devices was based on avoiding a quality gate and improving the way to press a ferrite in a hole of a plastic component. So, those improvements are: (1) a manual press is designed to ensure an easier procedure to press the ferrite into the hole and also improve the accuracy of the ferrite position, as well as reduce the force required by the operator to correctly press the ferrite (Figure 41), (2) another optimization was to eliminate the use of a quality fixture to check the previous assembly. The reason for this is that this quality check is already done by functional testing.

5.3.1.6. Workstation 07 – Tilt Center I Assembly

In this workstation, the challenge was greater comparing the rest of the workstations. The designed tools play a very critical role in relation to the accuracy of this assembled product, so this fixture must meet the following characteristics: (1) the design has reduced the need to use three fixtures to correctly run the right amount of pressure fitting on this workstation. Therefore, only one mounting device is designed as shown in Figure 42, (2) all attach of the adapters used in this tool are made by pins, guaranteeing good geometric position and fast fixing, (3) the changeover of this tool to a different version of the product is faster, because one only needs to make small adjustments.

Lean Manufacturing Applied to a Complex Electronic Assembly Line 323

Figure 41. New tool the press fit the ferrite.

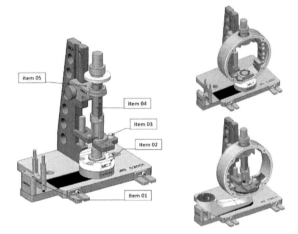

Figure 42. New assembly fixture to workstation 07.

Figure 43. New assembly fixture to workstation 07 for tight the arms.

After all the pressure fit operations performed, the next task is to assemble the product arms. This tool needs to ensure stability during the arms tightening process, and also this workstation has a workbench similar to the aluminium plate made for the workstation 04. This fixture makes easier to rotate the arm assembly (Figure 43).

5.3.1.7. Workstation 08 – Tilt Center II Assembly

In the workstation the only design developed was a fixture to maintain the previous assembly in vertical position. The concept of the tool is the same observed in the previous fixtures.

5.3.1.8. Workstation 09 – Camera Module Assembly

In this station, the new fixture design took into consideration the following aspect: the changeover of this fixture is to change the item 02, 03 and item 04, and the setup of those parts is simple and quick (Figure 44).

5.3.1.9. Workstation 10 – Front Shell Assembly

In this station, the new fixture design took into consideration the following aspects: (1) a vertical press-fit toggle clamp was added to this fixture intending to eliminate the movement by the operators to another station to press fit the bushing (Figure 45), (2) the changeover for a different product version is to take off the item 02, 04 and sweep the item 03 to another hole. All connections are carried out by fast plugs.

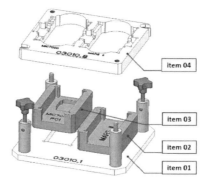

Figure 44. New assembly fixture to workstation 09.

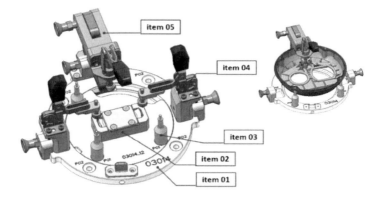

Figure 45. New assembly fixture to workstation 10.

5.3.1.10. Workstation 11 – Rear Shell Assembly

In the workstation 11, the only design developed was a fixture to maintain the previous assembly in vertical position. The concept of the tool is the same observed in the previous fixtures.

5.3.1.11. Workstation 12 – Final Assembly

In this station, the new fixture design took into consideration the following aspects: (1) the design has reduced the need to use two fixtures quite similar to this workstation. Therefore, only one mounting device is designed as shown in Figure 46. The real gain is the reduction of the occupied area, due to the size of the fixture, (2) eliminate the ergonomic issue related to handling this assembly fixture and the complete unit. Thus, the idea is to reduce the weight of the fixture and design a structure to minimize the need for handling the product, (3) the new design of this fixture contains a feature that permits adjusting the inclination of item 02 illustrated in Figure 46. The advantage of this is to have the better configuration of the fixture, (4) the changeover for a different product version is to take off the item 03, 04, 05 and item 06. All connections are carried out by fast plugs.

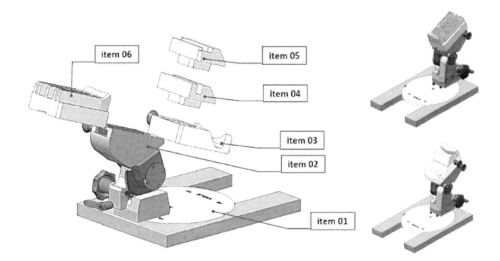

Figure 46. New assembly fixture to workstation 12.

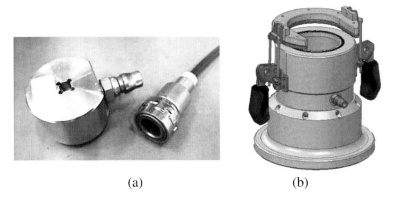

(a) (b)

Figure 47. (a) Fast air plug; (b) New fixture to test leak.

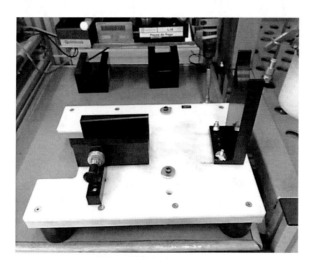

Figure 48. The adapted fixture to attach the yoke caps.

5.3.1.12. Workstation 15 – Leak Test

In this station, the new fixture design took into consideration the following aspects: (1) a new way to plug the air connection was created, eliminating the operation of tightening the air adaptor. Thus, the new design permits a faster plugging of the air connection as shown in Figure 47a, (2) in Figure 47b is shown the new fixture for performing the leak test. With this new design, it is possible to reduce the time needed for this operation, (3) after the leak test it is necessary to close the unit. Therefore, to avoid the need for the development of new fixtures, it was decided to adopt the current fixture with the objective of assuming the same characteristics as the previous assembly fixtures. In Figure 48 is possible to see the optimization done.

6. RESULTS AND DISCUSSION

6.1. Results and Discussion of the LLD Implementation

As found in the matrix calculation method, the option with the highest performance was option four, with a result of 91.33, thus being the best layout to implement.

The re-layout of this manual assembly line is planned to preserve almost all the stations from the previous assembly line layout, in order to reduce reorganization cost. Therefore, the key notion of this layout planning was to generate the minimum possible area required for the occupation of this assembly line, ensuring that all workstations were close to each other, without increasing station setup costs. The new plan suggests an assembly line with a U-shaped configuration, being the main proposal the reduction of the distance between the stations and also the improvement of the material flow.

Theoretically, balancing and layout planning of an assembly line are two dissociated procedures that do not influence each other while designing a production system from scratch. It must be taken into consideration that one of the objectives of using a Lean line design approach is to increase the production capacity of the assembly system. Thus, it is relevant to assume that in this case the balancing technique and the layout planning are strongly related, so a re-layout of the current line needs to be performed.

Regardless of the cycle times known from the current layout for each workstation, the purpose of Option 4 layout was to increase the production capacity of this manual assembly line, taking into account some balancing operations, transferring some tasks from station L19-05 to the previous workstation, and dividing station L19-09 into two. In Figure 49 is detailed the cycle time for each station.

After the cycle time distribution improvements shown in Figure 48, it is time to represent the new stack workflow for each operator obtained in the layout option 4, considering the scenarios of operator 6 and 8.

In Table 11 is shown the future state flow assembly sequence with the condition of having 8 operators working in the line. The walking time for each operator gives a total of 101 seconds moving across stations. Comparing with the actual state this reduction of loop time needed to cross stations, it represents a gain of 45.41%.

PROCESS STEPS		Manual	Machine
Description	Detailed	seconds	
L19-01 - Pan Motor Preparation		160,0	
L19-02 - Tilt Motor Preparation + Nexus Soldering + Power Soldering		265,0	
L19-03 - Wipper Motor Preparation		217,0	
L19-04 - Pan Stage I Assembly		363,0	
L19-05 - Pan Stage II Assembly		474,0	
L19-06 - ARM´s Preparation		296,0	
L19-07 - Tilt Center I Assembly		370,0	
L19-08 - Tilt Center II Assembly		310,0	
L19-09a - Camera Module I Assembly		321,0	
L19-09b - Camera Module II Assembly		200,0	
L19-10 - Front Shell Assembly		321,0	
L19-11 - Rear Shell Assembly		386,0	
L19-12 - Final Assembly		592,0	
L19-13A - Final Test		926,5	463,3
	Place on test	51,1	
	auto		80,5
	Manual	23,5	
	auto		77,7
	Manual	14,2	
	auto		126,6
	Manual	124,3	
	auto		409,4
	Manual	19,3	
L19-15 - Leackage Test		608,0	
L19-15.1	Place on the Leak	68,0	
L19-15.2	Auto		358,0
L19-15.3	Remove from Leak	112,0	
L19-15.4	Place on Soak Test	70,0	
L19-16 - Packaging station		665,0	

Figure 49. Cycle time for each workstation for the proposed layout selected.

Table 11. 4 Option flow assembly sequence for 8 operators

Operator	Flow Assembly Sequence			Walking [s]	Operator Time [s]
Operator 1	L19-01	L19-02	L19-03	8	**650,0**
Operator 2	L19-07	L19-09a		8	**699,0**
Operator 3	L19-04	L19-08		20	**693,0**
Operator 4	L19-05	L19-06		13	**783,0**
Operator 5	L19-10	L19-11		4	**711,0**
Operator 6	L19-12	L19-15.3	L19-15.4	8	**782,0**
Operator 7	L19-09b	L19-13	L19-15.1	9	**740,3**
Operator 8	L19-16			31	**696,0**
			Total =	**101**	

The future condition of the layout with the cycle times of all the operations is plotted in the operator balance chart, as shown in Figure 50.

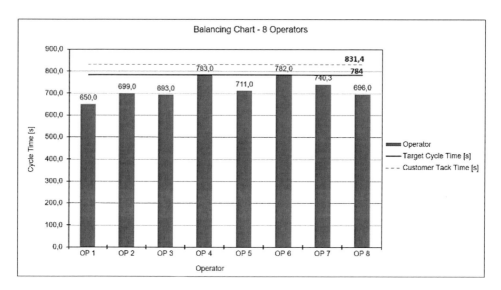

Figure 50. Four option Line Balance chart for 8 operators.

Table 12. 4 Option flow assembly sequence for 6 operators

Operator	Flow Assembly Sequence			Walking [s]	Operator Time [s]
Operator 1	L19-02	L19-03	L19-04	12	**857,0**
Operator 2	L19-01	L19-05	L19-08	24	**968,0**
Operator 3	L19-06	L19-07	L19-09a	16	**1003,0**
Operator 4	L19-09b	L19-10	L19-13	14	**998,3**
Operator 5	L19-11	L19-12	L19-15.1	14	**1060,0**
Operator 6	L19-15.3	L19-15.4	L19-16	44	**891,0**
			Total =	**124**	

The operator line balancing chart plotted for the future condition has a bottleneck loop which cycle time is close to the Customer Takt Time: the loop made by the operator 4. The cycle times of all operations are less than Cycle Time Target. This line is unbalanced by 2%

and the production capacity is 35 pcs/wd. Comparing with the current layout, this version permits an unbalanced decrease from 7% to 2%, and also an increase of the daily production capacity from 32 pcs/wd till 35 pcs/wd.

In Table 12 is shown the future state flow assembly sequence with the condition of having 6 operators working in the line. The walking time for each operator gives a total of 124 seconds moving across stations. Comparing with the actual state, this reduction of loop time needed to cross stations represents a gain of 30.34%.

The future condition of the layout with the cycle times of all the operations is plotted in the operator balance chart, as shown in Figure 51.

The operator line balancing chart plotted for the future condition has a bottleneck loop which cycle time is close to the Customer Takt Time: the loop made by the operator 5. The cycle times of all operations are less than the Targetted Cycle Time. This line is unbalanced by 3% and the production capacity is 26 pcs/wd. Comparing with the current layout, this version permits an unbalanced decrease from 7% to 3%, and also an increase of the daily production capacity from 24 pcs/wd till 26 pcs/wd.

6.2. Conclusion and Discussion Production Process Optimization

Generally, regarding the production process improvements, gains in time reduction of the assembly process to build one product were noticed.

Concerning the designing of new assembly fixtures, the optimization done brought good results as:

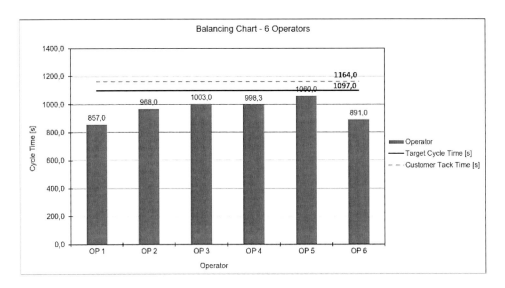

Figure 51. 4 option line balance chart for 6 operators.

- The handling of fixtures is simpler and faster;
- All the quality fixtures used were eliminated. It is important to reveal that all the quality issues continue to be verified.

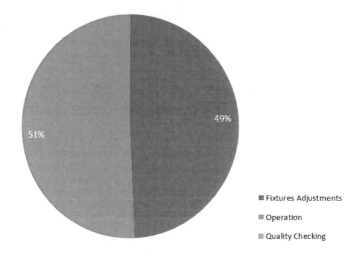

Figure 52. Mounting jigs - new fixtures usage.

Table 13. Time obtained from improving actions

Classification	Time [min]	[%]
Fixtures Adjustments	1,1	49
Operation	1,1	51
Quality Checking	Not needed	Not needed

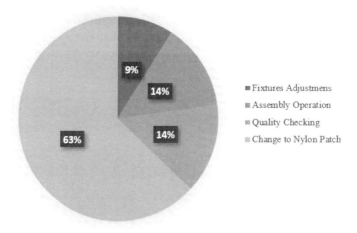

Figure 53. The different types of improvements.

Table 14. Time gain with process improvements

Improvement Classification	Time [s]	[%]
Fixtures Adjustments	26	9
Assembly Operation	40	14
Quality Checking	39	14
Change to Nylon Patch	182,8	64

Comparing the current state with the new reality, it was observed that with the new solutions implemented, some time reductions were achieved.

As showed in Table 13 and Figure 52, the need for fixtures adjustments during the assembly process for building one unit, a decrease of 1.1 minutes was verified. This is a reduction of 70% relative to the current state. The time reduction of the worker regarding the operation of the fixtures is not significant, which was decreased by 30%. For last, the time needed to check the quality of the product was eliminated.

Analyzing Figure 53 and Table 14, it is possible to see that a decrease of 289,8 seconds is verified in the duration of the entire process sequences necessary to accomplish a final product. One of the improvements actions with more impact was changing the practice of applying thread lock glue to the screws for a solution of nylon patch screws. However, this improvement is ongoing by the development team. Nevertheless, these changes represent an improvement of 3% when compared to the results achieved through the LLD implementation, with the following results: 0% of unbalanced; a decrease of takt-time to 808,3 sec/pc comparing with 831,4 sec/pc, with a production capacity of 36 units per day. The number the operators in the assembly line are eight.

CONCLUSION

This work is focused on the application of Lean methodologies, with the objective of implementing productive improvements in a manual assembly line. In order to be able to recognize the current reality in a more effective way, an initial VSM analysis was incorporated to detect the improvements needed to be implemented through the help of other Lean methods, such as assembly optimizations and the LLD technique, involving the design of line balancing layout and walking paths.

The main goal of the Option 4 layout was to increase the production capacity of this manual assembly line. The selection of this option was achieved taking into account the highest performance obtained by this option in the selection matrix used in this work, with a final result of 91.33. On the other hand, for a stream assembly sequence for 8 operators, and comparing with the current state the reduction of the loop time required to cross stations, represents a gain of 45.41%.

As depicted in Figure 53, the improvements in the production process can be summarized as follows: 9% fixtures adjustments, 14% assembly operations, 14% quality checking and finally the improvement that represents 63% is changing to nylon patch.

With regard to the improvement of the mounting fixtures systems, it should be pointed out that the application of the modular fixtures system concept is fundamental for production environments which require a very manual process since they increase the speed and flexibility in the preparation of the line for new product variants. In addition, by standardizing the type of fixtures systems, the setup times and the number of tools used have been reduced.

The implementation of Lean methodologies also allowed to increase productivity in approximately 10% with the same workforce, modifying the workstations and work methodologies. With the redesign of the accessory systems, there was a reduction in line setup times, improved product quality and reduced number of tools needed to produce these products.

Production based on a Lean philosophy guarantees enough benefits in manual production/assembly systems. This manufacturing concept allows to obtain a functional scenario for efficient and competitive production by means of: stock reduction/elimination, avoiding quality errors and inefficient processes, reducing production space, avoiding overproduction and reducing transportation routes.

REFERENCES

Araújo, W. F. S., Silva, F. J. G., Campilho, R. D. S. G. & Matos, J. A. (2017). "Manufacturing cushions and suspension mats for vehicle seats: a novel cell concept." *International Journal of Advanced Manufacturing Technology*, *90* (5-8), 1539-1545. doi: 10.1007/s00170-016-9475-6.

Bortolini M., Faccio, M., Gamberi, M. & Pilati, F. (2016). "Including Material Exposure and Part Attributes in the Manual Assembly Line Balancing Problem," *IFAC-PapersOnLine* 49(12), 926–931. doi:10.1016/j.ifacol.2016.07.894.

Bosch Group. (2003). Bosch - QCO, *Quick Changeover*. Internal procedure/brochure.

Bosch Group. (2015). *Bosch Production System Always. Doing. Better*. Internal procedure/brochure.

Castro, T. A. M., Silva, F. J. G. & Campilho, R. D. S. G. (2017). "Optimising a specific tool for electrical terminals crimping process." *Procedia Manufacturing*, *11*, 1438 – 1447. doi: 10.1016/j.promfg.2017.07.274.

Correia, D., Silva, F. J. G., Gouveia, R., Pereira, T. & Ferreira, L. P. (2018). "Improving Manual Assembly Lines Devoted to Complex Electronic Devices by Applying Lean Tools." *Procedia Manufacturing*, *17*, 663-671. doi: 10.1016/j.promfg.2018.10.115.

Costa, C., Silva, F. J. G., Gouveia, R. M. & Martinho, R. P. (2018). "Development of Hydraulic Clamping Tools for the Machining of Complex Shape Mechanical Components." *Procedia Manufactureing*, *17*, 563-570. Doi:10.1016/j.promfg. 2018.10.097.

Costa, M. J. R., Gouveia, R. M., Silva, F. J. G. & Campilho, R. D. S. G. (2018). "How to solve quality problems by advanced fully-automated manufacturing systems." *International Journal of Advanced Manufacturing Technology*, *94*, 3041–3063. doi: 10.1007/s00170-017-0158-8.

Costa, T., Silva, F. J. G. & Pinto Ferreira, L. (2017). "Improve the extrusion process in tire production using Six Sigma methodology." *Procedia Manufacturing*, *13*, 1104-1111. doi:10.1016/j.promfg.2017.09.171.

Costa, R. J. S., Silva, F. J. G. & Campilho, R. D. S. G. (2017). "A novel concept of agile assembly machine for sets applied in the automotive industry." *International Journal of Advanced Manufacturing Technology*, *91* (9-12), 4043-4054. 10.1007/s00170-017-0109-4.

Dombrowski U., Ebentreich, D. & Krenkel, P. (2016). "Impact Analyses of Lean Production Systems," *Procedia CIRP*, *57*, 607–612. doi:10.1016/j.procir.2016.11.105.

El Maraghy, H. & El Maraghy, W. (2016). "Smart Adaptable Assembly Systems." *Procedia CIRP*, *44*, 4–13. doi:10.1016/j.procir.2016.04.107.

Fernando M. D. D. & Rivera Cadavid, L. (2007). "Lean manufacturing measurement: the relationship between Lean activities and Lean metrics," *Estudios Gerenciales*, *23* (105), 69–83. doi: 10.1016/S0123-5923(07)70026-8.

Koppal, V. & Arunkumar, N. S. (2013). "Line Balancing and Lean Line Design for Common Rail Pump Housing Manufacturing", 74–77. *International Conference on Mechanical and Industrial Engineering (ICMIE'2013)* August 28-29, 2013 Penang (Malaysia).

Lam, N. T., Toi, L. M., Tuyen, V. T. T. & Hien, D. N. (2016). "Lean Line Balancing for an Electronics Assembly Line." *Procedia CIRP*, *40*(1), 437–442. doi:10.1016/j.procir.2016.01.089.

Locher, D. A. (2008). "Value stream mapping for Lean Development." *CRC Press*, Boca Raton, FL, USA. ISBN-13: 978-1563273728.

Magalhães, A. J. A., Silva, F. J. G. & Campilho, R. D. S. G. (2019). "A novel concept of bent wires sorting operation between workstations in the production of automotive parts." *Journal of the Brazilian Society of Mechanical Sciences and Engineering*, *41*, 25-34. doi: 10.1007/s40430-018-1522-9.

Martins, M. Radu Godina., Carina, Pimentel., Silva, F. J. G. & João, C. O. Matias. (2018). "A Practical Study of the Application of SMED to Electron-beam in the Automotive Industry." *Procedia Manufacturing*, *17*, 647-654. doi: 10.1016/j.promfg.2018.10.113.

Moreira, B. M. D. N., Ronny, M. Gouveia., Silva, F. J. G. & Campilho, R. D. S. G. (2017). "A Novel Concept Of Production And Assembly Processes Integration." *Procedia Manufacturing*, *11*, 1385-1395. doi: 10.1016/j.promfg.2017.07.268.

Nagi, M., Chen, F. F. & Da Wan, H. (2017). "Throughput Rate Improvement in a Multiproduct Assembly Line Using Lean and Simulation Modeling and Analysis," *Procedia Manufacturing*, *11*, 593–601. Doi: 10.1016/j.promfg.2017.07.153.

Neves, P., Silva, F. J. G., Ferreira, L. P., Pereira, M. T., Gouveia, A. & Pimentel, C. (2018). "Implementing Lean Tools in the Manufacturing Process of Trimmings Products." *Procedia Manufacturing*, *17*, 696-704. doi: 10.1016/j.promfg.2018.10.119.

Nunes, P. & Silva, F. J. G. (2013). "Increasing Flexibility and Productivity in Small Assembly Operations: A Case Study." In: *Advances in Sustainable and Competitive Manufacturing Systems*, Azevedo, A. (Eds.), Springer, 329-340. doi: 10.1007/978-3-319-00557-7.

Rocha, H. T., Luís, Pinto Ferreira. & Silva, F. J. G. (2018). "Analysis and Improvement of Process in Jewelry Industry." *Procedia Manufacturing*, *17*, 640-646. doi: 10.1016/j.promfg.2018.10.110.

Rohani J. M. & Zahraee, S. M. (2015). "Production Line Analysis via Value Stream Mapping: A Lean Manufacturing Process of Color Industry." In *Procedia Manufacturing*, *2*, 6–10. doi: 10.1016/j.promfg.2015.07.002.

Rosa, C., Silva, F. J. G., Ferreira, L. P. & Campilho, R. (2017). "SMED Methodology: The Reduction of Setup Times for Steel Wire-Rope Assembly Lines in the Automotive Industry." *Procedia Manufacturing*, *13*, 1034-1042. doi: 10.1016/j.promfg.2017.09.110.

Rosa, C., Silva, F. J. G., Luís, Pinto Ferreira., Teresa, Pereira. & Ronny, Gouveia. (2018). "Establishing Standard Methodologies to Improve the Production Rate of Assembly Lines Used for Low Added-Value Products." *Procedia Manufacturing*, *17*, 555-562. doi: 10.1016/j.promfg.2018.10.096.

Rosa C., Silva, F. J. G. & Luís, Pinto Ferreira. (2017). "Improving the Quality and Productivity of Steel Wire-rope Assembly Lines for the Automotive Industry." *Procedia Manufacturing*, *11*, 1035-42. doi: 10.1016/j.promfg.2017.07.214.

Rother, M. & Shook, J. (1999). "Learning to See: Value Stream Mapping to Add Value and Eliminate Muda (Lean Enterprise Institute)." 1st edition, Boston, *Lean Enterprise Institute Brookline*. ISBN-13: 978-0966784305.

Sabuncuoglu, I., Erel, E. & Alp, A. (2009). "Ant colony optimization for the single model U-type assembly line balancing problem." *International Journal of Production Economics*, *120*(2), 287–300. doi:10.1016/j.ijpe.2008.11.017.

Sousa, E., Silva, F. J. G., Ferreira, L. P., Pereira, M. T., Gouveia, R. & Silva, R. P. (2018). "Applying SMED Methodology in Cork Stoppers Production." *Procedia Manufacturing*, *17*, 611-622. doi: 10.1016/j.promfg.2018.10.103.

Venkataraman, K., Ramnath, B. V., Kumar, V. M. & Elanchezhian, C. (2014). Application of Value Stream Mapping for Reduction of Cycle Time in a Machining Process." *Procedia Materials Science*, 6, 1187–1196. doi:org/10.1016/ j.mspro.2014.07.192.

Wilson, L. (2009). *How to Implement Lean Manufacturing*. 1st edition, McGraw-Hill Professional, New York. ISBN-13: 978-0071625074.

Womack, J. P. & Jones, D. T. (2003). *Lean Thinking-Banish Waste And Create Wealth In Your Corporation*. 2nd edition, Free Press, New York. ISBN-13: 978-0743249270.

In: Lean Manufacturing
Editors: F. J. G. Silva and L. Carlos Pinto Ferreira
ISBN: 978-1-53615-725-3
© 2019 Nova Science Publishers, Inc.

Chapter 13

KARAKURI: THE APPLICATION OF LEAN THINKING IN LOW-COST AUTOMATION

Stephanie D. Nascimento, Milena B. Alves, Julia O. Morais, Laryssa C. Carvalho, Robson F. Lima, PhD, Ricardo R. Alves and Robisom D. Calado[*], PhD*

LabDGE - Laboratório Design Thinking Gestão e Engenharia Industrial,
School of Production Engineering, Federal Fluminense University (UFF)
Rio das Ostras, Rio de Janeiro, Brazil

ABSTRACT

In the current conjuncture of new operation management approaches, a great deal of effort and knowledge has been directed towards industries in order to carry out implementations based on the industry 4.0 approach. To this end, expensive investments have emerged in new forms of technological innovation. Through Lean Thinking, solutions for mechanical movements using the Karakuri technique have appeared, producing mechanical structures that are designed with simplicity and creativity. Besides the low-cost involved, the technique offers a wide spectrum of benefits. The projects combine concepts of physics with mechanical movements and easy handling, explore the use of the principles of gravity, as well as levers, springs, cams and pulleys, with the purpose of reducing physical and cognitive effort during operations. Such projects are often designed and manufactured by the operators themselves, integrating the Kaizen method with their technical knowledge and creativity in favour of improvements in internal logistics, ergonomics, productivity and worker satisfaction. The purpose of this chapter is to expose the practicality and simplicity of projects for low-cost automation using the Karakuri method and thus promote the elimination or reduction of the seven types of waste associated to the Lean approach (Transportation, Inventory, Motion, Waiting, Over-Processing, Overproduction and Defects). Therefore, this chapter will discuss the benefits of using Karakuri in industries and will, additionally, present

---

[*] Corresponding Author's E-mail: robisomcalado@gmail.com.
[**] Grateful to State School Jacintho Xavier Martins and Trilogiq Brazil (Modular Storage and Handling Solutions) for the collaboration. Grateful to program PROAES/UFF for the provided of scientific scholarship grants.

examples of its relationship with Lean thinking. It also addresses an introduction to the knowledge of obtaining efficient and functional systems, demonstrating alternatives for the application of low-cost and affordable automation technologies through creativity and an understanding of physics. It is concluded that by not resorting to energy sources, such as electrical and pneumatic, the automation proposed is suitably set within the scenario of present-day challenges. As such, besides offering companies significant savings and greater employee participation, it will also heighten motivation and enhance the use of human creativity, thus collaborating in producing a greater sense of well-being and safety for the people involved.

Keywords: Karakuri technique, automation, low cost solutions, lean thinking, Kaizen methodology

1. PRINCIPLES OF LEAN PRODUCTION

Lean Production not only challenges production practices, but also reviews and vitalizes the concepts of production and quality in the supply chain. According to researchers Womack, Jones and Roos in their original 1990 publication "The machine that changed the world," the Japanese automotive industry has shown the world its superiority in manufacturing (Womack et al. 1998,11).

Through its production system, Toyota's performance has proved to be more competitive and has forced its competitors to relearn, from shareholders to even the most modest operators. Holweg (2007, 420) researched and summarized the historical record of Lean production, from 1930 to the beginning of the concept in 1990, ending his study with some applications and relevant literature up to the year 2006. In recent literature and articles pertaining to operations and logistics, there are often chapters or citations on the approaches and concepts of lean production.

Companies experience different levels of maturity and, as such, lean methods and tools can be structured and applied in accordance with the principles and adequacy of each company's reality. It is understood that the manufacture of a product emerges from a project, with the functional need to serve customers, which is based on parameters. Since its structure is not rigid, but evolves continuously, it is important to explain the five principles of lean thinking in the Toyota Production System: value; value chain; flow; pulled production and perfection.

According to Womack and Jones (1998, 10), the five principles considered to constitute the basis for lean production in the Toyota Production System are:

Value: the capacity offered to a customer at the right time at a suitable price, as defined by the customer.

Value Chain: specific activities required to design, order and offer a specific product, from product conception to the launching of the order for delivery, and from the raw material to the customer's hands.

Flow: the progressive completion of tasks along a value chain for a product to transit from design to launching; and raw material to the customer's hands without interruption, waste or retro flows.

Pulled Production: a production system with instructions for the delivery of downstream activities to upstream activities in which the upstream supplier produces nothing without the downstream customer signalling this need.

Perfection: the total elimination of any activity that consumes resources, but which does not create value; thus, over time, all of the activities in a value chain will generate value.

The principles of lean production can be implemented to meet better manufacturing results. The methods and tools involved are inherent to the lean production approach, derive from its principles, and address the objectives. These methods describe the processes to be implemented, constitute the means to achieve the objectives, and are represented by the tools. Figure 1, for example, describes the five basic principles applied to all the levels of an operations and logistics system in a consumer goods manufacturing industry.

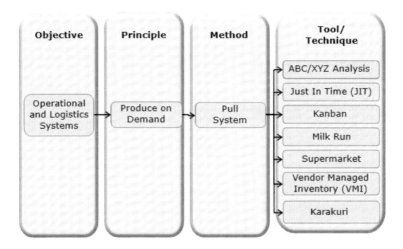

Figure 1. Tools for pull production systems.

2. PRINCIPLE OF PRODUCE ONLY ACCORDING TO DEMAND

The principle "to produce only according to demand" is translated into the continuous optimization of the process across the value aggregation chain, as well as orientation towards customer demand. Producing only on demand means producing according to customer demand and delivering on time, within the shortest time and inventory cycles. This effort to address customers promotes this group's satisfaction; besides others, it prevents inventory waste and generates simple and transparent processes. In this manner, the challenge for companies is to offer the right product, at the right moment, and in the location, quantity and quality desired by the customer, at the least cost.

Producing only when there is customer demand requires the application of knowledge and methods such as: the Pull System, takt time and supply chain management (Figure 1). It is worth mentioning that the principles presented are defined according to the strategy and needs presented by each business. In this chapter, the reference is to the Operations and Logistics System of a manufacturing industry, more specifically the supply chain of consumer goods. As such, the participants consist of external customers, wholesalers, retailers and subsidiaries; the internal customers may have distribution centres and warehouses that belong to or are under the company's control and management.

3. PRODUCE ON DEMAND

The Pull System (Figure 2) comprises a system where the production orders are pulled by manufacturing, contrary to what happens in the pushing principle. The impulse to execute manufacturing orders begins at the end of the supply chain (the customer). This means that the concrete order, and / or consumption by the customer, commands production.

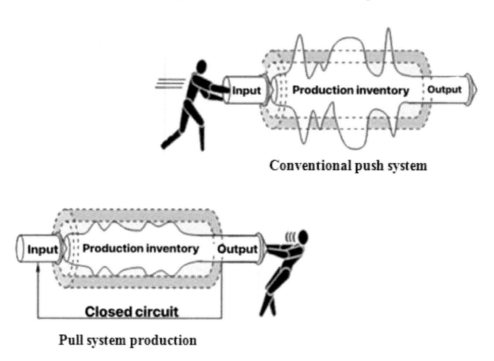

Figure 2. Analogy of pull and push systems. Source: Adapted from Slack et al. (2013, 300).

The purpose of the pull system is, whenever possible, to produce only part of the goods, in accordance with the consumer's needs, within a short time period. This thus dispenses with the stock usually required between the two stages of the process. The actions result in a reduction of inventory, less driving time, and an increase in delivery performance. There are also improvements in the process of planning and control, as well as better programming and production control.

The pull system generates advantages in the process chain (customer - production - supplier): since the customer's order triggers production at a defined point of the total process flow, the steps are connected to each other through the pull system during the entire process. Self-controlled processes thus replace those responsible for production control, and are based on a plan. Another advantage is that sales forecasts are used to plan capacities and demands in the medium and long-term.

4. PULL SYSTEM METHOD

The tools used in the pull system, in the context of manufacture, are described below:

4.1. ABC/XYZ Analysis

The ABC/XYZ Analysis tool comprises the combination of value / quantity (ABC) analysis with the forecast / consumption behaviour (XYZ) of the mix / product-mix parts. Thus, they constitute the basis for the classification of materials or products into supply / service classes.

Supply classes refer to standardized processes of material supply, be it raw material or the acquisition of parts from production suppliers. Service classes are understood as standardized supply processes, finished products, which transit from production to customers. The ABC/XYZ analysis serves as the basis for the selection and classification of supply classes and service classes. The following points are defined for each of the classes: delivery point and stock types (examples: warehouse, supermarket, production line, etc.), means of supply (examples: the milk run system), as well as control methods, such as Kanban and orders.

Materials and products are periodically controlled, at least on a quarterly basis, in accordance with the ABC/XYZ analyses. They are then classified into the proper supply classes, with the same materials and products being controlled according to the method determined. They are subsequently stored and made available if their prerequisites are fulfilled. The parameters are continuously revised and eventually adjusted when an Enterprise Resource Planning (ERP) system is in place. The ABC/XYZ analyses are executed automatically through a tool and are used for all products, semi-finished products, parts and raw materials.

4.2. Just in Time

Just In Time (JIT) is a concept which relates to the provision of material at a given point of delivery. JIT means that the right material is delivered at the right time, in the right amount and to the right place, based on a customer's concrete demand. Considering these requirements, supplier management is executed.

Just In Sequence (JIS) constitutes an additional step, which follows on from JIT: thus, in addition to the JIT requirements, the supplier must ensure that the sequence of deliveries is executed in accordance with the customer's request. This tool helps to reduce lead times, decreases inventory and space requirements, minimizes material movement and handling efforts, and enhances the transparency of the entire process involved.

The supplier and/or logistics and services provider implement JIT/JIS for each of the adequate material groups. The JIT/JIS supplier's capacity is checked or developed, with the constant concern of defining the delivery point. As an example: the frequency of the JIT/JIS supplier delivery is at least once per shift: i.e., the maximum stock required for one single shift. All the other parameters - surface, organization, technique, JIT/JIS contracts, and partnerships - are clarified, documented and made available, and no quality inspections are performed on the receipt of goods at the company. A favourable condition is when the supplier, under agreed and normal conditions, ensures the required quality.

4.3. Kanban

Kanban, a Japanese term defined as "card", is a platform which visually controls and manages the flow of information and regulates material supply between processes. It flags that the internal or external customer consumes a material or a product. The internal or external supplier is subsequently activated to re-supply or re-produce this product. When the client and the supplier communicate and act in accordance with this same procedure, external interventions are no longer used and there is then a self-controlled circle of regulation.

Kanban is a tool used in the application of the pull principle and control, which simplifies the flow of materials through the visualization of available materials and stocks. Besides preventing overproduction and minimizing defective parts, it is able to control the customer supplier relationship on its own. Furthermore, it detects process disturbances, which it uses to carry out continuous improvements.

Kanban control is applied and viewed according to the suitable groups of materials. Supplier/Customer control circles are defined and supply points (places of consumption) are established. Standard information contained on the kanban card includes: Name and part/material number, quantity per loader, source, image (optional), barcode (optional), customer, Kanban number, replenishing of lead-time, observation number and the Kanban total number.

4.4. Milk Run

The milk run system is a direct delivery derivative, similar to that of the old milk delivery truck: a milkman delivers bottles at a predefined time along a specific route, during which he replaces empty bottles with full ones. In the production process, through milk runs, the supply of materials runs along a defined route at predetermined times, and in fixed quantities and compositions.

The milk run reduces individual trips and minimizes supply costs for small groups of parts or materials. An additional advantage is that it is able to implement an efficient supply system by means of Kanban. Complex transport control systems and specialized systems are thus avoided, an example of which is the stackers control system.

This system allows for the standardization of material supply. For example, an internal circular cycle of less than two hours can be predefined within the factory, where the milk run train (Figure 3) makes several stops; the customer views these stops; there is direct delivery at the place of consumption; the packaging is standardized; and empty packaging is managed in an integrated manner.

Figure 3. Motorized car for an internal milk run.

4.5. Supermarkets

Supermarkets represent small buffers (understood as small stock) of material between workstations, in areas where it is difficult to ensure continuous material flow. Supermarkets maintain minimum and maximum quantities of raw materials, parts and the subassembly groups required in their respective processes. If the minimum quantity is not reached, new parts of the previous process are produced until the normal and predefined stock is filled.

Just as in a supermarket, the customer or the next process takes the necessary parts off the shelves and triggers the replacement of these parts through the seller, ie, the previous process. The supermarket is operated according to the FIFO principle (First In, First Out). The implementation of Kanban is only an auxiliary means and should be seen as an intermediate solution on the path to client-oriented continuous flow production.

Vendor processes are decoupled from customer demand, connecting batch-size processes with continuous flow, for example. This ensures the availability of finished and semi-finished products, but prevents overproduction. Consumption control, in line with the pull principle, eliminates production based on prognostics.

4.6. Vendor Managed Inventory

The Vendor Managed Inventory (VMI) is undertaken when the vendor assumes responsibility for the inventory of products sold and previously in possession of the customer. By complying with certain rules, the customer planner transfers the management of his inventory to the material and parts supplier. In addition, the supplier will have access to the stock data provided by the customer of the materials involved; the former will then begin the execution of supplies independently and assumes responsibility for the action.

This tool, in which the supplier is fully responsible for the availability of materials, has been highly recommended by the authors as a form of improving cooperation between companies. It also produces additional benefits, such as: ensuring the availability of components; speeding up responses; reducing the costs and interfaces of the order execution process; enhancing the transparency of supply relationships; building a stakeholder-based relationship with the supplier; and facilitating the planning and provision required for the entire supply chain by means of reliable inventory, consumption and demand data.

4.7. Karakuri

Karakuri is a Japanese term that can signify "mechanism" or "trick", the structure of which is based on the mechanization of mechanical clocks. At the time of its creation, it aroused curiosity in people, since it was described as a puppet which contained a hidden mechanism. These dolls are like puppets, but which move exclusively by means of mechanical devices.

Karakuri were produced in large numbers from the 17th to the 19th centuries in Japan, and are considered to be the forerunners of modern Japanese robots. Several festivals in Japan still use Karakuri to represent scenes relating to the country's history or myths.

These culturally rich and ancient artefacts have gained ground in industry and are being used to optimize processes. Due to the simplicity and mechanical practicality of the mechanism, this device has emerged today as an alternative of the applicability of the basic concepts of physics to solve daily problems. According to Katayama et al. (2013, 1895), Karakuri technology has led to environmentally friendly operations, such as energy savings, workload mitigation and operational simplification, favouring pull production and the continuous flow of parts and materials.

5. KARAKURI TECHNIQUE

In order to extend one's knowledge of Karakuri, one should revisit some of the concepts of physics as well those of ergonomics, which are fundamental to the understanding of the technique.

6. PHYSICS – MECHANICAL STRUCTURES DESIGNED WITH SIMPLICITY

Mechanical motions are known to science and have been put forward by physicists across the ages - Aristotle, Galileo, Newton and Kepler, amongst many others - up to the present Modern Era, with its quantum models for motion. However, if one is to come up with practical solutions by creatively using the concept of movement, one must invariably leave one's comfort zone and think of any scientific production before the existence of current reality. It is from this point of departure that one should try to create ideal solutions for answers to real problems.

The purpose of simple machines, such as pulleys, levers, hydraulic presses, or even inclined planes, is that of minimizing the force required to perform a given task. In this way, less human effort is required and ergonomic factors are ultimately improved, resulting in a reduction in energy costs. This feature is known as Mechanical Advantage.

Finding simple solutions through these machines will always be a challenge for the field of physical science and the integration of engineering in this process. Yet, by using the concepts of physics (gravity, inclined plane, levers, bearings and principles, as described by Blaise Pascal) this system has emerged as a significant alternative to the economically sustainable production process.

The use of levers, the processes of which were explained by Archimedes in the year 2 BC, has demonstrated the unique applicability of the concepts of force momentum, interspersed by three distinct handling operations. It should thus be noted that when a given force is applied outside a rotation axis, this improves efficacy and ensures greater energy efficiency, for example through the variation of vectors such as direction, orientation, speed and force.

6.1. Inclined Plane

The inclined plane is the most primordial of simple machine processes. By utilizing gravitational force, and coupled to roller systems, these mechanisms seek to minimize friction, thus ensuring greater efficiency. The inclined plane is observed through the movement of objects on a plane; depending on the angular variation, these objects will move upwards or downwards.

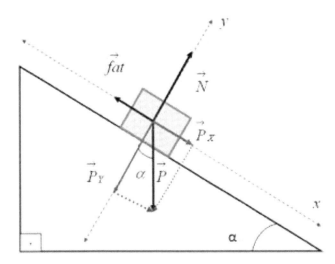

Figure 4. Inclined plane.

The motion generated by the inclined plane occurs in accordance with the Law of the Conservation of Mechanical Energy, which is responsible for the transformation of potential energy. At its maximum, it is transformed into kinetic energy through the speed gained as the object moves down the plane. Some practical examples of the applicability of the inclined plane are the ramp, the wedge and the screw.

Thus, in the inclined plane presented in Figure 4, several forces act upon this object. The frictional force (Fat) is ignored in an ideal system. However, in real systems the force of friction cannot be neglected and is equal to the force on the x-axis (Px), which guarantees the applicability of straight and uniform motion on an inclined plane. If this does not occur, then the object will accelerate to a speed which is proportional to the angle of inclination.

6.2. Levers

The fundamental physical concept inherent to the lever is that a rigid and extensive object is supported on a fixed point (o), where two forces will be applied: one is called the potent force (FP) and the other is named the resistant force (FR). The so-called potent force principle will be able to increase the force or displacement of the object you want to set in motion. In general, the operation of a force system applied to the lever is based on Figure 5. Its applications are contemplated in the first, second and third classes.

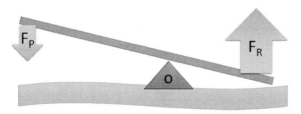

Figure 5. Simple machine – lever.

First Class Lever: In the first-class lever (Figure 6), the fulcrum is between the strength (FR) and the force (FP). When using this type of lever, a minimal force is required to lift heavy objects.

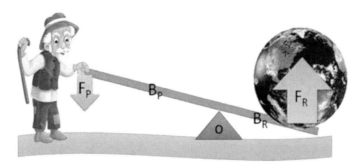

Figure 6. Simple machine - first class lever.

Second-Class Lever: In second-class levers (Figure 7), the resistive force (FR) is located between the powerful force (FP) and the fulcrum (o). In order to gain a mechanical advantage over this type of lever, the resistive force must be placed as close as possible to the fulcrum. One of the most practical examples of this type of lever is the wheelbarrow.

Figure 7. Simple machine - second-class lever.

Third Class Lever: In the third-class lever, the powerful force (FP) is located between the resistive force (FR) and the fulcrum. Characterized by a tweezer movement, this type of lever allows for the application of a certain force with a high degree of precision. Although it does not make use of the proper mechanical advantage, it enables movements of quality, when these are required.

Figure 8. Simple machine - third class lever.

Physical ergonomics is related to the characteristics of human anatomy, anthropometry, physiology and biomechanics in relation to physical activity. Relevant topics include the study of postures at work, material handling, repetitive movements, musculoskeletal injuries, workstation design, safety and health.

7. ERGONOMIC ASPECTS

Ergonomics is a scientific field which relates to the understanding of the interaction between humans and other elements or systems, as well as the application of theories, principles, data and methods to projects in order to optimize human well-being and overall system performance. Ergonomists contribute to the planning, design and evaluation of tasks, workplaces, products, environments and systems so as to make them compatible with people's needs, abilities and limitations.

The word Ergonomics derives from the Greek ergon (work) and nomos (norms, rules, laws). It is a subject which is directed at a systemic approach to all aspects of human activity. In order to account for the breadth of this dimension and to be able to intervene in work activities, ergonomists must take a holistic approach to the subject's entire field of action, both with regard to its physical and cognitive aspects, as well as in the social, organizational and environmental aspects at hand.

In everyday practice, fundamental physics is related to ergonomic factors, in which several studies investigate the impact of the worker's human and behavioural aspects on the company's performance and on the expected results. This makes ergonomics a critical factor in industry, acting directly to improve organizational performance.

The goal of ergonomics is to develop and adapt working conditions to the workers' psychological and physiological characteristics, so that they are able to perform the task safely, with maximum comfort and efficient performance.

In the workplace, some factors should be analysed ergonomically: furniture, equipment and transportation, as well as the loading and unloading of materials. Likewise, movements and postures should be considered when performing tasks which require the use of hands, arms and legs, whether sitting, half-sitting or standing. In addition to these, experts must consider the environmental conditions of the workstation itself (noise, vibration, lighting, humidity, climate and chemicals, etc.), as well as the work organization involved (visual information, hearing, other senses, controls for operation, dialogues, etc.).

Non-ergonomic factors are recurrent in the worker's daily life. Some relate to static work, a greater intensity of work rhythm, repetitive movements, productivity requirements, the absence of control over the work mode and rhythm, the lack of or insufficient pauses, as well as furniture and equipment which are uncomfortable and unsuitable for the execution of the tasks involved.

The adoption of the ergonomic concept reduces injury rates and absenteeism; it also improves productivity, quality and reliability. Studies have shown that Musculoskeletal Disorders (MSDs) and Repetitive Strain Injuries (RSI) lead to a significant loss of productivity.

RSI are the most common physical problems, which can cause limitations or lead to the inability to work. The use of ergonomic solutions in the workplace is an important initiative to develop the best prevention methods for these problems, which can significantly increase worker satisfaction and efficiency levels.

Brazilian regulation regarding ergonomics (NR-17) determines that, whenever the work can be performed in the seated position, the workstation must be planned or adapted to the position required. When in the standing position, as well as in the sitting position, the workstation position must possess the height and work surface characteristics which are compatible with the type of activity to be performed. Some requirements are: a suitable distance from the eyes to the work surface; work equipment and a work area which are easy to reach and can be visualized by the worker; and dimensional characteristics that will allow for the adequate positioning and movement of the body segments. For activities where work is to be carried out standing, resting zones must be provided in an area where all workers can use them during breaks.

For activities that also require the use of feet, and in addition to the requirements set forth above, foot pedals and/or other controls must possess the dimensions and position to ensure easy reach, as well as adequate angles between the various parts of the worker's body, depending on the characteristics of the work to be performed.

The seats used in work stations must meet the following requirements: the height of the seat and the armrests must be adjustable to the height of the worker and the nature of the function executed; characteristics of little or no conformation at the base of the seat; a rounded front edge; a backrest with adjustable angulation; and a shape which is slightly adapted to the body to ensure protection of the lumbar region. When the height of the seat is not adjustable, foot support may be required, which is adapted to the length of the worker's leg.

Failure to meet these regulated requirements can result in poor postural positioning in the development of an activity, which can lead to symptoms such as discomfort, stress and fatigue (Castilho, Barbirato and Sales 2015, 42).

Anthropometry constitutes the area of ergonomics which studies the measurements of the human body and how to use these in projects. The purpose of this study is to determine the population's mean values (weight and height), establish variations and the range of movements, as well as study the differences between groups and the influence of variables (gender, age, population, ethnic and climatic variations). Based on this analysis, heights, dimensions and angulations are then calculated and used in workstation projects and clothing manufacture, amongst others (Brogin, Merino and Batista 2014, 7).

All this research points to the importance of ergonomics in the design of work stations and processes, as well as to the effects of these on workers' safety and health, leading to the optimization of human performance and the overall work system.

When ergonomic factors are not adequately considered in the case of the production cells of assembly and manufacturing sectors, there is ultimately a greater occurrence of work absenteeism. These may be related to the inadequate characteristics of furniture, machinery, work and movement devices, which contribute to a greater physical effort required to carry out the activity. In addition, the organizational model of production could also contribute to the development of musculoskeletal injuries through exposure to repetitive and monotonous work.

On the recommendation of ergonomics specialists, for example, efforts have been directed towards a gymnastics programme at work, as well as the implementation of job rotation and changes in tools and work devices.

In order to prevent waste and improve aspects relating to internal logistics and the handling of materials, it has become increasingly common to apply changes in movement and in the handling of devices through the introduction of the Karakuri technique. This is implemented in the medium and long term, once the Lean Thinking approach has been adopted in the production system.

8. INTERNAL LOGISTICS AND THE HANDLING OF MATERIALS

Logistics is considered to be: "(...) the process of planning, implementing, controlling flow and undertaking the efficient low-cost storage of raw materials, in-process stocks, finished products and related information, from the source to consumption, in order to meet customer requirements."

According to Ballou (1973, 16), internal logistics studies how management can provide better profitability in the distribution of services to clients and consumers through planning, organization and effective control of handling and warehousing activities aimed at facilitating the flow of products.

An internal logistics system, which is well designed and used correctly, positively influences competitiveness, since it is linked to the planning and control of production, thus enhancing the company's efficiency. The crucial impact of this system on production efficiency, according to Alizon et al. (2009, 3854), makes supply chain design and planning a key factor in the reduction of costs and in the maximization of product availability. The continuous application of the Lean approach in the area of internal logistics is essential when reducing those activities that add no value to the product, thus improving processes in the production chain.

According to Ohno (1988, 30), Toyota identified seven types of wastes (Table 1) which add no value to activities in processes. One should, however, include an eighth form of waste, which was proposed by Liker (2004, 36): non-utilized talent, namely, underutilizing human capabilities such as creativity.

The lean logistics approach presents relevant collaboration in the reduction of waste throughout all of the stages involved, as well as in the improvement of operational performance. The application of the tools and techniques is based on two pillars: respect for people and continuous improvement. Karakuri contributes to these two pillars by motivating employees to propose improvements in the production processes at hand; namely, it contributes to the reduction or elimination of waste at a low cost.

Table 1. Eight types of wastes in the Lean approach (Ohno, 1988 and Liker, 2004)

Waste	Description
Overproduction	More production, sooner or faster than necessary.
Waiting	Operators or processes are inoperative due to delays in the arrival of material or unavailability of other resources, including information.
Transportation	Parts or products are moved unnecessarily.
Extra-Processing	Unnecessary or incorrect processes are performed, which do not add value to the product or service provided.
Inventory	More than the minimum requirements are available for a controlled system.
Motion	Unnecessary operator movements when executing a task.
Defects	Need for inspection, reworking or defective products.
Non-utilized Talent	This occurs when the company does not know of the employees' intellectual knowledge or skills, which does enable their use.

A lean thinking organization supports employee capacity building through interrelated factors, such as training and organizational infrastructure, which in turn sustain improvements over time.

It is common knowledge that the effectiveness of operational practices depends on human resources. The adoption of training and extensive training constitutes a strategy which leads to the improvement of productivity, quality, cost and safety. One must bear in mind that productivity is related to issues such as well-being: only workers who are satisfied and healthy can be productive. High productivity is, consequently, essential and the best way to stay in business.

Socio-technical interaction - constituted by a particular class of complex systems composed of people and technology, which interact to produce a desirable result - will therefore affect the speed of change, ensure the achievement of goals and develop a capacity for improvement, thus culminating in heightened productivity.

8.1. Productivity

Productivity is the ability to produce more by using fewer resources, and can be measured by analyzing the relationship between Output (the quantitative measure of what was produced) and Input (the quantitative measure of resources). It is, above all, an attitude of the mind: it seeks to improve what already exists and seeks to eliminate waste. Productivity is based on the belief that it is always possible to do better today what was done yesterday, and that tomorrow will be better than today. Through the participative spirit of Kaizen, continuous improvement is sought, and its immediate implementation in the productive process on the factory floor is valued.

Since it connects human talent with organizational performance, productivity should reflect on the impact of implementing management practices.

8.2. Kaizen Tool

Kaizen events have been widely reported to produce positive change in business results and human resource outcomes.

It has been observed that, when combined with the pulled production system, the Kaizen tool represents the elimination of waste, which occurs continuously through minimal improvements. "Kaizen" is a Japanese word which means "continuous gradual and organized improvement". It constitutes an endless improvement process for quality and productivity. In numerous case studies, where Kaizen was applied in accordance with lean thinking in small and medium-sized enterprises in several countries, it was revealed that such an application brings about effective improvements in combating hidden inefficiencies in organizations.

The various studies undertaken in several countries show that the Kaizen tool has been approved worldwide, and that it is being combined with various waste disposal tools and techniques in an easy and effective manner.

Positive attitudinal outcomes from specific lean implementation activities, such as Kaizen events, could increase employee commitment to the lean program as a whole, ultimately improving the program's success and sustainability.

Kaizen culture needs to be rooted in the minds of managers and workers. This term includes: team building, personal commitment, the improvement of employee morale, pre-defined quality circles, and suggestions for the improvement of the system. Through Kaizen, Karakuri applications have been shown to be effective.

9. KARAKURI CONCEPTS IN LEAN INTERNAL LOGISTICS

The solution for problems faced during the processes is often in the hands of the workers. Therefore, the Karakuri technique, as well as Kaizen and QCC - Quality Control Circle, provide employees with the opportunity to contribute to the improvement of the production flow. This is achieved through the design and construction of devices and equipment to eliminate waste and solve ergonomic problems by means of low-cost automation, and is known in the industry as Karakuri.

At an operational level, employees will seek to propose alterations of what needs to be improved or solved; in this way, the organization will thus be in possession of an important technique which, besides motivating the workers, will reduce investment in automation and contribute to the improvement of productivity and safety.

The following three practical examples of equipment used in companies will be presented to provide a better illustration of Karakuri. Noteworthy is the fact that the Karakuri in companies such as Toyota are thought up by the employees themselves by harnessing their own creativity. Even without any knowledge of the term Karakuri, some companies already apply simple solutions for the improvement of internal logistics. These can be used as examples of the way materials are stored and moved on the production lines, as is shown below in Figures 9 and 10.

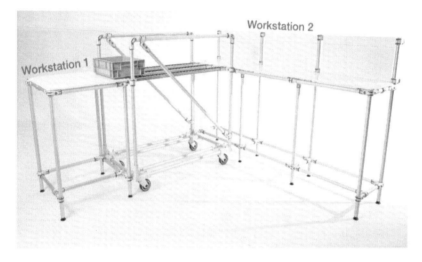

Figure 9. Karakuri example for an inclined plane between two stations (courtesy of TRILOGIQ).

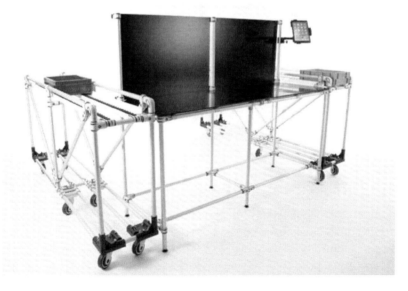

Figure 10. Karakuri example of inter-workstation transport with a rail conveyor (courtesy of TRILOGIQ).

Companies providing solutions for internal logistics and ergonomics have produced solutions which are in line with the Karakuri technique; the equipment and devices designed to reduce times do not add value and ensure a better organization of the manufacturing plant.

There are also examples of companies which acquire the components, and the employees themselves subsequently create and assemble the solutions. This is undoubtedly a way of encouraging and praising workers' motivation as participants in the continuous improvement process, in addition to enabling the use of workers' technical knowledge in the development of solutions in line with lean thinking.

Below are three basic examples of applications, which resort to pre-ordered commercial equipment using the Karakuri technique. In the first example (Figure 9), one is presented with two workstations, which have been adapted according to process needs. In a conventional situation, for example in the first station, the worker performs his routine activity and the box is then transported to station 2.

In this case, internal logistics equipment and human resources are used. In the Karakuri application, which aims to eliminate unnecessary movements of no added value, a conveyor was placed between the two stations; the work is thus carried out in optimized time and with greater efficiency. In this manner, Karakuri can be used to transfer boxes between the two workstations. In order to do so, the two stations are purposely positioned near each other and this transfer is carried out by a rail which possesses a mechanical interface (Karakuri), allowing the boxes to move by means of a simple pushing movement.

Similarly, in the second example (see Figure 10), Karakuri was used for the rail conveyor. There is a conveyor which transports the boxes to the workstation; another conveyor returns them once the process is completed.

It is notable that Karakuri has become a facilitating factor for employees working in industries due to its adaptability to each work environment. In turn, Karakuri with conveyors has facilitated a speedier movement of the most varied objects, weights and volumes. It has additionally assisted in the elimination of waste relating to waiting and transportation, thus providing greater dynamics in the work environment.

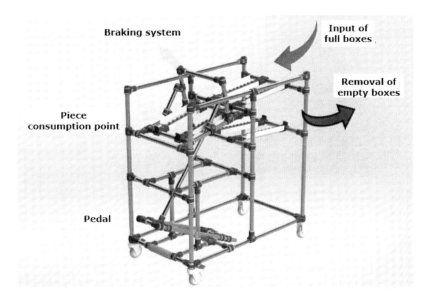

Figure 11. Karakuri dump rack example (courtesy of TRILOGIQ).

In the third example, a Dump Rack (Figure 11) is used. The structure of the Rack consists of two levels: one for supply (indicated with the green arrow) and another for return (indicated by the red arrow). There is also a footswitch pedal to return the empty box. Operation is initiated through full box supply (green dot), which allows the operator to consume the box parts. In this manner, once the box is empty, the operator presses the pedal to return the empty box to the lower level. This return is only possible by means of a pedal, which is responsible for the automatic return to the starting position. As a result, the empty box is removed from the rack and acts as a trigger so that logistics can replenish the component required, which subsequently restarts the operation.

In conventional means, the operator has to take the empty box and place it on the return section, thus performing several movements that add no value. When one examines this isolated example with only one box, the value of this Karakuri solution might not be detected. However, if one were to imagine an entire production line, where any lost second could affect the volume produced at the end of the day, the gains generated by movement, and even by labour, are considerable. In this model, the operator who is at the point of consumption has no physical contact with the packaging; he only deals with the parts used in the assembly of his product. Another benefit is related to ergonomics, as the operator does not have to lower or move his body to return empty packaging.

10. ADVANTAGES OF KARAKURI APPLICATION

The benefits of applying Karakuri have proven the efficiency and effectiveness of this tool as a solid path to the elimination of waste. Above all, in the context of today's exhaustion of resources, Karakuri technology has innovated by demonstrating that it is not dependent on electric or pneumatic energy: it utilizes natural physical phenomena such as counterweights, gravity, levers and pulley mechanisms, amongst others. Thus, the technique considerably minimizes the costs associated to implementation, maintenance and operations.

Through the implementation of Karakuri technology, one can use the operators' creativity by creating Kaizen teams to present simple and practical solutions. This will ultimately reduce costs and contribute towards simplicity in manufacture. Once Kaizen has been implemented, in association with Karakuri applications, one often observes a shorter processing time due to the elimination of activities or operation elements which add no value. This shorter processing time invariably results in lower process and product costs (Alves and Calado, 2018, 372).

Karakuri changes can produce benefits not only in the industry itself but also in the case of employees. Once workers begin to use their own perception in problem identification and resolution, the reduction of the eighth waste of lean thinking then ensues, during the development and / or proposal of new Karakuri devices to improve processes.

Karakuri devices may exclude the need for the transportation and unloading of materials by the worker, thus eliminating the non-ergonomic postures associated to carrying weight, such as squatting and twisting of the body. The correct implementation of Karakuri meets the basic requirements of employee ergonomics and eliminates repetitive movements. It also improves the efficiency of internal logistics on the shop floor by eliminating waste at a low cost, thus allowing for a real reduction of process times.

Karakuri not only offers an approach of method engineering but also motivates the workers' sense of creativity. Grounded on assumptions of simplicity, it constitutes an attractive alternative for industries wishing to increase their productivity, safety and automation at a low cost.

CONCLUSION

The Karakuri technique constitutes an appealing alternative in the improvement of lean and internal logistics. It is able to assist in enhancing productivity through mechanical automation and ergonomic layouts, thus eliminating waste from the project phase to the production process. In addition to reducing times that do not add value, the technique also minimizes the need for resources.

However, while Karakuri technology undoubtedly contributes to increased productivity, reduced automation costs, satisfaction, and employee morale, it also has its limitations in that it cannot be applied as a solution to all processes and to inadequate working conditions.

It is concluded that automation which does not depend on energy sources, such as electrical and pneumatic, is ideally set within the scenario of present-day challenges. As such, the technique will invariably offer companies significant savings and greater employee participation, as well as increased motivation and the enhanced use of human creativity, thus collaborating towards a greater sense of well-being and security for the staff involved.

REFERENCES

Alizon, F., Dallery, Y., Essafi, I. and Feillet, D. (2009). Optimising material handling costs in an assembly workshop. *International Journal of Production Research,* 47: 3853-3866.

Alves, R. R., and Calado, R. D. (2018). Karakuri: uma alternativa de automação de baixo custo [Karakuri: a low-cost automation alternative]. In *Lean na Prática*, edited by Spagnol, G. S., Calado, R. D., Sarantopoulos, A. and Li, L. M. 1: 367-378. Rockville: *Global South Press.*

Ballou, R. H. (1973) *Business Logistics Supply Chain Management.* New Jersey: Prentice Hall, 1973.

Brogin, B., Merino, E. A. D. and Batista, V. J. (2014). Contribution of Ergonomics and Anthropometry in Clothing Design to Children with Physical Disability. *Design & Tecnologia,* 8: 1-10.

Castilho, J. B. S., Barbirato, J. M. R. C. and Sales, C. M. R. (2016). Postural and ergonomic analysis: study of productive activities in a Dairy Cooperative located in the city of Itaperuna - RJ. *GEPROS* 11: 39-56.

Farris, J. A.; Van Aken, E. M.; Doolen, T. L.; Worley, J. (2009). Critical success factors for human resource outcomes in Kaizen events: An empirical study. *International Journal of Production Economics,* 117 (1): 42-65.

Glover, W. J., Farris, J. A., Van Aken, E. M. and Doolen, T. L. (2011). Critical success factors for the sustainability of Kaizen event human resource outcomes: An empirical study. *International Journal of Production Economics,* 132 (2): 197-213.

Holweg, M. (2007). The genealogy of lean production. *Journal of Operations Management.* 25: 420-437.

Ohno, T. (1973). Toyota Production System: Beyond Large-Scale Production. *Boca Raton: Productivity Press.*

Liker, J. (2004). The Toyota way: 14 Management Principles from the World's Greatest Manufacturer. *New York: McGraw-Hill.*

Slack, N., Alistair, B.J. and Johnston, R. (2013). *Operations Management.* London: Pearson. Seventh edition.

Womack, J. P. and Jones, D. T. (1998). *Lean Thinking: Banish Waste and Create Wealth in Your Corporation.* New York: Free Press.

Chapter 14

LEAN AND ERGONOMICS: HOW TO INCREASE THE PRODUCTIVITY IMPROVING THE WELLBEING OF THE WORKERS – A CASE STUDY

J. Santos, F. J. G. Silva, G. Pinto and A. Baptista*
DME – Departement of Mechanical Engineering, ISEP – School of Engineering, Polytechnic of Porto, Porto, Portugal

ABSTRACT

The principles of Lean manufacturing, in addition to addressing the reduction of waste throughout manufacturing processes, are also highly concerned about the conditions that can favor the productivity of operators, which is achieved through their well-being while performing tasks. This work was elaborated with a view to the analysis of a production line capable of producing different electronic devices and which presented deficiencies in terms of operator ergonomics, demanding undesired and exhausting efforts, having as aim the elimination of such conditions. Thus, a detailed study of the movements carried out by the operators along the assembly line was elaborated, and possible solutions were considered to eliminate the situations that were perceived as more tiring and capable of generating health problems for the operators, when repeated several times. The study was successfully developed, and solutions were found that completely eliminated actions that required exaggerated efforts or unnecessary movements, clearly optimizing the assembly flow and the well-being of the operators in the line.

Keywords: lean manufacturing, ergonomics, assembly line, workstation, well-being, increased productivity, labour health problems

---

* Corresponding Author's E-mail: fgs@isep.ipp.pt.

1. INTRODUCTION

The principles of Lean Manufacturing arose based on the TPS (Toyota Production System), which was created aiming at cutting as much existing waste as possible in production processes, due to the scarcity of material and human resources in Japan after WWII, as well as the singularity of the market where Toyota actuated at that time. The high productivity and efficiency achieved has allowed Toyota to overcome and even prosper under those difficult times (Cirjaliu and Draghici 2016, 106). The Lean Manufacturing techniques allow to combine the advantages of the flexibility typical of the artisan production, with the rigor and automation characteristic of mass production, allowing it to become more adjusted to the different needs of each customer (Rosa, Silva & Ferreira 2017, 1036). These tools require a strong involvement of all the people in the organization in order to identify wastes and communicate them with a view to their rapid elimination (Antoniolli et al. 2018, 1121). Due to its simple approach to problems and how effectively it can solve them, Lean Manufacturing has become a philosophy intensively adopted by many industrial sectors (Sousa et al. 2018, 612) (Neves et al. 2018, 697) (Rocha, Ferreira and Silva 2018, 641) (Moreira et al. 2018, 625), with a particular emphasis on the automotive sector (Rosa et al. 2017, 1035) (Martins et al. 2018, 648) as well as the production of electronic devices (Nguyen and Do 2016, 596) (Correia et al. 2018, 671). Moreover, the design of processes and equipment can also be performed following Lean principles, making easier the fulfillment of tasks carried out by workers (Magalhães, Silva and Campilho, 2019, 26) (Moreira et al. 2017, 1394) (Nunes and Silva, 2015, 333-334) (Araújo et al. 2017, 1544-1545) (Costa, Silva and Campilho 2017, 1052) (Costa et al. 2018, 3057). However, in order for Lean tools to be applied to produce the expected success, stakeholders need to be fully aware of the contribution they can bring to the process and respect the work of others through effective and efficient communication (Pearce, Pons and Neitzert 2018, 96). In fact, the Lean philosophy also promotes a sustainable work climate, and not all organizations have the capacity to strategically be able to integrate productivity growth with the care to be taken in the management of human resources (Souza and Alves 2018, 2668). In order to access the maturity that organizations present regarding the implementation of Lean tools, a model has even been developed, which helps organizations to understand to what extent the main concepts are internalized and in practice (Maasouman and Demirli 2015, 1880). The application of Lean tools may require the re-design of work systems, as well as monitoring before, during and after these tools, to ensure the best results. This tool is known as Resilient Engineering and is defined as the "ability of the system under study to adjust its functioning prior to, during, or following changes and disturbances, so that it sustains required performance under both expected and unexpected conditions" (Rosso and Saurin 2018, 45). That work has resulted in the elaboration of guidelines that can be used to reformulate jobs that need to be redesigned, taking into account the health problems that can result from excessive movements or efforts caused by the operators, and the development of guidelines is based on principles of Lean Manufacturing and Resilient Engineering. However, the problem of redesigning production or assembly lines cannot be seen in isolation from the other problems that need to be addressed, namely line balancing. Another situation that is critical in designing or redesigning production or assembly lines is component feeding, a problem that worsens if the weight of one or some of the components is significant. In these cases, Lean

Manufacturing techniques can be used for the supply of components to the line, and fatigue factors are granted to the work of the operators, extending the time dedicated to certain manual tasks when they exceed the recommended limit of effort assigned to the job to be performed (Battini et al. 2016, 195).

The study of labor, when appropriately associated with Lean techniques, allows any company to increase its productivity and increase the quality of the products produced (Rosa et al., 2018, 556). However, the implementation of Lean tools can also lead to stress or musculoskeletal problems of operators, if the focus is only put on increasing productivity, without taking into account the well-being of operators. It has been noted that in the automotive industry a higher stress is felt in the assembly lines by the operators, even when certain Lean tools are applied, which is not common to other sectors of activity (Koukoulaki 2014, 207). In a study carried out in the past decade, some situations typical of some work environments that generate stress in the workplace were identified. These were intensity/rhythm of work, lack of human or material resources, excess of working hours, short cycle time to carry out the assigned tasks, substitution of absent workers, guilt feelings about defects produced by the operator himself and ergonomics difficulty (Conti et al. 2006, 1027-1032). In fact, when all the waste is removed, an intensification of the workflow is verified, which minimizes the possibility of operator relaxation, thus contributing to the operator's increased stress. This will be all the more serious as higher loads and greater displacements are performed repetitively by the operator within his area of operation (Brito et al. 2017, 1113). Another factor that causes stress in operators is the work in the Just-In-Time system, which requires an intensification of the work and an increased responsibility for the work pace imposed on the production or assembly line. Considering a recently published work (Wyrwicka and Mrugalska, 2017, 784), the application of Lean tools is said to be able to reduce waste along the production flow, not only at specific points, designing processes that require less human effort, less space and invested capital, allowing the production of products with higher quality and at a lower price.

In order for the main purposes of the Lean philosophy to be achieved, ie greater productivity and flexibility, it is essential that the workstations be designed correctly, taking into account the well-being of the operator. The main requirements stated as important to allow for a better efficiency, effectiveness, functionality and operators' satisfaction, based on the "Hierarchy of workstation needs" are: Health and safety, Cleanliness and orderliness in the work environment, Wastes elimination, Inventory and logistics, Flexibility, Visual management and Quality (Gonçalves and Salonitis 2017, 391). Taking into account these requirements, a model based on a check-list was developed by those authors to guide the workstation designer and evaluators through best practices, allowing the design or evaluation of production lines and comfortable workstations for the operator and able to generate the productivity and flexibility required for the development of any industrial activity in the today's market. However, the model developed showed some weaknesses, such as the lack of consensus in the previous research about the main requirements to assess to the workstation design, the assessment method can be improved because it is based on other assessment tools which can lead to forgetfulness and errors, the reliable results only can be achieved if the tool can count on the participation of a large group of people in the company, and if the factors attributed to the different issues have the correct weight, depending on its importance in the global results. One technique commonly used to avoid monotony and displeasure on production or assembly lines is to frequently switch workstation operators to avoid distractions and loss of performance. However, an appropriate design of the production or assembly line may have a similar effect, provided that certain assumptions essential of the well-being of the operators are safeguarded. Several models have been developed to optimize the design of production and assembly lines, in order to achieve productivity gains and increase operator satisfaction levels, without jeopardizing their level of alertness to each of the operations to accomplish (Gnanavel, Balasubramanian, and Narendran 2015, 580) (Botti, Mora, and Regattieri 2015, 363).

Ergonomics can be defined as the study of the interaction between operators, equipment and the dynamics that affect the interaction, aiming in improving the performance of production systems by enhancing human machine interaction (Vukadinovic 2019). In simplified terms, the main objective of ergonomics is to improve the performance of the production system by eliminating the undesirable aspects usually associated with them. These aspects usually include the inefficiency, fatigue, accidents and injuries, user difficulties, and low morale.

Performing manual tasks involving high effort, long duration, incorrect or static postures, performing similar repeated movements over and over during a work shift, or even more severely, a combination of the above factors, increases significantly the risk of musculoskeletal disorders in the regions of the body involved. Eliminating the manual task that presents this risk or the redesign of the movements or the task to reduce the exposure to these risk factors provoked by the task becomes an objective to be attained, in order to avoid major evils in the near future (Burguess-Limerick 2018, 293).

The use of hybrid production or assembly lines, in which the pace is imposed by automatic equipment, usually creates stress problems for workers. In such cases, in order to avoid creating ergonomic problems for workers, it is necessary that the pace is imposed by man in order to minimize risks at work, according to the model developed by Botti et al. (Botti et al., 2017, 486). The design of the workstations also assumes a crucial role. There are some general practical guidelines for the proper designing of workstations. These guidelines can be translated into seven rules (Grandjean, 2004), as follows: (a) the first rule states that any bent or unnatural posture of the body should be always avoided; (b) the second rule suggests to avoid immobility of the arms in the extended position, forward or to the side, as this leads to fatigue and decreased accuracy and dexterity of movement; (c) the third rule says that, whenever possible, standing positions should be avoided; (d) the fourth rule refers that the operators' arm movement must be in the opposite direction of each one or in symmetrical direction; (e) the fifth rule denotes that the height of the work surface should allow optimal visual observation with the most natural body posture possible; (f) the sixth rule states that handles, levers, tools and work materials should be arranged on the workplaces in such a way that the movements more frequently done are with the elbows bent and close to the body. This position allows the arms to find a support to rest on the work surface. If the arms are always away from the trunk, tension in the muscles will be created, therefore causing fatigue; (g) the seventh rule refers that in continuous manual labor (in seated position), the workstation should have supports for the arms and elbows to rest. These supports should be covered in soft material and height adjustable.

An important concept in ergonomics is the reach zone, which defines the area possible to cover by an operator when in its workplace. Thus, containers, equipment, and operating elements distributed into the workstation must be in the reach zone. They must be easily accessible, and operators should not have to rotate the torso to reach them. The reach zone is constituted by three different areas, with the following characteristics:

- **Center of work, two-handed zone**, which is optimum for operating with two hands, as each hand will reach this zone and are under the worker's field of view, allowing for accurate movements, making possible to handle lighter weights and additionally permits the improvement of review and coordination activities, where only the forearm movements are needed and just smaller muscles groups are used.
- **Large reach zone,** which corresponds to gross motor movements, being usually reserved for tools and components that are typically grabbed with one hand, and where upper and lower arm movements without required use of shoulders or torso rotation can be necessary.
- **Extended one-hand zone**, which should be used for occasional handling, empty containers or transferring components to a position where the next employee can reach them without difficulty. Usually, shoulder and torso are used in movement into this area.

In Table 1, the maximum distances reached by men and women are shown considering three different percentiles (5th, 50th and 95th).

Table 1. Maximum distances reached by males and females

		Percentile	Range height (mm)
Female	Tallest	95	2060
	Medium	50	1930
	Shorter	5	1800
Male	Tallest	95	2180
	Medium	50	2060
	Shorter	5	1950

Source: Grandjean, 2004

The concept of range is related to the movements necessary to make in order to perform a given task. It is usually related to tasks of grabbing and/or operating manual controls or pedals. Thus, it is intended to determine the maximum height that can exist, so that people with smaller physical dimensions can reach a certain object or control it. Therefore, the 5th percentile should be used as reference in this case because if these people are accommodated, all the others will be. It is also considered an one-way limitation since only one extreme of the population are considered (Rebelo 2004). Depending on whether it is male or female, maximum distances are established around the axis of rotation of the shoulder that can be reached systematically or sporadically. Moreover, in a seated position, it was established that the highest shelves position should be between 1500 mm and 1600 mm for men, or 1400 mm and 1500 mm for women. At this height, the shelves can be accessible up to a depth of 600 mm.

In standing position, the shelves recommended height can be calculated using the following expression (Grandjean, 2004):

Maximum reach height $=1.24 \times$ height (1)

As the strength of an individual can vary according to various factors such as age, genre, and training, it is necessary to define the limits for the handling of loads required for the operators. Thus, when studying the maximum loads limits allowed, the individuals with less force should have priority because if they do not have problems, the remaining people will not have problems either. The maximum strength of a muscle or group of muscles depend on age, genre, constitution, degree of fitness, and motivation (Grandjean, 2004). It is also important to note that, women have about 2/3 of men's strength. Thus, some conclusions related to working in a standing position can be drawn, as follows:

- When standing in most arm positions, the pushing force (pressure) is greater than the pulling force;
- The pushing and pulling forces in the vertical position are the highest and in the horizontal position the lowest;
- The pushing and pulling forces in the sagittal position is of the same order as with the arm extended to the sides;
- The force of pushing in the horizontal position in men reaches 160-170 N and in women 80-90 N.

The handling of loads (especially lifting loads) should be considered heavy work. The main problem with handling heavy loads is not so much the requirement of the muscles but the wear of the intervertebral discs (Grandjean, 2004). The position adopted when lifting heavy loads directly influences the wear of the intervertebral discs, and for this reason some care is essential. Some basic consists of having the legs bent and the back straight. The wear of the intervertebral discs can result in problems in the spine and legs causing pain and limiting the mobility of operators. Diseases of the spine may lead to a extended absence from work and today are considered one of the main causes of premature disability.

Some other ways of avoiding accidents when handling loads are:

- Using mechanical equipment for lifting heavy loads;
- Ask for support, when moving heavy and bulky loads operators should ask for help whenever possible;
- Avoid trunk rotation, one should always rotate the entire body;
- Team coordination, one person should assume the coordinating role and give the team clear instructions for lifting and transporting.
- Carrying loads close to the body and well-balanced;
- Avoid obstructed vision, it is important to have unobstructed vision on the entire carrying course;
- Using Personal Protective Equipment (PPE);
- Whenever possible, one should push rather than pull;
- Using transport equipment with appropriate tires and bearings, and do not overload transport equipment;

Despite the existing rules on how to improve workers' health in the workplace, solving ergonomic problems should be a participatory process in which workers play a proactive role in identifying problems and in indicating possible solutions (Burgess-Limerick 2018, 289). Moreover, knowledge, skills, encouragement, tools and resources, it is legitimate to say that workers are the best placed people to identify the problems that may affect them, find the best solutions and know how to implement them, with a view to reduce the degree of accidents and risk of occupational diseases that may affect them in the short or medium term, without affecting productivity (Brown 2005). Regarding the participatory ergonomics process, a framework has been developed for defining the variations of the programs and dimensions found in this field, as described in the works of Heines and Hignett (Heines et al. 2002, 309; Hignett et al. 2005, 200), being pointed out that the most important one is the location of the decision making power. In a study carried out on the application of Lean Manufacturing taking into account the ergonomic and well-being aspects for workers in the automobile industry, it was found that reducing absenteeism was one of the main consequences, with a decrease in the non-quality of manufactured products, an increase in the effective time of production, an enhancement in productivity and the elimination of industrial accidents (dos Santos et al., 2015, 5952). Therefore, the combined application of Lean and Ergonomics, produces results that can be excellent for the company and for its workers.

Although the subject of ergonomics is crucial to a good work environment, its applicability is sometimes very hard. One of the barriers to its applicability is due to by the considerable differences between people from different regions and even of different genres. To overcome this challenge, it is necessary to develop a set of solutions able to cover the larger number possible of people. This can be achieved by analyzing anthropometry statistical data of the population. As also mentioned, the lack of vision of top management can also be a barrier to the implementation of ergonomic measures, as some managers may see the implementation of ergonomics in the workplace as an expense rather than as an investment (Brito et al. 2017, 1118). In fact, some researchers have already concluded that the ergonomic aspects should be integrated in a gentle way in other works being carried out in the company, as Lean improvements, not very focused, so that they are not exposed to some disruption of the evolutionary process, since it is usual that these processes can take 3 to 5 years to have practical and effective implementation on the factory floor, besides being necessary to involve all those who have the power to make decisions, as well as all others who are stakeholders in this process (Neumann et al., 2009, 535). The dialogue between all stakeholders should also be present so that there are no problems between employees of the company with different levels of knowledge. The dialogue should also be extended beyond the physical boundaries of the company, so as to add more knowledge and involvement of all.

2. PROBLEM DESCRIPTION

The assembly line studied consists of four workstations and produces five different types of products. All the products manufactured in this assembly line are similar regarding the flow of production seen in Figure 1.

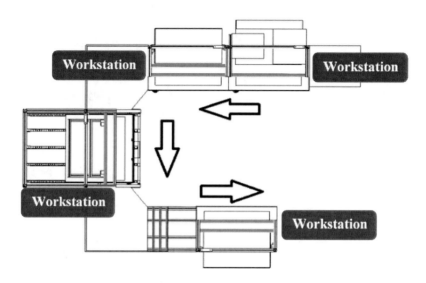

Figure 1. Workstations and production flow.

All products pass at least three different workstations where three different types of operations are performed. These operations are:

- Encasing, is the main assembly of parts. This operation is often divided into more than one workstation;
- Tests, where the product is subjected to a Functional Test (FcT) and High Voltage Test (HVT);
- Packing, the last stage of production. Here the product is prepared to export.

2.1. Products: General Description

The products manufactured in this assembly line are: (a) Dual Delegate Interface (DDI); (b) Radiator Medium Power (M); (c) Radiator High Power (H); (d) Integrus Charging Case; and (e) Integrus Cabinet. Table 2 shows some relevant characteristics of all products referred before and manufactured in this assembly line.

2.2. Products: Specifications

2.2.1. Radiators M and H

The only differences in both these products are the height (model M 200 mm and model H 300 mm), and the weight. Regarding the manufacturing process, they are very similar. In the first assembly stage of both radiators, the operators need to unload the extrusions into the working platforms. Since the weight of these extrusions are different, and the supply heights can be of 315 mm and 550 mm, different results were obtained when evaluating ergonomic deviations. Table 3 shows the results obtained when evaluating the actions of unloading the extrusions into the working platforms, as well as manually carrying them to the next workstation.

Table 2. Products manufactured

Description/Picture	Main dimensions (mm)	Weight (kg)
DDI	200×100×35	0.69
Radiator M	477.2×200×178	9.74
Radiator H	477.2×300×178	13.1
Integrus Case	688×514×224.4	17.38
Integrus Cabinet	659×521×113	14.2

Table 3. Ergonomic analyses of the first stage of the radiators

Task	Result
Picking model M from first car shelf (555 mm).	🟡
Picking model M from second car shelf (315 mm).	🟡
Picking model H from first car shelf (555 mm).	🔴
Picking model H from second car shelf (315 mm).	🔴
Carrying extrusion of model M from workstation 4 to 1.	🟡
Carrying extrusion of model H from workstation 4 to 1.	🟡
Carrying semifinal assembled extrusion model M from workstation 1 to 2.	🟡
Carrying semifinal assembled extrusion model H from workstation 1 to 2.	🟡

After finishing all tests required in workstation 2 (please see Figure 1), the radiators are carried to workstation 3 for packing. This action was evaluated, and the results are shown in Table 4.

Table 4. Ergonomic analyses of the second stage of the radiators

Task	Result
Carrying fully assembled model M from workstation 2 to 3.	🟡
Carrying fully assembled model H from workstation 2 to 3.	🔴

While the Radiator M is easy to carry by the operator, the superior weight of the Radiator H implies a higher load to the operator, which will need to be properly dealt in the future, because it exceeds the maximum load usually permitted in repetitive tasks. After the packing is completed, the final box is placed in the trolley using a vacuum lifter, as depicted in Figure 2.

Figure 2. Placing a radiator H already packed in the finished product car.

This means that despite the difference in heights, the operator is not subjected to any kind of stress.

2.3. Integrus (Case/Cabinet)

In the case of Integrus cases and cabinets, the process is also very similar, being the only difference the operation of placing the product in the milk run trolley. In the first assembly stage, the cases/cabinets are supplied under workstations 4 and 1 at the same height. The bottom part of the case/cabinet is transported directly to workstation 2, the top part of the case remains in the supply car and the textplate is placed manually on workstation 4 for assembling. Table 3 shows the results obtained after evaluating these actions, as well as carrying the fully assembled textplate to workstation 2.

After all tests and remaining assembly operations are finished in workstation 2, the products are carried manually to the final workstation for packing. The loading of the finished product trolleys is done in two different ways depending on the product. For the case model, the loading is done using a lifter and for the cabinet model this is done manually. The actions of carrying the products from workstation 2 to 3 and the loading of the cabinet model into the finished product car were evaluated, and the results are shown in Table 4.

The cumulative load was also evaluated for each product. Table 55 shows the results obtained through that analysis.

Table 3. Ergonomic analyses of the first stage of the Integrus assembly process

Task	Result
Carrying bottom part of Integrus case from workstation 4 to workstation 2.	🔴
Carrying bottom part of Integrus case from workstation 1 to workstation 2.	🔴
Carrying cabinet from workstation 4 to workstation 2.	🔴
Carrying cabinet from workstation 1 to workstation 2.	🔴
Carrying fully assembled textplate from workstation 1 to workstation 2.	🔴

Table 4. Ergonomic analyses of the second stage of the Integrus assembly process

Task	Result
Carrying bottom case and textplate from workstation 2 to workstation 3.	🔴
Carrying cabinet and textplate from workstation 2 to workstation 3.	🔴
Carrying cabinet from workstation 3 to milk run trolley and placing it inside the packing box.	🔴

Table 5. Cumulative load results

Product	Cumulative Load (kg)	Result
Radiator H	1734.51	🟡
Radiator M	1250.91	🔴
Integrus Case	618.80	🟡
Integrus Cabinet	1111.76	🟡

3. METHODOLOGY

The methodology adopted in the development of this work consisted of several different steps, as follows:

1) Background study of the different subjects involved;
2) Analysis of the current manufacturing process;
3) Problem identification;
4) Setting the main goals of the presented work;
5) Establishing a hierarchy between the goals that were set;
6) Collecting feedback among more experienced professionals regarding possible solutions;
7) Selection of the best suitable solution;
8) Elaboration of the 3D models of the chosen solution;
9) Estimating costs of the solution presented.

4. BRAINSTORMING

After analyzing the present process, the time came to think about the possible solutions in order to overcome the problems found. For this, suggestions were collected among a multidisciplinary team and the suggested scenarios were weighted. The team was formed by all elements of the three different groups, such as: process engineering department, industrial engineering department, and logistics department. The process of collecting the different inputs of each subgroup consisted of several meetings about the assembly line studied with different elements of each subgroup. Throughout these meetings, it was possible to gather different solutions for each individual problem and possible issues that could arise from each one. The possible solutions obtained after these meetings, as well as advantages and drawbacks associated with each one, can be obsserved in Tables 6 to 9.

For helping the decision amongst the solutions presented before, a selection matrix was performed. In this matrix, all solutions presented have got a score, depending on how well they fulfil the objectives of this study. However, before the actual building of the matrix, the objectives were compared face to face in pairs, leading to establish a hierarchy for a more precise comparison between the different solutions. Table 10 shows the ponderation matrix performed considering each objective.

Table 6. Cumulative load results

Fully Automated Assembly	
Advantages	Drawbacks
Would fit in a smaller space than the available; Ergonomic issues would be solved; Greater line output.	High cost; Inflexibility in moving the assembly line; High programming complexity; High maintenance.

Table 7. Manual assembly with the product suspended

Manual assembly with the product suspended	
Advantages	**Drawbacks**
The operators would not have to carry the products by hand; Would not affect the current line output.	Hard to apply in the products manufactured; High cost; Despite solving some ergonomic issues, others could raise; It would be harder to handle the products, especially Radiators.

Table 8. Automatic conveyor work stands

Automatic conveyor work stands	
Advantages	**Drawbacks**
Some ergonomic issues would be solved; The line output would not be affected; Area available would be enough.	Issues regarding product handling would not be solved; It would not be possible to establish a work height that would be suitable for all products; Automation needed would require special technician for day to day use.

Table 9. Hybrid solution

Hybrid solution	
Advantages	**Drawbacks**
Every issue can be analyzed individually; Adjustments for each product are possible; Estimated costs lower than other options.	Capacity of the line could not be improved; Space required may be larger than other solutions; May require some pieces of equipment that are only used in some family of products.

Table 10. Ponderation matrix

	1-2	1-3	1-4	1-5	2-3	2-4	2-5	3-4	3-5	4-5	Σ
Ergonomics	1	0.7	0.7	1							3.4
Line output	0				0.5	0.3	0.3				1.1
Cost		0.3			0.5			1	0.7		2.5
Area			0.3			0.7		0		0.5	1.5
Flexibility				0			0.7		0.3	0.5	1.5

The main objective is to assure that the exposure of operators to elevated risk factors ceases to exist and by doing so, guarantee that no more work-related injuries will happen again. Thus, between all different objectives, the line ergonomics is considered a priority. Following ergonomics, the most important factor is the cost. Since the available funds for the restructuration of the assembly line are limited, it is imperative that the solution chosen not have a high associated cost. Required area and line flexibility came third in the hierarchy. These two factors are equally important in the analyses. And the least important factor to take in consideration is the line output. However, despite improving the line output is not a fundamental requirement, the solution chosen should not reduce the current line output. Thus, in the selection matrix different values for each factor were given, depending on the solution. The values are shown in Table 11.

Table 11. Values of each factor according to the different solutions

Solution	Ergonomics	Line output	Cost	Area	Flexibility
Fully automated assembly	100	100	100	60	40
Manual assembly with the product suspended	50	90	80	80	30
Automatic conveyor work stands	50	90	50	80	50
Hybrid solution	90	90	40	70	90

The values shown were empirically obtained, based on the opinion and experience of the different members of the department. Table 12 shows the performance index calculated for each solution in the selection matrix.

Table 12. Calculated performance index

Solution	Performance Index
Fully automated assembly	71.67
Manual assembly with the product suspended	50.65
Automatic conveyor work stands	59.61
Hybrid solution	84.96

Based on the index performance obtained, the selected solution is the hybrid solution. In this option, it will be possible to analyze each deviation individually. However, all solutions chosen must be implemented in one assembly line. Also, whenever its possible all materials from the old layout should be reutilized, reducing costs and waste in this project.

5. IDEAS IMPLEMENTATION

In this section, the new manufacturing process according to the new layout will be explained. This section will be divided into four parts, where a detailed explanation of the new manufacturing process regarding the different products will be presented. The parts in which this section will be divided are:

- Radiator (H/M);
- Integrus Case;
- Integrus Cabinet;
- DDI.

The assembly line (L18) will have two fixed workstations, one for testing and another for packing of Radiators and Integrus cases/cabinets. This station will also be used for all operations needed in the DDI. One fixed supermarket will also be needed for accommodating all components of DDI and some components used in packing of Radiators and Integrus.

However, there will be more components in the assembly line at any given time. Figure 3 shows the assembly line with all components that are part of the everyday use of the assembly line. Only the supply cars were omitted because only one type of car will be in the assembly line when producing.

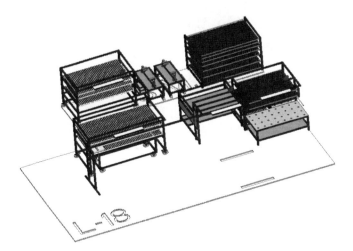

Figure 3. L18 with all elements to be used.

The remaining components represented are:

- One mobile workstation for Radiators;
- One mobile workstation for Integrus Case/Cabinets;
- Two trolleys adapted with jigs for Radiators;
- One mobile work table to be used when producing Integrus Case/Cabinets;
- One supermarket;
- One electric suspensor;
- One electric scissor lifter;
- One-foot operated scissor lifter trolley.

5.1. Radiators

In the next sub-sections it will be explained, step by step, the the ideas studied and implemented in order to overcome the problems initially identified regarding the Radiators assembly process.

5.1.1. Supply

After the supply cars are placed on the assembly line, the first task to be carried out is the unloading of the first extrusion into the adapted radiator trolley. To eliminate the deviation in the old layout, two measures were implemented. The first was to elevate the height at which the extrusions are supplied, and the second was the use of a compressed air suspensor that picks the extrusion. Figure 4 shows a 3D simulation of this task.

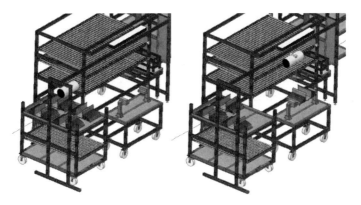

Figure 4. Extrusion being picked from supply car and being placed on the JIG.

Regarding the product, there are two treaded holes where two eyebolts will be screwed, and then connected to the suspensor with a cable. After the extrusion is loaded in the jig, the eyebolts will be disconnected from the cables, but will remain in the extrusion until the end of the process.

5.1.2. Encasing

With the extrusion fixed on the jig, the operators assemble all components required. As this jig allows the easy handling of the extrusion, the operators do not have the need to pick it. Therefore, the cumulative load associated with this task is significantly lower than before. Another deviation in the old layout was the working height, when producing both models of the extrusion. As it was explained before, by performing all encasing operations in the adapted radiator trolley, the working heights are always respected independently of the extrusions position.

5.1.3. Testing

After the encasing stage is done, the operator will push the trolley to the test workstation. Posteriorly to aligning the trolley, the operator pushes it to pass the HVT safety barriers and starts preparing the packing box while the test is being performed. When the test is finished, the operator needs to perform a visual inspection at two different distances (0,6 m and 2 m). Figure 5 shows the lines where the operators should be placed to performe this task.

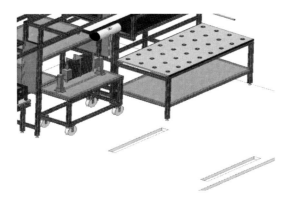

Figure 5. Radiators testing stage.

5.1.4. Packing

Concluding the visual inspection required, the next step is to place the final assembled Radiator inside the packing box. For this, the operators will once again use the compressed air suspensor, avoiding the need to pick the radiators by hand. Using the suspensor to perform this task, eliminates the ergonomic deviation felt in the previous layout. Figure 6 is a 3D simulation of this task.

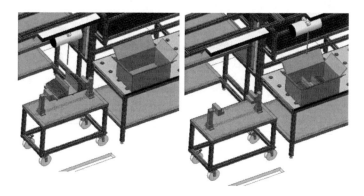

Figure 6. Radiators packing stage, before and after the radiator put into the package.

After placing the Radiator and remaining accessories inside the package, the eyebolts are removed, and the box is closed.

5.1.5. Loading the Supply Car

The last step of the process is the loading of the supply car (which also serves as the finished product car) with the packed product. In the previous layout, this action was done with the help of a vacuum lifter and did not subject the operators to any kind of physical stress. However, since one of the main objectives of this study was to make the assembly line more flexible, it was concluded that this equipment should be replaced by another system. The solution found was the use of scissor lifter trolleys adapted with rollers.

The operators will push the packed product to this trolley, and since the cover of the packing workstation is covered with plastic ball transfer units, the force necessary is very low. Figure 7 is a 3D representation of this task.

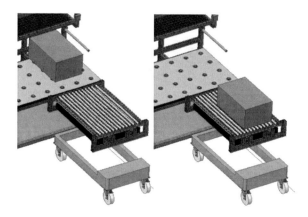

Figure 7. Transferring a packed radiator from the packing workstation to the trolley.

After the trolley is loaded with the package, the operator will carry it back to the beginning of the assembly line and unload the package back in the supply car (Figure 8).

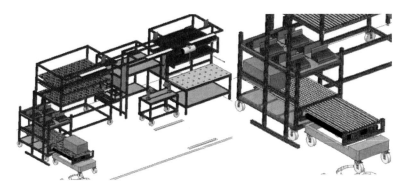

Figure 8. Transferring a packed radiator from the trolley to the finished product car.

After the production of four Radiators, the milk run will bring another supply car and take the one in the assembly line back to the warehouse.

Regarding any possible ergonomic concerns, the only action that can be evaluated by the current methods used, is the carrying of the loaded trolley from the packing workstation to the trolley. This action was evaluated using IGEL software (ISO 11228-2:2007), taking into account the parameters shown in Table 13. This software is an analysis tool as other developed by some researchers taking into account industrial ergonomic concerns (Labuttis, 2015, 4169).

Table 13. Values used in IGEL analyses (Radiators)

Parameters	Values
Shift Duration (h)	8
Pushing Force (N)	50
Pulling Force (N)	50
Distance (m)	5.7
Handle Height (m)	1
Force Angle (degrees)	30
Frequency (units per hour)	3.1
Gender mix (male: female)	25:75

Based on the parameters shown in Table , a maximum pushing force of 87.36 N and pulling force of 71.68 N was calculated. Since the maximum force that the operators will have to do is lower than the calculated values, there is no ergonomic concerns regarding this task.

5.2. Integrus Case

In the next sub-sections it will be explained, step by step, the the ideas studied and implemented in order to overcome the problems initially identified regarding the Integrus Case assembly process.

5.2.1. Unloading the Case onto the Electric Scissor Lifter

The first stage of production of the Integrus Case is the unloading of the empty case onto the top of the electric scissor lifter (Figure 9). This action is done by simply pushing the case to the lifter. Since the supply car shelves are made of rollers, and the top of the lifter is equipped with plastic ball transfer units, this action has almost none physical impact on the operators. In addition, because the lifter is mounted on a mobile structure, its positioning and alignment is very simple.

Figure 9. Transferring bottom part of case to test trolley.

Only the bottom part of the case is necessary at this point, so the top part can be placed back at the supply car.

5.2.2. Textplate/Chargers Assemble

The disassembled textplate is supplied to the assembly line in the Integrus mobile workstation. It is placed on a trolley work stand and, after the chargers are aligned and other components are assembled to the textplate, the operators pick it and place it on top of the chargers (Figure 10).

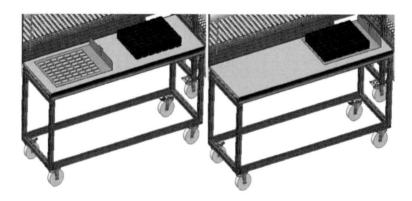

Figure 10. Textplate assembly process.

After the textplate is fully assembled, the operator carries it to the test workstation, where it is assembled in the bottom part of the case. This task was evaluated using IGEL software (ISO 11228-1:2003), taking into account the parameters shown in Table 14.

Table 14. Parameters and values of carrying the assembled textplate to the test workstation

Parameters	Values
Load weight (kg):	7.5
Carrying distance (m):	2.5
Number of load transfers (per hour):	1.3
Origin grip height (mm):	950
Destination grip height (mm):	950
Origin horizontal grip distance (mm):	300
Destination horizontal grip distance (mm):	300
Grip conditions:	Poor

The result of the carried out IGEL analysis can be seen in Figure 11.

Low risk - recommended, no actions necessary.
The risk of disease or injury is negligible or at an acceptable level for all operators in question.

Figure 11. Result of the IGEL analysis about the operation of carrying the textplate.

5.2.3. Testing

In this stage of the process, the operator only has to push the trolley pass the HVT safety barriers and wait for the test to be finished (Figure 12).

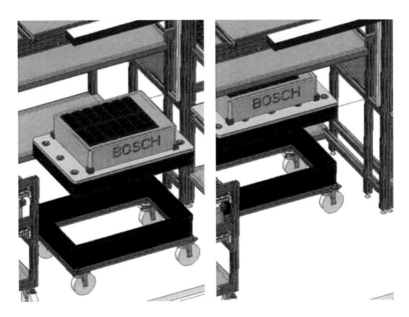

Figure 12. Integrus case testing stage.

While the test is being performed, the operator prepares the packing box of the product in the packing workstation. After the test is completed, the operator assembles the top part of the case and slides it into the packing box (Figure 13).

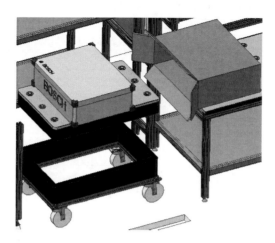

Figure 13. Integrus case packing stage.

The ball transfer units allow the easy alignment between the case and the box.

5.2.4. Loading the Supply Car

This stage of the process is very similar to what happens when producing Radiators. The Integrus cases were also loaded in the car with the help of the old lifter, but for the same reasons as Radiators (line flexibility), this equipment needed to be replaced. Thus, the same trolley used for transporting the packed radiators will transport the Integrus cases to the finished product car (Figure 14).

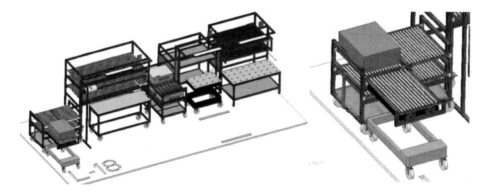

Figure 14. Transferring a packed Integrus case from the trolley to the finished product car.

Table 15. Values used in IGEL analyses (Integrus)

Parameters	Values
Shift Duration (h)	8
Pushing Force (N)	60
Pulling Force (N)	60
Distance (m)	7.2
Handle Height (m)	1
Force Angle (degrees)	30
Frequency (units per hour)	1.3
Gender mix (male: female)	25:75

As it happens with the Radiators, the only action that can be evaluated is the carrying of the trolley to the finished product car. Thus, using again the IGEL software (ISO 11228-2), a similar analysis was made taking into account the parameters shown in Table 15.

With the parameters shown in Table 15, a maximum pushing force of 90.48 N and pulling force of 74.24 N was calculated. As the maximum force necessary is 60 N, there are no ergonomic concerns in this task.

5.3. Integrus Cabinet

The manufacturing process of the Integrus cabinet is identical to the Integrus case. However, some differences need to be addressed.

5.3.1. Unloading the Case onto the Electric Scissor Lifter

The first stage of manufacturing the Integrus cabinets is also to unload them on to the scissor lifter. However, before unloading the cabinet it is necessary to assemble the packing box on top of the scissor lifter, and after this is done, the cabinet is placed directly inside the box (Figure 15).

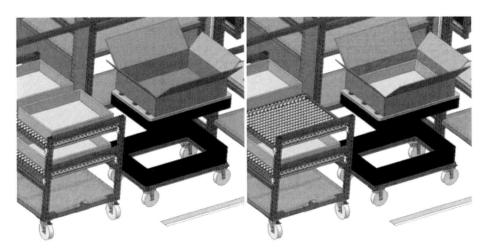

Figure 15. Placing bottom part of the cabinet inside the packing box.

This early action will eliminate the last deviation in the packing of the product regarding the old layout.

5.3.2. Textplate/Chargers Assemble

The textplate assemble stage in Integrus cabinet is exactly the same as in the Integrus case. The operator has to assemble the textplate and then place it in the empty cabinet in the test workstation (Figure 16).

The same ergonomic analyses made for the Integrus cases is valid when producing cabinets, and therefore there are no ergonomic concerns regarding this task.

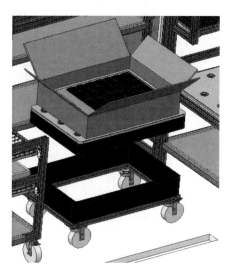

Figure 16. Textplate assemble in bottom part of cabinet.

5.3.3. Testing

The tests are made in exactly the same way as when producing Integrus cases. However, since the product is already inside of the final box, the task of sliding the product inside the box does not exist. Instead, after the tests are finished, the entire assemble (product and box) is pushed to the packing workstation were the remaining accessories are added (Figure 17).

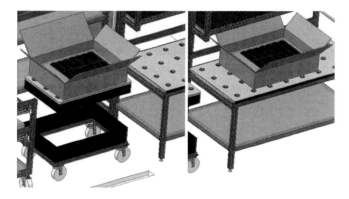

Figure 17. Product transferred from test trolleys to packing workstation.

With the remaining accessories already placed inside the box, it is then closed and placed in the trolley used in both Radiators and Integrus cases.

5.3.4. Loading the Supply Car

This task is also similar to the one performed when producing Integrus Cabinets. The scissor lifter trolley loaded with one packed product is carried to the finished product car, were the package is unloaded (Figure 18).

Once again, the same ergonomic analyses made in the Integrus Cases is valid for the Cabinets and, therefore, no ergonomic concerns were found in this scenario.

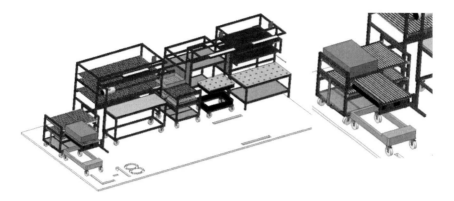

Figure 18. Transferring a packed Integrus cabinet from the trolley to the finished product car.

5.4. DDI

Regarding the DDI product, all stages of the manufacturing process are performed in the packing workstation. The only ergonomic concern that could arise is the height of the work surface. To avoid any problems, the same platform that was used in the old layout will also be used in the new one (Figure 19).

Figure 19. Packing workstation when producing DDI.

By using this platform, the working height is 1.05 m, which falls into the recommended limits. One other thing to consider when producing this product is that the equipment associated with the remaining products needed to be allocated somewhere within the assembly line or the supermarket area. Figure 20 shows one possible way to accommodate the equipment.

The equipment elements represented in the image above are as follows:

1) Test workstation;
2) Radiators mobile workstation;
3) Integrus mobile workstation;
4) Supermarket;

5) Packing workstation;
6) Compressed air suspensor;
7) Trolley work stand for Integrus;
8) Two adapted radiator trolleys;
9) Scissor lifter trolley;
10) Electric scissor lifter;
11) DDI working platform.

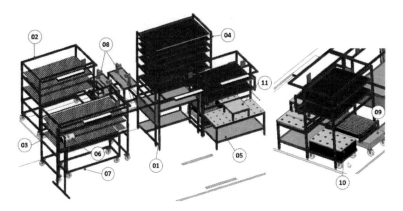

Figure 20. Equipment accommodation.

6. RESULTS AND DISCUSSION

With the implementation of the solution described, it will be possible to respond to all initial requirements and goals. However, some created actions need to be evaluated before the validation of the described project. These actions are:

- Pushing the trolley with the packed Radiator, from the packing workstation to the finished product car;
- Pushing the trolley with the packed Integrus Case/Cabinet, from the packing workstation to the finished product cart.
- Carrying the Integrus assembled textplate from the encasing workstation to the test workstation.

After identifying these actions, they were evaluated and none of them represented any threat to the operators well being therefore, no actions are needed.

Regarding ergonomics, the implementation of the compressed air lifter and trolleys will eliminate all physical stresses imposed to the operators. In addition, since the different pieces of equipment are light (compressed air lifter) and mobile (trolleys), it means that the line can be assembled in any location of the factory.

In terms of costs, with this new layout it is possible to minimize the investment by reusing many materials from the old layout. However, some investment in new equipment is inevitable. Table 16 shows the amount of different materials required for this assembly line and their respective costs.

Table 16. Main materials, quantities and respective prices needed to implement the ergonomic improvements

Material	Quantity	Price
Aluminum profile 45 mm x 45 mm (m)	259,034	1 618,96 €
Scissor lifter trolley	1	1 395,00 €
Electric scissor lifter	1	2 720,00 €
Compressed air suspensor	1	2 289,00 €
Ball transfer units	50	650,00 €
Roller conveyor with flaps (m)	51,06	368,91 €
Roller conveyor without flaps (m)	919,72	6 553,01 €
Articulated wheel set (Large - 4 units)	6	1 333,20 €
Articulated wheel set (Small - 4 units)	6	228,00 €
Corners	394	394,00 €
Replicated jig	1	1 500,00 €

The price of acquiring every new material, would represent an investment of 19050,08 €. However, since many materials can be reused from the old layout, the investment associated with acquiring all materials needed for this new layout is, approximately, 9500 €.

Although that, in this chapter the line balancing has not been studied, it is possible affirm based on the number of products to be supplied on the line, that the current line output will not be affected.

Lastly, this new layout will fit in the area of the previously one, thus eliminating the need for the expansion of the available area.

Table 17 summarizes the final results obtained.

Table 17. Initial objectives and final results summary

Initial Objectives	Final Result
Elimination of ergonomic concerns	✓
Area required	The same
Assembly line output	Expected to be the same
Costs	€ 9500
Flexibility	✓

CONCLUSION

Looking back to the initial line design and comparing it to the solution described, some conclusions can be drawn. All ergonomic deviations identified where related to one of two issues. The first issue was that some components (the heavier ones) where supplied to the line at a height that was not the ideal working height. Although this alone could be manageable by the operators, the fact that these components where heavy, made this situation a threat to the long-term health of operators. Another problem with the original layout was that operators needed to carry these heavy products by hand, from one workstation to another. Again, the weight of these products promoted these actions into threats. To solve these problems, a solution was studied where the need for these actions was eliminated. By using pieces of equipment such as lifters and trolleys, the operators are not subjected to any substantial physical stress. In this example, identifying all ergonomic concerns of the line, analyzing them individually, finding a solution to each one and finally implement those solutions in one assembly line, proved to be a successful way of solving the initial problems. Thus, the operators' health can be safeguarded, increasing well-being and job atmosphere, which are main factors that can lead to better productivity, eliminating wates of time in production and assembling operations.

REFERENCES

Antoniolli I., P. Guariente, T. Pereira, L. Pinto Ferreira, and F. J. G. Silva. 2017. "Standardization and Optimization of an Automotive components production line." *Procedia Manufacturing* 13, 1120-1127. doi: 10.1016/j.promfg.2017.09.173.

Araújo, W. F. S., F. J. G. Silva, R. D. S. G. Campilho, and J. A.Matos. 2017. "Manufacturing cushions and suspension mats for vehicle seats: a novel cell concept." *International Journal of Advanced Manufacturing Technology* 90 (5-8), 1539-1545. doi: 10.1007/s00170-016-9475-6.

Battini, D., M. Calzavara, A. Otto, and F. Sgarbossa. 2016. "The Integrated Assembly Line Balancing and Parts Feeding Problem with Ergonomic Considerations." *IFAC PapersOnline* 49 (12), 191-196. doi: 10.1016/j.ifacol.2016.07.594.

Botti, L., C. Mora, and A. Regattieri. 2017. "Application of a mathematical model for ergonomics in lean manufacturing." *Data in Brief* 14, 360-365. doi: 10.1016/j.dib.2017.06.050.

Botti, L., C. Mora, and A. Regattieri. 2017. "Integrating ergonomics and lean manufacturing principles in a hybrid assembly line." *Computers & Industrial Engineering* 111, 481-491. doi: 10.1016/j.cie.2017.05.011.

Bridger, Robert. 2004. *Introduction to Ergonomics*. Singapore: McGraw-Hill International Editions. ISBN 978-0849373060.

Brito, M., A. L. Ramos, P. Carneiro, and M. A. Gonçalves. 2017. "Combining SMED Methodology and Ergonomics for Reduction of Setup Time in a Turning Area." *Procedia Manufacturing* 13, 1112-1119. doi: 10.1016/j.promfg.2017.09.172.

Brown, O., 2005. "Participatory ergonomics." In: Stanton, N., Hedge, A., Brookhuis, K., Salas, E., Hendrick, H. (Eds.), *Handbook of Human Factors and Ergonomics Methods*. CRC Press, U.S.A.: Boca Raton FL.

Burgess-Limerick, R. 2018. "Participatory ergonomics: Evidence and implementation lessons." *Applied Ergonomics* 68, 289-293. doi: 10.1016/j.apergo.2017.12.009.

Cirjaliu B., and Anca Draghici. 2016. "Ergonomic Issues in Lean Manufacturing." *Procedia Socian and Behavioral Sciences* 221, 105-110. doi: 10.1016/j.sbspro.2016.05.095.

Conti, R., J. Angelis, C. Cooper, B. Faragher, and C. Gill. 2006. "The effects of just in time/lean production on worker job stress." *International Journal of Operations and Production Management* 26 (9), 1013-1038. doi: 10.1108/01443570610682616.

Correia, D., F. J. G. Silva, R. Gouveia, T. Pereira, and L. P. Ferreira. 2018. "Improving Manual Assembly Lines Devoted to Complex Electronic Devices by Applying Lean Tools." *Procedia Manufacturing* 17, 663-671. doi: 10.1016/j.promfg.2018.10.115.

Costa, M. J. R., R. M. Gouveia, F. J. G. Silva, and R. D. S. G. Campilho. 2018. How to solve quality problems by advanced fully-automated manufacturing systems." *International Journal of Advanced Manufacturing Technology* 94, 3041-3063. doi: 10.1007/s00170-017-0158-8.

Costa, R. J. S., F. J. G. Silva, and R. D. S. G. Campilho. 2017. "A novel concept of agile assembly machine for sets applied in the automotive industry." *International Journal of Advanced Manufacturing Technology* 91 (9-12), 4043-4054. 10.1007/s00170-017-0109-4.

dos Santos, Z. G., L. Vieira, G. Balbinotti. 2015. "Lean Manufacturing and ergonomic working conditions in the automotive industry." *Procedia Manufacturing* 3, 5947-5954. doi: 10.1016/j.promfg.2015.07.687.

Gnanavel, S. S., V. Balasubramanian, and T. T. Narendran. 2015. "Suzhal - An alternative layout to improve productivity and worker well-being in labor demanded lean environment." *Procedia Manufacturing* 3, 574-580. doi: 10.1016/j.promfg.2015.07.268.

Gonçalves, M. T., and K. Salonitis. 2017. "Lean assessment tool for workstation design of assembly lines." *Procedia CIRP* 60, 386-391. doi: 10.1016/j.procir.2017.02.002.

Grandjean, Etienne. 2004. *Manual de Ergonomia - Adaptando o trabalho ao homem* [*Manual of Ergonomics - Adapting work to man*], Brasil: Bookman, 2004, ISBN 978-8536304373.

Haines, H., J. R. Wilson, P. Vink, E. Koningsveld. 2002. "Validating a framework for participatory ergonomics (the PEF)." *Ergonomics* 45, 309-327. doi: 10.1080/00140130210123516.

Hignett, S., J. R. Wilson, W. Morris, 2005. "Finding ergonomic solutions - participatory approaches." *Occupational Medicine* 55, 200-207. doi: 10.1093/occmed/kqi084.

Koukoulaki, T. 2014. "The impact of lean production on musculoskeletal and psychosocial risks: An examination of sociotechnical trends over 20 years." *Applied Ergonomics* 45, 198-212. doi: 10.1016/j.apergo.2013.07.018.

Labuttis, J. 2015. "Ergonomics as element of process and production optimization." *Procedia Manufacturing* 3, 4168-4172. doi: 10.1016/j.promfg.2015.07.391.

Maasouman, M. A., and K. Demirli. 2015. "Assessment of Lean Maturity Level in Manufacturing Cells." *IFAC Papers Online* 48 (3), 1976-1981. doi: 10.1016/j.ifacol.2015.06.360.

Magalhães, A. J. A., Silva, F. J. G., and Campilho, R. D. S. G. 2019. "A novel concept of bent wires sorting operation between workstations in the production of automotive parts." *Journal of the Brazilian Society of Mechanical Sciences and Engineering* 41, 25-34. doi: 10.1007/s40430-018-1522-9.

Martins, M. Radu Godina, Carina Pimentel, F. J. G. Silva, and João C. O. Matias. 2018. "A Practical Study of the Application of SMED to Electron-beam in the Automotive Industry." *Procedia Manufacturing* 17, 647-654. doi: 10.1016/j.promfg.2018.10.113.

Moreira A., F. J. G. Silva, A. I. Correia, T. Pereira, L. P. Ferreira, and F. de Almeida. 2018. "Cost Reduction and Quality Improvements in the Printing Industry." *Procedia Manufacturing* 18, 623-630. doi: 10.1016/j.promfg.2018.10.107.

Moreira B. M. D. N., Ronny M. Gouveia, F. J. G. Silva, and R. D. S. G. Campilho. 2017. "A Novel Concept Of Production And Assembly Processes Integration." *Procedia Manufacturing* 11, 1385-1395. doi:10.1016/j.promfg.2017.07.268.

Neumann, W. P., M. Ekman, J. Winkel. 2009. "Integrating ergonomics into production system development - The Volvo Powertrain case." *Applied Ergonomics* 40, 527-537. doi: 10.1016/j.apergo.2008.09.010.

Neves, P., F. J. G. Silva, L. P. Ferreira, M. T. Pereira, A. Gouveia, and C. Pimentel. 2018. "Implementing Lean Tools in the Manufacturing Process of Trimmings Products." *Procedia Manufacturing* 17, 696-704. doi: 10.1016/j.promfg.2018.10.119.

Nguyen, M.-N., and N.-H. Do. 2016. "Re-engineering Assembly line with Lean Techniques." *Procedia CIRP* 40, 591-596. doi: 10.1016/j.procir.2016.01.139.

Nunes, P., and F. J. G. Silva. 2013. "Increasing Flexibility and Productivity in Small Assembly Operations: A Case Study." in: *Advances in Sustainable and Competitive Manufacturing Systems*, Azevedo, A. (Eds.), Springer, 329-340. doi: 10.1007/978-3-319-00557-7.

Rebelo, Francisco. 2004. *Ergonomia no dia a dia [Ergonomics in everyday life]*. Lisboa: Edições Sílabo. ISBN: 978-972-618-867-4.

Rexroth. 2012. *Ergonomics Guidebook for Manual Production Systems*. Retrieved online on October 18th, 2018. https:// www.valin.com/ documents/ pdf/ Bosch- Ergonomic-Guidebook.pdf.

Rocha, H. T., Luís Pinto Ferreira, and F. J. G. Silva. 2018. "Analysis and Improvement of Process in Jewelry Industry." *Procedia Manufacturing* 17, 640-646. doi: 10.1016/j.promfg.2018.10.110.

Rosa C., F. J. G. Silva, and Luís Pinto Ferreira. 2017. "Improving the Quality and Productivity of Steel Wire-rope Assembly Lines for the Automotive Industry." *Procedia Manufacturing* 11, 1035-42. doi: 10.1016/j.promfg.2017.07.214.

Rosa C., F. J. G. Silva, Luís Pinto Ferreira, Teresa Pereira, and Ronny Gouveia. 2018. "Establishing Standard Methodologies to Improve the Production Rate of Assembly Lines Used for Low Added-Value Products." *Procedia Manufacturing* 17, 555-562. doi: 10.1016/j.promfg.2018.10.096.

Rosa, C., F. J. G. Silva, L. P. Ferreira, and R. Campilho. 2017. "SMED Methodology: The Reduction of Setup Times for Steel Wire-Rope Assembly Lines in the Automotive Industry." *Procedia Manufacturing* 13, 1034-1042. doi: 10.1016/j.promfg.2017.09.110.

Rosso, C. B., and T. A. Saurin. 2018. "The joint use of resilience engineering and lean production for work system design: A study in healthcare." *Applied Ergonomics* 71, 45-56. doi: 10.1016/j.apergo.2018.04.004.

Sousa, E., F. J. G. Silva, L. P. Ferreira, M. T. Pereira, R. Gouveia, and R. P. Silva. 2018. "Applying SMED Methodology in Cork Stoppers Production." *Procedia Manufacturing* 17, 611-622. doi: 10.1016/j.promfg.2018.10.103.

Souza, J., P. E., and J. M. Alves. 2018. "Lean-integrated management system: A model for sustainability improvement." *Journal of Cleaner Production* 172, 2667-2682. doi: 10.1016/j.jclepro.2017.11.144.

Vukadinovic, S., I. Macuzic, M. Djapan, M. Milosevic. 2019. "Early management of human factors in lean industrial systems." *Safety Science*, In Press. doi: 10.1016/j.ssci.2018.10.008.

Wyrwicka, M. K., and B. Mrugalska. 2017. "Mirages of Lean Manufacturing in Practice." *Procedia Manufacturing* 182, 780-785. doi: 10.1016/j.proeng.2017.03.200.

In: Lean Manufacturing
Editors: F. J. G. Silva and L. Carlos Pinto Ferreira
ISBN: 978-1-53615-725-3
© 2019 Nova Science Publishers, Inc.

Chapter 15

MEASUREMENT OF THE LEVEL OF IMPLEMENTATION OF SOCIOTECHNICAL AND ERGONOMIC PRACTICES AND LEAN PRODUCTION PRACTICES: CONSIDERATIONS FROM A SYSTEMATIC REVIEW PROCESS

E. P. Ferreira[1], PhD, J. Schmitt[2], PhD, L. G. L. Vergara[2,*], PhD, D. F. de Andrade[2], PhD and G. L. Tortorella[2], PhD

[1]UNIPAMPA – Universidade Federal do Pampa, Brazil
[2]UFSC – Universidade Federal de Santa Catarina, Santa Catarina, Brazil

ABSTRACT

Increasing competition requires organizations to make an effort to improve the performance on production systems. It is important that any organization wishing to remain competitive in the market must has good working conditions. However, there is still another big challenge, which is becoming progressively a priority: the human element. In addition, their participation contributes to the continuous improvement of the work process, which in the long term can provide a remarkable organizational performance. Considering the human element as a key factor in lean implementation, organizations should seek continuous improvement of their human and technological systems. Thus, the objective of this study was to carry out an systematic review of the literature on the subject Lean Ergonomics, using the intervention instrument denominated Knowledge Development Process – Constructivist (Proknow-C). It is an exploratory study with a qualitative approach, using technical procedures as action research and bibliographic and use of primary and secondary data. As a result, it identified the existence of 51 articles, its represent the interest literature fragment. One of the major challenges of this study was to establish which practices, Sociotechnical and Ergonomic Practices (SE) and Lean Production Practices (LP), corroborate with the improvement of

---

[*] Corresponding Author's E-mail: l.vergara@ufsc.br.

the processes in order to provide health, well-being and safety to the workers, accordingly to the reality of each organization. For the researchers, it has become a challenge, because studies conducted so far indicate several adverse factors to the workers resulting from this implementation. However, through the systemic analysis of the literature with the respective constitutive and operational definitions it was possible to elaborate a conceptual model for the SE and LP.

Keywords: ergonomics, lean production, systematic review process

1. INTRODUCTION

Increasing competition requires organizations to make an effort to improve the performance of their production systems. It is undeniable that any organization that wants to remain competitive in the market must have good working conditions (Silva, Tortorella, and Amaral 2016, 1-2; Maia et al. 2015, 50; Santos, Vieira, and Balbinotti 2015, 5947-5948; Marksberry, Church, and Schmidt 2014, 29-30).

This competitive scenario requires a reduction of production costs, however, it demands better levels of productivity and quality. Together with the incentive to remain competitive, organizations still have another major challenge, which has progressively become a priority: the human element. Its participation contributes to the continuous improvement of the work process, which in the long run can provide a remarkable organizational performance (Silva, Tortorella, and Amaral 2016, 2-3; Tajini and Elhaq 2014, 473; Perez Toralla, Falzon, and Morais 2012, 2711; Figueira, Machado, and Nunes 2012, 1713-1714; Genaidy and Karwowski 2003, 317; Niepce and Molleman 1998, 87-89).

Considering the human element as a key factor in lean implementation, organizations should seek the continuous improvement of their human and technological systems (Paez et al. 2004, 289; Genaidy and Karwowski 2003, 319). Gnanavel, Balasubramanian, and Narendran (2015), 575 and Ferreira and Gurgueira (2013), 40 describe the importance of a holistic view of the process, especially emphasizing the human factor as a driver for higher productivity. However, organizations commonly consider the worker only as one more factor of production that they have to obtain the maximum utilization (Alves, Dinis-Carvalho, and Sousa 2012, 219).

This said, this research aims to establish which are the practices that allow to classify a lean manufacturing industry in relation to the level of implementation of SE and LP. However, for the construction of these practices, the establishment of the SE and LP constructs was first carried out through a systematic review process using the Knowledge Development Process - Constructivist (Proknow-C) instrument.

2. METHODOLOGICAL PROCEDURES

2.1. Methodological Framework

This study is characterized, when to the object of research as an exploratory research, because it explores a fragment of the literature (Gil 2002, 41) - Lean Ergonomics.

It is classified as a qualitative research, regarding the approach of the problem, since it seeks to respond to a questioning (Gil 2002, 193): "What are the characteristics that allow to classify a lean manufacturing industry in relation to the level of implementation of SE and LP practices?." In order to answer it a systemic analysis of the literature is carried out, analyzing the articles of the selected Bibliographic Portfolio (PB).

In terms of the technical procedures, it is classified as an action research and bibliographical (Gil 2002, 44-55), because it involves the analysis of published scientific articles and there is an interaction of the researchers with the result of the research throughout the analysis of the articles in the PB.

Finally, as for data collection, primary and secondary data are used, in which the first category refers to the PB selection stage, in which the researchers, through their interaction, perception and delimitations, identify which articles should be part of the PB. And secondary data because all analysis is performed based on the articles in the PB.

2.2. Knowledge Development Process – Construtivist (Proknow-C)

The main objective of the Proknow-C intervention instrument is to describe and present a process that builds on the researchers the required knowledge to enable them to investigate and analyze a specific theme (Valmorbida et al. 2014, 4; Ensslin, Ensslin, and Pacheco 2012, 71). Since its inception in 2007 it has been applied in many national and international papers, such as: Dutra et al. 2015, 250; Cardoso et al. 2015, 6; Sartori et al. 2014, 79; Valmorbida et al. 2014, 10; Ensslin, Ensslin, and Pacheco 2012, 78.

It consists of four stages: (a) selection of the bibliographic portfolio that will provide the literature review; (b) bibliometric analysis of the bibliographic portfolio; (c) systemic analysis of the bibliographic portfolio; and, (d) elaboration of the research objectives, which in this research case is the description of SE and LP practices.

The employment of this instrument is justified by presenting a complete structured process for selection and analysis of the literature, which recognizes research opportunities and provides future researches (Valmorbida et al. 2014, 6), achieving in this way, scientific publications that are representative for the issue of lean ergonomics.

2.3. Procedures for Collecting and Analyzing Data

The procedures for collecting and analyzing the data, as presented in Figure 1, are comprised of three steps: (i) selection of the raw database of articles, (ii) filtering of the article database and (iii) test of representativeness of articles in the BP.

2.4. Bibliometric Analysis of the Bibliographic Portfolio (BP)

The bibliometric analysis identifies and underlines the basic variables and characteristics that stand out by counting their occurrences for all the articles belonging to the BP and their references (Cardoso et al. 2015, 7; Valmorbida et al. 2014, 10; Afonso et al. 2012, 57).

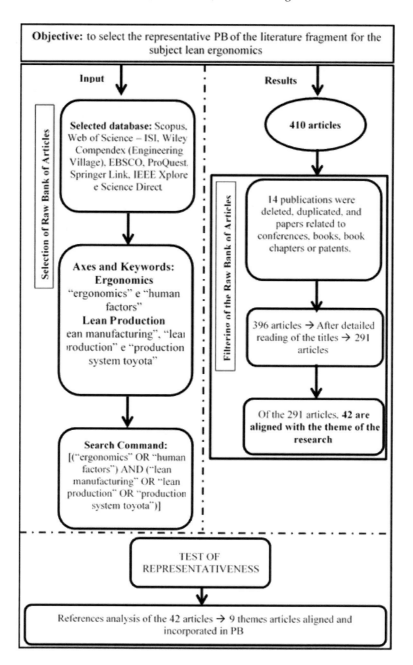

Figure 1. Selection process of the bibliographic portfolio.

As proposed by the Proknow-C method, it is considered as basic characteristics to be identified: (i) who are the researchers with trajectory in the area of knowledge; (ii) which journals publish the most on the subject; (iii) which articles have the highest scientific recognition; (iv) what are the most representative keywords of this subject; and (v) which is the impact factor of journals that publish on this subject.

For this research, the bibliometric analysis was sectioned into three main stages: (i) analysis of the articles belonging to the Bibliographic Portfolio; (ii) analysis of the references of the articles in the Bibliographic Portfolio; and (iii) joint analysis of the articles and references of the articles belonging to the BP.

2.5. Systemic Analysis of the Articles in the Bibliographic Portfolio

The systemic analysis of this research was carried out by analyzing the content of the articles in the Bibliographic Portfolio. It consists of a set of techniques that uses systematic and objective procedures to describe the content of messages with the aim of obtaining indicators, whether quantitative or not, that allow the inference of knowledge regarding the conditions of production and/or reception of these messages (Bardin 2011, 112).

For this analysis were elaborated the constitutive and operational definitions of the construct and its operationalization, as presented below.

3. RESULTS

3.1. PB Bibliometric Analysis

In the analysis of the articles belonging to the BP, five basic variables were identified: authors, periodicals, articles, keywords and the impact factor of the journals, considered by the authors as 'highlight'. In this research will be presented a joint analysis of the articles and references of the BP articles, by a cross-referencing of the information about the basic variables in relation to authors, periodicals and articles.

For the analysis of the most prolific authors it was verified that the articles of the BP and the references of these articles were written by a total of 153 authors. The prominent authors of BP and the references are presented in Figure 2.

Through the authors' intersection, the researcher Patrick W. Neumann was highlighted with two articles in the Bibliographic Portfolio and three in the references. He is an associate professor in the Department of Mechanical and Industrial Engineering at Ryerson University, Canada. His current research, carried out in the Laboratory of Human Factors Engineering, includes the study of the design of work systems from a human and technical approach. His main areas of study are: human factors and corporate strategy, industrial systems design processes, organizational design and change management, simulation and virtual performance modeling, and performance and exposure measurement.

With two articles in the Bibliographic Portfolio and two in the references, the researchers Tarcísio Abreu Saurin and Jörgen Winkel deserve special mention. Professor Saurin holds a postdoctoral degree from the University of Salford and a PhD in Production Engineering from the Federal University of Rio Grande do Sul. He is currently associate professor at the Federal University of Rio Grande do Sul in the Department of Production and Transportation Engineering. His research area is especially related to the themes: security and production management in complex systems, lean production systems and resilience engineering.

In turn, Winkel was Professor of Applied Work Physics and Production Ergonomics at the National Institute for Life at Work, based in Sweden (Karolinska Institute, Stockholm) from 1989 to 2007. He served as visiting professor from 2008 to 2017 at the National Center for Research for the Work Environment, and at the Technical University of Denmark, Department of Management Engineering. The author has published more than 450 studies in the field of Production Ergonomics, and among the main themes, the following stand out: ergonomic intervention research, ergonomic epidemiology, quantification and prediction of mechanical exposure, as well as physiological issues related to occupational work. He also received several international awards, such as the best article by Liberty Mutual Insurance Company, in 1997 with the study 'Ergonomic intervention research for improved musculoskeletal health: A critical review' published in the International Journal of Industrial Ergonomics, and by the Institute of Ergonomics and Human Factors Applied Ergonomics with the article entitled 'Occupational musculoskeletal and mental health: significance of rationalization and opportunities to create sustainable production systems - a systematic review' published in 2011 in the journal Applied Ergonomics.

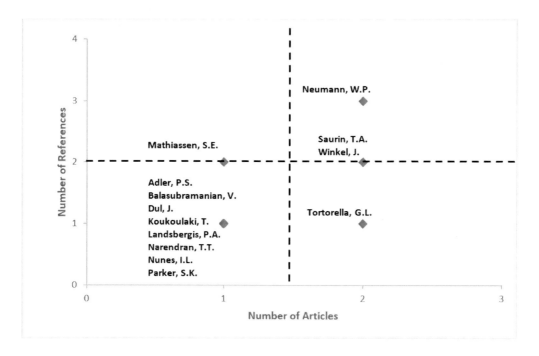

Figure 2. Cross-referencing of prominent authors in PB.

Also worthy of mention is the researcher Guilherme Luz Tortorella with two articles in the Bibliographic Portfolio and one in the references. Tortorella is an adjunct professor of Production Engineering at the Federal University of Santa Catarina, with a full degree from the Federal University of Rio Grande do Sul - UFRGS. He worked for 15 years in the automotive industry and has experience in the academic area in Production Systems and Quality Systems. He is currently developing his projects and researches in two laboratories: Laboratory of Productivity and Continuous Improvement, and Laboratory of Simulation of Production Systems. In addition, he is leader of the CNPq research group entitled Product, Process and Service Management.

Researcher Svend Erik Mathiassen had one article in the Bibliographic Portfolio and two in the references. Mathiassen holds a PhD in Occupational Physiology from the National Institute of Occupational Health in Stockholm, 1993. He conducts his research at the

Department of Medical Sciences, Occupational and Environmental Medicine at the University of Gävle, Sweden, and between his main area of interest it is the study of work-related musculoskeletal disorder. In addition to the cited authors, eight more stood out for having an article in the Bibliographic Portfolio and one in the references of this BP, which were: Paul S. Adler, Sai Venkatesh Balasubramanian, Jan Dul, Theoni Koukoulaki, Paul A. Landsbergis, TT Narendran, Isabel L. Nunes and Sharon K. Parker.

It was also observed that the BP articles and their references were published in 37 journals. The cross-referencing of journals with more emphasis in the Bibliographic Portfolio and its references is presented in Figure 3.

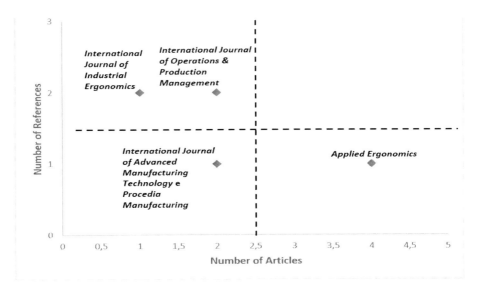

Figure 3. Cross-referencing of prominent journals of the PB.

With the intersection of the periodicals in which the articles of the Bibliographic Portfolio and their respective references were published, the journal Applied Ergonomics was highlighted, with four articles in the BP and one in the BP references. This journal deals with subjects related to ergonomics (human factors) including the conception, planning and management of technical and social systems at work. As having this broad selection of topics, several professionals can contribute academically and, at the same time, exploit their publications, such as: ergonomists, designers, health and safety engineers, among others.

In a broader context, i.e., in the field of Supply Chain Management and Operations, the International Journal of Operations & Production Management (IJOPM) was emphasized with two articles in the BP and two in references. It investigates opportunities, challenges and frontiers in the development and implementation of strategies, systems, processes and practices in operations and supply chain management.

Another important journal for the ergonomics area is the International Journal of Industrial Ergonomics which stood out with an article in the BP and two in the references of the Bibliographic Portfolio. Its main contribution relates to the understanding of the role of human beings in current systems and their interactions with various components of the system. It has as main themes: industrial and occupational ergonomics, systems design, tools and equipment, human performance measurement and modeling, and human productivity.

Two other periodicals stood out for having two articles in the BP and one in the references, that were: International Journal of Advanced Manufacturing Technology and Procedia Manufacturing. These journals aim at a more practical application of the studies, especially when related to manufacturing engineering and advanced systems. With the analysis of the most receptive journals to the subject studied, it is perceived that they are well aligned to such research, as they provide studies with theoretical advances in the literature, empirical studies, such as case studies, and launching of new methodologies and procedures.

The last analysis regarding the basic variables of the joint analysis of the articles belonging to the Bibliographic Portfolio and its references refers to the scientific recognition of the articles. Table 1 presents the two articles with the highest recognition of the BP.

Table 1. Scientific recognition of articles in the Bibliographic Portfolio

Author(s)	Title	Year	Journal	Number of citations
Landsbergis; Cahill; Schnall	The impact of lean production and related new systems of work organization on worker health	1999	Journal of Occupational Health Psychology	555
Parker, S. K.	Longitudinal effects of lean production on employee outcomes and the mediating role of work characteristics	2003	Journal of Applied Psychology	400

The most cited articles, although their journals are not among the leading journals, are directly linked to an interdisciplinary field that is psychology. In these articles it can be seen the connection with the theme of this research, especially regarding the ergonomic aspects, as they aim at improving the quality of professional life, in order to protect and promote safety, health and the well-being of workers.

The two articles with the highest scientific recognition of the references of the Bibliographic Portfolio are presented in Table 2.

Table 2. Scientific recognition articles

Author(s)	Title	Year	Journal	Number of citations
Boudreau, J.; Hopp, W.; McLain, J.O.; Thomas, L.J.	On the interface between operations management and human resources management	2003	Manufacturing & Service Operations Management	355
Neumann, W.P.; Winkel, J.; Medbo, L.; Mathiassen, S.E.; Magneberg, R.	Production system design elements influencing productivity and ergonomics – a case study of parallel and serial flow strategies	2006	International Journal of Operations & Production Management	92

Regarding the most cited article of the BP references, its journal was also not highlighted as prominent. However, Manufacturing & Service Operations Management (M & SOM) is a journal that concentrates a range of research focused on the management of production and operations of goods and services. In its turns, the second most cited article stands out by its authors, especially Patrick W. Neumann, Winkel, J., Yang and Svend Erik Mathiassen, placed among the most influential authors in this area, as well by the journal in which the article was published, considered as a highlight, with two articles in BP and two in references. Thus, we note the recognition of these articles in the area.

3.2. PB Systemic Analysis - Constitutive and Operational Definitions

For establishing the constitutive and operational definitions of this research, a conceptual model was elaborated for the sociotechnical and ergonomic construct - SE (Figure 4). It is worth mentioning that the study proposal is the elaboration of two constructs, SE and LP. However, it was noticed, after the systematic review of the literature, that LP practices are widespread and already established by 19 practices in the literature, which is not the case with sociotechnical and ergonomic practices.

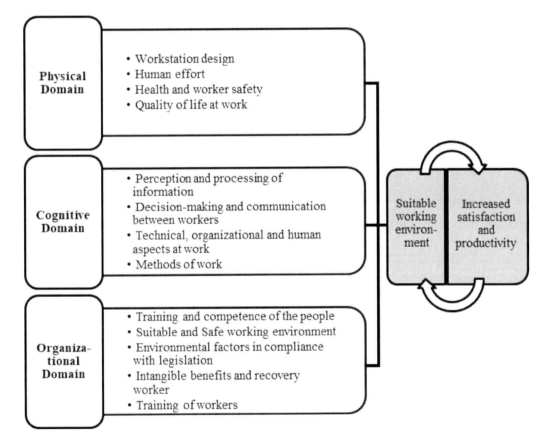

Figure 4. Conceptual model of the socio-technical and ergonomic construct for industrial environments of lean production.

In this context, a conceptual model for SE was proposed, based on the ergonomic domains: physical, cognitive and organizational. In addition, the classification of socio-technical and ergonomic principles for the management of projects, which was developed by Balbinotti 2013, 26 in a study of the automobile industry, was evidenced in the literature review and has contributed to the present study. In this way, the elaboration of the conceptual model was based on the ergonomic domains found in classical studies and lean ergonomics concepts related to the study context.

According to the International Ergonomics Association (2018), Ergonomics aims to apply theories, principles, data and methods to projects with the aim of optimizing human well-being and the overall performance of a productive system, in order to make them compatible with the needs, abilities and limitations of the workers. Based on these assumptions, the constitutive definition of the construct "socio-technical and ergonomic principles in lean manufacturing environments" was elaborated. The definition of this construct was based on the physical, cognitive and organizational domains in an ergonomic context, as presented in Table 3.

Table 3. Definition of construct socio-technical and ergonomic principles (SE) in lean manufacturing environments

Construct	Constitutive Definition
SE in lean manufacturing environments	Physical attributes - relationship between the project and the physical arrangement of the work place, aspects related to human effort, such as: work rhythm, intensity, overload, repetitiveness and postures, health care (levels of work sickness distances, such as LER/DORT) and worker safety (accident/incident indexes, human error) and quality of life at work;
	Cognitive attributes - relationship between perception and processing of information, decision-making and communication between stakeholders are effective, attention to technical, organizational and human aspects at work and understanding of working methods;
	Organizational attributes - relationship that occurs between the formation and competence of people; the work environment is safe and adequate; the environmental factors (especially lighting, noise, vibration and temperature) are in accordance with the legislation; the intangible benefits (such as motivation, satisfaction, stress levels - low or no, fatigue reduction and monotony), employee appreciation and worker empowerment.

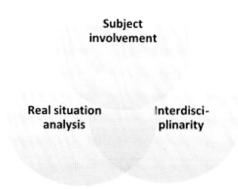

Figure 5. Main assumptions of ergonomics.

After the constitutive definition of the construct for the SE practices, its operational definition was carried out, which means that it was defined which actions should be performed by the lean manufacturing industries so that the work environment is adequate and provides better health conditions, safety, comfort and well-being to workers.

It is important to emphasize that, according to Abrahão et al. 2009, 158, for the accomplishment of an ergonomic action and its methodological choices, it is first necessary to know what are the bases that underlie this activity. Ergonomics presents three assumptions: interdisciplinarity, real situation analysis and subject involvement (Figure 5). The interdisciplinarity in ergonomics makes it possible to analyze a real work situation on different perspectives, which makes possible the evaluation of complex work situations.

In this context, ergonomics is specially required to improve work conditions, contributing mainly to the design of efficient and safe systems. However, to enable the defined attributes for the construct of this study to be efficient and effective, it is essential that they be adapted to the characteristics and limits of the workers.

In the next step, the operationalization of the construct was carried out, that is, the elaboration of the items that will form the instrument of measurement on the socio-technical and ergonomic practices in industrial environments of lean production (Abrahão et al. 2009). The construct operationalization can be defined as the behavioral representation of the construct, that is, the tasks - set of items of the instrument - that the respondents must carry out in order to evaluate the magnitude of presence of the construct - attribute - (Pasquali 1998, 4), which in this study is the definition of SE and LP practices. The items of the instrument were established based on the definitions presented and the criteria recommended by Pasquali 1998, 7-9 for its elaboration.

3.2.1. Elements that Compose the Construct for Lean Production Practices (LP)

Organizations that are in the lean implementation process need to transition from traditional mass production models to new models, specifically by organizing their work systems and management practices (Longoni et al. 2013, 5). The challenge is to justify and examine the motive and under what conditions the lean practices have competitive value and contextual relevance (Ketokivi and Schroeder 2004, 176).

Consequently, the selection of appropriate practices to improve processes and the identification of their applicability in operations presents an additional problem for industrial managers (Herron and Braiden 2006, 146). This problem is reinforced by the large number of available lean practices, totaling more than 100 (Pavnaskar, Gershenson, and Jambekar 2003, 3089). However Bhasin and Burcher (2006), 66 argue that organizations often begin lean implementation using one or two practices, implementing them across the enterprise, but soon realize that such practices do not lead to systemic improvements in the value chain. Marodin and Saurin (2013), 12 comment that regardless of the fact that lean practices have been used for decades, the implementation steps should be specific to each organization according to its objectives.

In this sense, Table 4 consolidates LP practices based on scientific research on lean production and consolidated philosophies in the field, such as: JIT - Just-In-Time, TQM - Total Quality Management, TPM - Total Preventive Maintenance, HRM - Human Resource Management and CI - Continuous Improvement.

It was possible to establish 19 lean practices among which 'standardized work' and 'problem solving methods' are the most frequent in the researched literature. The first is to be applied under different motivational reasons: (i) to create basic stability in production processes, mitigating the process variability (Doolen and Hacker 2005, 61; Arlbjørn and

Freytag 2013, 187); (ii) to balance workload among employees, as described by Shah and Ward (2007), 799 and Bortolotti, Boscari, and Danese (2015), 182, and (iii) to emphasize quality procedures and daily routines (Furlan, Vinelli, and Pont 2011, 841; Bhamu and Sangwan 2014, 891).

Table 4. Elements that compose the construct for Lean Production practices (LP)

Philosophies of Lean Thinking	Elements that compose the construct for Lean Production practices (LP)
JIT	Flexible manpower
	Pull system
	Takt time
	Continuous flow
	Material supply
TQM	Zero defects
	Quality assurance
	Product/process quality planning
TPM	Standardized work
	Production leveling
	Maintenance system
	5S, visual management and housekeeping
HRM	Goal oriented teams
	Simultaneous engineering
	Multifunction Working Times
CI	Organizational design
	Problem solving methods
	Unfolding guidelines (Hoshin Kanri)
	Small Group Activities

Despite their relevance, the research efforts of the second practice were recently associated in the academic literature (Marodin et al. 2015, 1389). This fact can be justified by the evolutionary understanding of lean production and its practices, which have reached different patterns of understanding over time. Overall, all nineteen practices have been consistently studied in the literature and therefore may be representative to characterize lean implementation.

3.2.2 Elements of the Construct for Sociotechnical and Ergonomic Practices (SE)

Any successful enterprise must have an effective and efficient work organization in the management level in order to balance work demands and workforce conditions and hence establish best sociotechnical and ergonomic practices conducive to the maximum human health, productivity and quality of work (Jaworek et al. 2010, 368; Genaidy and Karwowski 2003, 317; Karwowski et al. 1994, 3) The concepts underlying SE practices can be considered in the planning and execution of operational activities, establishing adequate conditions and better results in the interaction between workers and the work environment (Ferreira and Gurgueira 2013, 41).

Based on the literature review, this study highlight 20 SE practices as the most frequent ones at a level of ergonomic management, as shown in Table 5. Of these, the practice "regulation of technical, organizational and human aspects" is the most cited in the literature.

This practice is generally associated with the internal procedures of the organizations that compound the management routines (Dul and Neumann 2009, 745; Nunes 2015, 896). These procedures aim to reinforce and provide adequate guidance of management's expectations regarding technical, organizational and human aspects on a daily basis for all productive activities within the company (Arezes, Dinis-Carvalho, and Alves 2015, 57). Some researchers also relate it to the norms of behavior in a work environment (Camarotto and Vanalle 2015, 1; Koukoulaki 2010, 940).

In contrast, the "clarity in setting goals" presents a low percentage of agreement among the studied authors. Despite its clear relevance within an organizational context, this practice is rarely indicated as influencing the implementation of SE factors (Koukoulaki 2010, 936) or even considered as an SE practice (Saurin and Ferreira 2009, 404). In general, all SE practices have been evidenced in the literature and may represent, for the purpose of this study, the implementation of SE factors within a company.

Table 5. Elements that compose the construct for Sociotechnical and Ergonomic Practices (SE)

Domain	Elements that compose the construct for Sociotechnical and Ergonomic Practices (SE)
Organizational	Communication and information system
	Problem solving indicators exposure
	Management of staff turnover
	Workers' recognition and reward
	Teamwork and coaching
	Clarity in targets definition
	Clarity in defining the role of workers
	Search for good organizational climate
	Balancing among quality, scope, time and cost
	Meetings for communication of projects
Physical	Ergonomics criteria for workstation design
	Workstations appropriated to workers
	Risk alerts utilization
	Search for the health and safety of workers
	Anticipating and reducing the risk of incidents
	Ergonomics recommendations as regulations
Cognitive	Overload for achievement of goals
	Appreciation for workers training
	Clear strategies, symbols and methods
	Regulation of technical, organizational and human aspects

4. FINAL REMARKS

The aim of this study was to establish which practices can be used to classify a lean manufacturing industry in relation to the level of implementation of SE and PE. In order to reach the proposed goal, the Knowledge Development Process - Constructivist (Proknow-C) tool was used, which, due to its constructivist vision, allowed the selection of the 51 articles in the BP, the bibliometric analysis of the basic and advanced variables and the systemic analysis of these publications.

The bibliometric analysis of the publications concerning the basic variables evidenced the researchers: Patrick W. Neumann with two articles in the Bibliographic Portfolio and three in the references, in turn the researchers Tarcísio Abreu Saurin and Jörgen Winkel stood out with two articles in the BP and two in the references, and in Brazil it is worth mentioning the researcher Guilherme Luz Tortorella with two articles in the Bibliographic Portfolio and one in the references. According to the information presented on these researchers, it is evident that they are specialists in production systems and have the ergonomics of work as one of their fields of study. Thus, these authors are characterized as prolific authors of this fragment of literature.

The journal with most highlight was Applied Ergonomics, with four articles in the BP and one in the BP references. As for the scientific recognition of the articles in the BP, the most cited article is entitled 'The impact of lean production and related systems of work organization on worker health' written by the authors Landsbergis, Cahill and Schnall, with 555 citations. From the articles in the BP references the highlight was 'On the interface between management and human resources management' written by Boudreau, Hopp, McLain and Thomas with 355 citations. However, these last authors are not among the researchers who stood out, as well as the published journal, that did not stand out between those in the BP.

This stage of the bibliometric analysis revealed the main characteristics of the publications on the subject under investigation, i.e., ergonomics and lean production. In this way, it allowed the researcher to know relevant information about the subject under study and about who are the prominent researchers, as well as indicated the main channels where current and/or future research can be published (Dutra et al. 2015, 265; Ensslin et al. 2014, 19; Ensslin, Ensslin, and Pacheco 2012, 86).

One of the major challenges of this approach was to establish which practices corroborate with the improvement of the processes in order to provide health, well-being and safety to the workers, accordingly to the reality of each organization. For the researchers it has become a challenge, because studies conducted so far indicate several adverse factors to the workers resulting from this implementation. However, through the systemic analysis of the literature with the respective constitutive and operational definitions it was possible to elaborate a conceptual model for the sociotechnical and ergonomic construct - SE and later, the establishment of two constructs, both for Sociotechnical and Ergonomic Practices (SE) and for Lean Production Practices (LP).

The present research established the constructs, however, its continuation will be accomplished with the elaboration of two scales to measure the level of implementation of these practices, deploying the Gradual Response Model of Samejima, through the application of a valuable tool, the Item Response Theory (IRT). This tool is responsible for validating the proposed instrument.

In demonstrating the relationship between SE and LP, the present research will evidence contributions to theory and practice. Academically, it will provide a method that combines the assessment of lean and ergonomic practices in a single approach. Such combination fills the gaps identified in the literature, since it integrates aspects of lean implementation from the point of view of ergonomics. The proposal of the method is not to provide an optimal solution, but to point out alternatives for improvement that can be developed consecutively in organizations.

In conclusion, this instrument aims to enable the integration of such practices and contribute to the construction of a synergistic approach, which will support a process of change and improvement of the ergonomic aspects in organizations, allowing to glimpse the problems in the long term, besides a clear vision of the current shortcomings in lean implementation.

APPENDIX A– BIBLIOGRAPHIC PORTFOLIO – PB

Authors	Article (Title)	Year	Periodical
Tortorella; Vergara; Ferreira	Lean manufacturing implementation: an assessment method with regards to socio-technical and ergonomics practices adoption	2017	*The International Journal of Advanced Manufacturing Technology*
Botti; Mora; Regattieri	Integrating ergonomics and lean manufacturing principles in a hybrid assembly line	2017	*Computers & Industrial Engineering*
Botti; Mora; Regattieri	Application of a mathematical model for ergonomics in lean manufacturing	2017	*Data in Brief*
Zare; Croq; Hossein-Arabi; Brunet; Roquelaure	Does Ergonomics Improve Product Quality and Reduce Costs? A Review Article	2016	*Human Factors and Ergonomics in Manufacturing & Service Industries*
Jarebrant; Winkel; Hanse; Mathiassen; Ojmertz	ErgoVSM: A Tool for Integrating Value Stream Mapping and Ergonomics in Manufacturing	2016	*Human Factors and Ergonomics in Manufacturing & Service Industries*
Silva; Tortorella; Amaral	Psychophysical Demands and Perceived Workload An Ergonomics Standpoint for Lean Production in Assembly Cells	2016	*Human Factors and Ergonomics in Manufacturing & Service Industries*
Mouayni; Etienne; Siadat; Dantan; Lux	A simulation based approach for enhancing health aspects in production systems by integrating work margins	2016	*IFAC-PapersOnLine*
Arezes; Dinis-Carvalho; Alves	Workplace ergonomics in lean production environments: A literature review	2015	*Work*
Santos; Vieira; Balbinotti	Lean Manufacturing and Ergonomic Working Conditions in the Automotive Industry	2015	*Procedia Manufacturing*
Charalambous; Fletcher; Webb	Identifying the key organisational human factors for introducing human-robot collaboration in industry: an exploratory study	2015	*The International Journal of Advanced Manufacturing Technology*
Gnanavel; Balasubramanian; Narendran	Suzhal – An Alternative Layout to Improve Productivity and Worker Well-being in Labor Demanded Lean Environment	2015	*Procedia Manufacturing*

Appendix A. (Continued)

Authors	Article (Title)	Year	Periodical
Camarotto; Vanalle	Production organization and work aspects in companies of the automotive sector in Spain and Brazil	2015	Revista Espacios
Maia; Eira; Alves; Leão	The organizational improvement as trigger for better working conditions	2015	Revista Ibérica de Sistemas e Tecnologias de Informação
Koukoulaki	The impact of lean production on musculoskeletal and psychosocial risks: An examination of sociotechnical trends over 20 years	2014	Applied Ergonomics
Lesková	Principles of lean production to designing manual assembly workstations	2013	International Journal of Engineering
Ferreira L, Gurgueira G	Ergonomia como fator econômico no pensamento Enxuto: uma análise crítica bibliográfica	2013	Gepros: Gestão da Produção, Operações e Sistemas
Toralla; Falzon; Morais	Participatory design in lean production: which contribution from employees? for what end?	2012	Work
Vieira; Balbinotti; Varasquin; Gontijo	Ergonomics and Kaizen as strategies for competitiveness: a theoretical and practical in an automotive industry	2012	Work
Yang; Yang	An integrated model of the toyota production system with total quality management and people factors	2012	Human Factors and Ergonomics in Manufacturing & Service Industries
Figueira; Machado; Nunes	Integration of human factors principles in LARG organizations - a conceptual model	2012	Work
Marksberry; Church; Schmidt	The employee suggestion system: A new approach using latent semantic analysis	2012	Human Factors and Ergonomics in Manufacturing & Service Industries
Silva; Bento	The regulation of work activity and the new labor and production contexts	2012	Work
Brännmark M, Håkansson M.	Lean production and workrelated musculoskeletal disorders: overviews of international and Swedish studies	2012	Work
Westgaard; Winkel	Occupational musculoskeletal and mental health: Significance of rationalization and opportunities to create sustainable production systems - A systematic review	2011	Applied Ergonomics
Hasle	Lean production - An evaluation of the possibilities for an employee supportive lean practice	2011	Human Factors and Ergonomics in Manufacturing & Service Industries
Finnsgård; Wänströ; Medbo; Neumann	Impact of materials exposure on assembly workstation performance	2011	International Journal of Production

Authors	Article (Title)	Year	Periodical
Yang; Yeh; Yang	The implementation of technical practices and human factors of the toyota production system in different industries	2011	*Research*
Wong; Richardson	Assessment of working conditions in two different semiconductor manufacturing lines: Effective ergonomics interventions	2010	*Human Factors and Ergonomics in Manufacturing & Service Industries*
Eswaramoorthi; Rajagopal; Prasad; Mohanram	Redesigning assembly stations using ergonomic methods as a lean tool	2010	*Human Factors and Ergonomics in Manufacturing & Service Industries*
Dul; Neumann	Ergonomics contributions to company strategies	2009	*Work*
Wong; Wong; Ali	A study on lean manufacturing implementation in the Malaysian electrical and electronics industry	2009	*Applied Ergonomics*
Saurin; Ferreira	The impacts of lean production on working conditions: A case study of a harvester assembly line in Brazil	2009	*European Journal of Scientific Research*
Womack; Armstrong; Liker	Lean Job Design and Musculoskeletal Disorder Risk: A Two Plant Comparison	2009	*International Journal of Industrial Ergonomics*
Hunter	The toyota production system applied to the upholstery furniture manufacturing industry	2008	*Human Factors and Ergonomics in Manufacturing*
Saurin; Ferreira	Guidelines to evaluate the impacts of lean production on working conditions	2008	*Materials and Manufacturing Processes*
Brown; O'Rourke	Lean manufacturing comes to China: A case study of its impact on workplace health and safety	2007	*Revista Produção*
Conti, R., Angelis, J., Cooper, C.	The effects of lean production on worker job stress	2006	*International Journal of Occupational and Environmental Health*
Seppälä; Klemola	How Do Employees Perceive Their Organization and Job When Companies Adopt Principles of Lean Production?	2004	*International Journal of Operations & Production Management*
Paez, O., Dewees, J., Genaidy, A., Tuncel, S., Karwowski, W., Zurada, J.,	The lean manufacturing enterprise: an emerging sociotechnological system integration	2004	*Human Factors and Ergonomics in Manufacturing*
Genaidy; Karwowski	Human performance in lean production environment: Critical assessment and research framework	2003	*Human Factors and Ergonomics in Manufacturing*
Smith	Growing an ergonomics culture in manufacturing	2003	*Human Factors and Ergonomics in Manufacturing*
Parker, S. K.	Longitudinal effects of lean production on employee outcomes and the mediating role of work characteristics	2003	*Journal of Engineering Manufacture*

Appendix A. (Continued)

Authors	Article (Title)	Year	Periodic
Hunter	Ergonomic evaluation of manufacturing system designs	2002	*Journal of Applied Psychology*
Anderson-Connolly, R. Grunberg L, Greenberg E, Moore S.	Is lean mean? Workplace transformation and employee wellbeing	2002	Journal of Manufacturing Systems
Landsbergis; Cahill; Schnall	The impact of lean production and related new systems of work organization on worker health	1999	Work, Employment and Society
Niepce; Molleman	Work Design Issues in Lean Production from a Sociotechnical Systems Perspective: Neo-Taylorism or the Next Step in Sociotechnical Design?	1998	Journal of Occupational Health Psychology
Adler, P., Goldoftas, B., & Levine, D.	Ergonomic, employee involvement, and the Toyota production system: A case study of NUMMI's 1993 model introduction	1997	Human Relations
Niepce; Molleman	A case study - Characteristics of work organization in lean production and sociotechnical systems	1996	Industrial and Labor Relations Review
Bjorkman	The rationalisation movement in perspective and some ergonomic implications	1996	International Journal of Operations
Lewchuk, W., Robertson, D.	Working Conditions under Lean Production:	1996	& Production Management
Jackson, P. R., & Martin, R.	A Worker-based Benchmarking Study	1996	Applied Ergonomics

REFERENCES

Abrahão, Júlia, Laerte Sznelwar, Alexandre Silvino, Maurício Sarmet, and Diana Pinho. 2009. *Introdução à Ergonomia: Da Prática à Teoria*. [*Introduction to Ergonomics: From Practice to Theory*] Editora Blucher.

Afonso, Michele Hartmann Feyh, Juliane Vieira de Souza, Sandra Rolim Ensslin, and Leonardo Ensslin. 2012. "Como Construir Conhecimento Sobre O Tema De Pesquisa? Aplicação Do Processo Proknow-C Na Busca De Literatura Sobre Avaliação Do Desenvolvimento Sustentável." ["How to Build Knowledge on the Research Theme? Application of the Proknow-C Process in the Search for Literature on Sustainable Development Assessment"] *Revista de Gestão Social e Ambiental* 5 (2): 47–62. https://doi.org/10.5773/rgsa.v5i2.424.

Alves, Anabela C., José Dinis-Carvalho, and Rui M. Sousa. 2012. "Lean Production as Promoter of Thinkers to Achieve Companies' Agility" *Learning Organization* 19 (3): 219–37. https://doi.org/10.1108/09696471211219930.

Arezes, Pedro M., José Dinis-Carvalho, and Anabela Carvalho Alves. 2015. "Workplace Ergonomics in Lean Production Environments: A Literature Review." *Work* 52 (1): 57–70. https://doi.org/10.3233/WOR-141941.

Arlbjørn, Jan Stentoft, and Per Vagn Freytag. 2013. "Evidence of Lean: A Review of International Peer-Reviewed Journal Articles." *European Business Review* 25 (2): 174–205. https://doi.org/10.1108/09555341311302675.

Balbinotti, Giles Cesar. 2013. *O Gerenciamento Dos Aspectos Humanos Nas Atividades de Projetos de Processo Produtivo Na Indústria Automotiva: Princípios Com Abordagem Sociotécnica e Ergonômica.* [*The Management of Human Aspects in the Activities of Productive Process Projects in the Automotive Industry: Principles with a Sociotechnical and Ergonomic Approach.*] PhD thesis., Universidade Federal de Santa Catarina.

Bardin, Laurence. 2011. *Análise de Conteúdo.* Lisboa: Edições.

Bhamu, Jaiprakash, and Kuldip Singh Sangwan. 2014. "Lean Manufacturing: Literature Review and Research Issues." *International Journal of Operations and Production Management* 34 (7): 876–940. https://doi.org/10.1108/IJOPM-08-2012-0315.

Bhasin, Sanjay, and Peter Burcher. 2006. "Lean Viewed as a Philosophy." *Journal of Manufacturing Technology Management.* 17 (1): 56 -72. https://doi.org/10.1108/17410380610639506.

Bortolotti, Thomas, Stefania Boscari, and Pamela Danese. 2015. "Successful Lean Implementation: Organizational Culture and Soft Lean Practices." *International Journal of Production Economics*, 160 (1): 182–201. https://doi.org/10.1016/j.ijpe.2014.10.013

Camarotto, João Alberto, and Rosangela Maria Vanalle. 2015. "Production Organization and Work Aspects in Companies of the Automotive Sector in Spain and Brazil." *Espacios*, 36 (18): 1-10. http://www.revistaespacios.com/a15v36n18/15361810.html.

Cardoso, Thuine Lopes., Sandra Rolim Ensslin, Leonardo Ensslin, Vicente Mateo Ripoll-Feliu, and Ademar Dutra. (2015). *Reflexões para avanço na área de Avaliação e Gestão do Desempenho das Universidades: uma análise da literatura científica.* [*Reflections for advancement in the area of Evaluation and Management of University Performance: an analysis of the scientific literature.*] Anais do Seminários em Administração (XVIII SEMEAD) São Paulo (SP), 4(1): 1-17.

Doolen, Toni L., and Marla E. Hacker. 2005. "A Review of Lean Assessment in Organizations: An Exploratory Study of Lean Practices by Electronics Manufacturers." *Journal of Manufacturing Systems* 24 (1): 55–67. https://doi.org/10.1016/S0278-6125(05)80007-X.

Dul, Jan, and W. Patrick Neumann. 2009. "Ergonomics Contributions to Company Strategies." *Applied Ergonomics* 40 (4): 745–52. https:// doi. org/ 10. 1016/ j. apergo. 2008.07.001.

Dutra, Ademar, Vicente Mateo Ripoll-Feliu, Arturo Giner Fillol, Sandra Rolim Ensslin, and Leonardo Ensslin. 2015. "The Construction of Knowledge from the Scientific Literature about the Theme Seaport Performance Evaluation." *International Journal of Productivity and Performance Management* 64 (2): 243–69. https://doi.org/10.1108/IJPPM-01-2014-0015.

Ensslin, Leonardo, Sandra Rolim Ensslin, and Giovanni Cardoso Pacheco. 2012. "Um Estudo Sobre Segurança Em Estádios de Futebol Baseado Na Análise Bibliométrica Da Literatura Internacional." ["A Study on Safety in Football Stadiums Based on Bibliometric Analysis of International Literature."] *Perspectivas Em Ciência Da Informação* 17 (2): 71–91. https://doi.org/10.1590/S1413-99362012000200006.

Ensslin, Sandra Rolim, Leonardo Ensslin, Aline Willemann Kremer, Altair Borgert, and Leonardo Correa Chaves. 2014. "Comportamentos Dos Custos: Seleção de Referencial

Teórico e Análise Bibliométrica." ["Costs Behaviors: Selection of Theoretical Referential and Bibliometric Analysis."] *Revista de Contabilidade Do Mestrado Em Ciências Contábeis Da UERJ* 19 (3): 2–25.

Ferreira, Leonardo, and Giovana Pimentel Gurgueira. 2013. "Ergonomia Como Fator Ergonômico No Pensamento Enxuto: Uma Análise Crítica Bibliográfica." ["Ergonomics as an Ergonomic Factor in Lean Thinking: A Critical Bibliographical Analysis."] *Gestão Da Produção, Operações e Sistemas* 8 (3): 39–51. http://repositorium.sdum.uminho.pt/bitstream/1822/18865/1/CLME2011WB_AA_PA.pdf.

Figueira, Sara, V. Cruz MacHado, and Isabel L. Nunes. 2012. "Integration of Human Factors Principles in LARG Organizations - A Conceptual Model." *Work* 41 (1): 1712–19. https://doi.org/10.3233/WOR-2012-0374-1712.

Furlan, Andrea, Andreia Vinelli, and Giorgia Dal Pont. 2011. "Complementarity and Lean Manufacturing Bundles: An Empirical Analysis." *International Journal of Operations and Production Management* 31 (8): 835–850. https://doi.org/10.1108/01443571 111153067.

Genaidy, Ash M., and Waldemar Karwowski. 2003. "Human Performance in Lean Production Environment: Critical Assessment and Research Framework." *Human Factors and Ergonomics In Manufacturing* 13 (4): 317–30. https://doi.org/10.1002/hfm.10047.

GIL, Antonio Carlos. 2002. *Como Elaborar Projetos de Pesquisa. [How to design research projects.]* Editora Atlas.

Gnanavel, S. S., Venkatesh Balasubramanian, and T. T. Narendran. 2015. "Suzhal – An Alternative Layout to Improve Productivity and Worker Well-Being in Labor Demanded Lean Environment." *Procedia Manufacturing* 3 (1): 574–580. https://doi.org/10.1016/j.promfg.2015.07.268.

Herron, Colin, and Paul M. Braiden. 2006. "A Methodology for Developing Sustainable Quantifiable Productivity Improvement in Manufacturing Companies." *International Journal of Production Economics* 104 (1): 143–53. https://doi.org/10.1016/j.ijpe.2005.10.004.

Jaworek, Magdalena, Tadeusz Marek, Waldemar Karwowski, Chris Andrzejczak, and Ash M. Genaidy. 2010. "Burnout Syndrome as a Mediator for the Effect of Work-Related Factors on Musculoskeletal Complaints among Hospital Nurses." *International Journal of Industrial Ergonomics* 40 (3): 368–75. https://doi.org/10.1016/j.ergon.2010.01.006.

Karwowski, W., G. Salvendy, R. Badham, P. Brodner, C. Clegg, S. L. Hwang, J. Iwasawa, et al. 1994. "Integrating People, Organization, and Technology in Advanced Manufacturing: A Position Paper Based on the Joint View of Industrial Managers, Engineers, Consultants, and Researchers." *The International Journal of Human Factors in Manufacturing* 4 (1): 1–19. https://doi.org/10.1002/hfm.4530040102.

Ketokivi, Mikko, and Roger Schroeder. 2004. "Manufacturing Practices, Strategic Fit and Performance." *International Journal of Operations & Production Management* 24 (2): 171–91. https://doi.org/10.1108/01443570410514876.

Koukoulaki, Theoni. 2010. "New Trends in Work Environment - New Effects on Safety." *Safety Science* 48 (8): 936–942. https://doi.org/10.1016/j.ssci.2009.04.003.

Longoni, Annachiara, Mark Pagell, David Johnston, and Anthony Veltri. 2013. "When Does Lean Hurt? – an Exploration of Lean Practices and Worker Health and Safety

Outcomes." *International Journal of Production Research* 51 (11): 3300–3320. https://doi.org/10.1080/00207543.2013.765072.

Maia, Laura C., Rúben Eira, Anabela C. Alves, and Celina P. Leão. 2015. "A Melhoria Organizacional Como Alavanca Para Melhores Condições de Trabalho." ["Organizational Improvement as Leverage for Better Working Conditions."] *RISTI - Revista Iberica de Sistemas e Tecnologias de Informacao*, 4(1): 50–65. https://doi.org/10.17013/risti.e4.50-65.

Marksberry, Phillip, Joshua Church, and Michael Schmidt. 2014. "The Employee Suggestion System: A New Approach Using Latent Semantic Analysis." *Human Factors and Ergonomics in Manufacturing & Service Industries* 24 (1): 29–39. https://doi.org/10.1002/hfm.20351.

Marodin, Giuliano Almeida, and Tarcisio Abreu Saurin. 2013. "Implementing Lean Production Systems: Research Areas and Opportunities for Future Studies." *International Journal of Production Research* 51 (22): 6663–80. https://doi.org/10.1080/00207543.2013.826831.

Marodin, Giuliano Almeida, Tarcísio Abreu Saurin, Guilherme Luz Tortorella, and Juliano Denicol. 2015. "How Context Factors Influence Lean Production Practices in Manufacturing Cells." *The International Journal of Advanced Manufacturing Technology* 79 (5–8): 1389–99. https://doi.org/10.1007/s00170-015-6944-2.

Niepce, W., and E. Molleman. 1998. "Work Design Issues in Lean Production Form a Sociotechnical Systems Perspective: Neo-Taylorism or Th next Step in Sociotechnical Design?" *Human* 51 (3): 259–287. https://doi.org/10.1177/07399863870092005.

Nunes, Isabel L. 2015. "Integration of Ergonomics and Lean Six Sigma. A Model Proposal." *Procedia Manufacturing* 3 (1): 890–897. https://doi.org/10.1016/j.promfg.2015.07.124.

Paez, O., J. Dewees, A. Genaidy, S. Tuncel, W. Karwowski, and J. Zurada. 2004. "The Lean Manufacturing Enterprise: An Emerging Sociotechnological System Integration." *Human Factors and Ergonomics In Manufacturing* 14 (3): 285–306. https://doi.org/10.1002/hfm.10067.

Pasquali, Luiz. 1998. "Princípios de Elaboração de Escalas Psicológicas." ["Principles of Elaboration of Psychological Scales."] *Revista de Psiquiatria Clínica* 25 (5): 206–13.

Pavnaskar, S. J., J. K. Gershenson, and A. B. Jambekar. 2003. "Classification Scheme for Lean Manufacturing Tools." *International Journal of Production Research* 41 (13): 3075–90. https://doi.org/10.1080/0020754021000049817.

Perez Toralla, M. S., P. Falzon, and A. Morais. 2012. "Participatory Design in Lean Production: Which Contribution from Employees? For What End?" *Work* 41 (1): 2706–2712. https://doi.org/10.3233/WOR-2012-0514-2706.

Santos, Zélio Geraldo dos, Leandro Vieira, and Giles Balbinotti. 2015. "Lean Manufacturing and Ergonomic Working Conditions in the Automotive Industry." *Procedia Manufacturing* 3 (1): 5947–5954. https://doi.org/10.1016/j.promfg.2015.07.687.

Sartori, Simone, Leonardo Ensslin, Lucila Maria De Souza Campos, and Sandra Rolim Ensslin. 2014. "Mapeamento Do Estado Da Arte Do Tema Sustentabilidade Ambiental Direcionado Para a Tecnologia de Informação." ["State Mapping Of The Theme Art Environmental Sustainability Targeted For Information Technology."] *Transinformacao* 26 (1): 77–89. https://doi.org/10.1590/S0103-37862014000100008.

Saurin, Tarcisio Abreu, and Cléber Fabricio Ferreira. 2009. "The Impacts of Lean Production on Working Conditions: A Case Study of a Harvester Assembly Line in Brazil."

International Journal of Industrial Ergonomics 39 (2): 403–412. https://doi.org/10.1016/j.ergon.2008.08.003.

Shah, Rachna, and Peter T. Ward. 2007. "Defining and Developing Measures of Lean Production." *Journal of Operations Management* 25 (4): 785–805. https://doi.org/10.1016/j.jom.2007.01.019.

Silva, Marcelo Pereira da, Guilherme Luz Tortorella, and Fernando Gonçalves Amaral. 2016. "Psychophysical Demands and Perceived Workload—An Ergonomics Standpoint for Lean Production in Assembly Cells." *Human Factors and Ergonomics in Manufacturing & Service Industries* 26 (6): 643–654. https://doi.org/10.1002/hfm.20404.

Tajini, Reda, and Saâd Lissane Elhaq. 2014. "Methodology for Work Measurement of the Human Factor in Industry." *International Journal of Industrial and Systems Engineering* 16 (4): 472-492. https://doi.org/10.1504/IJISE.2014.060655.

Valmorbida, Sandra Mara Iesbik, Sandra Rolim Ensslin, Leonardo Ensslin, and Vicente Mateo Ripoll-Feliu. 2014. "Avaliação de Desempenho Para Auxílio Na Gestão de Universidades Públicas: Análise Da Literatura Para Identificação de Oportunidades de Pesquisas." ["Evaluation of Performance for Aid in the Management of Public Universities: Analysis of Literature for Identification of Research Opportunities."] *Contabilidade, Gestão e Governança* 17 (3): 4–28. https://cgg-amg.unb.br/index.php/contabil/article/view/520.

In: Lean Manufacturing
Editors: F. J. G. Silva and L. Carlos Pinto Ferreira
ISBN: 978-1-53615-725-3
© 2019 Nova Science Publishers, Inc.

Chapter 16

LEAN MANUFACTURING AND INDUSTRY 4.0: FACING NEW CHALLENGES FROM A SHOP-FLOOR PERSPECTIVE

Antonio Sartal[1,2,*] *and Helena Navas*[2]

[1]University of Vigo, Spain
[2]Universidade Nova de Lisboa, Portugal

ABSTRACT

In today's environments, which are characterized by great volatility and customization, the lean practices and tools conceived for High-Volume & Low-Variability (HVLV) manufacturing systems, seem to show important difficulties to adapt efficiently. Thus, it is necessary to consider the limitations of lean thinking in this new environment and to analyze how they can be overcome. Many researchers and practitioners hold the idea that Industry 4.0 can offer solutions to these current challenges; however, it is not clear which of these solutions is the most appropriate or which technologies can actually generate a sustainable competitive advantage. We attempt to evaluate how lean manufacturing establishes the right conditions for developing technology-intensive environments for Industry 4.0, as well as how shop-floor technologies, particularly clean technology (CT) and Information technology (IT), can be leveraged to address the weaknesses of the lean approach and to achieve improved industrial performance and a better competitive position.

Keywords: lean manufacturing, Industry 4.0, information technologies, cleaner technologies, shopfloor

---

[*] Corresponding Author's E-mail: antoniosartal@uvigo.es.

1. OBJECTIVES OF THIS CHAPTER

In today's environments, which are characterized by great volatility and customization, the lean principles and tools conceived for High-Volume & Low-Variability (HVLV) manufacturing systems, seem to show important difficulties to adapt efficiently. In fact, some of these difficulties have already led to criticisms that have been used to justify new paradigms of operations management (Sartal et al. 2017a, 260). These authors have considered it necessary to complement the lean philosophy with contributions from other models to mitigate these limitations.

Thus, it is necessary to consider the limitations of lean thinking in this new environment and to analyze how they can be overcome. Clearly, firms must consider sudden changes in preferences, shorter product life cycles, and environmental concerns. In this context, many researchers and practitioners hold the idea that Industry 4.0 can offer solutions to these current challenges (Sartal et al. 2018a, 1). However, it is not clear which of these solutions is the most appropriate or which technologies can actually generate a sustainable competitive advantage. To meet the demands of today's consumers (e.g., shorter product life cycles, reduced cost, more customizable products, and greater consideration of environmental concerns), of all the technologies that fall under the Industry 4.0 umbrella, clean technology (CT) and Information technology (IT) are best able to resolve the weaknesses of the lean approach and to achieve improved industrial performance and a better competitive position.

With this in mind, this chapter seeks to reflect on the debate about the relationship between lean manufacturing and both CT and IT in Industry 4.0 shop floors. We attempt to evaluate how lean manufacturing establishes the right conditions for developing technology-intensive environments for Industry 4.0, as well as how shop-floor technologies can be leveraged to enhance the contributions of lean practices to industrial performance.

2. OPERATIONAL EXCELLENCE AND LEAN MANUFACTURING: FROM JAPANESE ORIGINS TO THE AMERICAN REINTERPRETATION

The term operational excellence describes a "specific strategic approach to the production and delivery of products and services" (Treacy and Wiersema 1993, 2). Companies pursuing operational excellence are indefatigable in seeking ways to minimize overhead costs, to eliminate intermediate production steps, to reduce transaction and other 'friction' costs, and to optimize business processes across functional and organizational boundaries" (Treacy and Wiersema 1993, 2). Although there are various labels for these strategies (Six Sigma, agile manufacturing, world-class manufacturing, etc.), the Toyota Production System (TPS), which was reinterpreted later as lean manufacturing, has been one the most influential new paradigm in manufacturing.

The lean manufacturing philosophy brings together the Japanese conventional wisdom and various tools that aim to identify and systematically eliminate waste - that is defined as activities with low or zero added value - based on ongoing improvement and a pull flow. Many of the ideas and tools within this philosophy of waste reduction are a matter of common sense; some stem from Taylor's optimizing legacy; and others originate in the industrial heritage of Henry Ford. However, the TPS has undoubtedly had the greatest influence on the generation of this philosophy's tools.

Lean manufacturing includes the teachings that J applied to Toyota (the automobile company), which, in the 1950s, aimed to serve markets with low volumes and a greater variety of vehicles. Toyota reached the United States in the 1980s and its alternative to the traditional system of mass production - in spite of its limitations, which are discussed below - has gradually become the orthodoxy in operations management.

The International Motor Vehicle Program, which was created by researchers at the Massachusetts Institute of Technology in the 1980s, popularized the term lean manufacturing to describe this alternative approach to mass production. However, it was the subsequent analysis in the book The Machine that Changed the World (Womack et al. 1990) that sparked the TPS's eventual adoption in the West. Today, the concepts of TPS and lean manufacturing are interchangeable.

The consensus in the literature is that the main goal of this philosophy is the systematic elimination of any activities that do not add value. This ties in with Ohno's (1988, 23) statement that the final aim of the TPS was the absolute elimination of waste, which he defined as anything in excess of the necessary minimum of materials, equipment, parts, space, or time. However, in spite of this unanimity, agreement is lacking regarding how to measure a company's degree of leanness and regarding the practices and tools involved in the process.

The eclectic nature of lean manufacturing has generated a multitude of approaches in the literature (e.g., Shah and Ward 2007, 786). Here we frame lean manufacturing using two main perspectives. On the one hand, the practical lean toolbox, which includes a set of management practices and tools associated with just in time (JIT), total productive maintenance, total quality management, advanced human resources practices, and a set of external practices related to the degree of integration with suppliers and customers. On the other, the more theoretical lean culture, which emphasizes the main principles of this philosophy.

In the opinions of several authors, both this dual perspective and the lack of connection between lean tools and lean culture in transformation processes help explain why the success rate of lean manufacturing transformations in non-Japanese industries has been very low. In fact, according to estimates, just 10% of companies manage to conclude such a transformation (Bhasin and Burcher 2006, 56). Although many non-Japanese companies continue to perceive lean manufacturing as an aggregation of tools for eliminating waste - the lean toolbox approach there is also a tradition linked to the Japanese roots of lean manufacturing, in which all attention tends to be focused on the organization's cultural transformation - the lean culture approach.

To build a model based on permanent aspects, the analysis of lean manufacturing should perhaps be approached from the viewpoint of the pillars of TPS: JIT, Jidoka, and the third pillar: Respect for People (RfP) which is very much present in the TPS but which is less prominent in the subsequent dissemination of lean manufacturing (Emiliani 2008, 1). Figure 1 is an adaptation of the well-known "TPS house" (Liker 2004, 33); it includes these pillars and their main characteristics. We explain each of them below in detail.

The first pillar, JIT, is perhaps the most widely recognized aspect of the Toyota system. This principle is a response to a set of tools and interrelated management practices that aim to produce only the required products (in line with customers' expectations), at the right time and in the right amount (Sugimori 1977, 553). This implies two fundamental changes in perception: on the one hand, a change from a push system to a pull one (in which each process uses the products or parts from the previous processes, as required); on the other, the elimination of stocks between processes and production at the rate imposed by demand (takt time).

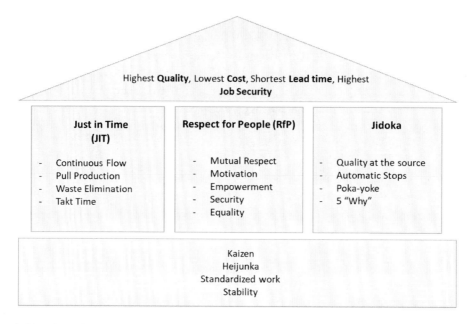

Figure 1. The three pillars of the TPS, based on the "TPS house" (Liker 2004).

The second pillar, Jidoka, is usually conceived as involving the automation of quality control with an added human touch. This principle covers a set of tools and practices that are meant to ensure that workers self-inspect and that provides them with the means, in the case of an anomaly, to stop production to prevent defective parts from advancing. This well-known "zero defects" philosophy is intended to correct problems as they arise (built-in quality). This principle breaks with the Taylorist tradition, in which only the plant manager could stop the production line. The TPS gives workers not just this capability but also the responsibility to do so, thus adding new allies to the problem-solving process. Moreover, this principle replaces end-of-line quality inspection, which is characteristic of the Taylorist and Fordist systems, with self-inspecting systems during the actual production process (Sugimori, 1977, 554).

The third and final pillar is RfP. Although certain authors have argued that this pillar is the key to making the lean system work (Emiliani 2008, 1), and although it was explicitly included in the foundations of the TPS (Sugimori et al. 1977, 557), it has received practically no dissemination within the lean thinking literature. However, its importance to Toyota is reflected in the very first English-language article on the TPS, in which Sugimori et al. (1977) - in their capacity as Toyota managers - claimed that the TPS was based on two essential concepts: ongoing improvement and the full use of workers' capabilities (the RfP system). In the same way, in the preface to his work, Ohno (1988, 15) stated that waste elimination and RfP are, with equal importance, the basis for the TPS. In addition, in the Toyota 4P model, Liker (2004, 66) explicitly mentions this pillar as one of the four components (which he calls "People and Partners" and which includes respect, challenge, and trust).

Over the years, however, even as the first two pillars of ongoing improvement (JIT and jidoka) have received increasing attention, academics and practitioners alike seem to have underestimated this third pillar. There are few documented examples of implementations that include both ongoing improvement and the practices of this third pillar (Emiliani 2008, 1).

Undoubtedly, the fact that it is easier to generate results in the short term with JIT and jidoka than with RfP is one of the reasons for this discrepancy. In addition, RfP is not easy to understand or to apply because it has to be interpreted from the same viewpoint as Toyota. In the TPS, the meaning of RfP is related to workers' capability at work rather than to the actual individuals. Sugimori et al. (1977, 557) mentioned it, for example, when they stressed that full use of workers' capabilities requires a system of RfP, which would improve their working conditions and give them greater responsibility and authority.

We can, therefore, observe the dual nature of this pillar within the TPS. On the one hand, Toyota revealed a particular form of respect that is more practical than moral and that is based on mutual support between supervisors and employees. In this case, respect refers to workers and managers cooperating in the problem-solving process, as well as helping and trusting each other. The aim is therefore to ensure that workers are well-trained, that they actively participate, and that they are granted decision-making capacity (Sugimori 1977, 557).

This form of cooperation reflects a deal in which workers and managers act as part-owners, with the aim of minimizing costs and maximizing the possibilities for success. Moreover, also in line with this explanation, we observe that RfP complements the initial empowerment that is inherent in JIT and jidoka by adding an explicit concern regarding the improvement of work conditions and workers' motivation by balancing out the burnout that improvement processes can generate (Emiliani 2008, 2).

Even in an industrial environment that has technically perfect coordination and that features learning systems to resolve any problems that arise, decision-makers still have distinct - and often diverging - interests, so a stable agreement is needed to avoid constant conflict in the organization of operations. From this point of view, the goal is thus to guarantee a certain stability, which facilitates the correct deployment of routines for the transmission of information and for problem-solving. This is what actually happened within the TPS, a model of human resources management that respects people more than Taylorism or Fordism, at least to the extent that it involves putting them on equal footing in problem-solving. However, this management style could end up generating even more stress than the systematic repetition of a few simple tasks. In the 1980s - in case there was any doubt about the meaning of RfP - Toyota began to see very high levels of staff turnover and to find it difficult to attract the most capable workers. It then started to consider the need for a balanced form of lean manufacturing that would restore the balance of employee satisfaction and customer satisfaction.

3. EVOLUTION OF LEAN MANUFACTURING AND TODAY'S MAIN CHALLENGES

Lean practices have been associated with very satisfactory results in terms of productivity, quality, and workplace safety, as they act to reduce the causes of variability, waste, and anything else that does not add value from the customer's point of view. Many authors have described the success of lean implementation in aspects such as reducing production and/or lead time, minimizing inventory, ensuring delivery reliability, and eliminating waste. Therefore, certain lean capabilities are difficult to imitate and are potential sources of sustainable competitive advantage.

These successful results have led to significant expansion of lean manufacturing throughout the value chain, from product design to distribution and marketing. The principles originally developed for use in production plants are now being used in the general management of all organizational functions, as well as in operations. Even the terminology reflects this evolution, with lean manufacturing or lean production changing to lean management or lean thinking today.

This change is also reflected in the sectors in which lean management is applied. Although it originated in the automobile industry, it is now also used in other industrial sectors (e.g., Sartal and Vazquez 2017, 4) and in services (e.g., Arfmann and Federico 2014, 2). In all cases, the basic premise of the systematic elimination of non-value-added activities has been welcomed.

However, in spite of its advantages and increasing implementation, the lean concept is relatively new in the West, especially in certain sectors (e.g., food) and in small and medium firms (SMEs). In fact, in the several decades since the first English-language article on the TPS (Sugimori 1977), lean manufacturing - the name that this approach eventually received outside Japan - has been adopted as the dominant paradigm for the organization of operations in a broad range of sectors the world over. Even today, however, the success rate of lean manufacturing transformations in non-Japanese industries is very low. These transformations' often have limited impact and fall short of the desired improvements for the overall system.

It would also be wrong to consider this approach to be a panacea. Some authors, despite defending its applicability to any area, have suggested that certain circumstances are necessary for this approach to be successful - for instance, a particular type of process, a certain firm size, or a given context. Doubtless, because this approach is so difficult to implement, especially regarding aspects of work organization - which are more difficult to replicate than technical processes are - and because its integral nature makes it difficult to maintain over time, it is essential that the lean principles be correctly articulated if the expected results are to be achieved.

Although lean practices have been associated with high business performance in terms of productivity, quality, and workplace safety, much remains to be done regarding the dissemination of these practices in SMEs and in activities that support the value chain. However, this diagnosis does not prevent the conclusion that lean thinking (as a management paradigm) seems to have certain weaknesses with regard to the structural trends that stem from environmental evolution, including clients that demand greater personalization and shorter response times; environmental concerns that stem from issues such as insufficient resources and climate change; economic uncertainty; and the intensity, impact, and speed of technological change, especially with the information technology (IT) evolution (Sartal and Vazquez 2017, 1). The principles and tools that are at the core of the paradigm have shown increasing weakness in three main areas.

- First, although the standard waste-elimination philosophy of lean manufacturing should be consistent with growing environmental concerns and with the need to reduce energy consumption and low-carbon production (Quintás et al. 2018, 2); this is not actually the case in practice because those who are engaged in the development and application of lean tools do not seem to have internalized the increasing demands for sustainable operations (Rothenberg et al. 2001, 229). In fact, the truth is quite the opposite. If we avoid (for now) a full analysis of the product life cycle, in efforts focused on the manufacturing area, these ideas have been considered: the production and transport of raw materials and components, as well as distribution and waste management. In other words, although both philosophies refer to waste, they have different, and often conflicting, motivations (e.g., Rothenberg et al. 2001, 235).

- Second, various authors have suggested that the main weakness of this model is the role that it assigns workers. Although much of the legacy of lean manufacturing has to do with the importance of people in the daily management of processes, the mass introduction of IT could be undermining the core role that human factors play in the TPS. However, IT is here to stay. It has evolved to include more benefits and to be easier to use, and the digitalization of society has allowed IT to find its way into even industrial environments and all kinds of classic logistic (e.g., Groba et al. 2018, 55) and optimization applications that have little human capital (e.g., Sartal and Vazquez 2017, 3).

- Finally, in environments in which customers' requirements give rise to sudden changes in preferences or in which there are shorter product life cycles, the benefits of adopting efficient operations management are largely dependent on a firm's management of its innovation process. This is an activity is underestimated in lean thinking. In fact, "fat" (not lean) products are often industrialized in lean environments, and real-world lean thinking has placed greater weight on ongoing process improvement than on disruptive product innovation - even though lean concepts can, and should, also be applied to this area.

Therefore, in today's environments, which are characterized by great volatility and customization, the lean principles and tools conceived for High-Volume & Low-Variability (HVLV) manufacturing environments seem unable to adapt quickly (Sartal et al. 2017a, 260). In fact, some of these difficulties have led to criticisms that have already been covered in the literature and that have been used to justify new paradigms of operations management. Several authors have pointed to the need to integrate lean principles into new hybrid models such as lean Six Sigma (Sheridan 2000, 1) and lean, agile, resilient, and green (LARG) supply-chain management (Carvalho and Cruz-Machado 2011); or new perspectives (Vazquez et al. 2016, 321). These authors have considered it necessary to complement the lean philosophy with contributions from other models to mitigate the lean philosophy's limitations. Other authors have even proposed that lean thinking should evolve toward new systems such as agile manufacturing (Dubey and Gunasekaran 2015, 2147), as these systems adapt better to current market requirements.

However, in many analysts' opinions, these new approaches are essentially descriptive and educational. More importantly, these approaches' real contributions have been criticized because they are based on principles and practices that have always formed part of the lean philosophy's theoretical basis within the TPS (although they bore other names in that context).

Thus, it is objectively necessary to consider the limitations of lean thinking in this new environment and to analyze how they can be overcome. Clearly, firms must consider sudden changes in preferences, shorter product life cycles, and environmental challenges in their day-to-day work (Sartal et al. 2017b, 16). In addition, higher standards of living have made it necessary for firms to go beyond management philosophies that are based exclusively on efficiency. Without questioning the core of the lean philosophy, it is possible to act on its limitations and to evaluate the effects of the incorporation of complementary resources (tools and technologies), which can be used to more effectively respond to today's various challenges.

With this in mind, the roles of all technologies linked to the Industry 4.0 concept should be analyzed to determine how they can help resolve the limitations of lean manufacturing. In fact, many researchers and practitioners have followed the idea that Industry 4.0 can offer solutions to the challenges discussed above (Sartal et al. 2018a). However, it is not clear which of these solutions is the most appropriate or which technologies can actually generate a sustainable competitive advantage. To meet the demands of today's consumers (e.g., shorter product life cycles, reduced cost, more customizable products, and greater consideration of environmental concerns), of all the technologies that fall under the Industry 4.0 umbrella, clean technology (CT) and IT are best able to resolve the weaknesses of the lean approach and to achieve improved industrial performance and a better competitive position. This chapter seeks to reflect on the debate about the relationship between lean manufacturing and both CT and IT in Industry 4.0 shop floors. We attempt to evaluate how lean manufacturing establishes the right conditions for developing technology-intensive environments for Industry 4.0, as well as how shop-floor technologies can be leveraged to enhance the contributions of lean practices to industrial performance.

We analyzed these resources because they are two of the most widespread technologies on shop floors and because they are closely related to the fulfilment of customer demands (Sartal et al. 2017a, 260). The technology-enabled capabilities of both resources can therefore constitute powerful mechanisms through which lean routines can further contribute to manufacturing efficiency. In addition, because of their nature, they can help meet changing client requirements and overcome concerns (e.g., environmental challenges) for which lean principles seem to have a weakness (Sartal et al. 2017a, 260).

On the one hand, although the principles of IT and lean manufacturing have long been seen as mutually exclusive (Piszczalski 2000, 26), authors have increasingly claimed that they are interdependent and complementary (Riezebos et al. 2009, 237). Although lean practices can be simply but adequately conducted without using IT, recent academic and business evidence has indicated the increasing importance of technology in managing today's huge amounts of real-time data, as well as in enhancing a firm's absorptive capacity to respond to variations in client requirements (Martínez-Senra et al. 2015, 205). Scholars have increasingly considered IT resources to be decisive in leveraging organizational practices, which then leads to further improvements in industrial performance (Moyano-Fuentes et al. 2012, 108). Similarly, many lean manufacturers have begun adopting IT to support daily process management and new product development so as to respond with greater speed and flexibility to changes in customer demands (Ghobakhloo and Hong 2014, 5371).

On the other hand, social awareness of global warming, and of water and land contamination, also makes a difference in how operations should be run. According to a recent Nielsen survey, 55% of consumers will pay extra for products and services from companies that are committed to pursuing positive social and environmental impacts. For instance, in the design of tense flows inspired by JIT, it is important to consider how these flows will affect CO2 emissions. Shorter product life cycles and sudden changes in preferences, analogously, can influence how product design, industrialization, and logistics affect a firm's environmental impact. The issue at hand is that the efficiency and sustainability approaches to operations are not always aligned, so additional resources can be required to achieve green goals. CT thus comes into play as a valuable resource that helps develop and complement lean initiatives. The CT approach to waste and energy reduction, together with lean principles and tools, should enable firms to find new opportunities for waste elimination and thereby improve their industrial performance. However, despite the enormous development of lean and green topics in recent years, the absence of studies that analyze the relationships among lean principles, CT, and industrial performance is striking (Garza-Reyes 2015, 27).

4. INFORMATION TECHNOLOGY AS A DRIVER OF CHANGE

Although the a priori the introduction of IT in business seems perfectly aligned with the objectives of operational excellence, the early authors in this area had doubts as to how their research would affect the core role of the human factor; this is still holding back IT's mass deployment in lean environments. Toyoda (1983, 173) affirmed for instance that, if it were possible to receive a huge amount of information by pressing a button, this could damage the workers' capacity to think. This corroborated Toyoda's idea that, when wrongly used, IT would damage workers' problem-solving skills.

If this reflection was pertinent in the early 1980s, today it become a critical factor in any company's strategy because the risk of excess information has increased, meaning that new sources of waste have opened up. However, we stress that IT's benefits and ease of use have evolved and that the digitalization of society has facilitated IT's progressive incorporation into industrial environments. Perhaps for this reason, even though IT and lean principles have traditionally been considered antagonistic, authors have gradually moved toward positions that favor, not only their joint deployment, but also their complementarity and interdependence (Moyano Fuentes et al. 2012, 108).

The initial group of analysts - including the abovementioned classic authors - argued that IT, without a management philosophy to support it, is no more than a set of tools, meaning that it is doomed to fail because it only generates value for itself. However, a second, growing group of authors has defended both the complementarity and interdependence of these resources. Thus, for IT projects, lean manufacturing could be considered a necessary prior stage that avoids the muda and that leads to efficient and successful results. Finally, in a third position, halfway between the others, authors have defended the use of IT in lean projects but have argued that the use of this resource should be limited "only to reliable and totally proven technology," relegating IT to "serving the processes" (Liker 2004, 237).

However, apart from these various trends, the three major positions agree that failure is likely when either lean practices or IT is applied in isolation. In the case of IT, for example, authors have found that investment in technology is an important factor in optimizing firms' internal processes and in generating profits; however, if IT is implemented alone, it is not be sufficient to improve competitiveness. The other set of factors is necessary in improving efficiency, obtaining better business results, and generating a competitive advantage. Here, we use the productivity paradox to describe the difficulty of translating the execution of such projects into a real improvement in productivity.

Similarly, when lean transformations are carried out without any type of support, authors have found that only 10% of them actually achieve the expected results. The conclusion is clear: Although technology cannot be an end in itself (because it hinders the capacity to think), it is also impossible for lean tools to efficiently collect and interpret the huge amount of data required today. Using only lean tools would lead to inefficiency - which goes against the lean philosophy - because it would concentrate efforts on collecting data rather than on analyzing them - which is how value is generated.

In relation to the Industry 4.0 concept, IT can therefore serve as a necessary catalyst in the successful implementation of various lean programs and in ensuring their sustainability. The advantages of IT are because of its intrinsic characteristics - which increase the amount and speed of managed data, the reliability of real-time information, and the capacity for integration. IT can transform lean routines, giving rise to a new type of evolution in the production system that will be much more efficient and flexible, as well as more adaptable to fast-changing environments. From the philosophical and practical points of view, lean thinking and IT pursue the same objective: improving business results. Moyano Fuentes et al. (2012, 107) empirically showed that there is a direct link between the level at which internal IT is used and the degree of implementation of lean initiatives.

Lean shop floors, therefore, can be considered ideal for the adoption and deployment of IT, with a view to continuing to increase the efficiency of processes that might have been left behind with traditional lean routines. In fact, many authors have described IT solutions that arose on the shop floor from the use of various lean tools but that were insufficient because of new requirements for flexibility and agility. Kanban cards, Heijunka boards, and visual inventory management, among others, become unmanageable when there is a huge amount of data to be managed on a daily basis; this is why information processing, management, and real-time analysis using IT are essential.

Kotani (2008, 5790), for example, described the appearance of electronic Kanban (even within Toyota), expanding upon traditional functions to include real-time monitoring and performance indicators. IT extends throughout the plant, and inventory monitoring applications based on Radio Frequency Identification (RFID) Systems can allow for efficient inventory management or can be used for statistical process control in support of total quality management programs (Sánchez-Rodríguez et al. 2006, 487).

Other IT systems are present across an entire manufacturing plant. For example, Sartal et al. (2008, 76) described Manufacturing Execution Systems (MES) and Overall Equipment Effectiveness (OEE) as the ideal solutions for standardizing and monitoring lean practices. Similarly, GMAO systems allow for real-time monitoring of equipment, serving as the basis for total productive maintenance programs (Riezebos et al. 2009, 243). Finally, Powell (2013, 1490) described ERP (Enterprise Resource Planning) systems as ideal for transferring the results of lean practices from the plant to the business area.

Therefore, it seems obvious that, even though (from a conceptual point of view) lean practices can be adopted without the use of IT, current market determinants require organizations, at least in the short term, to apply a higher-level Industry 4.0 technology if they aim to become leaner. The actual lean initiatives themselves can even reveal the need for more, and better, information. However, this also works the other way around. Bruun and Mefford (2004, 248), for example, found that the Internet and e-commerce facilitate the development of lean practices. Similarly, Moyano Fuentes et al. (2012, 113) empirically showed that there is a direct relation between the degree to which internal IT is used and the degree to which lean production is adopted. Ward and Zhou (2006, 181) also found that, when lean practices are adopted, ERP systems have improved effects on certain results (in this case, by reducing lead time).

The existence of an appropriate IT infrastructure therefore can promote the appearance of synergies between business units, but only when the organization makes effective use of IT characteristics - increasing the amount and speed of data, the reliability of real-time information, and the capacity for integration. Thus, there is a symbiotic relationship between lean manufacturing and IT. Each continues the path that the other begins regarding cost and time reduction, conflict resolution, and other organizational functions that are directly involved in the production process.

5. CLEANER TECHNOLOGIES AS A DRIVER OF CHANGE

In a world characterized by a shortage of energy and resources, new production strategies should embrace both the search for economic efficiency (lean) and sustainability of operations (green).

However, the industrial sector has traditionally seen an important trade-off between environmental improvements and economic development. In fact, it was only in 1987, with the Brundtland Report, that the need for sustainable development was first considered to meet present needs without compromising future needs. This report and subsequent treaties[1] to fight climate change awoke environmental awareness among nations; consumers started to demand environmentally friendly products, and firms started to see new business opportunities.

Lean manufacturing, which did not initially include the idea of protecting the environment, is now perceived, based on one of its principles (the "systematic elimination of waste") as the paradigm that could internalize "green considerations" and balance the trade-off between firm performance and sustainable development. Although we understand the appropriateness of lean routines, we stress that they need to evolve and adopt the resources provided by the Industry 4.0 revolution in order to adapt to today's environment.

There is still no consensus in the literature on this. The minority position questions the lean-green complementarity based on possible trade-offs that might arise between environmental goals and the ongoing search for efficiency of lean programs (Dües et al., 2013, 98). For example, Rothenberg et al. (2001, 231) showed in their case studies that the lean cost reduction goals are not necessarily more environmentally friendly. Similarly, Dues et al. (2013, 98) described CO_2 emissions in the supply chain as the main lean-green conflict. In this area, in addition to combining the two paradigms, it seems necessary to develop more efficient means of transport in order to reduce emissions. Although these studies were very useful for explaining the lean-green debate, they have an important weakness: They evaluate the "greenness" of certain lean practices (JIT or batch reduction, among others) in isolation, assuming that the results can be generalized without evaluating the lean conceptual framework.

---

[1] The latest Climate Summit took place in September 2014 in New York.

Following the main positive trend (e.g., Yang et al. 2011, 258), we believe that even though the two paradigms (lean and green) start out from different objectives for the elimination of waste, because they focus on "similar waste" - inventory, transport, by-products, pollutants, etc. - they end up converging. Therefore, "green behavior" can be seen as a natural extension of lean thinking. In fact, lean principles often reveal environmental needs. Several studies have confirmed that firms that adopt lean management also show greener behavior, intensifying environmental practices, adopting standards (ISO1400, EMAS, etc.), or adopting CT.

If we consider the pillars of the TPS - JIT, jidoka, and RfP and follow research such as, for example, Yang et al. (2011, 258), we find that the "concern for greenness" appears in each of them. In the case of JIT (e.g., zero-buffer principles, "no work-in-process") and jidoka (e.g., "zero defects," self-inspecting, quality circles), we find that both aspects are directly related to organizational efforts to reduce environmental waste. Similarly, in terms of RfP, worker participation (through programs such as staff training, regular individual assessment interviews, and financial incentives) leads to greater worker engagement in the organization and creates the ideal scenario for the adoption of environmental technologies and practices.

Therefore, experiences with lean manufacturing encourage organizations to adopt cleaner technologies and practices. We believe that environmental considerations could be added immediately to the basic principles. However, the lean and green principles could also be promoted by incorporating the best available technologies (BATs) and replacing current equipment with more efficient equipment. Some green technologies have proven to be successful, such as high-efficiency pumps, electric motors with speed regulation, or energy recovery systems. However, environmental management involves many other tools that could amount to new sources of opportunities for lean-green enhancement. Life-cycle assessment and sustainable Value Stream Mapping (VSM) are two clear examples of tools that, when adopted in combination, allow for a holistic approach and for the consideration of several forms of lean and green wastes at the same time.

Therefore, although from a conceptual viewpoint lean principles and practices can be adopted separately from IT, the more lean practices there are in an organization, the more it will be possible to adopt technologies of this type, and vice versa. CT can build bridges between lean manufacturing systems and green systems because many of the environmental programs and reasons for using such technologies are based on lean thinking. Moreover, CT can be considered a powerful mechanism allowing lean practices to make a stronger contribution to production competitiveness in mature or even over-mature lean environments. The fact that CT makes it possible to go one step further in lean environments that have already gone as far as they can might allow managers to discover new opportunities for improvement because of the reduction in search costs and marginal costs. And increasing, therefore, the possibility that this type of initiative will be adopted and the results will be maximized.

6. FINAL REFLECTIONS

In this section, some of the conceptual bases of new industrial businesses are considered, bearing in mind customer demands in terms of customization, response times, and greater environmental concerns. It is not only a matter of cleverly combining autonomous machines and efficient processes. We aim to consider some of the characteristics of an efficient, sustainable firm in which the introduction of advanced technologies and efficient organization of industrial processes are as important as social and environmental considerations.

All these ideas can be summarized in the following question: What are the challenges faced by lean thinking in view of the new requirements regarding the environment, and in what direction should lean routines evolve?

Just as Fordism continued to use the bases of Taylorism and lean thinking started out by questioning the foundations of Fordism but retaining its strong points; this new management paradigm that we pursue, should also start from the permanent knowledge that has been gathered on management (Taylorism, Fordism, and lean thinking) and from the new requirements and opportunities offered by the technological environment. While the lean principles are considered ideal for 1) inclusion within the plant's information systems and 2) balancing the trade-off between efficiency and sustainability, we stress that lean routines should be adapted to the changing conditions of the environment, thus generating a sustainable competitive advantage (Figure 2).

Therefore, IT and CT seem to have become new challenges to be internalized in management models, acting as drivers for change that will enable old lean routines to be transformed, generating new capabilities that adapt and respond efficiently to current requirements of the environment. Exclusive consideration of economic efficiency should be complemented by sustainability of operations, and adaptation for survival should take place quickly.

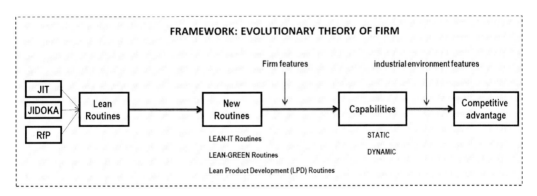

Figure 2. Theoretical framework from an evolutionary viewpoint (Own elaboration from the literature review).

Even though today, lean manufacturing is considered the orthodoxy in operations management, its principles and tools seem to be suffering from increasing weakness with respect to current trends: more demanding customers in terms of customization and response times, and greater environmental concern. In developed countries especially, rising standards of living have forced firms to go beyond the deployment of management philosophies based exclusively on efficiency. Sudden changes in preferences, shorter product life cycles, and environmental challenges now have to be considered. If we question this in lean manufacturing, we see that it needs to explicitly internalize in its principles some of the resources that are traditionally associated with the Industry 4.0 transformation, such as IT and green technologies. Starting from the principles of the TPS, the basis of lean and permanent knowledge, we consider it essential to propose alternatives based on what is permanent in organizational phenomena in order to overcome any limitations of lean thinking in this area. We evaluate how lean principles make it possible to establish the right conditions for developing technology-intensive Industry 4.0 environments, as well as how shop-floor technologies can subsequently be leveraged to enhance the contribution of lean practices to industrial performance.

It is thus possible to see how IT can become an important asset for improving business performance, or how new CT and lean manufacturing can be complementary or even synergic. There is no reason the new Industry 4.0 technologies should negatively affect business performance. In fact, they can help preserve the goal of efficiency of lean manufacturing and serve as a support in situations in which the lean principles seem weak. Because of their intrinsic characteristics, both IT and CT can generate other types of unique, competitively valuable organizational capabilities that are exclusive to them, becoming a source of competitive advantage.

IT can be considered essential for competing in most industries. While supporting other organizational resources (lean routines), it will develop its full potential, becoming especially useful in environments with the most variable requirements and therefore also the lowest volume, in line with its own characteristics: an increase in the quantity and speed of data, reliability, in-time information, and capacity for integration. However, implementing it merely as a strictly technological challenge will not be sufficient for achieving a more competitive position. Its complementarity with other resources in the firm and with its organizational processes must be considered. Otherwise, investment in IT will end up being no more than expenditure that brings in no returns.

Another key aspect of IT, which is sometimes forgotten, is that it improves horizontal and vertical communication within the organization. While lean practices imply a certain empowerment in that they grant the worker greater responsibility and decision-making capacity, they also make the actual work more stressful. Therefore, the introduction of various IT systems (MES, ERP, RFID, etc.) has an immediate effect on the reduction of information asymmetries (both horizontal and vertical) among workers, stimulating departmental cooperation and helping to make the various links in the value chain more fluid. In addition, the integration of ERP and MES can encourage the introduction or improvement of other functionalities, such as alarm management or product traceability, and can lead to better understanding of the expected benefits. This is a key aspect for creating a positive perception of the initiative and for maximizing the possibilities of success.

CT should be focused on reducing waste so that new opportunities for elimination of waste can be discovered and exploited, leading to improved productivity and improving the plant's preparedness for environmental initiatives, thus balancing the conflict between firm performance and sustainable development. Thus, there is no reason green technologies should negatively influence business performance. They may also serve as a support in situations in which the lean principles seem weak, generating other types of unique and valuable organizational capabilities (exclusive of this type of resources) as a source of competitive advantage.

This should lead professionals to consider the integration of both systems, but as with IT, efforts should be focused on cases in which success is maximized because not all the lean pillars will necessarily show this positive behavior. For example, JIT practices may lead to severe incompatibilities between environmental requirements and the ongoing search for efficiency (Sartal et al. 2018b). This is what happens, for example, when supply chain synchronization intensifies CO2 emissions, or when batch size reduction requires more frequent product changes with more frequent maintenance on production equipment and the creation of waste (Dues et al. 2012). Therefore, in view of the need to optimize work teams and technical resources to work on projects for improvement, priority should be placed on technologies or practices that are complementary, and alternatives should be sought where conflicts exist.

The reflections in this chapter might also be of interest for those responsible for public policies because they consider some of the conflicts that arise between an industry's efficiency requirements and the environmental interests of institutions.

As stated above, IT and CT may make it possible to generate unique and competitively valuable organizational capabilities that will be converted into competitive advantages. However, it is also true that the capabilities generated by these resources in support of lean principles do not necessarily have to be in line with market requirements and may therefore not lead to sustainable competitive advantages for the firm. Thus, provided that it is of interest for society, public policy designers can intervene by introducing specific policies allowing entrepreneurs to use such technologies or directly supporting projects that might have a positive impact on the organization's results.

Public administrations wishing to increase the competitiveness of certain strategic industrial sectors therefore need to develop policies that will help overcome the structural inertia that is inherent in the adoption of lean projects, especially in the presence of new technologies. Obviously, we are not suggesting that public policies should aim only to promote research in lean-IT-green, but, considering that public resources are scarce, it is important for public aid and incentives to focus on relevant projects (which otherwise might not go ahead) so that the returns on investment in regions can be maximized and sustainable competitive advantages generated.

REFERENCES

Arfmann, David, and G. Topolansky Barbe. 2014. "The Value of Lean in the Service Sector: a Critique of Theory & Practice." *International Journal of Business and Social Sciences* 5, no. 2: 2-18.

Bhasin, Sanjay, and Peter Burcher. 2006. "Lean viewed as a philosophy." *Journal of Manufacturing Technology Management* 17, no. 1: 56-72.

Bruun, Peter, and Robert N. Mefford. 2004. "Lean production and the Internet." *International Journal of Production Economics* 89, no. 3: 247-60.

Carvalho, Helena, and V. Cruz-Machado. 2011. "Integrating Lean, Agile, Resilience and Green Paradigms (LARG_SCM)." In Supply Chain Management, edited by P. Li, 27-36.

Dubey, Rameshwar, and Angappa Gunasekaran. 2015. "Agile manufacturing: framework and its empirical validation." *The International Journal of Advanced Manufacturing Technology* 76, no. 9-12: 2147-2157.

Dües, Christina Maria, Kim Hua Tan, and Ming Lim. 2013. "Green as the new Lean: how to use Lean practices as a catalyst to greening your supply chain." *Journal of Cleaner Production* 40: 93-100.

Emiliani, Bob. 2008. *The Equally Important "Respect for People" Principle Real Lean: The Keys to Sustaining Lean Management*. Vol. 3. The CLBM, LLC, Wethersfield, CT, available at: www. bobemiliani. com/goodies/respect_for_people. pdf (accessed August 15, 2017).

Garza-Reyes, Jose Arturo. 2015. "Lean and green-a systematic review of the state of the art literature." *Journal of Cleaner Production* 102: 18-29.

Ghobakhloo, Morteza, and Tang Sai Hong. 2014. "IT investments and business performance improvement: the mediating role of lean manufacturing implementation." *International Journal of Production Research* 52, no. 18: 5367-5384.

Groba, Carlos, Antonio Sartal, and Xosé H. Vázquez. 2018. "Integrating forecasting in metaheuristic methods to solve dynamic routing problems: Evidence from the logistic processes of tuna vessels." *Engineering Applications of Artificial Intelligence* 76: 55-66.

Kotani, Shigenori. 2008. "Optimal method for changing the number of kanbans in the e-Kanban system and its applications." *International Journal of Production Research* 45, no. 24: 5789-809.

Liker, Jeffrey K. 2004. *The Toyota Way: 14 Management Principles from the World's Greatest Manufacturer*. Madrid: McGraw Hill.

Martínez-Senra, Ana Isabel, Quintás, María de los Angeles, Sartal, Antonio, & Vázquez, Xosé Henrique. 2015. How can firms' basic research turn into product innovation? The role of absorptive capacity and industry appropriability. *IEEE Transactions on Engineering Management* 62, no 2, 205-216.

Moyano-Fuentes, José, Pedro Martínez-Jurado, Jose Manuel Maqueira Marín, and Sebastian Bruque Cámara. 2012. "El papel de las tecnologías de la información y las comunicaciones (TIC) en la búsqueda de la eficiencia: un análisis desde Lean Production y la integración electrónica de la cadena de suministro." ["The role of information and communication technologies (ICT) in the search for efficiency: an analysis from Lean Production and the electronic integration of the supply chain."] *Cuadernos de Economía y Dirección de la Empresa* 15, no. 3: 105-16.

Moyano-Fuentes, José and Macarena Sacristán- Díaz. 2012. "Learning on lean: a review of thinking and research." *International Journal of Operations & Production Management* 32, no. 5: 551-82.

Ohno, Taichi. 1988. *Toyota Production System: Beyond Large-Scale Production*. New York: Productivity Press.

Piszczalski, Melville. 2000. "Lean vs. information systems." *Automotive Manufacturing and Production* 112, no. 8: 26-8.

Powell, Daryl. 2013. "ERP systems in lean production: new insights from a review of lean and ERP literature." *International Journal of Operations & Production Management* 33, no. 11/12: 1490-1510.

Quintás, María A., Martínez-Senra, Ana I., & Sartal, Antonio. 2018. The Role of SMEs' Green Business Models in the Transition to a Low-Carbon Economy: Differences in Their Design and Degree of Adoption Stemming from Business Size. *Sustainability* 10, no 6: 1-20.

Riezebos, Jan, Warse Klingenberg, and Christian Hicks. 2009. "Lean Production and information technology: Connection or contradiction?" *Computers in Industry* 60, no. 4: 237-47.

Rothenberg, Sandra, Frits K. Pil, and James Maxwell. 2001. "Lean, green, and the quest for superior environmental performance." *Production and Operations Management* 10, no. 3: 228-43.

Sánchez-Rodríguez, Cristóbal, Frank W. Dewhurst, and Angel Rafael Martínez-Lorente. 2006. "IT use in supporting TQM initiatives: an empirical investigation." *International Journal of Operations and Production Management* 26, no. 5: 486-504.

Sartal, Antonio, Carou, Diego, Dorado-Vicente, Raul, & Mandayo, Lorenzo. 2018a. Facing the challenges of the food industry: Might additive manufacturing be the answer?. *Proceedings of the Institution of Mechanical Engineers, Part B: Journal of Engineering Manufacture,* 1-5.

Sartal, Antonio, González, Miguel Fernando, & Quiroga, José Ignacio A. 2008. OEE: una propuesta para definir la productividad real de planta. *Automática e instrumentación,* no 396: 76-80.

Sartal, Antonio, Llach, Josep, Vázquez, Xosé Henrique, & de Castro, Rodolfo. 2017a. How much does Lean Manufacturing need environmental and information technologies. *Journal of Manufacturing Systems* 45, no 1: 260-272.

Sartal, Antonio, Martinez-Senra, Ana Isabel, & Cruz-Machado, Virgilio. 2018b. Are all lean principles equally eco-friendly? A panel data study. *Journal of Cleaner Production*, 177: 362-370.

Sartal, Antonio, Martínez-Senra, Ana Isabel, & García, José Manuel. 2017b. Balancing Offshoring and Agility in the Apparel Industry: Lessons from Benetton and Inditex. Fibres & Textiles in Eastern Europe 3: 4.

Sartal, Antonio, & Vázquez, Xosé Henrique. 2017. Implementing Information technologies and Operational Excellence: planning, emergence and randomness in the survival of adaptive manufacturing systems. *Journal of Manufacturing Systems* 45, no 1: 1-16.

Shah, Rachna, and Peter T. Ward. 2007. "Defining and developing measures of lean production." *Journal of Operations Management* 25, no. 4: 785-805.

Sheridan, Jhon. H. 2000. "'Lean Sigma' synergy." *Industry Week* 249, no. 17: 81-8.

Sugimori, Y., K. Kusunoki, F. Cho, and S. Uchikawa. 1977. "Toyota production system and kanban system materialization of just-in-time and respect-for-human system." *The International Journal of Production Research* 15, no. 6: 553-64.

Toyoda, Eiji. 1983. *Creativity, Challenge and Courage, Toyota Motor Corporation* (cited in Liker, Jeffrey K. 2004).

Treacy, Michael, and Fred Wiersema. 1993. "Customer intimacy and other value disciplines." *Harvard Business Review* 71, no. 1: 84-93.

Vazquez, Xosé H., Sartal, Antonio, & Lozano-Lozano, Luis M. 2016. Watch the working capital of tier-two suppliers: a financial perspective of supply chain collaboration in the automotive industry. Supply Chain Management: *An International Journal* 21, no 3: 321-333.

Ward, Peter, and Honggeng Zhou. 2006. "Impact of information technology integration and lean/ just-in-time practices on lead-time performance." *Decision Sciences* 37, no. 2: 177-203.

Womack, Daniel T. Jones, Daniel Roos. 1990. *The Machine That Changed the World*. New York: MacMillan/Rawson Associates.

Yang, Ma Ga Mark, Paul Hong, and Sachin B. Modi. 2011. "Impact of lean manufacturing and environmental management on Industrial performance: An empirical study of manufacturing firms." *International Journal of Production Economics* 129, no. 2: 251-61.

In: Lean Manufacturing
Editors: F. J. G. Silva and L. Carlos Pinto Ferreira
ISBN: 978-1-53615-725-3
© 2019 Nova Science Publishers, Inc.

Chapter 17

LEAN HEALTHCARE IN A CANCER CHEMOTHERAPY UNIT: IMPLEMENTATION AND RESULTS

T. M. Bertani[1], A. F. Rentes[1],
M. Godinho Filho[2],* and R. Mardegan[3]

[1]EESC - University of São Paulo, São Paulo, Brazil
[2]DEP - Federal University of São Carlos, São Carlos, Brazil
[3]Hominiss Consulting, São Carlos, Brazil

ABSTRACT

Lean manufacturing is derived from the Toyota Production System (TPS) and focuses on the identification and combating of wastes that pertain to production with a series of tools to make lead time reductions possible and permit greater flexibility. Although the lean philosophy was developed and continues to be used in manufacturing, Womack et al. (2005) defends the idea that the applications of lean thinking are much broader. One of such application is healthcare. Lean healthcare appears to be an effective solution to generate improvements in healthcare organizations. This book chapter presents an action research to show the implementation and results from the application of the lean healthcare concept to a stream of patients who were undergoing chemotherapy at a hospital located in Brazil. The lean healthcare implementation method used in the studied hospital was based on the DMAIC method. The implementations led to the following improvements in the studied hospital: an increase of 24% in the number of patients served in triage, a reduction of up to reductions of up to 42% (from 65 days to 38 days) in the average patient lead time (time between triage and patient treatment), a reduction of up to 90% in the exam results lead time, and an increase of up to 23% in productivity (patients served) in chemotherapy. Additionally, the financial area was improved, with an approximately 33% increase in billing. It is believed that the application shown in this chapter can serve as motivation and reference for other hospitals to implement lean healthcare in their chemotherapy operations.

Keywords: lean healthcare, cancer chemotherapy, implementation

---

* Corresponding Author's E-mail: moacir@dep.ufscar.br.

1. INTRODUCTION

Lean manufacturing is derived from the Toyota Production System (TPS) and focuses on the identification and combating of wastes that pertain to production with a series of tools to make lead time reductions possible and permit greater flexibility. Although the lean philosophy was developed and continues to be used (see for example Liker and Morgan, 2011)) in manufacturing, Womack et al. (2005) defend the idea that the applications of lean thinking are much broader:

> [...] "Lean thought is not a manufacturing tactic or a cost reduction program, but rather a management strategy that is applicable to all organizations because it involves the improvement of processes. All organizations - including health sector organizations - are composed of a series of processes, or sets of actions for the creation of value for those that use or depend on them (clients/patients)."

Actually, a lot of operations management paradigms, such as Theory of Constraints and Total Quality Management are being implemented in services industries (for example Reid, 2007 and Yasin et al, 2004). Lean Manufacturing is one of such theory and has begun to draw the attention of the goods and services sector. Examples of this trend are increasingly common and can be observed in civil construction, credit businesses, insurers, training processes, pharmaceutical industries, and hospitals, among others.

Considerable attention has been given to the healthcare area because, according to Graban (2011), hospitals around the world are suffering under new market demands. The hospitals' costs are increasing—often at faster rates than revenues—and, with increasing frequency, patients are harmed or killed due to preventable errors. Today, operations and Supply chain management researchers are giving a great emphasis to healthcare field. For example, see the work of Modig and Ahlstrom (2012) and Su et al. (2012). For a review about operations and supply chain management research in healthcare, see Dobrzykowski, Saboori Deilami, and Kim (2014).

According to Souza (2009), lean healthcare appears to be an effective solution with which to generate improvements in healthcare organizations. Many cases have been reported in the literature with respect to the application of lean healthcare in healthcare organizations in the United States. ThedaCare Inc. of Wisconsin can be cited as an example; this organization developed the ThedaCare Improvement System (TIS) and, by reducing costs and increasing productivity, reported gains on the order of US$10 million per year (Womack et al. 2005). Similar to ThedaCare (USA), the Virginia Mason Medical Center of Seattle (USA) created the Virginia Mason Production System (VMPS), which is based on lean concepts. Relevant studies about lean health care have also emerged from the United Kingdom, Australia, and Canada, such as the study by Souza and Archibald (2008).

According to Panchak (2004), manufacturing executives are also interested in the application of lean concepts to healthcare. For example, [Companies] "have to demand this, because you're paying insurance premiums, paying when errors to your staff [keep them out of work longer than necessary]..." To the author, even small delays in medical consultation generate costs for companies: "To see a doctor for seven minutes, it's common to wait 15 or 20 minutes -- that's not care, that's all non-value-added waste".

Regarding the Brazilian reality, the report titled "Brazilian Hospital Performance" (2006) (La Forgia and Couttolenc 2008), created by specialists from the World Bank (Bird), disapproved both, public and private Brazilian hospitals. According to the report, the Brazilian hospital network is inefficient and uses resources poorly, thus increasing hospital costs. The authors defend the idea that the healthcare sector management model must undergo profound reforms. Guandalini and Borsato (2008) cite data about the cost increases in Brazilian healthcare in recent decades; for example, the cost for a day of ICU care increased by more than 90% in the last 10 years, waste from the Brazilian healthcare system increased from 5% to 10% of the gross domestic product (GDP), and medicine prices increased by an average of more than 170% over the last 10 years.

Our paper is motivated by the following research question: "How lean healthcare can be implemented to improve the patients flow in a cancer chemotherapy unit in a Brazilian hospital?"

Therefore, the main goal of our paper is presenting an action research to show the implementation and results from the application of the lean healthcare concept to a stream of patients who were undergoing chemotherapy at a hospital located in the state of São Paulo. Action research is an empirically based research method that is conceived and performed in strict association with an action or the resolution of a collective problem and in which the researchers and representative participants of the situation or problem are involved in a cooperative or participative manner (Thiollent 1997).

Although the literature on lean healthcare has been increasing a lot in the last years, it lacks research presenting implementation studies, especially in the chemotherapy sector (Costa and Godinho Filho, 2016). Our study aims to contribute to fill out this gap.

The study structure is as follows: the theoretical references that form the basis of the study are presented in section 2, the action research itself is presented in section 3, and the study conclusions are presented in section 4.

2. Literature Review

2.1. Lean Healthcare: Fundamental Concepts

The philosophy of lean healthcare is supported by a set of concepts, techniques, and tools that improve hospital organization and management (Graban 2011). Lean thinking is an approach derived from methods that were developed in the manufacturing sector in an attempt to increase client satisfaction through better uses of resources. Successful manufacturers such as the Toyota Motor Corporation are concerned with the efficient and effective production of a diversified range of high quality cars or other goods in large volumes. Similar to the automotive industry, the healthcare sector also faces challenges to meet elevated demand and thus requires speed and flexibility with elevated safety and quality standards (Ben-Tovim et al. 2008).

Drawing on various hypotheses, Womack et al. (2005) report that

> [...] "Lean management is not a new concept, but it is relatively new to the health sector. While skeptics are right when they say, "Patients are not cars", medical care is, in fact, delivered in extraordinarily complex organizations, with thousands of interacting

processes, much like the manufacturing industry. Many aspects of the Toyota Production System and other lean tools therefore can and do apply to the processes of delivering care."

According to Souza (2009), although some publications in the 1990s considered the possibility of applying some lean concepts in the healthcare industry, including a report by Heinbuch (1995), none of them characterized lean thought as a methodology in their studies. According to the author, the first studies that made consistent speculations about the use of lean philosophy as an approach to improve the healthcare services sector were those published by the British National Health Service (NHS) in 2001 and 2002.

The main characteristic of lean thought is the reduction/elimination of wastes whenever possible. In the healthcare industry, this is a delicate subject because the industry's activities are centered on people, both on those who provide services such as nurses, doctors and other employees, as well as those who receive services, such as patients. In this sense, Aherne and Whelton (2010) emphasize that the objective of lean philosophy is not to eliminate or reduce the number of people, but rather to tackle the wastes, according to the principles developed by Toyota.

Silberstein (2006) emphasizes this aspect by highlighting the importance of training involved professionals to identify the wastes that develop in the processes in order to continuously improve their work. One example given by the author is a nurse who spends part of her time looking for medications; she does this to serve the patient and for this reason might not see this activity as a waste. However, if the medication is in the correct location, the nurse can spend more time with the patient because the latter is an activity that adds value.

Some important points can be emphasized by the following 5 principles of lean thought, proposed by Womack and Jones (1996), when applied to the health care industry. These are to:

1. *Specify what is valued.* According to Aherne and Whelton (2010), value in healthcare can be defined as anything that improves the patient's mental or physical state. According to the authors, patients usually value fast treatment and organized and orderly processes.
2. *Identify the value stream.* By considering the patients as "products" in a hospital who arrive at the establishment with some problem and expect to receive appropriate treatment to resolve the problem, the value stream in this type of organization can be characterized as the sequences of events that comprise the patients' journey within (Aherne and Whelton 2010).
3. *Establish a continuous flow.* The ideal state in a hospital, according to Aherne and Whelton (2010), is achieved when flow is provided between the client's treatment stages in the smoothest manner possible so that the client receives the highest possible level of attention and quality without delays.
4. *Use pull logic.* The organization of activities in a hospital should be such that it provides services for patients' needs at the moment that they occur. One example of this is the planning and execution of tests on samples according to the need for patient treatment, rather than accumulating many tests to achieve the maximum efficiency of the machines that perform the tests (Aherne and Whelton 2010).

5. *Seek perfection.* This principle acts in healthcare services in the same manner as in other sectors. By establishing of the first 4 principles, it creates in the involved persons a positive feeling related to continuous improvement, without ever accepting the current state as ideal and always seeking perfection in the processes developed by the organization.

2.2. The Literature Related to Lean Healthcare

The literature related to the application of lean healthcare concepts in hospital environments has grown exponentially in the last decade. In this study, a literature review was conducted with respect to the application of lean healthcare concepts. To this end, the following sources were used: the University of São Paulo (USP) Library in Healthcare Sciences; Portal Emerald; Google Books; the ISI Web of Knowledge; IEEE Explore; the Brazilian Federal Agency for the Support and Evaluation of Graduate Education Coordenação de Aperfeiçoamento de Pessoal de Nível Superior (CAPES) Periodicals Portal in Health Sciences; Scielo; and Google Scholar. The keywords used in the study were lean healthcare, Healthcare Toyota, lean hospitals, lean healthcare methodology, Kaizen Healthcare, and TPS Healthcare. This literature review yielded 50 studies.

It could be concluded from an analysis of this literature review that the majority of the studies that described the implementation of lean healthcare accomplished this through small improvement cycles that were based on mapping the current situation, developing of the future situation, and generating an implementation plan for improvements. Implementations of improvements or kaizen events and employee involvement were also frequently cited. As mentioned by Rich and Piercy (2013), research on improvement in healthcare is extensive, but focuses on distinct characteristics of the system rather the system as a whole. The following main topics can be found in the literature.

1. *Definition of the client:* according to Womack and Jones (1996), the first principle of lean production is to determine what represents value to the client. However, in the hospital complex, defining the client might not be an easy task (Young and McClean 2008). According to Shah and Robinson (2008), in a hospital system, various users such as supporters and healthcare professionals can be identified, and each exhibits characteristics that can make them analogous to clients. However, Womack et al. (2005), Filingham (2007), and McGrath et al. (2008) state that it is extremely important to define value according to the primary client, the patient.
2. *Implementation structure:* Dickson et al. (2009), Bush (2007), Filingham (2007), Taninecz (2007), Nelson et al. (2009), and Trilling et al. (2010) cite the formation of ad-hoc teams to accomplish kaizen events. In addition to the teams formed for the timely implementation of improvements, Ben-Tovim et al. (2008) describe the formation of a fixed internal team to manage and direct efforts during the lean journey. McGrath et al. (2008) complement this by emphasizing the presence of a member of upper management on the team.
3. *Establishment of objectives and goals:* Langabeer et al. (2009) emphasize that the majority of hospitals do not clearly establish goals and objectives during the initial phases of the projects. Laursen, Gertsen and Johansen (2003), Womack et al. (2005),

Amirahmadi et al. (2007), and Trilling et al. (2010) emphasize the definition of clear objectives such as the starting point for introducing lean concepts into hospitals.

4. *Involvement of people:* Womack et al. (2005), Dickson et al. (2009), Taninecz (2007), Lefteroff and Graban (2008), Pexton (2007), and Graban (2009) highlight the importance of involving employees who act within the operation to successful implementation. Taninecz (2007), Lefteroff and Graban (2011), and McGrath et al. (2008) emphasize that the involvement of employees from other areas contributes to paradigm disruption. Although doctors are cited by Langabeer et al. (2009) as the group that is most resistant to changes, only Ben-Tovim et al. (2008) and McGrath et al. (2008) clearly cite the involvement of doctors when developing improvements. Womack et al. (2005), McGrath et al. (2008), and Trilling et al. (2010) describe the involvement of upper management as an essential factor for the success of lean healthcare projects.

5. *Training:* According to Womack et al. (2005), Amirahmadi et al. (2007), Ben-Tovim et al. (2008), and McGrath et al. (2008), employee training administered by professionals with lean experience is essential.

6. *Mapping of the current situation and development of the future situation:* the majority of implementations use value stream mapping (VSM) as a tool for mapping processes and identifying wastes.

7. *Implementation of improvements:* Dickson et al. (2009), Bush (2007), Fillingham (2007), Taninecz (2007), Nelson et al. (2009), Trilling et al. (2010) and Meredith et al. (2011) cite examples of lean healthcare implementations through kaizen events.

8. *Sustainability of implemented improvements:* Activity standardization is a practice commonly found in lean healthcare publications. In addition to the creation of standards, McGrath et al. (2008) highlight the importance of system maintenance by defining an owner for each stream worked, with the responsibility to frequently maintain and review the work.

9. *Continuous Improvement:* Kim et al. (2006) emphasize that once a future situation is implemented, it becomes the current situation that must be worked in order to achieve a new future situation. In texts from Womack et al. (2005), King, Ben-Tovim, and Bassham (2006), Rickard (2007), Taninecz (2007), Ben-Tovim et al. (2008), and McGrath et al. (2008), the culture of continuous improvement is evident in the cyclical forms of the methods employed.

The methods used for implementation are predominantly based on the plan-do-check-act (PDCA) or define-measure-analyze-improve-control (DMAIC) methodologies. Despite this, particularities are observed for each implementation. Apparently, there is no single standard method for introducing lean concepts in hospitals.

Table 1 shows countries where lean healthcare applications can be found. As shown in this table, the majority of studies were performed in the United States. Table 2 shows that the application of lean healthcare concepts can be observed in the most diverse healthcare sectors. This table shows 18 healthcare areas (including hospital as a whole) in which lean healthcare studies have already been conducted.

Table 1. Countries where lean healthcare studies were conducted

Country	References	Number of Publications
United States	Hinckley (2003), Panchak (2004), Jimmerson et al. (2005), Womack et al. (2005), Culbertson (2005), Dickson et al. (2009), Kim et al. (2006), Bush (2007), Amirahmadi et al. (2007), Rickard and Mustapha (2007), Heidger, Lambert and Mourshed (2008), Zarbo and Angelo (2007), Cottington (2008), Pexton et al. (2007), Graban (2011), Langabeer et al. (2009), Tucker (2009), Mazur and Chen (2009), Gastineau et al. (2009), Nelson et al. (2009), Al-Araidah et al. (2010) and Graban (2011)	22
United Kingdom	Jones (2006), Fillingham (2007), Bamford and Griffin (2008), Souza (2009), Castro, Dorgan and Richardson (2008), Proudlove, Moxham and Boaden (2008), Shah and Robinson (2008), Castle and Harvey (2009), Papadopoulos, Radnor, and Merali (2011)	9
Brazil	Figueiredo (2006), Silberstein (2006), Zanchet, Saurin and Missel (2007), Costa (2008), Ferro (2009) and Selau et al. (2009), Bertani (2012), Godinho Filho et al. (2015), Henrique et al. (2016)	9
Australia	King et al. (2006), Young and Mcclean (2008), Ben-Tovim et al. (2008) and McGrath et al. (2008)	4
France	Ballé and Régnier (2007) and Trilling et al. (2010)	2
Canada	Taninecz (2007), Lefteroff and Graban (2008)	2
Sweden	Trägårdh and Lindberg (2004) and Larsson et al. (2012)	2
Holland	Pieters et al. (2010)	1
Denmark	Laursen, Gertsen and Johansen (2003)	1
New Zealand	Oldeshaw (2009)	1

Lean practices and tools are broadly employed in manufacturing. In this study, the practices and tools employed in lean healthcare were surveyed. This survey is found in Table 3. Activity standardization is the most common practice in the recorded texts. As the second most studied tool, VSM appears to be a powerful instrument in the hospital environment.

In total, 14 tools were found, thus demonstrating that lean tools can be broadly employed in hospitals. Regarding the relationship between Hospital Areas and lean healthcare tools, it can be inferred that:

- 5S, Kaizen Events, VSM, and Standard Work are used in the most diverse hospital areas;
- The Emergency Room and Laboratory are the hospital areas with the largest number of applied tools; and
- Layout changes are found predominantly in the Laboratories.

For a complete literature review about lean healthcare, see Costa and Godinho Filho (2016).

Table 2. Hospital areas in which lean healthcare studies were performed

Area	References	Number of Publications
Emergency Room	Dickson et al. (2006), Figueiredo (2006), Filingham (2007, King et al. (2006), Silberstein (2006), Taininecz (2007), Bush (2007), Bem-Tovim (2008), Castle and Harvey (2009, Young and Mcclean (2008), Ferro (2009) and Larsson et al. (2012)	12
Laboratory	Panchak (2003), Culberston (2005), Rickard and Mustapha (2007), Zarbo and Angelo (2007), Lefteroff and Graban (2008), Pexton (2008), Cayou et al. (2009), Gastineau et al. (2009) and Nelson et al. (2009)	9
Operating Room	Panchak (2003), Laursen, Gertsen and Johansen (2003), Silberstein (2006), Bush (2007), Bem-Tovim (2008), Lefteroff and Graban (2008), Young and Macclean (2008), Castle and Harvey (2209), Selau et al. (2009) and Matos et al. (2016)	10
Hospital	Panchak (2003), Laursen, Gertsen and Johansen (2003), Womach et al. (2005), Filingham (2007), Bush (2007), Bem-Tovim (2008), Cottington (2008), Young and Macclean (2008), Bamford and Griffin (2008) and Graban (2010)	10
Hospital Pharmacy	Silberstein (2006), Lefteroff and Graban (2008) and Mazur and Chen (2009)	3
Sterilized Material Center (SMC)	Taninecz (2007), Zanchet, Saurin ans Missel (2007) and Castle and Harvey (2009)	3
Internment Ward	Filingham (2007), Bush (2007) and Bem-Tovim (2008)	3
Diagnostic and Therapeutic Ancillary Services	Laursen, Gertsen and Johansen (2003), Filingham (2007) and Taninecz (2007)	3
Hospital Ward	Kim et al. (2006), Ballé and Régnier (2007)	2
Chemotherapy	Bush (2007) and Bertani (2012)	2
Orthopedics	Taninecz (2007)	1
Pathology unit	Papadopoulos et al. (2011)	1
Obstetric Ward	Pieters et al. (2010)	1
Psychiatry	Taninecz (2007)	1
Clinical Teaching Unit	Taninecz (2007)	1
Radiation Therapy	Trilling et al. (2010)	1
Clinical Ophthalmology	Castle and Harvey (2009)	1
Endoscopy Ward	Bush (2007)	1

Table 3. Lean healthcare practices and tools

Practices and Tools	References	Number of Publications
Standard Work	Santos and Maçada (1998), Culbertson (2005), Figueiredo (2006), Filingham (2007), Kim et al. (2006), Silberstein (2006), Amirahmadi et al. (2007), Ballé and Régnier (2007), Rickard and Mustapha (2007), Taninecz (2007), Bem-Tovim (2008), Cottington (2008), Lefteroff and Graban (2008), Ferro (2009), Gastineau et al. (2009) and Bertani (2012)	23
Value Stream Mapping	Laursen, Gertsen and Johansen (2003), Womack et al. (2005), Dickson et al. (2006), Filingham (2007), Kim et al. (2006), Rickard and Mustapha (2007), Taninecz (2007), Saurin and Missel (2007), Ferro (2009), Selau et al. (2009), Trilling et al. (2010) and Bertani (2012)	21
5S	Culbertson (2005), Womack et al. (2005), Filingham (2007), Bem-Tovim (2008), Cottington (2008), Lefteroff and Graban (2008), Gastineau et al. (2009), Al-Araidah et al. (2010) and Trilling et al. (2010)	16
Kaisen Event	Womack et al. (2005), Dickson et al. (2006), Filingham (2007), Bush (2007), Taninecz (2007), Tucker (2009), Trilling et al. (2010) and Bertani (2012)	13
Production Leveling	Laursen, Gertsen and Johansen (2003), Rickard and Mustapha (2007), Zarbo and Angelo (2007), Bem-Tovim (2008), Pexton (2008), Gastineau et al. (2009) and Bertani (2009)	11
Continuous Flow	Culbertson (2005), Dickson et al. (2006), Filingham (2007), Amirahmadi et al. (2007), Bush (2007), Taninecz (2007), Zarbo and Angelo (2007) and Nelson et al. (2009)	9
Pulled System	Culbertson (2005), Filingham (2007), Silberstein (2006), Taninecz (2007), Zarbo and Angelo (2007), Bem-Tovim (2008), Lefteroff and Graban (2008) and Selau et al. (2009)	8
A3	Jimmerson (2004), Ballé and Régnier (2007), Cottington (2008), Cayou et al. (2009), Ferro (2009) and Ghosh (2012)	8
Value Stream Redesign	Womack et al. (2005), Kim et al. (2006), Zarbo and Angelo (2007), Pexton (2008), Castle and Harvey (2009) and Graban (2009)	7
Cellular Layout	Culbertson (2005), Bush (2007), Cayou et al. (2009), Gastineau et al. (2009), Pieters et al. (2010), Trilling et al. (2010) and Larsson et al. (2012)	8
Work Leveling	Amirahmadi et al. (2007), Taninecz (2007) and Lefteroff and Graban (2008)	4
SMED	Kim et al. (2006), Silberstein (2006), Bush (2007) and Selau et al. (2009)	4
Visual Management	Amirahmadi et al. (2007) and Papadopoulos et al. (2011)	3
Team working	Bamford and Griffin (2008) and Larsson et al. (2012)	2

3. THE IMPLEMENTATION OF LEAN HEALTHCARE IN THE STUDIED HOSPITAL

3.1. The Implementation Method

The lean healthcare implementation method used in the studied hospital was based on the DMAIC method. This method is systematic and fact-based and facilitates the management of results-oriented projects (Pyzdek 2003). The DMAIC improvement cycle is the central process used to conduct Six Sigma projects. However, DMAIC is not exclusive of Six Sigma and can be used as a structure for other improvement applications (Sokovic, Pavletic, and Pipan 2010). The implementation method employed in this study is therefore based on 5 phases, as shown in Figure 1.

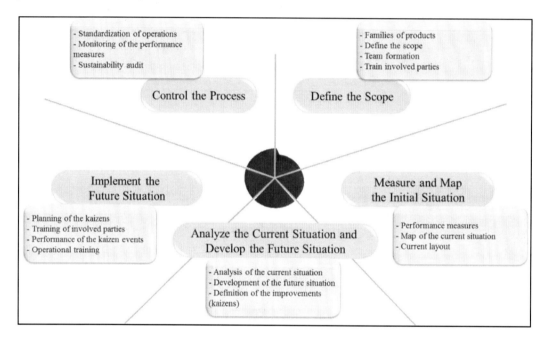

Figure 1. Lean healthcare implementation phases in the studied hospital.

3.2. The Studied Hospital

The hospital that served as the object of this study is a philanthropic organization that operates with 95% of its internments destined for the Brazilian Unified Health System and was founded in 1920 in the city of São Paulo, SP, Brazil. It includes a structure composed of 2 buildings, the outpatient clinic and the hospital. The outpatient clinic houses the administrative organization and outpatient operations such as consultations, while the procedures defined by the outpatient service such as surgeries, chemotherapy, radiation therapy, and exams are mostly conducted in the hospital. The outpatient clinic has 10 floors, of which 5 are reserved for outpatient care. The hospital structure is primarily composed of an internment ward with 77 beds; an intensive care unit (ICU) with 7 beds; an operating area with four rooms for minor, medium, and major surgeries; a diagnostics center (SADT); a radiation therapy center; a chemotherapy center; and a pharmacy and storage area. The upper management of the institute comprises a council and presidency with a 4-year term, as well as 3 boards (administrative, technical and clinical).

3.3. Description of the Implementation Phases

In this section each of the 5 phases of implementation is described in detail.

3.3.1. Phase 1: Defining the Scope

Definition of the scope began with the project objectives established by the project team and by members of the hospital board. The initially objectives were to improve patient assistance, thus reducing the time for patient access to treatment from the hospital and to improve the financial condition of the hospital (although the hospital is a philanthropic organization, the financial outcome is extremely important because the hospital accumulated negative results that made new investments impossible and even threatened the existence of the hospital).

For definition of the scope, the main hospital streams were initially surveyed. Figure 2 shows a macro flow of the primary treatments performed in the hospital. The 3 main patient streams, surgical, chemotherapy and radiation therapy, are shown in this figure.

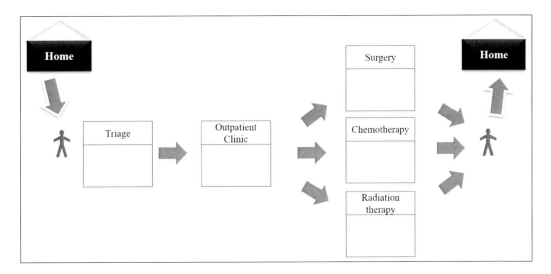

Figure 2. Macro flows of the studied hospital.

Additionally, to define the scope, the streams were analyzed with respect to their financial representativeness and the difficulty of implementing the improvements. The surgical stream, despite being representative with regard to demand and having a negative financial result, is highly complex with regard to implementing the improvements because of the need for investments and the strong involvement of medical professionals.

The chemotherapy and radiation therapy streams show elevated financial representativeness. Both streams require moderate involvement of the medical professionals. Of the 2 streams, the radiation therapy sector is administered through a service provider, thus increasing the difficulty of implementing improvements. The chemotherapy patient stream was therefore defined as the scope of the project, while emphasizing that the triage and outpatient clinics relative to these patients were also within this scope.

Various patient family members are present in a chemotherapy patient's macro stream. The patient families are defined by the similarities of the processes that are performed to treat the patient. The types of patients and the processes of this stream are shown in Table 4.

Table 4. Patient families in the chemotherapy patient stream

Patient \ Process	Triage	1st Consultation	Tomography	Ultrasound	Blood test	Consultation	APAC	Chemotherapy
New patients without exams	x	x	x	x	x	x	x	x
New patients without ultrasound	x	x	x		x	x	x	x
New patients without tomography	x	x		x	x	x	x	x
New patients with exams	x	x			x	x	x	x
Recurring patients without exams			x	x		x	x	x
Recurring patients						x	x	x

Among the types of patients listed in the table, the new patients without exams family is defined when detailing the other phases of the project. This decision was made with the intent to reduce the time to the beginning of treatment for new patients and to guarantee a set of actions that would impact all of the patient streams, as this family would comprise all of the listed processes.

Continuing within the phase of defining the scope, the project team must be defined. In the present study, a team composed of 2 lean healthcare specialists, the authors of this study (with dedications to the project of 2 days per week and 4 hours per week, respectively) and a hospital employee (with a dedication of 2 days per week to the project) was defined. The Institute of Cancêr Dr. Arnaldo (Instituto de Cancêr Dr. Arnaldo (ICAVC)) employee present on the team performs the function of a quality coordinator. Professionals from this sector are good candidates to participate within the teams, as they generally are familiar with the hospital's processes. Additionally, the active participation of upper management is essential to support and collaborate with the decisions of the team and provide guidelines for the progress of the project. To assume this responsibility, a board member was defined as the project sponsor, who dedicated 2 hours per week to participate in team meetings. The project sponsor performs the role of the technical director of the hospital. In addition to being a professional with broad autonomy in the institution, the director is a physician, which makes it possible to assist with the elimination of barriers to implementation.

During the initial project phase, technical training regarding general lean concepts was administered only to the lean team. The primary hospital members were only made aware of the concepts with brief presentations given by the lean team during leadership meetings.

3.3.2. Phase 2: Measuring and Mapping the Initial Situation

The tool used to diagnose the situation prior to the implementation of the project was VSM. The mapped processes were those that belonged to the value streams from the family of new patients without exams, which was the study focus. At the beginning of the mapping processes, meetings were held with sector supervisors to initiate an understanding of the processes streams. Additionally, various visits were performed for data collection and in loco verification of the current process conditions.

The selected patient stream represents all trajectories of the chemotherapy patients in the hospital. This means that the mapping is not limited to the chemotherapy department, which is responsible only for the administration of chemotherapy to patients. Thus, the mapping involved employees from various areas through the participation of hierarchical levels that varied from operations to the hospital board during meetings and unstructured interviews. It is worth emphasizing the importance of the involvement of employees from both the assistance and the administrative sectors. The interviewed healthcare professionals included the nursing staff, pharmacists, and doctors. Among the doctors, care was taken to primarily involve the doctors with a greater interest in the global performance of the institute. Thus, the doctors that manage the hospital, including the technical director and the clinical director responsible for emergency care and the ICU, were preferred over the clinicians. With the intent to ease understanding, a simplified chemotherapy patient stream is shown in Figure 3.

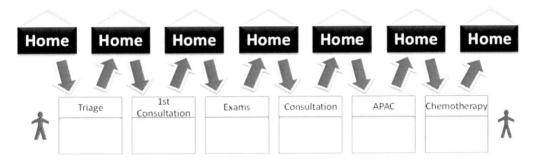

Figure 3. Simplified chemotherapy patient flow.

During mapping, some modifications were made to the VSM pattern so that this tool could be adapted to the reality of this case. In the chemotherapy patient stream, material streams (such as exam processing), information streams (such as treatment approvals), and patient streams (such as consultations) can be observed in parallel.

The initial VSM for the chemotherapy patient without exams, represented in a simplified form in Figure 3, is detailed in Figures 4 and 5. The detailed map shows the relationship between all of the value stream processes for this family and the "stock" points of the patients. The patient stream flows from the left to the right and shows the patients' waiting points, which are represented by triangles, and the processes with their data boxes. Notably, there are 2 symbols for the patient stocks. This adaptation was performed to emphasize the quantity of patient travel to their homes, a factor that increases costs to the patient and waiting times. The line below the value stream is the timeline that provides the lead time for the family in question.

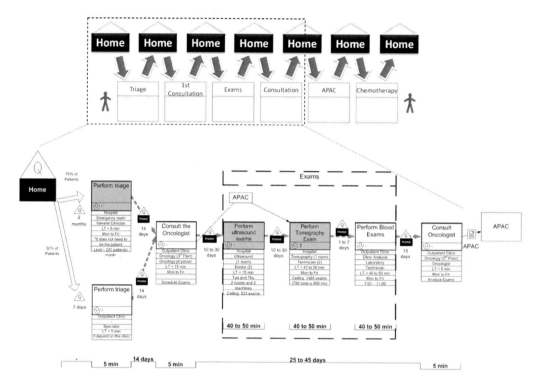

Figure 4. Details of the current situation, part 1 (treatment definition).

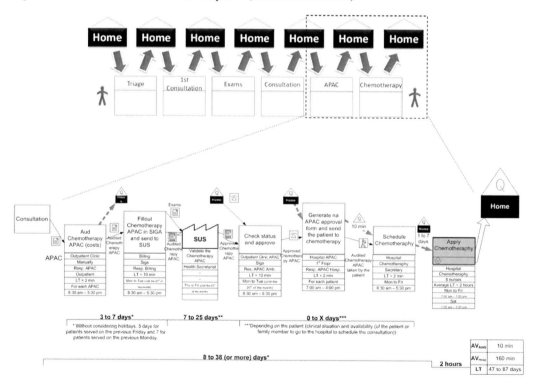

Figure 5. Details of the current situation, part 2 (authorization and treatment).

The first process in the chemotherapy patient stream is triage, in which a doctor analyzes whether the case is a cancer patient and what the possible treatment might be (chemotherapy, radiation therapy, surgery, or a combination of these). After triage, the patient who will be treated with chemotherapy is directed to the first consultation with the oncologist. During this consultation, the doctor analyzes the patient's exams and, if there is a need, requests pending exams. When necessary, the patient submits to the exams and, after the results are available, they return for another consultation with the oncologist. During this consultation, the doctor determines how the chemotherapy treatment will be conducted (number of sessions, type of medication, and dosage). This activity can be performed during the first consultation for patients with all of the necessary exams. The treatment determined by the doctor must be approved by the health department. To this end, the hospital prepares an Authorization for High Complexity Procedures (Autorização de Procedimentos de Alta Complexidade (APAC)) for each patient, with an average duration of 3 months. Upon receiving APAC approval, the patient schedules the chemotherapy sessions and undergoes the chemotherapy treatment.

The initial VSM (Figure 4 and 5) for the chemotherapy patient shows a lead time of 47 to 87 days between triage and chemotherapy, and value added times of 20 minutes in the outpatient clinic and 180 minutes in the hospital. The current hospital layout was not surveyed at this stage of the project. Due to the complexity of this task, the team decided to perform analyses of the specific layouts that were directed toward planned improvements.

3.3.3. Phase 3: Analyzing the Current Situation and Developing the Future Situation

In this stage, the team analyzed the situation that was found and mapped during the previous stage in order to design the future situation. Using the VSM of the initial situation and the interviews performed, the team elaborated the relationships of the main problems identified, as listed below:

- Triage in the hospital: the patient schedules the procedure for triage in the hospital. This type of hospital input has a limited capacity of 10 daily appointments that are performed by a doctor who has to be removed from his/her functions and without specific times, resulting in a long wait for triage in the hospital.
- Triage in the outpatient clinic: for triage in the outpatient clinic, the patient is directed from another care institution to the SUS, which schedules the procedure through the Integrated System of Care Management (SIGA). Through this entry door, the patient is directed directly to the correct specialty, which in turn can request the necessary exams to analyze the case. The openings made available by the hospital through the SIGA were outdated, and the descriptions from the specialties were entered in an incorrect manner, resulting in a low use of triage in the outpatient clinic, in contrast with the long wait for triage in the hospital. Another means of patient input into triage in the outpatient clinic occurs through the indications of doctors who provide care to the hospital. Thus, the patient is also directed to the correct specialty.
- Ultrasound exams: the reports from these exams are prepared in lots. The doctor performs various exams and later prepares the reports for the exams. Lot processing results in greater a lead time for the exam results and increases the possibility of errors in the reports.

- Tomography exams: tomography has a long waiting line due to the high volume of exams requested. In addition to the high use, the state provides a "ceiling" quota of exams per month. When analyzing the requested exams with the aid of the technical and medical directors, it was verified that various incorrect or unnecessary requests contributes to the overload of this sector.
- Blood exams: blood samples are collected in the hospital and sent to a laboratory that subsequently performs the analyses. For each patient, 1 or more analyses are required, with different processing times. In general, the longest exams have the objective of documenting the state of the patient for later analyses of treatment evolution, while the fastest exams are necessary to define the patient's treatment. The results from the analyses are returned to the hospital only at the end of the analysis with the longer lead time, thus increasing the waiting time for the patient.
- APAC: various problems were identified during the preparation and request for authorization to perform chemotherapy, especially the incorrect filling out of the documents by the doctors and the uneven submission of documents for approval. The latter is caused by the rules for submission, which were established by the Health Secretariat. The new APACs can only be sent on Mondays and Tuesdays until the twentieth day of the month, with a response time for authorization until Friday of the week in which the document was sent. It is important to note that the uneven submission of APACs results in the uneven entry of patients into chemotherapy, especially in the first week of the month, thus resulting in an increase in the patients' waiting times.
- Scheduling of chemotherapy: with the approved APAC, the patient or a companion must go to the hospital to obtain an approval form and then go to the chemotherapy reception desk, where the chemotherapy session scheduling occurs. Because chemotherapy serves patients according to the order of their arrival, scheduling does not provide a specific time for the patients and does not consider the sector's capacity. This causes patients to arrive at chemotherapy during the first hour of the day, thus making the workload uneven and impacting patient satisfaction.
- Administration of chemotherapy - the sector has 24 seats and operates from Monday to Friday, 7:00 AM to 7:00 PM, and on Saturday, 7:00 AM to 1:00 PM. The uneven workload throughout the day (higher percentage of patients at the beginning of the shift) and the month (greater quantity of patients in the last weeks of the month) directly impacts the sector's activities. Additionally, the lack of activity standardization reduces the employees' efficiency at moments of high demand. For example, despite the elevated number of patients present in the sector at the beginning of the day, on average, the seats are not completely occupied until approximately 9:00 AM, 2 hours after the beginning of activities.

From the initial VSM analysis, the lean team developed the future VSM for the chemotherapy patient. The steps followed are shown below:

- Structure the stream by eliminating and altering the activity sequence;
- Use continuous flow where possible by balancing the work load according to the task time;

- Improve the processes by standardizing documents and activities and defining the manner of communication between the processes and necessary resources; and
- Level the activities by defining the performance frequencies of the activities and leveling the stream input according to the projected capacity.

With the objectives of reducing the lead time for patient treatment and improving the financial condition of the hospital, the main solutions were directed to elevate the production capacity, thus permitting the treatment of more patients and eliminating wastes and subsequently making a shorter response time possible. The suggestions for improvement or kaizens are described below:

- Kaizen in triage: the studied hospital only performs cancer treatment; that is, the institute patients are directed to the hospital after a cancer diagnosis is made at another hospital. The Health Secretariat, through SIGA, directs cancer patients not only to the hospital but also to the correct medical specialty. For these patients, the triage performed in the hospital is redundant. For future situations, the patients will be directed through SIGA directly to the first consultation. To make triage in the Health Secretariat possible, it is necessary to update the openings available in the hospital through SIGA;
- Kaizen in the exams: the balancing of activities, structuring of sectors, and redesigning the processes, thus making exam results possible in the stream;
- Kaizen in the information stream: to reduce reworking during the APAC auditing process, reduce the lead time for APAC approval, level patient entry into chemotherapy and reduce patient movement, it is necessary to train the medical professionals regarding filling out of the APAC, leveling the sending of requests, and redesigning of the process of scheduling, thus permitting chemotherapy scheduling after the consultation;
- Kaizen in chemotherapy - In chemotherapy, the leveling of production is proposed through patient load-based scheduling, activity balancing, and work standardization for nursing and pharmacy, as well as the establishment of a continuous flow between medicine manipulation and chemotherapy administration to patients, thus permitting reductions in wait times and increasing the sector's productivity.

3.3.4. Phase 4: Implementing the Future Situation

The process of implementing improvements is guided by planning the proposed kaizens. This planning consists of elaborating a chronogram while considering the improvement implementation order. This sequence considers the impact of the improvements on both the chemotherapy patient stream and the other streams present in the hospital, such that it results in the projected gains with a lower system impact. Thus, as increases in the quantity of triaged patients reflects the increases in demand on all of the streams from the ICAVC, a decision is made to begin the improvements through the intermediary APAC and Exams processes, with the objective of organizing the hospital according to the current demand. Thus, new triaged patients could pass through the stream with greater ease until the final processes of radiation therapy, surgery, and chemotherapy. With more efficient intermediary processes, triage can be adjusted to permit an increase in the quantity of patients in the hospital. To assume the increase in demand, chemotherapy finally receives the attention of the team in order to organize the sector and guarantee a quality of care to the patient.

The first step for implementing the kaizens is training of the involved parties. This process of education, recycling, and behavioral changes is essential to conducting improvements. With the objective of generating a uniform understanding for all involved, the lean team was trained relative to the main concepts and tools of lean production such as continuous flow, standardized work, kaizen events, pulled systems, and 5S. The other involved parties received kaizen-related training. During the initial implementation phase, the professionals were instructed about the concepts used and, later, about the new standards and manners for sustaining the implemented improvements. Kaizen implementation is described below.

3.3.4.1. Kaizen in the APAC Information Stream

The APAC information stream improvement project was performed from April 2011 to June 2011 with the participation of the lean team and the individuals responsible for the outpatient clinic APAC, billing, and the hospital APAC, as well as the collaboration of the Health Secretariat. As project requirements, the team defined a reduction in lead time for APAC approval of up to 50%, an increase of the frequency of APAC submission for approval from weekly to daily, and the elimination of the number of patient movements to the hospital to resolve bureaucratic subjects such as obtaining an approval form and scheduling chemotherapy.

Note that in this stream, a third party (Health Secretariat) performs the APAC validation process. The problem of uneven APAC approval is the result of the submission rules for APAC validation (new APACs sent on Mondays and Tuesdays until the twentieth day of the month, with a response period for authorization until Friday of the week in which the APAC was sent). A new proposal for the submission rules for APAC validation was prepared by foreseeing the daily delivery of requests. To approve the proposal and reduce the APAC reworking index, the team, with the participation of the Health Secretariat, scheduled a meeting to discuss the submission rules and training regarding filling out of the APACs. In May 2011, the clinical group was trained by a representative of the Health Secretariat, and the new submission rules for the APACs were established. Beginning in June 2011, daily APAC submission began until the twenty-fifth day of the month, with a response period for authorization of up to 3 business days. To eliminate the patient's return to the hospital to schedule chemotherapy, the processes were redesigned, thus allowing the patient to schedule the session date soon after the auditing of the APAC by the ICAVC.

The actions of this improvement project resulted in a 74% or 28-day reduction in the patient lead time, the elimination of the patient's return to schedule chemotherapy, and the leveling of patient entry into chemotherapy.

3.3.4.2. Kaizen in the Blood Exams

The project to improve the blood exams was performed from April 2011 to June 2011 with the participation of the lean team and the financial director, as well as the collaboration of the laboratory responsible for analyzing the exams. In this project, only the blood exams were addressed. Improvements in the tomography and ultrasound processes were postponed by the need for investments in those sectors. As a requirement for this project, the team defined a reduction in the lead time of 80% for the patients undergo exams to document their statuses and define treatments.

As cited previously, the transmission of results from the laboratory to the ICAVC was performed by the patient. As the analyses needed to document the patient's status have longer processing times, while those needed to define the treatment are rapid, the system that sends the results thus generates an unnecessary waiting time for the patient. Therefore, the initial system was altered with the consent of the involved parties. The results began to be sent for analysis and the responsible laboratory made available the schematic for printing the exams to the ICAVC, thus permitting the sending of results via the Internet. These actions permitted up to a 90% reduction in the exam lead time.

3.3.4.3. Kaizen in Triage (SIGA and Indications)

The triage improvement project was performed from May 2011 to September 2011 with the participation of the lean team, the doctor responsible for triages performed in the hospital, the administrative director of the ICAVC, and the individual responsible for the Medical Care and Statistics Service (Serviço de Atendimento Médico e Estatística (SAME)).

As cited previously, triage can be performed in the following 3 ways: triage scheduled in the hospital (the patient seeks out the ICAVC and schedules a triage with a general clinician), triage scheduled via the SIGA (the patient is diagnosed with cancer in another care unit and the responsible doctor schedules the triage for the patient in the ICAVC), and triage scheduled via indication (ICAVC partner doctors indicate patients who are diagnosed with cancer in other care units to the medical clinic of the ICAVC that specialized in the patient's case).

As project requirements, the team defined eliminating hospital-scheduled triage, adapting and increasing the occupation of available triage openings as scheduled by SIGA by 90% and, for triage via indication, simply monitoring the monthly demand and taking specific actions as needed. These requirements were derived from the larger objective of reducing the chemotherapy patient lead time. For patients directed via SIGA or an indication, the first process in the ICAVC is performed by the doctor designated to the case. Therefore, the doctor in question performs the first consultation in triage, thus reducing the patient lead time.

Triage is a point that must be treated with caution. Once triaged, the hospital is responsible for the patient, and this patient can then pass through various streams within the institute. For example, a patient who is now classified as a chemotherapy patient might at another time pass through the emergency room stream and internment. Therefore, an increase in triaged patients might reflect an increase in demand throughout the hospital and lead to overloading of the various sectors. To minimize the overloading of other sectors of the ICAVC, the triage improvement project analyzed the patients from the medical specialties with regard to the patient complexity and the clinical care capacity. In addition to this analysis, the projects for APAC and exam improvement were performed.

The main points that were addressed were reception activity standardization and the adaptation of the openings made available through the ICAVC in SIGA. The team also established a monthly meeting between the technical director and the SAME to update the openings offered through the ICAVC.

Initially, the triage scheduled directly in the hospital was not eliminated. It was preferred to adapt the triage scheduled via SIGA, thus permitting a later reduction of triage in the hospital. Adaptation of the openings scheduled via SIGA began in June 2011. For the month of July 2011, openings were only made available for the Oncology specialty, thus reducing the number of openings. For the following months, the other specialties were inserted to increase the quantity of available openings. After the analysis, the occupation rate of the openings jumped from 49% in May 2011 to 98% in October 2011. The number of openings available through SIGA in the second semester of 2011 was 42% lower than in the first semester of the same year (from 2,039 to 1,188 openings available); however, the number of patients served increased by 24% in the same period (from 889 to 1,102 patients served). Figure 6 shows this evolution.

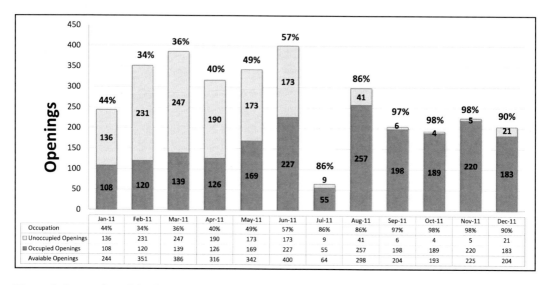

Figure 6. Occupation of the SIGA schedule.

3.3.4.4. Kaizen in Chemotherapy

The chemotherapy improvement project was performed from June 2011 to October 2011 with the participation of the lean team and the employees connected directly to the chemotherapy sector, including the responsible nurse, sector nurses, responsible pharmacist, and clerks.

As project requirements, the team defined increasing the chair occupation from 62% to 70% and reducing the patient waiting time in chemotherapy. The chemotherapy sector of the ICAVC has 24 chairs (16 for SUS patients and 8 for health plan patients) and reported an average monthly demand of 1650 applications in the first semester of 2011.

In the initial situation, chemotherapy operated from 7:00 AM to 7:00 PM from Monday to Friday and from 7:00 AM to 1:00 PM on Saturdays and served patients according to their order of arrival. This rule of service prioritization conditioned the patients to arrive at the sector at the beginning of the shift, thus generating an excess of patients during the morning period. Added to this factor, the lack of activity standardization and the work schedule of the pharmacy team resulted in an elevated loading time for the chairs from the beginning of the shift and a long patient wait time.

The time for medication infusion into patients varies according to the patient condition, medication, and dosage. Because chemotherapy scheduling did not have access to the chemotherapy infusion times and the sector ceased activities at 7:00 PM, the patients were directed to avoid arriving at the sector in the final hours of the evening period. In contrast to the situation in the morning period, the evening period had low patient demand.

After analyzing the situation, the team designed the future situation, using the kaizens. The main solutions adopted were the leveling of production with previous scheduling, based on the infusion time and sector capacity; chemotherapy scheduling after consultation; activity balancing and standardization for nursing and the pharmacy; and a continuous flow between medication preparation and chemotherapy administration.

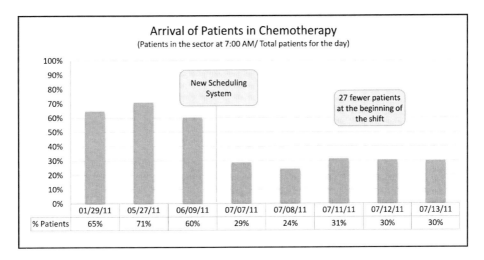

Figure 7. Patients at the beginning of the chemotherapy shift.

Work Pattern											
Chemotherapy	Pharmacy	Period Morning and Afternoon				Validity : __/__ to __/__					
Resp.	Elements of Work		Beginning of the Period								
			6:00	6:15	6:30	6:45	7:00	7:15	7:30	7:45	8:00
	Sterilize the hood		x								
	Prepare Medicine from Sao Caetano			x							
	Prepare Medicine from ICAVC				x	x	x	x	x	x	x
	Separate the prescriptions that are continuous			---							
	Distribute passwords to the patients (w/conditions)			---							
	Affix password to the prescription			---							
	Prepare trays for the ICAVC patients				---	---	---	---	---	---	---
End-of-shift Activities											
Prepare trays for the Sao Caetano patients											
Prepare medicine labels for the patients for the next day											
Complete the Serum Buffer											
Individual responsible for turning on the hood (5:30): _____											
Last Update:16/06/2011		Legend	x		Hood		---			Outside of the Hood	

Figure 8. Chemotherapy pharmacy work pattern.

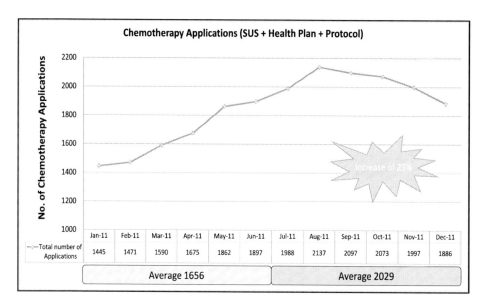

Figure 9. Chemotherapy applications.

The implementation of improvements was performed through a kaizen event during the period from July 4 to 8, 2011. The sector employees were trained in general lean concepts and involved in the implementation of the improvements. Figures 7, 8, and 9 show the analysis of the quantity of patients at the beginning of the shift, the work pattern of the chemotherapy pharmacy employees, and the number of chemotherapy applications in 2011, respectively.

As shown in Figures 7 and 9, the number of applications increased by 23% between the first and second semesters of 2011, with a 50% reduction in the number of patients present in chemotherapy at the beginning of the shift. Because the sector team remained constant throughout the analysis period, it can be stated that the improvements yielded a 23% increase in sector productivity.

3.3.5. Phase 5: Controlling the Process

Control of the process begins in parallel with the implementation of improvements. In this phase of the ICAVC project, the lean team, in conjunction with those involved in the chemotherapy patient stream, controlled the project indicators and developed control plans and sustainability audits.

Regarding the performance indicators, which were guided by the project objectives of improving patient care and the financial health of the hospital, the following performance indicators were selected and analyzed: the lead time between triage and the first chemotherapy session and chemotherapy billing.

The treatment time for a single cancer patient profoundly impacts the success of the disease treatment. The lead time for a chemotherapy patient was analyzed during 2011. Due to the quantity of patients and, consequently, the volume of information, this analysis involved an indicator that was difficult to compile. The data were removed from the ICAVC server, worked, and compiled in Figure 10. The average lead times of the patients who began chemotherapy were 65 days in January 2011, 50 days at the end of the year, and 38 days, the lowest value, in August. Despite the obtained values falling below the expectations (the future

VSM shows the lead time of the patient varying between 25 and 52), the indicator shows a positive evolution and a reduction of up to 42%.

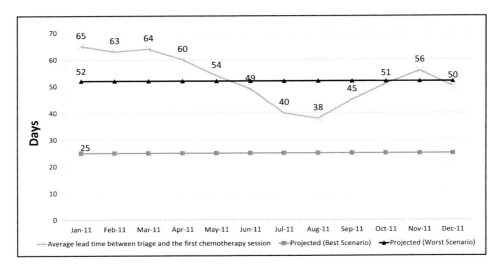

Figure 10. Lead-time for a chemotherapy patient.

The chemotherapy billing indicator is extremely important when analyzing the efficacy of the implemented improvements. Figure 11 shows the evolution of this indicator during 2011. Using the chemotherapy billing indicator, a 33% increase can be observed in the average monthly billing in the second semester of 2011, relative to that of the first semester of 2011, with a greater value realized in December 2011. Because it involves a high profitability sector, the increase in the billing of this sector reflects positively on the financial health of the hospital.

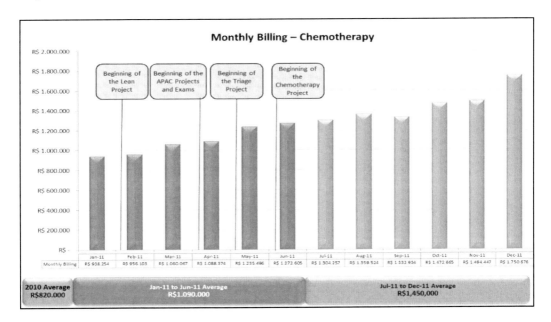

Figure 11. Chemotherapy billing.

In addition to evaluating the project performance indicators, the control phase of the improvements involves developing the control plans and sustainability audits. Sustainability is a very important element for the dissemination of lean production concepts in a company (Bateman 2005). To Womack (2007), a large part of the lean production implementation efforts is centered on avoiding the return to old work patterns. To promote the maintenance and evolution of the implemented improvements, periodic activities for sustainability reviews and evaluation were developed, as follows:

- Occupation of the SIGA schedule by specialty: The quantity of available openings for triage via SIGA is defined to be addressed in monthly meetings with the SAME leadership and the technical board, known as the SIGA triage meetings. In these meetings, the reasons for the scheduling impediment and the previous month's results by specialty are discussed. The openings to be made available result from an analysis of the medical clinics' availability and the schedule adherence during the previous month.
- Chemotherapy loading: At the beginning of the chemotherapy morning shift, the patients awaiting treatment are assigned to chairs according to the work pace of the sector employees. An efficient initial patient loading permits greater use of the chairs and reduces the patient waiting time. To follow up on the efficiency of the patient allocation to the chairs, a daily record plan of the loading and the problems encountered was generated. In chemotherapy, nursing, the pharmacy, and scheduling participate in daily meetings to discuss the recorded data and, if necessary, generate action plans for the pertinent problems. The chemotherapy loading indicator is also shown monthly to the technical board of the ICAVC.

CONCLUSION

This study had the objective of presenting action research to demonstrate the implementation and results from the application of lean healthcare concepts and tools in the chemotherapy patient care stream in a hospital located in the state of São Paulo, Brazil. In addition to the chemotherapy sector itself, the implementation included the triage sector, the exams and information streams, and chemotherapy scheduling. Table 5 summarizes the main lean healthcare tools implemented in these sectors.

Table 5. Lean healthcare tools used in this action research

Hospital Area	Tools	Number of Tools
Triage	Value stream mapping, standardized work, and levelling of work.	3
Laboratory (Blood Tests)	Value stream mapping and continuous flow.	2
Approval of Treatment (APAC)	Value stream mapping, levelling of work, and redesign of the value stream.	3
Chemotherapy	Value stream mapping, standardized work, levelling of work, balancing of activities, continuous flow, and kaisen events.	6

Beliefs and paradigms must be confronted to successfully implement these tools. Phrases that demonstrate initial concern such as "hospitals are different from industry" and "patients are not parts" must be addressed naturally. Although correct, these affirmations do not restrict the applications of the cited tools and must not serve as "excuses" against the application of lean healthcare concepts and tools. Some observed points such as lean training and the participation of operations people in project development facilitate the change management process conduction (for a study about lean learning and behaviors in hospitals see Mazur, McCreery, and Rothenberg, (2012)).

This study could possibly contribute information regarding the applications of lean concepts in the chemotherapy area. Among the cases analyzed in the bibliographic review of this study, only Bush (2007) comments briefly about the application of lean concepts to chemotherapy. According to the author, through a kaizen event, the Virginia Mason Medical Center (USA) was able to reduce the time from breast cancer diagnosis to treatment initiation from 21 days to 11 days and the patient wait time in chemotherapy from 240 minutes to 90 minutes. Similar to the case described by Bush (2007), the current action research also obtained positive results. In the 4 areas in which the project was implemented, the following improvements were achieved: an increase of 24% in the number of patients served in triage, a reduction of up to 28 days in the time to treatment approval, a reduction of up to 90% in the exam results lead time, and an increase of up to 23% in productivity (patients served) in chemotherapy. This led to gains throughout the hospital, including reductions of up to 42% (from 65 days to 38 days) in the average patient lead time (time between triage and patient treatment). Additionally, the financial area was improved, with an approximately 33% increase in billing.

Finally, in addition to the practical contributions, this study also contributes to theory in this area, which remains in need of studies and empirical evidence from hospitals located in developing countries, such as Brazil.

REFERENCES

Aherne, Joe., and John. Whelton. 2010. *Applying Lean in Healthcare: A Collection of International Case Studies*. Productivity Press.

Al-Araidah, O. et al., 2010. Lead-Time Reduction Utilizing Lean Tools Applied to Healthcare: The Inpatient Pharmacy at a Local Hospital. *Journal for Healthcare Quality* 32(1), 59-66.

Al-Araidah, Omar, Amer Momani, Mohammad Khasawneh, and Mohammed Momani. 2010. "Lead-Time Reduction Utilizing Lean Tools Applied to Healthcare: The Inpatient Pharmacy at a Local Hospital." *Journal For Healthcare Quality* 32 (1): 59-66. doi:10.1111/j.1945-1474.2009.00065.x.

Amirahmadi, F. et al., 2007. *Innovations in the Clinical Laboratory: An Overview of Lean Principles in the Laboratory*. EUA: Mayo Clinic.

Amirahmadi, F., A. Dalbello, D. Gronseth, and J. McCarthy. 2007. *Innovations in the Clinical Laboratory - An Overview of Lean Principles in the Laboratory*. Mayo Clinic.

Baesler, Felipe F, and Jost A Sepúlveda. 2006. "Multi-Objective Simulation Optimization: A Case-Study in Health Care Management." *International Journal of Industrial Engineering: Theory, Applications and Practice* 13 (2): 156-65.

Ballé, M.; Régnier, A., 2007. Lean as a Learning System in a Hospital Ward. *Emerald* 20 (1), 33-41.

Ballé, Michael, and Anne Régnier. 2007. "Lean as a Learning System in a Hospital Ward." *Leadership in Health Services* 20 (1): 33-41. doi:10.1108/17511870710721471.

Bamford, D., Griffin, M. 2008. A case study into operational team-working within a UK hospital. *International Journal of Operations and Production Management* 28 (3), 215-237.

Bamford, David, and Michael Griffin. 2008. "A Case Study into Operational Team-working within a UK Hospital." *International Journal of Operations & Production Management* 28 (3): 215-37. doi:10.1108/01443570810856161.

Bateman, N., 2005. Sustainability: the elusive element of process improvement. *International Journal of Operations & Production Management* 25 (3), 261-276.

Bateman, Nicola. 2005. "Sustainability: The Elusive Element of Process Improvement." *International Journal of Operations and Production Management* 25 (3): 261-76. doi:10.1108/01443570510581862.

Bem-Tovim, D., 2008. Redesigning care at the Flinders Medical Centre: clinical process redesign using "lean thinking". *The Medical Journal of Australia* 188 (6), 27-31.

Ben-Tovim, David I., Jane E. Bassham, Denise M. Bennett, Melissa L. Dougherty, Margaret A. Martin, Susan J. O'Neill, Jackie L. Sincock, and Michael G. Szwarcbord. 2008. "Redesigning Care at the Flinders Medical Centre: Clinical Process Redesign Using 'Lean Thinking.'" *The Medical Journal of Australia* 188 (6 Suppl). doi:ben11046_fm [pii].

Bertani, T. M., 2012. *Lean Healthcare: recommendations for implementations of the concepts of Lean Production in hospital environments*. Dissertation, University of São Paulo, Brazil.

Bertani, Thiago Moreno. 2012. *Lean Healthcare: Recomendações Para Implantações Dos Conceitos de Produção Enxuta Em Ambientes Hospitalares*. [*Lean Healthcare: Recommendations for Deployments of Lean Manufacturing Concepts in Hospital Environments.*] Universidade de São Paulo. doi:10.11606/D.18.2012.tde-29102012-235205.

Bush, R. D., 2007. Reducing Waste in US Health Care Systems. *The Journal of American Medical Association* 297(8), 871-874.

Bush, Roger W. 2007. "Reducing Waste in US Health Care Systems." *JAMA* 297 (8): 871. doi:10.1001/jama.297.8.871.

Castle, A., Harvey, R., 2009. Lean information management: the use of observational data in health care. *International Journal of Productivity and Performance Management*, 58 (3) 280-299.

Castle, Andrew, and Rachel Harvey. 2009. "Lean Information Management: The Use of Observational Data in Health Care." *International Journal of Productivity and Performance Management*. doi:10.1108/17410400910938878.

Castro, P.J., S.J. Dorgan, and B. Richardson. 2008. "A Healthier Health Care System for the United Kingdom." *The Mckinsey Quarterly*, 2008.

Cayou, J. et al. 2009. "Special Coagulation Laboratory: Layout Improvements." *Mayo Clinic*.

Cayou, J. et al., 2009. *Special Coagulation Laboratory: Layout Improvements*. EUA: Mayo Clinic.

Costa, Daniela Akemi. 2008. *Aplicação Do Lean Sigma No Centro Cirúrgico de Um Hospital Privado de São Paulo* [Application of Lean Sigma in the Surgical Center of a Private Hospital of São Paulo]. Escola Politécnica da Universidade de São Paulo, EPUSP.

Costa, L.B., Godinho Filho, M. 2016. Lean healthcare: review, classification and analysis of literature. *Production Planning and Control* 27 (10), 823-836.

Costa, L.B.M. and M. Godinho Filho. 2016. "Lean healthcare: review, classification and analysis of literature". *Production Planning & Control* 27 (10):823-36.

Cottington, S., 2011. *Lean healthcare processes at Pella Regional Health Center, CDC*. Available in: <http://www.cdc.gov/ncidod/eid/vol5no1/rubin.htm>.

Culbertson, G., 2005. *Iowa Health Des Moines Laboratory*. EUA: Ortho-Clinical Diagnostics.

Dickson, E. W. et al., 2006. *Application of Lean Manufactoring Techniques in the ED*. Available in: <http://www.leanhealth.net.au/content/links/>.

Dickson, Eric W., Sabi Singh, Dickson S. Cheung, Christopher C. Wyatt, and Andrew S. Nugent. 2009. "Application of Lean Manufacturing Techniques in the Emergency Department." *Journal of Emergency Medicine* 37 (2): 177-82. doi:10.1016/j.jemermed.2007.11.108.

Dobrzykowski, D., Deilami, V.S., Hong, P., Kim, S. 2013. A structured analysis of operations and supply chain management research in healthcare (1982-2011). *International Journal of Production Economics* (2013), http://dx.doi.org/10.1016/j.ijpe.2013.04.055.

Dobrzykowski, David, Vafa Saboori Deilami, and Seung-Chul Kim. 2014. "A Structured Analysis of Operations and Supply Chain Management Research in Healthcare (1982-2011)." *International Journal of Production Economics* 147: 514-30. doi:10.1016/J.IJPE.2013.04.055.

Ferro, M., 2009. *Sistema Lean na reorganização de Pronto Socorro hospitalar*. Available in: <http://www.*lean*.org.br>.

Ferro, Mara. 2009. *Sistema Lean Na Reorganização de Pronto Socorro Hospitalar* [Lean System In Hospital Outpatient Reorganization]. Lean Institute Brasil.

Figueiredo, K., 2006. A logística enxuta [Lean logistics]. *Centro de Estudos em Logística, COPPEAD*, UFRJ, Brazil.

Figueiredo, Kleber. 2006. "A Logística Enxuta." ["The Lean Logistics"]. *Centro de Estudos Em Logística - COPPEAD/UFRJ*, 2006.

Fillingham, D., 2007. Can lean save lives? *Leadership in Health Services* 20 (4), 231-241.

Fillingham, David. 2007. "Can Lean Save Lives?" *Leadership in Health Services* 20 (4): 231-41. doi:10.1108/17511870710829346.

Forgia, Gerard M. La, and Bernard F. Couttolenc. 2008. "*Hospital Performance in Brazil.*" Washington, DC. doi:10.1596/978-0-8213-7358-3.

Gastineau, D. et al., 2009. *Human Cell Therapy Laboratory: Improvement Project*. EUA: Mayo Clinic.

Ghosh, M., 2012. A3 Process - A pragmatic problem-solving technique for process improvement in healthcare. *Journal of Health Management* 14 (1), 1-11.

Godinho Filho, M., Boschi, A., Rentes, A.F., Thurer, M., Bertani, T.M., 2015. Improving Hospital Performance by Use of *Lean Techniques: An Action Research Project in Brazil*. Quality Engineering, 27:2, 196-211.

Godinho Filho, Moacir, Artur Boschi, Antonio Freitas Rentes, Matthias Thurer, and Thiago Moreno Bertani. 2015. "Improving Hospital Performance by Use of Lean Techniques: An Action Research Project in Brazil." *Quality Engineering* 27 (2): 196-211. doi:10.1080/08982112.2014.942039.

Graban, M., 2009. *Lean Hospitals - Improving Quality, Patient Safety, and Employee Satisfaction.* Nova Iorque: Taylor & Francis Group.

Graban, Mark. 2011. *Lean Hospitals: Improving Quality, Patient Safety, and Employee Engagement.* 89. New York: CRC Press. doi:10.1016/S0001-2092(09)00184-7.

Guandalini, G., Borsato, C., 2008. A inflação da saúde [Health inflation]. Revista Veja, 2060.

Guandalini, Giuliano, and Cíntia Borsato. 2008. "A Inflação Da Saúde." ["Health Inflation"]. *Revista VEJA*, 2008.

Hediger, Iktor, Toby MH Lambert, and Mona Mourshed. 2008. "Private Solutions for Health Care in the Gulf." *McKinsey Quarterly*, 2008.

Heinbuch, Susan E. 1995. "A Case of Successful Technology Transfer to Health Care." *Journal of Management in Medicine* 9 (2): 48-56. doi:10.1108/02689239510086524.

Henrique, D.B., Rentes, A.F., Godinho Filho, M., Esposto, K.F. 2016. A new value stream mapping approach for healthcare environments. Production *Planning and Control* 27 (1), 24-48.

Henrique, Daniel Barberato, Antonio Freitas Rentes, Moacir Godinho Filho, and Kleber Francisco Esposto. 2016. "A New Value Stream Mapping Approach for Healthcare Environments." *Production Planning & Control* 27 (1): 24-48. doi:10.1080/09537287.2015.1051159.

Hinckley, C Martin. 2003. "Make No Mistake - Errors Can Be Controlled." *Quality and Safety in Health Care* 12 (5): 359-65. doi:10.1136/qhc.12.5.359.

Hinckley, C.M., 2003. *Make No Mistake—Errors Can Be Controlled.* Portland, EUA: Productivity Press.

Jimmerson, Cindy, Dorothy Weber, and Durward K. Sobek. 2005. "Reducing Waste and Errors: Piloting Lean Principles at Intermountain Healthcare." *The Joint Commission Journal on Quality and Patient Safety* 31 (5): 249-57. doi:10.1016/S1553-7250(05)31032-4.

Jones, D., 2006. *Leaning Healthcare.* Lean Enterprise Academy.

Jones, Daniel T. 2006. "*Leaning Healthcare.*" Lean Enterprise Academy. 2006. http://www.leanuk.org/article-pages/articles/2006/may/07/leaning-healthcare.aspx.

Kim, C., Spahlinger, D.A., Kin, J.M., Billi, J.E., 2006. Lean Health Care: What Can Hospitals Learn from a World-Class Automaker? *Journal of Hospital Medicine* 1 (3), 191-199.

Kim, Christopher S., David A. Spahlinger, Jeanne M. Kin, and John E. Billi. 2006. "Lean Health Care: What Can Hospitals Learn from a World-Class Automaker?" *Journal of Hospital Medicine (Online)* 1 (3): 191-99. doi:10.1002/jhm.68.

King, D.L., Ben-Tovim, D.I., Bassham, J., 2006. Redesigning emergency department patient flows: Application of Lean Thinking to health care. *Emergency Medicine Australasia* 18, 391-397.

King, Diane L., David I. Ben-Tovim, and Jane Bassham. 2006. "Redesigning Emergency Department Patient Flows: Application of Lean Thinking to Health Care." *Emergency Medicine Australasia* 18 (4): 391-97. doi:10.1111/j.1742-6723.2006.00872.x.

Langabeer, J., Dellifraine, J., Heineke, J., Abbass, I., 2009. Implementation of Lean and Six Sigma quality initiatives in hospitals: A goal theoretic perspective. *Operations Management Research* 2 (1-4), 13-27.

Langabeer, James R., Jami L. DelliFraine, Janelle Heineke, and Ibrahim Abbass. 2009. "Implementation of Lean and Six Sigma Quality Initiatives in Hospitals: A Goal Theoretic Perspective." *Operations Management Research* 2 (1): 13-27. doi:10.1007/s12063-009-0021-7.

Larsson, A., Johansson, M., Baathe, F., Neselius, S. 2012. Reducing throughput time in a service organization by introducing cross-functional teams. *Production Planning and Control* 23 (7), 571-580.

Larsson, Agneta, Mats Johansson, Fredrik Bååthe, and Sanna Neselius. 2012. "Reducing Throughput Time in a Service Organisation by Introducing Cross-Functional Teams." *Production Planning & Control* 23 (7): 571-80. doi:10.1080/09537287.2011.640074.

Laursen, M.L., Gertsen, F., Johansen, J., 2003. *Applying Lean Thinking in hospitals: exploring implementation difficulties.* Aalborg: Aalborg University, Center for Industrial Production.

Laursen, Martin Lindgård, Frank Gertsen, and John Johansen. 2003. "Applying Lean Thinking in Hospitals - Exploring Implementation Difficulties." In *3rd International Conference on the Management of Healthcare and Medical Technology.* Warwick, United Kingdom: Aalborg Universitet.

Lefteroff, Lewis, and Mark Graban. 2008. "*Lean and Process Excellence at Kingston General.*" Lean Manufacturing, 2008.

Liker, J., Morgan, J., 2011. Lean product development as a system: a case study of body and stamping development at Ford. *Engineering Management Journal* 23 (1), 16 - 28.

Liker, Jeffrey K, and James Morgan. 2011. "Lean Product Development as a System: A Case Study of Body and Stamping Development at Ford." *Engineering Management Journal* 23 (1): 16-28. doi:10.1080/10429247.2011.11431884.

Matos, I.A., Alves, A.C., Tereso, A.P., 2016. Lean principles in na operating room environment: na action research study. *Journal of Health Management* 18 (2), 239-257.

Mazur, L., McCreery, J., Rothenberg, L., 2012. Facilitating lean learning and behaviors in hospitals during the early stages of lean implementation. *Engineering Management Journal* 24 (1) 11-22.

Mazur, Lukasz M, and Shi-Jie Chen. 2009. "An Empirical Study for Medication Delivery Improvement Based on Healthcare Professionals' Perceptions of Medication Delivery System." *Health Care Management Science* 12 (1): 56-66.

Mazur, Lukasz, John McCreery, and Lori Rothenberg. 2012. "Facilitating Lean Learning and Behaviors in Hospitals during the Early Stages of Lean Implementation." *EMJ - Engineering Management Journal* 24 (1): 11-22. doi:10.1080/10429247.2012.11431925.

McGrath, K., Bennett, D., Ben-Tovim, D., Boyages, S., Lyons, N., O'Connell, T., 2008. Implementing and sustaining transformational change in health care: lessons learnt about clinical process redesign. *The Medical Journal of Australia* 188 (6), 32-35.

McGrath, Katherine M., Denise M. Bennett, David I. Ben-Tovim, Steven C. Boyages, Nigel J. Lyons, and Tony J. O'Connell. 2008. "Implementing and Sustaining Transformational Change in Health Care: Lessons Learnt about Clinical Process Redesign." *The Medical Journal of Australia* 188 (6): 32-35.

Meredith, J.O., Grove, A.L., Walley, P., Young, F., Macintyre, M.B. 2011. Are we operating effectively? A lean analysis of operating theatre changeovers. *Operations Management Research* 4 (3-4), 89-98.

Meredith, James O., Amy L. Grove, Paul Walley, Fraser Young, and Mairi B. Macintyre. 2011. "Are We Operating Effectively? A Lean Analysis of Operating Theatre Changeovers." *Operations Management Research* 4 (3-4): 89-98. doi:10.1007/s12063-011-0054-6.

Modig, N., Ahlstrom, P., 2012. This is lean - resolving the efficiency paradox. *Rheologica Publisching.*

Nelson et al. 2009. *Lean Principles in the Laboratory: Efficient Laboratory Space Design.* Mayo Clinic.

Nelson, L. et al., 2009. *Lean Principles in the Laboratory: Efficient Laboratory Space Design.* EUA: Mayo Clinic.

Oldeshaw, M. 2009. *Health Policy: Tackling Hospital Waiting Lists.*

Oldeshaw, M., 2009. *Health Policy: Tackling Hospital Waiting Lists.* Available in: <http://www.national.org.nz/files/2008/HEALTH/Waiting_lists.pdf>.

Panchak, P., 2003. *Lean Health Care? It works!* EUA: Industry Week Publisher.

Panchak, Patricia. 2004. "Lean Health Care? It Works!" *Industry Week*, 2004.

Papadopoulos, T., Radnor, Z., Merali, Y. 2011. The role of actor associations in understanding the implementation of Lean thinking in healthcare. *International Journal of Operations and Production Management* 31 (2), 167-191.

Papadopoulos, Thanos, Zoe Radnor, and Yasmin Merali. 2011. "The Role of Actor Associations in Understanding the Implementation of Lean Thinking in Healthcare." *International Journal of Operations and Production Management* 31 (2): 167-91. doi:10.1108/01443571111104755.

Pexton, C., 2008. *Working to Eliminate Bottlenecks.* EUA: Cath Lab Digest.

Pexton, C., G. Nail, M. Donoghue, and B. Egolf. 2007. "Working to Eliminate Bottlenecks: Florida Hospital's Cardiac Cath Lab Achieves Greater Efficiency and Higher Satisfaction." *Cath Lab Digest* 15 (11): 1-4.

Pieters, A., Van Oirschot, C., Akkermans, H. 2010. No cure for all evils - Dutch obstetric care and limits to the applicability of the focused factory concept in health care. *International Journal of Operations and Production Management* 30 (11), 1112-1139.

Pieters, Angele, Charlotte van Oirschot, and Henk Akkermans. 2010. "No Cure for All Evils." *International Journal of Operations & Production Management* 30 (11): 1112-39. doi:10.1108/01443571011087350.

Proudlove, N., Moxham, C., Boaden, R., 2008. Lessons for Lean in Healthcare from Using Six Sigma in the NHS. *Public Money & Management, Chartered Institute of Public Finance and Accountancy* 28 (1), 27-34.

Proudlove, Nathan, Claire Moxham, and Ruth Boaden. 2008. "Lessons for Lean in Healthcare from Using Six Sigma in the NHS." *Public Money and Management* 28 (1): 27-34. doi:10.1111/j.1467-9302.2008.00615.x.

Pyzdek, T., 2003. *The Six Sigma Handbook: A Complete Guide for Green Belts, Black Belts and Managers at All Levels.* New York, EUA: McGraw-Hill.

Pyzdek, Thomas. 2003. *The Six Sigma Handbook - A Complete Guide for Green Belts, Black Belts, and Managers at All Levels.* 2nd ed. New York: McGraw-Hill. doi:10.1036/0071415963.

Reid, R.A., 2007. Applying the TOC five-step focusing process in the service sector: A banking subsystem, *Managing Service Quality: An International Journal* 17 (2), 209-234.

Reid, Richard A. 2007. "Applying the TOC Five-Step Focusing Process in the Service Sector: A Banking Subsystem." *Managing Service Quality* 17 (2): 209-34. doi:10.1108/09604520710735209.

Rich, N., Piercy, N. 2013. Losing patients: a systems view on healthcare improvement. *Production Planning and Control* 24 (10-11), 962-975.

Rich, Nick, and Niall Piercy. 2013. "Losing Patients: A Systems View on Healthcare Improvement." *Production Planning and Control* 24 (10-11): 962-75. doi:10.1080/09537287.2012.666911.

Rickard, T. 2007. "*Lean Principles in the Laboratory: Inpatient Phlebotomy.*" Mayo Clinic.

Rickard, T., 2007. *Lean Principles in the Laboratory: Inpatient Phlebotomy*. EUA: Mayo Clinic.

Selau, L. et al., 2009. Produção Enxuta no setor de serviços: caso do Hospital de Clínicas de Porto Alegre – HCPA [Lean Production in the services sector: case of the Hospital de Clínicas of Porto Alegre – HCPA]. *Revista de Gestão Industrial* 5 (1), 122-140.

Selau, Lisiane P. R., Mônica G. B. Pedó, Daniela Dos S. Senff, and Tarcisio a. Saurin. 2009. "Produção Enxuta No Setor de Serviços: Caso Do Hospital de Clínicas de Porto Alegre - HCPA." *Revista Gestão Industrial* 5 (1): 122-40. doi:10.3895/S1808-04482009000100008.

Shah, S., Robinson, I., 2008. Medical device technologies: who is the user? *International Journal of Healthcare Technology and Management* 9 (2), pp. 181-197.

Shah, Syed Ghulam Sarwar, and Ian Robinson. 2008. "Medical Device Technologies: Who Is the User?" *International Journal of Healthcare Technology and Management* 9 (2): 181-97. doi:10.1504/IJHTM.2008.017372.

Silberstein, Augusto. 2006. *Um Estudo de Casos Sobre a Aplicação de Princípios Enxutos Em Serviços de Saúde No Brasil*. [*A Case Study on the Application of Health Care Principles in Brazil*]. Universidade Federal do Rio de Janeiro.

Sokovic, M., D. Pavletic, and K. Kern Pipan. 2010. "Quality Improvement Methodologies- PDCA Cycle, RADAR Matrix, DMAIC and DFSS." *Journal of Achievements in Materials and Manufacturing Engineering* 43 (1): 476-83.

Sokovic, M., Pavletic, D., Pipapn, K., 2010. Quality Improvement Methodologies - PDCA Cycle, RADAR Matrix, DMAIC and DFSS. *Journal of Achievements in Materials and Manufacturing Engineering* 42 (1), 476-483.

Souza, L.B., 2008. Trends and approaches in lean healthcare. *Leadership in Health Services* 22 (2), 121-139.

Souza, L.B., Archibald, A., 2008. The use of lean thinking to reduce LOS in elderly care. *Proceedings of the Operational Research Applied to Health Services Conference*, Toronto, ON, 61.

Souza, Luciano Brandão de, and A. Archibald. 2008. "The Use of Lean Thinking to Reduce LOS in Elderly Care." In *Proceedings of the Operational Research Applied to Health Services Conference*, 61. Toronto, ON.

Souza, Luciano Brandao de. 2009. *Trends and Approaches in Lean Healthcare. Leadership in Health Services*. 22 (2): 121-39. doi:10.1108/17511870910953788.

Su, C., Chou, C., Hung, S., Wang, P. (2012). Adopting the Healthcare Failure Mode and Effect Analysis to Improve the Blood Transfusion Processes. *International Journal of Industrial Engineering: Theory, Applications and Practice* 19 (8).

Su, C.-T., C.-J. Chou, S.-H. Hung, and P.-C. Wang. 2012. "Adopting the Healthcare Failure Mode and Effect Analysis to Improve the Blood Transfusion Processes." *International Journal of Industrial Engineering : Theory Applications and Practice* 19 (8).

Taninecz, G., 2007. *Pulling Lean Through a Hospital - Hotel-Dieu Grace Success story*. Canada: Lean Institute.

Taninecz, George. 2007. *Pulling Lean Through a Hospital - Hotel-Dieu Grace Success Story*. Lean Entreprise Institute.

Thiollent, Michel. 1997. *Pesquisa-Ação Nas Organizações [Action Research In Organizations]*. São Paulo: Atlas.

Trägårdh, Björn, and Kajsa Lindberg. 2004. "Curing a Meagre Health Care System by Lean Methods-Translating 'Chains of Care' in the Swedish Health Care Sector." *The International Journal of Health Planning and Management* 19 (4): 383-98. doi:10.1002/hpm.767.

Trilling, L., Pellet, B., Delacroix, S., Marcon, E., 2010. *Improving care efficiency in a radiotherapy center using Lean philosophy: A case study of the proton therapy center of Institut Curie*. Orsay: IEEE Workshop, Health Care Management (WHCM).

Trilling, Lorraine, Bertrand Pellet, Sabine Delacroix, Hélène Colella-Fleury, and Eric Marcon. 2010. "Improving Care Efficiency in a Radiotherapy Center Using Lean Philosophy: A Case Study of the Proton Therapy Center of Institut Curie - Orsay." In *2010 IEEE Workshop on Health Care Management, WHCM 2010*, 1-6. Venice, Italy: IEEE. doi:10.1109/WHCM.2010.5441251.

Tucker, R., 2009. *Can lean healthcare help to improve shortage of nursing staff?* EUA: Healthcare Performace Partners. Available in: http://www.leanhealthcareexchange.com.

Tucker, Richard. 2009. *Can Lean Healthcare Help to Improve Shortage of Nursing Staff?* Lean Healthcare Exchange, 2009.

Womack, J.P., 2005. *Going Lean in Healthcare*. Innovation Series 2005, Institute for Healthcare Improvement.

Womack, J.P., 2007. *The Problem of Sustainability*. Available in: <http://www.*lean*uk.org/downloads/jim/jim_eletter_200706.pdf>.

Womack, J.P., Jones, D.T., 1996. *Lean Thinking - Banish Waste and Create Wealth in Your Corporation*. New York, EUA: Simon & Schuster.

Womack, James P. 2007. *The Problem of Sustaintability*. Jim Womack's ELetters & Columns. 2007. https://www.lean.org/womack/DisplayObject.cfm?o=752.

Womack, James P., and Daniel T. Jones. 1996. *Lean Thinking: Banish Waste and Create Wealth in Your Corporation*. New York: Simon & Schuster.

Womack, James P., Arthur P. Byrne, Orest J. Fiume, Gary S. Kaplan, and John Toussaint. 2005. "Innovation Series: Going Lean in Health Care." *Institute for Healthcare Improvement* 21.

Yasin, M.M., Alavi, J., Kunt, M., Zimmerer, T.W., 2004. TQM practices in service organizations: an exploratory study into the implementation, outcome and effectiveness", *Managing Service Quality: An International Journal* 14 (5), 377-389.

Yasin, Mahmoud M., Murat Kunt, Jafar Alavi, and Thomas W. Zimmerer. 2004. "TQM Practices in Service Organizations: An Exploratory Study into the Implementation, Outcome and Effectiveness." *Managing Service Quality: An International Journal* 14 (5): 377-89. doi:10.1108/09604520410557985.

Young, T P, and S I McClean. 2008. "A Critical Look at Lean Thinking in Healthcare." *Quality and Safety in Health Care* 17 (5): 382-86. doi:10.1136/qshc.2006.020131.

Young, T.P., McClean, S.I., 2008. A critical look at Lean Thinking in healthcare. *Quality and Safety in Health Care* 17, 382-386.

Zanchet, T., Saurin, T.A., Missel, E.C., 2007. *A Aplicação do Mapeamento de Fluxo de Valor em um centro de material em um complexo hospitalar.* VII SEPROSUL - Semana de Engenharia de Produção Sul-Americana, 2007.

Zanchet, Tiago, Tarcisio Abreu Saurin, and Elenara Consul Missel. 2007. "Aplicação Do Mapeamento de Fluxo de Valor Em Um Centro de Material e Esterilização de Um Complexo Hospitalar." In *VII SEPROSUL - Semana de Engenharia de Produção Sul-Americana*, 1-5. Salto, Uruguay: UDELAR.

Zarbo, R.J., Angelo, R., 2007. The Henry Ford Production System: Effective Reduction of Process Defects and Waste in Surgical Pathology. *American Journal of Clinical Pathology* 128, 1015-1022.

Zarbo, Richard J., and Rita D'Angelo. 2007. "The Henry Ford Production System: Effective Reduction of Process Defects and Waste in Surgical Pathology." *American Journal of Clinical Pathology* 128 (6): 1015-22. doi:10.1309/RGF6JD1NAP2DU88Q.

ABOUT THE EDITORS

Francisco J. G. Silva is PhD in Mechanical Engineering by Faculty of Engineering, University of Porto, MSc in Mechanical Engineering – Materials and Manufacturing Processes by Faculty of Engineering, University of Porto, BSc in Mechanical Engineering – Industrial Management by School of Engineering, Polytechnic of Porto. He has also a Post-Graduation in Materials and Manufacturing Processes by Faculty of Engineering, University of Porto. He is currently Head of the Master's Degree in Mechanical Engineering at School of Engineering, Polytechnic of Porto, position occupied since 2014. He also was Head of the Bachelor's Degree in Mechanical Engineering at Superior School of Industrial Studies and Management, Polytechnic of Porto, from 2003 to 2006. He has supervised some PhD students at Faculty of Engineering, University of Porto, as well as more than 100 students at School of Engineering, Polytechnic of Porto. Moreover, he co-supervised more than 30 MSc students at School of Engineering, Polytechnic of Porto and Faculty of Engineering, University of Porto. He has more than 120 papers published in peer-reviewed journals with impact factor and more than 100 papers presented in International Conferences. He is member of the Scientific Committee of some International Conferences, such as FAIM – Flexible Automation and Intelligent Manufacturing, among others. He has conducted some Special Issues in MDPI journals, such as Coatings and Metals. Moreover, he has conducted some national and international research projects involving other academic institutions and companies. He has a strong linkage to the Industry, due to his past linked to Industrial Sector, having been founder, owner and manager of a company in the electric sector for 18 years.

Luís Pinto Ferreira is PhD in Engineering by University of Vigo (Spain), MSc in Industrial Engineering - Logistics and Distribution by School of Engineering, University of Minho (Portugal) and Bachelor's Degree in Industrial Electronic Engineering by School of Engineering, University of Minho (Portugal). He taught in higher education since 2000. He is currently Sub-Director of Master's Degree in Engineering and Industrial Management at School of Engineering, Polytechnic of Porto, position occupied since September 2016. He also was Head of the Master's Degree in Engineering and Industrial Management at Superior School of Industrial Studies and Management, Polytechnic of Porto, from July 2012 to July 2016. He has supervised more than 25 MSc students at School of Engineering, Polytechnic of Porto and at Superior School of Industrial Studies and Management, Polytechnic of Porto. He has more than 40 papers published in peer-reviewed journals and proceedings of International Conferences. He is member of the Scientific Committee of some International Conferences, such as MESIC – Manufacturing Engineering Society International Conference, among others. In 2004 he was awarded with the APDIO/IO2004 prize, given by the Operation Research Portuguese Association for the best paper in Operations Research from a Master of Science Dissertation in the period 2002/2004.

INDEX

#

4.0 industry, 256
5S, v, 5, 13, 14, 15, 19, 21, 24, 25, 28, 95, 99, 101, 102, 106, 107, 108, 109, 111, 112, 114, 115, 116, 117, 118, 119, 120, 121, 155, 169, 171, 172, 181, 182, 185, 189, 199, 200, 201, 202, 212, 214, 215, 218, 219, 294, 296, 304, 319, 396, 431, 433, 442
6S, 101, 109, 117, 120, 121, 248

A

ABC/XYZ analysis, 339
acceptance, 41, 79, 82, 83, 85, 86, 87, 90, 91
aggregation, 22, 337, 409
algorithm, 77, 124, 127, 130, 131, 132, 133, 135, 140, 144, 145
assembly line, vi, 21, 22, 29, 59, 61, 63, 64, 65, 66, 67, 68, 69, 73, 76, 77, 78, 173, 189, 190, 194, 205, 206, 207, 209, 210, 211, 212, 213, 214, 215, 216, 217, 218, 220, 221, 223, 252, 254, 282, 285, 286, 287, 290, 292, 294, 296, 297, 298, 299, 300, 301, 303, 304, 309, 310, 311, 312, 315, 317, 318, 319, 326, 327, 331, 332, 333, 334, 355, 356, 357, 358, 361, 362, 367, 368, 369, 370, 372, 373, 374, 379, 380, 382, 383, 384, 399, 401, 405
assessment, 9, 31, 33, 47, 48, 50, 51, 52, 53, 102, 104, 112, 115, 187, 357, 383, 398, 399, 401, 418
audit, 12, 94, 171, 194, 441, 442, 446, 448
automation, vi, 6, 41, 71, 92, 93, 124, 134, 174, 189, 199, 280, 284, 289, 335, 336, 349, 350, 353, 356, 368, 410, 459
automatisms, 318
automobile, 26, 118, 185, 219, 220, 221, 222, 279
automotive, vi, 2, 10, 11, 21, 25, 29, 36, 42, 50, 53, 54, 76, 77, 98, 101, 102, 110, 119, 124, 163, 173, 189, 190, 191, 193, 194, 195, 200, 204, 212, 218, 219, 220, 221, 222, 223, 252, 253, 254, 278, 283, 332, 333, 334, 336, 356, 357, 382, 383, 384, 390, 399, 400, 403, 405, 422, 423, 427
automotive components industry, 190
automotive industry, 29, 54, 77, 98, 101, 102, 110, 119, 163, 173, 189, 190, 191, 193, 194, 200, 204, 212, 219, 220, 221, 223, 252, 253, 254, 283, 332, 333, 334, 336, 357, 383, 384, 390, 399, 400, 403, 405, 423, 427

B

Bibliographic Portfolio, 387, 388, 389, 390, 391, 392, 398, 399
bibliometric analysis, 387, 388, 389, 398, 403, 404
business environment, 31, 50, 53
business model, 52, 179, 188
business strategy, 184, 256
businesses, 35, 46, 153, 181, 194, 418, 426

C

cancer, vi, 88, 425, 427, 439, 441, 443, 446
cancer chemotherapy, vi, 425, 427
candidates, 58, 76, 152, 157, 436
capabilities, 31, 32, 45, 46, 51, 52, 53, 57, 58, 75, 77, 99, 146, 153, 155, 160, 253, 348, 410, 411, 412, 414, 419, 420, 421
case study, vi, vii, 1, 2, 10, 11, 14, 25, 26, 27, 29, 31, 48, 49, 52, 54, 57, 98, 101, 102, 109, 110, 115, 118, 119, 120, 121, 122, 148, 163, 165, 166, 168, 183, 184, 185, 187, 205, 218, 220, 221, 223, 253, 278, 280, 281, 282, 283, 287, 294, 310, 333, 349, 355, 384, 392, 401, 402, 405, 417, 450, 453, 455, 456
changeover, 25, 65, 78, 169, 175, 176, 177, 203, 225, 226, 227, 243, 244, 245, 246, 248, 249, 250, 251, 253, 287, 292, 300, 319, 320, 321, 322, 324, 325

chemotherapy, 88, 89, 425, 427, 434, 435, 436, 437, 439, 440, 441, 442, 443, 444, 445, 446, 447, 448, 449
clean technology (CT), 27, 134, 137, 138, 141, 142, 143, 145, 285, 407, 408, 414, 415, 418, 419, 420, 421, 422
cleaner technologies, 407, 417, 418
cleaning, 14, 15, 107, 113, 114, 127, 131, 132, 175, 178, 202
cognitive attributes, 394
collaboration, 44, 67, 88, 145, 179, 335, 348, 399, 423, 442
collective bargaining, 95, 96, 97
comparative case study, 31
competition, 32, 155, 166, 195, 257, 266, 267, 276, 385, 386
competitive advantage, 7, 8, 54, 151, 161, 166, 167, 407, 408, 412, 414, 415, 419, 420, 421
competitiveness, 23, 32, 34, 36, 59, 92, 96, 176, 179, 181, 183, 189, 194, 218, 225, 226, 256, 275, 286, 348, 400, 415, 418, 421
competitors, 32, 41, 93, 97, 159, 336
contextual assessment, 31
contextualization, 11, 12, 14, 15, 16, 17, 19, 21, 22, 23
continuous improvement, v, 4, 12, 15, 17, 19, 24, 27, 28, 43, 44, 51, 52, 53, 106, 107, 108, 151, 155, 157, 165, 166, 171, 172, 175, 177, 178, 179, 180, 183, 191, 200, 202, 204, 218, 219, 222, 223, 256, 259, 260, 274, 275, 310, 319, 340, 348, 349, 351, 385, 386, 390, 395, 429, 430
cork stoppers, vi, 223, 225, 226, 228, 229, 230, 231, 234, 252, 334, 384
creativity, 4, 37, 48, 81, 85, 158, 200, 335, 348, 350, 352, 353
critical thinking, 151, 157, 259

D

data analysis, 17, 22, 134, 145, 261
data collection, 10, 14, 15, 255, 297, 387, 437
data set, 243, 261
data-based assignment heuristics, 124
digital technologies, 179, 184, 188
DOWNTIME, 4, 7, 8, 9

E

earliest finish, 124, 125, 127, 135
e-commerce, 416
economic activity, 191
economic efficiency, 417, 419

education, 92, 93, 106, 153, 156, 187, 259, 442
energy, 79, 86, 178, 182, 276, 286, 336, 342, 343, 352, 353, 412, 415, 417, 418
engineering, 18, 32, 38, 42, 43, 44, 45, 52, 53, 92, 156, 220, 256, 305, 343, 353, 367, 384, 389, 392, 396
environment, 1, 2, 9, 15, 17, 40, 45, 50, 51, 59, 64, 66, 81, 96, 103, 104, 105, 107, 109, 117, 124, 151, 152, 171, 172, 178, 182, 183, 194, 200, 202, 251, 256, 267, 272, 293, 294, 331, 345, 351, 383, 394, 399, 401, 407, 408, 411, 413, 414, 415, 417, 418, 419, 420, 431, 450, 452, 453
environmental management, 109, 418, 424
equipment, 13, 14, 15, 65, 71, 104, 109, 110, 114, 155, 165, 168, 172, 174, 175, 177, 178, 179, 183, 191, 199, 200, 201, 203, 206, 211, 214, 215, 217, 218, 225, 227, 238, 239, 240, 243, 246, 248, 250, 251, 253, 270, 291, 297, 298, 305, 306, 307, 312, 319, 320, 322, 346, 349, 350, 351, 356, 358, 360, 368, 372, 376, 379, 380, 382, 391, 409, 416, 418, 421
ergonomic aspects, 23, 293, 295, 321, 345, 361, 392, 399
ergonomics, vi, 1, 6, 9, 11, 21, 22, 23, 25, 26, 27, 28, 29, 30, 76, 118, 119, 121, 163, 179, 181, 212, 215, 259, 277, 281, 282, 283, 295, 296, 304, 311, 335, 342, 345, 346, 347, 351, 352, 353, 355, 357, 358, 361, 368, 369, 380, 382, 383, 384, 385, 386, 387, 389, 391, 392, 394, 395, 397, 398, 399, 400, 401, 402, 403, 404, 405, 406

F

factories, 11, 173, 177
financial, 7, 33, 34, 36, 37, 39, 79, 80, 84, 105, 191, 250, 256, 257, 258, 418, 423, 425, 435, 441, 442, 446, 447, 449
financial condition, 435, 441
flexibility, vii, 46, 48, 57, 58, 59, 60, 61, 63, 64, 65, 67, 68, 69, 70, 71, 72, 73, 74, 75, 77, 89, 107, 130, 134, 189, 199, 203, 204, 225, 226, 227, 273, 289, 291, 294, 296, 312, 331, 356, 357, 368, 376, 414, 416, 425, 426, 427
flexible manufacturing, 41, 70
framework, 28, 31, 41, 45, 46, 54, 56, 67, 77, 80, 82, 97, 99, 119, 146, 148, 180, 185, 280, 281, 361, 383, 386, 401, 404, 417, 419, 421

G

goods and services, 256, 259, 392, 426
gravity, 111, 335, 343, 352

H

health, vii, 9, 92, 94, 95, 96, 102, 160, 163, 178, 179, 181, 182, 186, 187, 345, 347, 355, 356, 361, 381, 386, 390, 391, 392, 394, 395, 397, 398, 399, 401, 402, 426, 427, 428, 439, 444, 446, 447, 450, 452, 453, 454
health care, 181, 394, 426, 428, 450, 452, 453, 454
health problems, 355, 356
Healthcare Toyota, 429
High-Volume & Low-Variability (HVLV), 407, 408, 413
human, 2, 6, 19, 20, 21, 24, 25, 32, 74, 75, 77, 95, 103, 104, 105, 106, 152, 162, 163, 174, 188, 199, 200, 258, 259, 263, 274, 298, 322, 336, 343, 345, 346, 347, 348, 349, 351, 353, 356, 357, 358, 384, 385, 386, 389, 391, 392, 394, 396, 397, 398, 399, 400, 401, 409, 410, 411, 413, 415, 423
human activity, 6, 103, 345
human health, 104, 396
Human Resource Management, 162, 185, 395
human resources, 2, 19, 20, 21, 24, 77, 103, 105, 152, 188, 199, 258, 274, 348, 351, 356, 392, 398, 409, 411

I

identification, 6, 18, 20, 23, 42, 84, 97, 104, 155, 166, 170, 171, 200, 202, 210, 212, 214, 219, 226, 264, 266, 269, 288, 296, 305, 352, 367, 395, 425, 426
improvements, 2, 6, 13, 14, 23, 51, 52, 53, 59, 84, 91, 95, 96, 105, 123, 125, 134, 144, 165, 175, 176, 178, 182, 200, 202, 216, 217, 218, 219, 226, 227, 247, 249, 250, 251, 252, 256, 273, 287, 294, 295, 304, 310, 315, 319, 322, 327, 329, 330, 331, 335, 338, 340, 348, 349, 361, 381, 395, 412, 414, 417, 425, 426, 429, 430, 435, 436, 439, 441, 442, 446, 447, 448, 449
increased productivity, 60, 95, 212, 219, 256, 287, 291, 353, 355
increasing competition, 385, 386
industrial environments, 393, 395, 413, 415
industrial performance, 407, 408, 414, 415, 420, 424
industrial revolution, 125, 179
industrial sectors, 1, 2, 3, 356, 412, 421
industrialization, 211, 212, 293, 415
Industry 4.0, vi, 29, 57, 58, 70, 74, 75, 77, 148, 166, 179, 183, 255, 260, 274, 276, 278, 280, 281, 407, 408, 414, 416, 417, 419, 420
information flow, 1, 6, 7, 8, 11, 12, 16, 17, 19, 25, 29, 244, 287

information processing, 44, 416
information technology (IT), 26, 89, 90, 146, 162, 179, 260, 405, 407, 408, 412, 413, 414, 415, 416, 417, 418, 419, 420, 421, 422, 423
intervention instrument, 385, 387
investment, 33, 35, 39, 65, 93, 94, 145, 159, 162, 183, 218, 226, 256, 257, 258, 289, 296, 312, 321, 335, 350, 361, 380, 381, 415, 420, 421, 422, 435, 442
isolation, 356, 415, 417
Item Response Theory (IRT), 398

J

job position, 191, 193
job satisfaction, 153, 156, 159, 160, 163
Just in Time, 185, 339

K

Kaizen Healthcare, 429
Kaizen tool, 349
Kanban, 2, 5, 12, 15, 25, 27, 28, 41, 155, 173, 184, 185, 186, 201, 267, 283, 339, 340, 341, 416, 422
Knowledge Development Process – Constructivist (Proknow-C), 385, 386, 387, 397

L

labour health problems, 355
lean concept, 31, 33, 42, 46, 251, 412, 413, 417, 426, 428, 430, 436, 446, 449
lean healthcare, vi, 99, 425, 426, 427, 429, 430, 431, 432, 434, 436, 448, 449, 450, 455, 456
lean healthcare methodology, 429
lean hospitals, 429, 452
lean implementation, 40, 79, 80, 82, 84, 88, 89, 90, 95, 97, 182, 223, 276, 279, 282, 349, 385, 386, 395, 396, 398, 399, 403, 412, 453
lean line design (LLD), 286, 287, 289, 290, 292, 294, 296, 310, 326, 331, 333
lean manufacturing, v, vi, vii, 2, 29, 30, 40, 57, 58, 73, 75, 93, 98, 100, 101, 102, 108, 118, 120, 121, 144, 151, 152, 155, 157, 163, 167, 168, 170, 182, 184, 186, 187, 188, 189, 219, 220, 221, 222, 223, 254, 271, 276, 277, 278, 279, 281, 283, 285, 287, 333, 334, 355, 356, 361, 382, 383, 384, 386, 387, 394, 395, 397, 399, 401, 403, 404, 405, 407, 408, 409, 411, 412, 413, 414, 415, 417, 418, 419, 420, 422, 423, 424, 426, 450, 451, 453
lean production, 10, 39, 40, 54, 93, 154, 155, 166, 179, 181, 182, 190, 198, 200, 256, 270, 286, 287,

336, 337, 354, 382, 383, 384, 386, 389, 392, 393, 395, 396, 398, 399, 400, 401, 402, 412, 417, 421, 422, 423, 429, 442, 448
lean production practices, vi, 385, 395, 398, 405
lean thinking, v, vi, 1, 2, 3, 8, 19, 29, 30, 56, 86, 102, 105, 118, 120, 168, 185, 186, 188, 191, 198, 200, 222, 253, 256, 258, 259, 260, 276, 277, 283, 334, 335, 336, 347, 348, 349, 351, 352, 354, 404, 407, 408, 410, 412, 413, 416, 418, 419, 420, 425, 426, 427, 450, 452, 453, 454, 455, 456, 457
lean tools, v, vii, 4, 14, 16, 18, 20, 24, 27, 28, 30, 77, 107, 111, 165, 167, 170, 183, 187, 189, 190, 200, 201, 212, 213, 214, 215, 216, 218, 219, 220, 221, 222, 223, 225, 252, 253, 260, 286, 287, 288, 293, 332, 333, 356, 357, 382, 384, 449
level of implementation, vi, 255, 264, 385, 386, 387, 397, 398
line balancing, 216, 285, 286, 294, 296, 297, 298, 300, 302, 303, 310, 312, 328, 329, 331, 333, 356, 381
logistics, 1, 2, 5, 6, 7, 10, 11, 13, 22, 23, 24, 25, 26, 27, 28, 29, 54, 174, 184, 186, 194, 259, 260, 280, 296, 335, 336, 337, 339, 347, 348, 349, 350, 351, 352, 353, 357, 367, 415, 451, 459
low cost solutions, 219, 336

M

make-to-order, 12, 57, 58, 286
manufacturing companies, 31, 162
manufacturing industry, 2, 10, 337, 401, 428
mapping, 6, 16, 17, 20, 88, 155, 166, 169, 181, 182, 186, 281, 288, 289, 333, 429, 430, 437, 448, 452
market share, 11, 180
marketing, 45, 49, 412
marketplace, 32, 33, 34
medical, 61, 64, 65, 66, 426, 427, 435, 440, 441, 443, 448
medication, 428, 439, 444, 445
medicine, 88, 89, 102, 427, 441
mental frame, 81, 82
mental health, 390, 400
milk run, 339, 340, 341, 365, 366, 373
modelling, 127, 134
models, 45, 56, 60, 61, 63, 64, 65, 68, 69, 70, 74, 124, 159, 163, 178, 182, 189, 256, 266, 267, 268, 269, 272, 275, 276, 286, 296, 299, 304, 319, 342, 358, 367, 371, 395, 408, 413, 419
modifications, 38, 53, 108, 126, 143, 246, 294, 312, 437

N

new product development, v, 31, 33, 53, 414
nurses, 88, 89, 90, 91, 428, 444
nursing, 102, 437, 441, 445, 448, 456

O

occupational health, 94, 95, 96, 181
occupational safety, 101, 102, 103, 111, 118, 121, 163, 188, 278, 281, 282, 283
organizational attributes, 394
organizational change, 79, 81, 82, 94, 98, 99, 100, 181
organizational culture, 12, 60, 176
organizational learning, 40, 52
organizational paradox, 79, 80, 81, 84, 97, 98, 99
organizations, v, 2, 3, 9, 23, 24, 42, 52, 54, 79, 80, 81, 82, 95, 98, 99, 102, 109, 152, 155, 158, 159, 160, 161, 162, 165, 166, 167, 168, 170, 175, 176, 177, 178, 179, 180, 181, 183, 184, 187, 193, 221, 225, 226, 253, 279, 286, 349, 356, 385, 386, 395, 397, 399, 400, 403, 404, 416, 418, 425, 426, 427, 456, 457
outpatient, 88, 89, 90, 91, 434, 436, 439, 442

P

paradox, 79, 80, 81, 82, 83, 84, 85, 86, 87, 89, 90, 91, 92, 93, 94, 95, 96, 97, 98, 99, 100, 416, 454
performance indicator, 13, 171, 177, 201, 311, 416, 446, 448
performance measurement, 166, 391
philosophies of lean thinking, 396
physical attributes, 394
plan-do-check-act (PDCA), 8, 18, 19, 22, 25, 44, 170, 189, 190, 200, 204, 205, 212, 213, 214, 215, 218, 221, 223, 275, 292, 294, 310, 311, 319, 430, 455
poka-yoke, 5, 6, 21, 155, 174, 182, 188, 199, 206, 213, 286, 292, 304
Portuguese textile and clothing industry, 255, 256, 257
practices, 2, 3, 9, 14, 21, 29, 31, 32, 45, 47, 48, 51, 52, 53, 56, 80, 84, 88, 95, 96, 99, 102, 107, 108, 119, 157, 166, 167, 176, 180, 181, 186, 188, 201, 222, 234, 252, 256, 279, 282, 283, 336, 348, 349, 357, 385, 386, 387, 391, 393, 394, 395, 396, 397, 398, 399, 401, 403, 404, 407, 408, 409, 410, 411, 412, 413, 414, 415, 416, 417, 418, 420, 421, 423, 431, 433, 456, 457
principles of lean production, 163, 336, 401

produce on demand, 338
producers, 101, 192, 260
product design, 199, 212, 216, 286, 303, 412, 415
product life cycle, 286, 408, 413, 414, 415, 419
product performance, 34, 35, 298
production cell, 281, 347
production costs, 273, 286, 386
production systems, 2, 41, 59, 92, 107, 179, 181, 190, 191, 199, 220, 258, 267, 276, 277, 278, 280, 286, 312, 332, 358, 384, 385, 386, 389, 390, 398, 399, 400, 405
production targets, 61, 64, 66, 69
productive capacity, 11, 286
productivity growth, 356
pull system method, 339

Q

quality control, 6, 41, 110, 155, 174, 211, 239, 410
quality improvement, 155, 168, 176, 292
quality of life, 106, 394
quality standards, 215, 427
quality tools, 27, 286
quick changeover, 175, 176, 291, 332

R

reduction of production costs, 386
reframing, 82
relevance, 97, 124, 171, 395, 396, 397
reliability, 35, 175, 177, 178, 211, 213, 215, 218, 219, 261, 346, 412, 416, 417, 420
researchers, ix, 41, 46, 103, 167, 227, 336, 361, 373, 386, 387, 388, 389, 397, 398, 407, 408, 409, 414, 426, 427
resistance, 17, 79, 80, 84, 86, 89, 90, 91, 97, 167, 209, 210, 219, 227, 258, 273, 274, 275, 276
resolution, 20, 21, 82, 83, 90, 91, 96, 97, 100, 215, 352, 417, 427
resource flexibility, 57, 59, 78
resources, 3, 4, 7, 8, 9, 21, 33, 39, 43, 45, 48, 50, 57, 58, 75, 88, 92, 105, 143, 152, 154, 157, 160, 161, 167, 179, 181, 203, 206, 225, 260, 275, 286, 291, 337, 348, 349, 352, 353, 356, 361, 412, 414, 415, 417, 419, 420, 421, 427, 441
risk, 9, 12, 23, 24, 25, 32, 35, 36, 37, 40, 43, 66, 85, 94, 101, 102, 103, 104, 111, 112, 115, 117, 120, 171, 195, 358, 361, 368, 397, 415
risk assessment, 101, 102, 104, 111, 112, 115, 117, 120
risk factors, 9, 23, 24, 25, 358, 368
risk taking, 37, 40

S

safety, v, 2, 9, 10, 21, 22, 23, 28, 64, 66, 69, 89, 90, 92, 94, 95, 96, 100, 101, 102, 103, 104, 105, 106, 108, 109, 110, 111, 112, 113, 114, 115, 117, 118, 119, 120, 121, 122, 155, 156, 165, 166, 171, 177, 178, 181, 182, 183, 184, 185, 186, 187, 188, 194, 202, 259, 278, 283, 286, 292, 336, 345, 347, 348, 350, 353, 357, 371, 375, 384, 386, 391, 392, 394, 395, 397, 398, 401, 403, 404, 411, 412, 427, 452, 457
savings, 9, 38, 133, 250, 336, 342, 353
scheduling optimization, 124, 126
SE in lean manufacturing environments, 394
shopfloor, 185, 407
single minute exchange of die (SMED), vi, 5, 14, 25, 29, 169, 175, 176, 185, 189, 190, 199, 200, 201, 203, 204, 211, 212, 214, 215, 218, 219, 220, 221, 222, 223, 225, 226, 227, 245, 246, 249, 251, 253, 254, 291, 304, 319, 333, 334, 382, 383, 384, 433
smart operators, 57, 58, 74, 75
sociotechnical and ergonomic practices, vi, 385, 393, 396, 397, 398
sorting process, 101, 110
standard work, 6, 16, 172, 178, 189, 190, 200, 201, 202, 212, 215, 218, 219, 431, 433
standardization, 15, 43, 44, 52, 53, 81, 82, 83, 89, 90, 91, 155, 176, 177, 178, 184, 199, 202, 211, 218, 340, 430, 431, 440, 441, 443, 444, 445
supermarkets, 5, 7, 10, 26, 173, 282, 299, 341
systematic review, vi, 27, 279, 385, 386, 390, 393, 400, 422
systematic review process, vi, 385, 386

T

technology, 2, 11, 32, 33, 34, 36, 41, 42, 48, 49, 51, 70, 72, 75, 144, 165, 176, 178, 183, 190, 191, 194, 198, 257, 258, 260, 276, 336, 342, 348, 352, 353, 407, 408, 414, 415, 416, 417, 418, 419, 420, 421, 423, 455
technology-intensive, 407, 408, 414, 420
tensions, v, 9, 79, 80, 81, 82, 84, 85, 86, 87, 89, 90, 91, 94, 95, 96, 97, 99, 358
textile industry and clothing industry, 256
textiles, 258, 263, 279
total productive maintenance (TPM), 4, 5, 14, 169, 174, 176, 177, 178, 179, 182, 183, 188, 199, 226, 253, 304, 395, 396
total quality management (TQM), 121, 179, 180, 187, 395, 396, 423, 456, 457

Index

Toyota, 2, 3, 21, 28, 29, 30, 32, 40, 41, 45, 54, 55, 67, 98, 99, 105, 106, 109, 119, 120, 154, 155, 163, 166, 168, 170, 173, 175, 176, 184, 186, 187, 188, 190, 198, 200, 222, 223, 226, 256, 258, 267, 272, 273, 280, 281, 283, 287, 336, 348, 350, 354, 356, 402, 408, 409, 410, 411, 416, 422, 423, 425, 426, 427, 428, 429
TPS Healthcare, 429
training, 12, 13, 19, 41, 59, 60, 65, 66, 67, 68, 70, 74, 76, 77, 83, 84, 86, 88, 92, 93, 94, 96, 105, 106, 108, 151, 153, 155, 156, 162, 172, 175, 178, 186, 191, 199, 200, 202, 211, 215, 218, 219, 271, 272, 273, 276, 348, 360, 397, 418, 426, 428, 430, 436, 442, 449
training programs, 66, 92, 155
Trilogiq, 335

V

value stream mapping (VSM), 4, 5, 6, 8, 14, 16, 17, 18, 25, 27, 28, 29, 30, 38, 88, 107, 165, 166, 168, 169, 170, 181, 182, 184, 186, 187, 189, 190, 200, 201, 212, 213, 214, 215, 218, 220, 222, 225, 226, 227, 244, 245, 285, 286, 287, 288, 289, 292, 294, 295, 296, 331, 333, 334, 399, 418, 430, 431, 433, 437, 439, 440, 447, 452
vendor managed inventory, 341
visual management, 5, 7, 14, 24, 26, 107, 113, 172, 181, 189, 190, 201, 212, 215, 218, 219, 285, 396, 433
visualization, 166, 177, 289, 295, 312, 340

W

weaknesses of the lean, 407, 408, 414
well-being, 23, 106, 153, 160, 163, 179, 336, 345, 348, 353, 355, 357, 361, 382, 383, 386, 392, 394, 395, 398, 399, 404
wet etch station, 124, 126, 143, 147
work activities, 178, 345, 400
work environment, 9, 41, 89, 95, 102, 117, 172, 181, 205, 351, 357, 361, 394, 395, 396, 397
workers, vii, 9, 11, 13, 14, 23, 25, 40, 41, 57, 58, 59, 60, 61, 63, 64, 65, 66, 67, 68, 69, 73, 74, 75, 76, 78, 85, 89, 92, 93, 94, 95, 96, 97, 98, 105, 108, 152, 153, 155, 156, 157, 158, 159, 160, 161, 162, 163, 172, 176, 181, 199, 200, 201, 202, 203, 211, 218,229, 233, 235, 246, 247, 251, 257, 263, 269, 273, 276, 285, 289, 293, 304, 346, 347, 348, 349, 350, 351, 352, 353, 356, 357, 358, 361, 386, 392, 394, 395, 396, 397, 398, 410, 411, 413, 415, 420
workflow, 15, 17, 76, 109, 298, 301, 302, 303, 327, 357
workforce, 11, 57, 58, 59, 60, 66, 67, 68, 70, 76, 79, 84, 92, 93, 152, 153, 162, 167, 182, 251, 285, 331, 396
working conditions, vii, 9, 24, 27, 94, 95, 103, 111, 151, 181, 211, 294, 346, 353, 383, 385, 386, 400, 401, 411
working groups, 88, 268
working hours, 59, 357
workload, 10, 68, 75, 106, 135, 137, 145, 155, 173, 217, 289, 292, 298, 312, 342, 396, 440
workplace, 5, 9, 95, 96, 101, 102, 103, 105, 106, 107, 152, 156, 171, 179, 182, 202, 215, 251, 346, 357, 358, 361, 401, 411, 412
workstation, 9, 15, 24, 61, 64, 65, 66, 67, 68, 69, 73, 74, 75, 133, 171, 190, 202, 211, 214, 215, 218, 285, 286, 290, 293, 298, 299, 301, 303, 305, 306, 307, 308, 309, 318, 319, 320, 321, 322, 323, 324, 325, 326, 327, 345, 346, 347, 350, 351, 355, 357, 358, 362, 363, 364, 365, 366, 370, 371, 372, 373, 374, 375, 377, 378, 379, 380, 381, 383, 397, 400

Related Nova Publications

The Rise of Accounting, Auditing, and Finance: Key Issues and Events That Shaped These Professions for Over 200 Years since 1800

Author: Lal Balkaran

Series: Business, Technology and Finance

Book Description: With over 200 professional associations, 120 pieces of authoritative literature, 65 well-known fraud cases, 62 accounting firms (including the origins and growth of the "Big Four"), 55 regulatory statutes, 30 frameworks, and much more, this unique book shows in a chronological sequence a range of select issues and events that have impacted and led to the growth of the professions of accounting, auditing, and finance since 1800.

Hardcover ISBN: 978-1-53614-732-2
Retail Price: $160

Asset Management: Strategies, Opportunities and Challenges

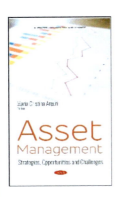

Editor: Maria Cristina Arcuri

Series: Business, Technology and Finance

Book Description: This book aims to provide an overview of asset management by focusing on some of the main issues in the sector. It gathers contributions on the system, strategies, opportunities and challenges.

Hardcover ISBN: 978-1-53614-246-4
Retail Price: $160

To see a complete list of Nova publications, please visit our website at www.novapublishers.com

Related Nova Publications

AUDITING: AN OVERVIEW

EDITORS: Timothy Cavenagh and Jacob Rymill

SERIES: Business, Technology and Finance

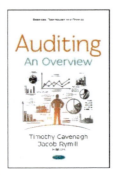

BOOK DESCRIPTION: In this compilation, critical aspects of the internal audit function are contrasted in order to provide an insight into the value of internal auditing and, within that, to submit arguments for the implementation and maintenance of an internal audit function.

SOFTCOVER ISBN: 978-1-53615-116-9
RETAIL PRICE: $95

DEVELOPMENTS AND PROSPECTS OF BUSINESS ECONOMICS AND FINANCE IN MUSLIM COUNTRIES

EDITOR: Eleftherios Thalassinos

SERIES: Business, Technology and Finance

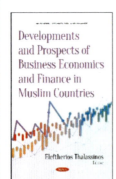

BOOK DESCRIPTION: As the title suggests, the Chapters are a mix of research studies on business economics, financial and managerial issues in Muslim countries with remarkable original research in some of them.

HARDCOVER ISBN: 978-1-53615-015-5
RETAIL PRICE: $195

To see a complete list of Nova publications, please visit our website at www.novapublishers.com